Environment and Ecology: Theories and Practices

Environment and Ecology: Theories and Practices

Edited by Jeffery Clarke

SYRAWOOD
PUBLISHING HOUSE

New York

Published by Syrawood Publishing House,
750 Third Avenue, 9th Floor,
New York, NY 10017, USA
www.syrawoodpublishinghouse.com

Environment and Ecology: Theories and Practices
Edited by Jeffery Clarke

© 2018 Syrawood Publishing House

International Standard Book Number: 978-1-68286-558-3 (Hardback)

Cataloging-in-Publication Data

Environment and ecology : theories and practices / edited by Jeffery Clarke.
    p. cm.
Includes bibliographical references and index.
ISBN 978-1-68286-558-3
1. Environmental sciences. 2. Ecology. I. Clarke, Jeffery.
GE105 .E58 2018
363.7--dc23

TABLE OF CONTENTS

PREFACE

This book aims to highlight the current researches and provides a platform to further the scope of innovations in this area. This book is a product of the combined efforts of many researchers and scientists, after going through thorough studies and analysis from different parts of the world. The objective of this book is to provide the readers with the latest information of the field.

The ecology of a region is determined by external as well as internal factors. External factors include the climate patterns and the geographical relief of a region. Internal factors are the flora and fauna that habitat an ecosystem. Ecological functioning also depends on physical, chemical, biological and anthropogenic processes. Long-term sustainability and propagation of biodiversity are the main goals of any ecological system. The extensive contents of this book provide the readers with a thorough understanding of the subject. It will be of great help to students and researchers in the fields of systems ecology, ecosystem management and biological diversity.

I would like to express my sincere thanks to the authors for their dedicated efforts in the completion of this book. I acknowledge the efforts of the publisher for providing constant support. Lastly, I would like to thank my family for their support in all academic endeavors.

Editor

Bioinformatic Approaches Reveal Metagenomic Characterization of Soil Microbial Community

Zhuofei Xu, Martin Asser Hansen, Lars H. Hansen, Samuel Jacquiod, Søren J. Sørensen*

Section of Microbiology, Department of Biology, University of Copenhagen, Copenhagen, Denmark

Abstract

As is well known, soil is a complex ecosystem harboring the most prokaryotic biodiversity on the Earth. In recent years, the advent of high-throughput sequencing techniques has greatly facilitated the progress of soil ecological studies. However, how to effectively understand the underlying biological features of large-scale sequencing data is a new challenge. In the present study, we used 33 publicly available metagenomes from diverse soil sites (i.e. grassland, forest soil, desert, Arctic soil, and mangrove sediment) and integrated some state-of-the-art computational tools to explore the phylogenetic and functional characterizations of the microbial communities in soil. Microbial composition and metabolic potential in soils were comprehensively illustrated at the metagenomic level. A spectrum of metagenomic biomarkers containing 46 taxa and 33 metabolic modules were detected to be significantly differential that could be used as indicators to distinguish at least one of five soil communities. The co-occurrence associations between complex microbial compositions and functions were inferred by network-based approaches. Our results together with the established bioinformatic pipelines should provide a foundation for future research into the relation between soil biodiversity and ecosystem function.

Editor: Raffaella Balestrini, Institute for Plant Protection (IPP), CNR, Italy

Funding: This work was partly funded by the EU ITN project TRAINBIODIVERSE and the Center for Environmental and Agricultural Microbiology (CREAM) funded by The Villum Foundation. The funders had no role in study design, data collection and analysis, decision to publish, or preparation of the manuscript.

Competing Interests: The authors have declared that no competing interests exist.

* E-mail: SJS@bio.ku.dk

Introduction

Soil is considered to be the most diverse natural environment on the Earth [1,2]. The soil microbial communities harbor thousands of different prokaryotic organisms that contain a substantial number of genetic information, ranging from 2,000 to 18,000 different genomes estimated in one gram of soil [3]. One of the most important issues in the field of soil ecology is to uncover the complex relationships between microbial compositions and functional diversity in soil.

Based on traditional approaches for cultivating and isolating soil microorganisms, early studies have focused on culturable bacteria which only account for less than 1% of soil microbial populations [4]. These studies have already discovered many novel genes encoding interesting enzymes and antimicrobials in soils via functional screens and clone-based Sanger sequencing [1,5,6]. Due to the recent advent of High-Throughput Sequencing (HTS) technologies, metagenomic sequencing approaches have been applied to investigate characterizations of diverse soil microbial communities, including target sequencing of the phylogenetic marker gene encoding 16S rRNA [7,8] and whole-metagenome shotgun sequencing [9–13]. However, the majority of 16S rRNA gene-based studies are committed to the interpretation of community composition but poorly focus on the functional and metabolic properties in a microbial community [14]. In addition, integrated bioinformatic analyses for microbial community-level taxonomic affiliation, metabolic reconstruction, and interaction network, seems to be less studied for the highly diverse soil ecosystems. Currently, MG-RAST [15], IMG/M [16], and CAMERA [17] are the major databases that can support

deposition and analysis of metagenomic datasets. Uploading large sequencing data and the subsequent analysis jobs on these web servers sometimes take long waiting time and even weeks. The computational pipelines implemented by these prominent platforms are capable of processing many analysis tasks, but some approaches for special biological inference and graphical visualization still need to be complemented [18].

Recently, together with the rapid development of the Human Microbiome Project, numerous computational tools and methodologies have been developed for effective interpretation and visualization of taxonomic and metabolic profiling of complex microbial communities [19,20] and could be applied to the analysis of the soil microbiota. Particularly, some outstanding computational techniques that could better explain the complexity and heterogeneity of microbial communities are still less applied in the study of the soil microbiota, e.g. prediction of metagenomic biomarkers and network-based correlation analyses [21,22]. In this study, we aim to explore the characterizations of the soil microbiota through integrating the current state-of-the-art bioinformatics tools. A collection dataset of 33 publicly available soil metagenomes was investigated in a custom metagenomic data mining pipeline for explaining and visualizing microbial compositions and metabolic potential. A full spectrum of metagenomic biomarkers and a network of taxon co-occurrence patterns were inferred to hopefully provide some new insights into the underlying mechanisms of complex ecological relationships in the soil microbial community.

Materials and Methods

Ethics statement

No specific permissions were required for the described field studies. The study locations are not privately owned and the field studies did not involve endangered or protected species.

Collection and quality control of metagenomic datasets

Thirty-three metagenomes sampled from five natural soil environments were publicly available and collected in the present study: 14 from grassland, seven from forest soil, nine from desert, two from Arctic soil, and one from mangrove sediment. The metagenomic datasets used can be downloadable according to the list of sequence accession numbers or web links shown in Table S1. All datasets have been produced by whole-metagenome shotgun sequencing using the Roche 454 or Illumina platforms. More reference information about these chosen metagenomes was listed in Table 1. For the datasets of FASTQ formatted sequence reads without quality control, we performed a quality check of bases by using the package Biopieces (http://www.biopieces.org). Low quality ends per read were trimmed by *trim_seq*. Trimming progressed until all bases in a 3-bp stretch with minimum quality score of 20. High quality reads were retained if satisfying the following criteria: minimum average quality score of 15 in a sliding window of 20 bp; minimum read length of 50 bp.

Estimation of microbial composition

MetaPhlAn v1.7 [24] and BLAST v2.2.22 [25] were employed for profiling the taxonomic clades in the metagenomic datasets. Briefly, metagenomic reads were firstly mapped to the MetaPhlAn reference database composed of unique clade-specific marker genes using BLASTN. The non-default parameters used for BLASTN sequencing similarity searching were as follows: E-value cutoff of 1e-10, word size of 12, and minimum alignment length of 75 nt. Relative abundance scores at all taxonomic levels from the domain level to the species level were then estimated by MetaPhlAn. In the text, mean values of abundances were shown for the mentioned taxon. To assess the compositional similarity among soil samples from different microbial communities, the Bray-Curtis measure of beta diversity [26] was employed to compare all pairwise taxonomic abundances between each sample-pair using a R function *vegdist* in the package vegan [27]. The permutation-based multivariate analysis of variance (PERMANOVA) and 2D stress value were then estimated. Based on the resultant Bray-Curtis similarity distance matrix, non-metric multidimensional scaling (NMDS) was adopted to visualize the dispersion of community structure. Multivariate analysis was carried out using vegan [27] and R (http://www.R-project.org) [28].

Metabolic reconstruction of metagenomes

Metabolic reconstruction was carried out using the HUMAnN methodology designed for the functional analyses of meta'omics [29]. High quality reads were initially mapped to the characterized protein functional database KEGG Orthology v54 [30] using the accelerated translated BLAST program USEARCH v6.0.307 [31]. The cutoff E-value was set to 1e-6 and best hits were then used to estimate relative abundances of KEGG orthologous (KO) gene families by HUMAnN v0.98. Base on the resulting KO information, MinPath was used to calculate the coverage and relative abundances of KEGG modules that are manually defined functional units [32]. Circular cladograms representing microbial taxonomic compositions and metabolic modules were implement-ed by using a standalone graphical tool GraPhlAn v0.9.5 (http://huttenhower.sph.harvard.edu/GraPhlAn).

Detection of metagenomic biomarkers

In order to further test whether some taxa/metabolic modules are significantly overrepresented in the individual soil habitat, statistical analyses were performed according to the inferred relative abundances. Differentially abundant features were identified by the approach of the linear discriminant analysis (LDA) effect size (LEfSe) and could be used as metagenomic biomarkers [21]. As the sample size is not very large in this test, the significance threshold of the alpha parameter for the Krushkal-Wallis (KW) test among classes was set to 0.01 and the cut-off logarithmic LDA score was 2.0. These analyses were performed through the Galaxy server [33]. Additionally, a non-parametric test of Spearman rank correlation between the relative abundances of each KO entry and taxonomic unit was employed to estimate co-variation of community composition and functional features using the R function *cor.test*.

Detection of microbial interactions

A recently developed computational methodology was used to investigate microbial co-occurrence and co-exclusion relationships within and between soil sites [22]. The microbial network of significant co-occurrence and co-exclusion interactions was built by a Cytoscape plugin CoNet 1.0b2 (http://psbweb05.psb.ugent.be/conet/). The taxonomic abundances estimated by MetaPhlAn were used to prepare an input matrix consisting of data from three sites (grasslands, deserts, and forest soils). The analysis was carried out with the non-default parameters listed below: 50 initial top and bottom edges; four similarity measures (Spearman, Pearson, Kullbackleibler, and Bray Curtis); edgeScores for the randomization routine; 1000 permutations and bootstraps. The resulting networks were merged based on the Simes method [34] and Benjamini-Hochberg false discovery rate (FDR) correction [35]. The FDR cutoff was set to 0.05. The ensemble co-occurrence network was visualized by Cytoscape 2.8 [36].

Results and Discussion

General characterization of soil community composition

To explore comprehensive characterizations of taxonomic compositions in the soil microbiota, 33 metagenomes sampled from five soil habitats (i.e. grassland, forest soil, desert, Arctic soil, and mangrove sediment) were included in this analysis (Table 1). Based on the assessment by MetaPhlAn, a total of 63 clades (11 phyla and 53 genera) were identified at $\geq 0.5\%$ abundance in at least one sample (Table S2). Proteobacteria was the most dominated phylum in the microbial community of soil, $\geq 70\%$ abundance detected in all soil sites except for the microbiota in the desert samples (Figure 1A). In desert, both phyla Proteobacteria and Actinobacteria exhibit almost identical abundance: 30% for Proteobacteria and 29% for Actinobacteria. In addition, Firmicutes and Bacteroidetes, which are the two major phyla dominating the human microbiome [37,38], were not frequently present in the soil microbial communities. Particularly, bacterial species within the Firmicutes rarely occurred in soil. As the taxonomic distribution of environmental metagenomic sequences are greatly affected by distinct reference databases [23], the 16S amplicon approach should provide more accurate taxonomic profiling than metagenome shotgun sequencing [12]. Previous amplicon surveys of 16S rRNA gene have pointed out that bacterial phyla Acidobacteria, Actinobacteria, Bacteroidetes, Proteobacteria, and Verrucomicorbia are often abundant and

Table 1. Summary of 33 soil metagenomes used in this study.

Biome type	ID	Sequencer	Original name	Location	Coordinates	No. of sequences	No. of bases (Mb)	Reference
Forest soil	F1	454	NA	Puerto Rican	18°18'N, 65°50'W	782,404	322	[9]
Forest soil	F2	454	NA	Massachusetts, USA	42°54'N, 72°18'W	1,439,445	742	[11]
Forest soil	F3	Illumina	AR3	Misiones, Argentina	26°44'S, 54°41'W	5,235,352	524	[12]
Forest soil	F4	Illumina	PE6	Manu National Park, Peru	12°38'S, 71°14'W	9,206,662	921	[12]
Forest soil	F5	Illumina	BZ1	Bonanza Creek, Alaska, USA	64°48'N, 148°15'W	6,543,903	654	[12]
Forest soil	F6	Illumina	CL1	South Carolina, USA	34°37'N, 81°40'W	6,402,940	640	[12]
Forest soil	F7	Illumina	DF1	North Carolina, USA	35°58'N, 79°5'W	3,890,044	389	[12]
Arctic soil	A1	454	Control	Alert, Canada	82°31'N, 62°17'W	495,998	254	[13]
Arctic soil	A2	Illumina	TL1	Toolik Lake LTER, Alaska, USA	68°38'N, 149°35'W	6,011,971	601	[12]
Grassland	G1	454	F1	Hertfordshire, UK	51°48'N, 0°14'E	976,268	358	[10]
Grassland	G2	454	F2a	Hertfordshire, UK	51°48'N, 0°14'E	1,094,883	471	[10]
Grassland	G3	454	F2b	Hertfordshire, UK	51°48'N, 0°14'E	890,966	321	[10]
Grassland	G4	454	F3	Hertfordshire, UK	51°48'N, 0°14'E	754,829	311	[10]
Grassland	G5	454	F4	Hertfordshire, UK	51°48'N, 0°14'E	946,839	391	[10]
Grassland	G6	454	F5	Hertfordshire, UK	51°48'N, 0°14'E	754,135	256	[10]
Grassland	G7	454	F6	Hertfordshire, UK	51°48'N, 0°14'E	782,342	306	[10]
Grassland	G8	454	J1	Hertfordshire, UK	51°48'N, 0°14'E	1,130,719	466	[10]
Grassland	G9	454	J1a	Hertfordshire, UK	51°48'N, 0°14'E	1,137,813	433	[10]
Grassland	G10	454	J1b	Hertfordshire, UK	51°48'N, 0°14'E	919,406	343	[10]
Grassland	G11	454	J1rhizo	Hertfordshire, UK	51°48'N, 0°14'E	1,025,699	369	[10]
Grassland	G12	454	J4	Hertfordshire, UK	51°48'N, 0°14'E	1,135,084	506	[10]
Grassland	G13	454	J7	Hertfordshire, UK	51°48'N, 0°14'E	938,860	339	[10]
Grassland	G14	Illumina	KP1	Kansas, USA	39°6'N, 96°36'W	5,348,832	535	[12]
Desert	D1	Illumina	EB017	Garwood Valley, Antarctica	78°2'S, 163°52'E	7,947,086	795	[12]
Desert	D2	Illumina	EB019	Lake Bonney Valley, Antarctica	77°44'S, 162°18'E	5,454,640	545	[12]
Desert	D3	Illumina	EB020	Lake Fryxell Valley, Antarctica	77°36'S, 163°15'E	9,446,684	945	[12]
Desert	D4	Illumina	EB021	Lake Hoare Valley, Antarctica	77°38'S, 162°53'E	6,543,681	654	[12]
Desert	D5	Illumina	EB024	Wright Valley, Antarctica	77°32'S, 161°42'E	10,863,646	1,086	[12]
Desert	D6	Illumina	EB026	Lake Bonney Valley, Antarctica	77°44'S, 162°19'E	5,951,684	595	[12]
Desert	D7	Illumina	MD3	Mojave Desert, California, USA	34°54'N, 115°39'W	5,899,497	590	[12]
Desert	D8	Illumina	SF2	Chihuahuan Desert, New Mexico, USA	35°23'N, 105°56'W	6,805,456	681	[12]
Desert	D9	Illumina	SV1	Chihuahuan Desert, New Mexico, USA	34°20'N, 106°44'W	11,122,546	1,112	[12]
Mangrove sediment	M1	454	NA	Bertioga, Brazil	23°53'S, 46°12'W	913,752	216	[23]

A

B

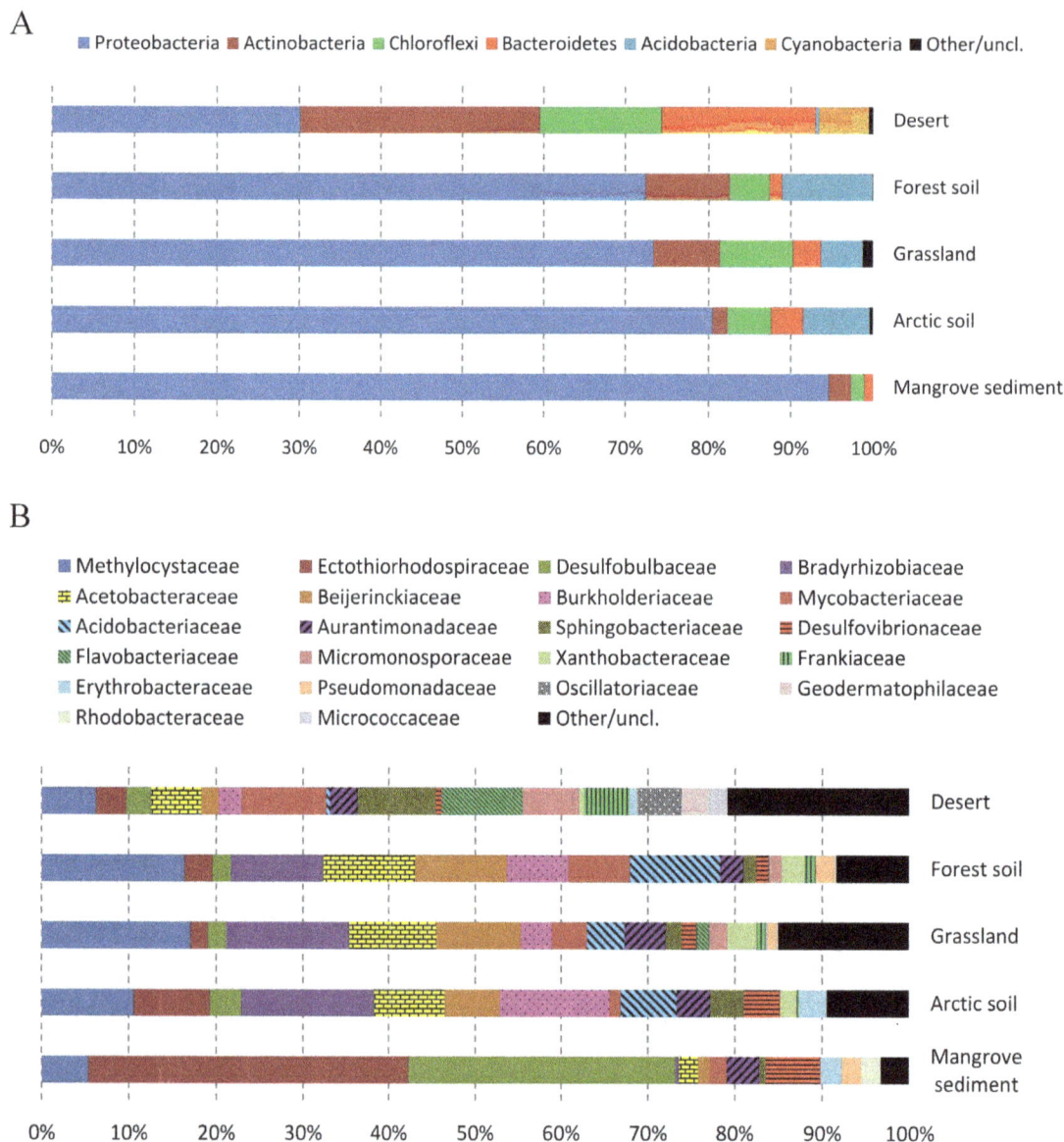

Figure 1. Taxonomic distribution of the soil microbial communities. A) Distribution at the phylum level; B) Distribution at the family level. Labels show the taxonomic units with average relative abundance >2% in at least one of five soil habitats: desert, forest soil, grassland, Arctic soil, and mangrove sediment.

ubiquitous in soil [12,39]. Although the clade-specific marker gene database in MetaPhlAn has successfully validated the composition of the human microbiome [24], it still needs to be updated with more genomes sequenced recently from various environments.

At the family level, several families were observed to be evidently more prevalent in and specific to one soil habitat or closely related habitats (Figure 1B). For instance, the family *Methylocystaceae* was dominated and almost equally present in both sites of forest soil (16.4% abundance) and grassland (17.1%) comparing with the other three soil habitats. Additionally, among five soil habitats, both families *Ectothiorhodospiraceae* (37.0%) and *Desulfobulbaceae* (30.8%) were found to be extremely abundant in the microbiota of mangrove sediment. The enrichment of these families could be reasonably explained by the selective pressures acting on certain ecological sites. For example, organisms within *Methylocystaceae* are usually methanotrophs that can metabolize

methane as their only carbon source and involved in methane oxidation [40,41]. The DNA-level evidence identified herein may support the oxidation of methane observed in forest soil and grasslands [42,43]. In addition, the microbiota of mangrove sediments is known to be sampled from anaerobic and hyperhaline seawater [23]. The corresponding environmental features should be beneficial for the dominance of *Ectothiorhodospiraceae* and *Desulfobulbaceae* in this particular habitat. The former comprises the most halophilic eubacteria [44] and bacteria in the latter family are strictly anaerobe sulphate reducers [45]. However, it is worth mentioning that taxonomic profiling of individual metagenome is visually distinguished from those of the other metagenomes within the same soil habitat (Figure S1). This is probably because publicly available soil metagenomes were generated by different research groups and varied in sampling strategies as well as sequencing methods. Thus, analysis of more soil metagenomes newly sequenced or coming soon is still required to statistically support the findings of soil biodiversity in the present study.

Structure similarity and taxonomic biomarkers of soil microbial communities

For a glimpse of structural similarity of soil microbial communities, ecological dissimilarity indices Bray-Curtis similarity scores were inferred and summarized in Table 2. The PERMA-NOVA test demonstrated that taxonomic compositions of microbial communities were significantly varied among soil habitats ($p = 0.001$). Meanwhile, the NMDS plot in Figure 2 further illustrated the compositional similarity among 33 samples from five soil sites. These results demonstrated that the microbiota from the same soil habitat should be more similar to each other. The community structure similarity is also influenced by varied geographical locations. E.g., the soil samples from grasslands were intensively clustered together and the corresponding similarity score (Bray-Curtis index 0.80 ± 0.07) is indeed the highest among all inter- or intra-group comparisons (Table 2). On the contrary, the Bray-Curtis similarity score between nine desert samples is the lowest (0.58 ± 0.16) among all intra-group comparisons, perhaps due to their sampling environments: three samples from hot deserts but the remaining ones from cold deserts [12]. Likewise, A2 sampled from the edge of the Arctic Circle is distant from A1 from high Arctic soil (Table 1). In addition, it was observed that the distances of most samples between forest soil and grasslands were closely clustered (Figure 2) and the Bray-Curtis similarity score was consistently high (0.76 ± 0.07). Whereas, the microbiota from two extreme conditions, desert and mangrove soil, respectively, exhibited the greatest compositional dissimilarity (0.37 ± 0.07) among all inter-group comparisons.

To further investigate the taxonomic distribution and differentially abundant clades of diverse soil ecosystems, we compared the abundances of microbial compositions at each taxonomic level. Figure 3A shows a cladogram visualizing all detected microbial compositions ($\geq 0.5\%$ abundance) from domain to species, respectively. Based on the inferred taxonomic profiling of all samples, a statistical strategy for discovering metagenomic biomarkers was carried out by LEfSe and determined 46 differentially abundant taxa (Table S2). Among these differentially abundant taxa, 10 and 12 were found to be family- and genus-level biomarkers, respectively (Figure 3A). These detected taxonomic biomarkers could be used as candidate indicators to distinguish at least one microbial community of five individual soil habitats. E.g., two families *Beijerinckiaceae* and *Methylocystaceae* that consist of methanotrophic taxa [46] were detected to be family-level biomarkers (P value <0.01) that were most abundant in the forest soil and grassland, respectively. The abundances of both families were found to be significantly decreased in desert and mangrove sediment (Figure 3B). The abundance differences of these methanotrophs might be positively associated with the expected capability of methane oxidation among distinct soil ecosystems. Although the organisms within the *Alphaproteobacteria* class were most differentially abundant in the grassland community, a genus-level biomarker within *Alphaproteobacteria* was specially enriched in the communities of forest soil and Arctic soil, respectively (Figure 3A). Intriguingly, the desert community had two phylum-level markers, Cyanobacteria and Chloroflexi, both of which showed the highest abundance in deserts comparing with other soil sites (Figure 3A). Bacteria in both phyla can produce their energy through photosynthesis [47]. It was worth noting that the family Oscillatoriaceae within Cyanobacteria was significantly enriched in the desert microbial community (Figure 3B). The enrichment of these bacterial groups should be consistent with the

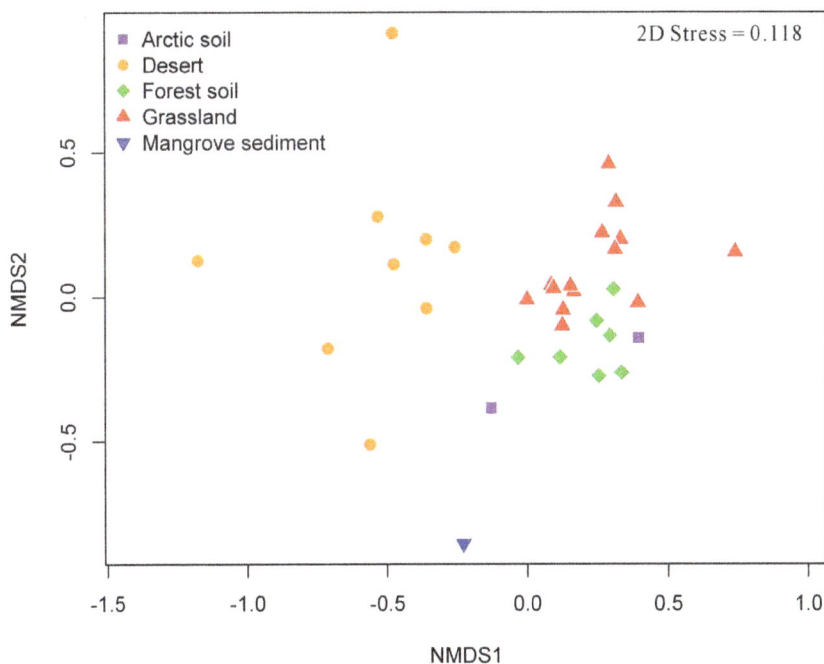

Figure 2. A nonmetric multidimensional scaling (NMDS) plot showing diversity of soil ecosystems. A Bray-Curtis distance similarity matrix was calculated based on the pairwise taxonomic profiles of 33 soil samples and used to generate NMDS coordinates of each sample. The distance linking two samples is shorter, indicating higher similarity between these samples. Samples from five soil sites were illustrated by different symbols and colors.

Table 2. Community structure similarity of the soil metagenomes within a habitat or between habitat pair.

Biome type	Grassland	Forest soil	Arctic soil	Desert	Mangrove soil
Grassland	**0.80±0.07**	0.76±0.07	0.68±0.09	0.50±0.12	0.45±0.02
Forest soil	0.76±0.07	**0.77±0.05**	0.68±0.10	0.48±0.12	0.45±0.05
Arctic soil	0.68±0.09	0.68±0.10	NA	0.45±0.11	0.52±0.15
Desert	0.50±0.12	0.48±0.12	0.45±0.11	**0.58±0.16**	0.37±0.07
Mangrove soil	0.45±0.02	0.45±0.05	0.52±0.15	0.37±0.07	NA

Mean and standard deviation of Bray-cutis similarities of all pairwise samples between any pair of soil habitats were shown herein. The number of sample combination less than 2 was denoted by NA.

following environmental features of deserts: extreme arid, strong light, and poor nutrient conditions. In addition, six species were found to be differentially abundant, some of which were uniquely present in the individual soil habitat. E.g., *Rubrobacter xylanophilus* only occurs in the microbiota of deserts. *R. xylanophilus* is the most thermophilic actinobacterium known and bears extreme tolerance to desiccation [48]. *Bradyrhizobium japonicum*, an agriculturally important species of legume-root nodulating [49], was found to be the most abundant in grassland and second in forest soil.

Metabolic potential and functional biomarkers of soil microbial communities

Besides microbial composition, metabolic potential of soil microbial communities was also investigated. In this study, we focused on the KEGG modules that are tight functional units composed of approximately 5 to 20 genes and beneficial for biological interpretation of metagenomes [29,30,38]. To further enhance the performance of statistical inference on the functional analysis, two soil sites with limited samples (two samples from Arctic soil and one sample from mangrove sediment) were excluded. After translated BLAST searching against the database of KO gene families, we found an average of ~33.6% of reads mapped to at least one KO entry (Table S3). Based on the metabolic reconstruction of 30 metagenomic datasets using HUMAnN, Figure 4 shows 119 functional modules detected in the microbial communities of grasslands, forest soils, and deserts (Table S4). Of these functional modules, we found 20 core metabolic modules that were almost entirely present at >90% coverage in all soil metagenomes tested (Table 3). Some of these core modules were essential for basic life activities of prokaryotic cells in soil, such as central carbon metabolism (M00002-3, M00007, M00009, M00011-12), nucleotide and amino acid metabolism (M00016, M00018, M00048, M00115, M00125), translation (M00178, M00359-360), and ATP synthesis (M00144). In addition, all the remaining core modules were involved in certain transport systems, three (M00207, M00222, M00239) of which are also detected in the core modules of the human microbiome [29]. On the other hand, three functional modules (M00026, M00133, M00319) were differentially covered among three soil sites (Figure 4; Table S5). It was worth noting that structural complex module manganese/zinc/iron transport system (M00319) was completely present only in the deserts but appeared to be absent in both grasslands and forest soils. It indicates that deserts microbiota is well-equipped with metal acquisition systems that play potential roles in the maintenance of metal homeostasis [50].

Furthermore, 33 functional modules were detected to be differentially abundant in at least one of three soil sites (Figure 4 and Table S4). Interestingly, two thirds of these modules were significantly enriched in the microbiota of deserts in comparison to the microbiota of grasslands and forest soils. Of them, three metabolic modules (M00165-167) are involved in the reductive pentose phosphate cycle (Benson-Calvin cycle), which is the main pathway for the conversion of atmospheric CO_2 to organic compounds [51]. It was worth noting that these overrepresented modules involved in carbon fixation might be consistent with high abundance of photosynthetic organisms Cyanobacteria present in the microbiota of desert. Additionally, eight structural complex modules detected to be functional biomarkers in deserts are responsible for the transport of metallic cation (M00317, M00319), mineral and organic ion (M00321, M00299), saccharide and polyol (M00201, M00199), glutamate (M00233), and urea (M00323). On the other hand, we found that two metabolic modules (M00022: Shikimate pathway and M00237: Branched chain amino acid transport system) were significantly overrepresented in grasslands and forest soils comparing with plant-free deserts (Figure 4 and Table S4). Both modules are associated with plant-derived metabolites [52,53]. These results showed that some modular metabolic activities are likely to be associated with the individual soil ecosystem. However, more metagenome samples from different sites are needed for accurately statistical validation of these characterized modules as promising biomarkers for diverse soil communities.

Correlation between microbial compositions and functions

Similar to the approach presented by Segata et al. [38], we assessed the correlations between microbial compositional and functional enrichment. The results showed that some significant associations between taxonomic clades and functional gene families were detected in the soil microbial communities (Spearman non-parametric test; Benjamini-Hochberg corrected p-value <0.01) (Figure S2). Notably, several taxonomic biomarkers possessed by individual microbial community mentioned above were further confirmed by the related strong associations between gene families and taxonomic clades. E.g. the gene *petA* (K02634) encoding apocytochrome f protein involved in photosynthesis was positively associated with the members of Cyanobacteria (Spearman test; q-value <0.001), one of the earliest prokaryotic organisms which can carry out oxygenic photosynthesis on Earth [47]. In addition, a significantly positive correlation (Spearman test; q-value <0.001) between methanotrophs *Methylocystaceae* and the gene *mcl* (K08691) coding for malyl-CoA lyase was observed in the microbial community of grassland. The enrichment of protein Mcl involved in both pathways of methane metabolism and carbon fixation, should be consistent with the featured metabolic activities of these methanotrophs.

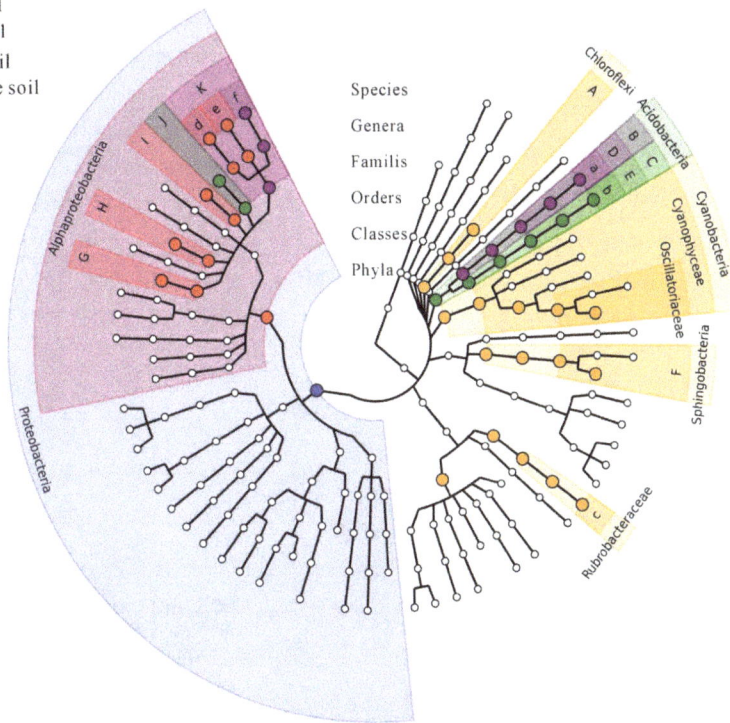

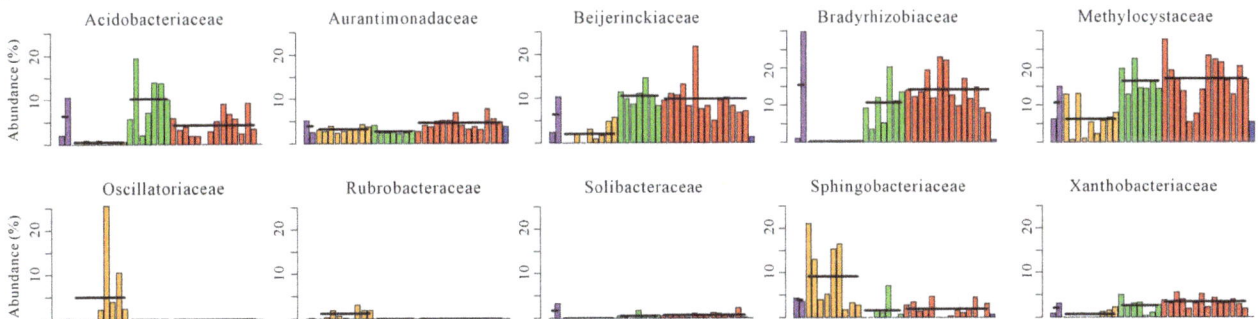

Figure 3. Taxonomic composition of soil microbial community based on the metagenomes from five soil habitats. A) Taxonomic cladogram showing all detected taxa (relative abundance ≥0.5%) in at least one sample. Taxonomic clades with more than five samples ≥0.5% abundance were used as inputs for LEfSe. Seven rings of the cladogram stand for domain (innermost), phylum, class, order, family, genus, and species (outermost), respectively. Enlarged circles in color are the differentially abundant taxa identified to be metagenomic biomarkers and the circle color is corresponding to the individual soil habitat in which the taxon is the most abundant among 5 soil ecosystems (Green for forest soil, red for grassland, purple for Arctic soil, blue for mangrove sediment, and orange for desert). B) The histograms of relative abundances of family-level biomarkers in each sample. Bacterial families significantly differential among all pairwise comparisons were illustrated. The average abundance of each family in the individual soil habitat was denoted by the horizontal line.

Soil microbial interaction network

To further decipher complex ecological relationships in the individual soil microbial community, microbial association networks were inferred based on the estimated taxonomic profiling. In this case, we intended to focus on the microbial associations within the single soil habitat, i.e. forest soil, grassland, and desert. The resultant metagenome-wide networks comprised 126 significant associations among 66 phylotypes at or above the genus level (Benjamini-Hochberg corrected p-value <0.05) (Figure 5). Of these significant phylotype correlations, 54% was detected to be co-present and the remaining was mutually excluded. Interestingly, we found that three quarters (~74%) of co-occurrence patterns

were constituted by the taxa within the same phyla; whereas nearly all co-exclusion patterns (~90%) consisted of the taxa from the distinct phyla. The evidence presented herein can again support the previous notion that phylotypes with closely evolutionary relationships usually tend to co-occur [8]. E.g., three families (*Bifidobacteriaceae*, *Mycobacteriaceae*, and *Frankiaceae*) belonging to the same class Actinobacteria showed pairwise positive correlation in the microbiota of desert (Figure 5). Similar taxon co-occurrence pattern was also found between *Bifidobacteriaceae* and *Frankiaceae* in the microbiota of grassland. Additionally, two genera within the family *Bradyrhizobiaceae* co-occurred in the grassland community: one is nitrogen-fixing bacteria *Nitrobacter* and the other is

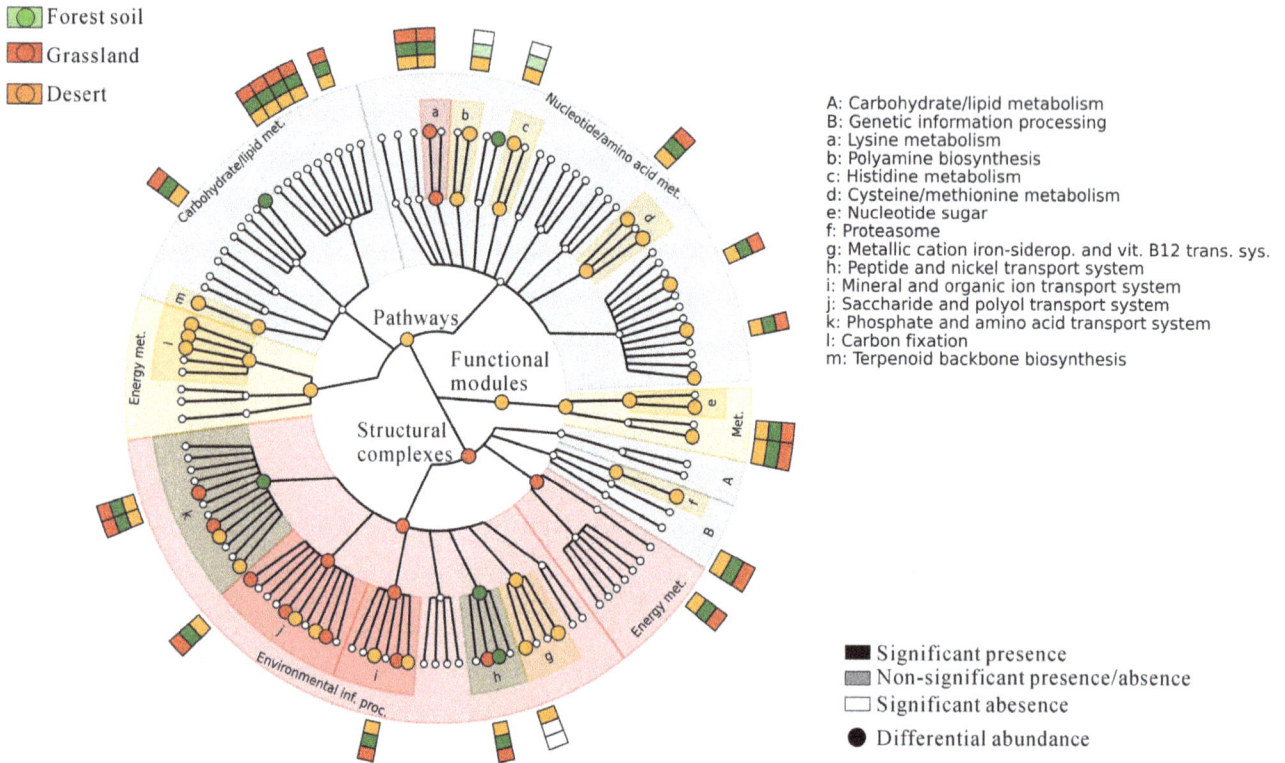

Figure 4. Metagenome-level metabolic reconstruction of the soil microbial community. KEGG BRITE hierarchical structures that are illustrated by the innermost four rings were used to cluster metabolic modules. The outermost ring composed of circles denotes KEGG functional modules detected in at least one of 30 metagenomes from three soil sites. Differentially abundant modules were inferred by LEfSe and illustrated by the enlarged circles in distinct colors: green stands for the modules most abundant in the forest soil, red for the grassland, and orange for the desert. The outermost rectangles denote core and differentially covered modules among three soil sites: ≥90% coverage stands for presence and ≤10% coverage for absence.

phototrophic bacteria *Rhodopseudomonas*. On the other hand, those mutually excluded bacteria were found to be evolutionarily unrelated. E.g., *Sphingobacteriaceae* belonging to the Bacteroidetes were negatively associated with *Desulfovibrionaceae* from the Proteobacteria and *Rubrobacteraceae* from the Actinobacteria in the microbiota of desert (Figure 5). Although most phylotype associations in the network lack empirical evidence to support their natural presence, it provides some promising targets at least to shed light on the complex cooperative or competitive mechanisms among soil microorganisms.

Conclusions

In this study, comparative metagenomic characterizations of divergent soil microbial communities were described in details by an integrated bioinformatics analysis pipeline. Complicated phylogenetic and metabolic networks with a spectrum of taxonomic and functional biomarkers were comprehensively illustrated at the metagenome level for soil. Cooperative or competitive associations among microbes from diverse soil ecosystems were also inferred to understand complex microbial interactions in the soil metagenome. This study provides new insights into the relation between soil biodiversity and ecosystem function, and provides applicable analysis and visualization approaches for studying soil microbial communities.

Supporting Information

Figure S1　Taxonomic distribution of 33 metagenomes from soil microbial communities. A) Distribution at the phylum level; B) Distribution at the family level. Labels show the taxonomic units with average relative abundance >2% in at least one of 33 samples.

Figure S2　Co-variation of bacterial clades and KEGG orthologous gene families in the desert microbiome. The spearman non-parametric correlation of each KEGG gene family against each taxonomic clade was assessed. After multiple testing corrections based on the Benjamini-Hochberg procedure, a network of significant correlations between gene families and taxonomic clades was shown herein (q-value <0.01). Ellipses denote taxa and rectangles stand for KEGG gene families. The edge linking taxonomic clade and gene family indicates that strong correlation was detected in the individual microbial community: green for forest soil, red for grassland, and orange for desert.

Table S1　Sequence data accession numbers and/or web links of soil metagenomes used in this study.

Table S2　Taxonomic profiling of the soil metagenomes estimated in this study. Relative abundances of taxa were inferred by MetaPhlAn. Differentially abundant clades among five soil habitats were detected by LEfSe and labeled by the soil site with

Table 3. Core metabolic modules shared by grasslands, deserts, and forest soils.

Module ID	Definition of modules in KEGG
M00002	Glycolysis, core module involving three-carbon compounds
M00003	Gluconeogenesis, oxaloacetate => fructose-6P
M00007	Pentose phosphate pathway, non-oxidative phase, fructose 6P => ribose 5P
M00009	Citrate cycle (TCA cycle, Krebs cycle)
M00011	Citrate cycle, second carbon oxidation
M00012	Glyoxylate cycle
M00016	Lysine biosynthesis, aspartate => lysine
M00018	Threonine biosynthesis, apartate => homoserine => threonine
M00048	Inosine monophosphate biosynthesis, PRPP + glutamine => IMP
M00115	NAD biosynthesis, aspartate => NAD
M00125	Riboflavin biosynthesis, GTP => riboflavin/FMN/FAD
M00144	Complex I (NADH dehydrogenase), NADH dehydrogenase I
M00178	Ribosome, bacteria
M00185	Sulfate transport system
M00207	Multiple sugar transport system
M00222	Phosphate transport system
M00237	Branched-chain amino acid transport system
M00239	Peptides/nickel transport system
M00359	Aminoacyl-tRNA biosynthesis, eukaryotes
M00360	Aminoacyl-tRNA biosynthesis, prokaryotes

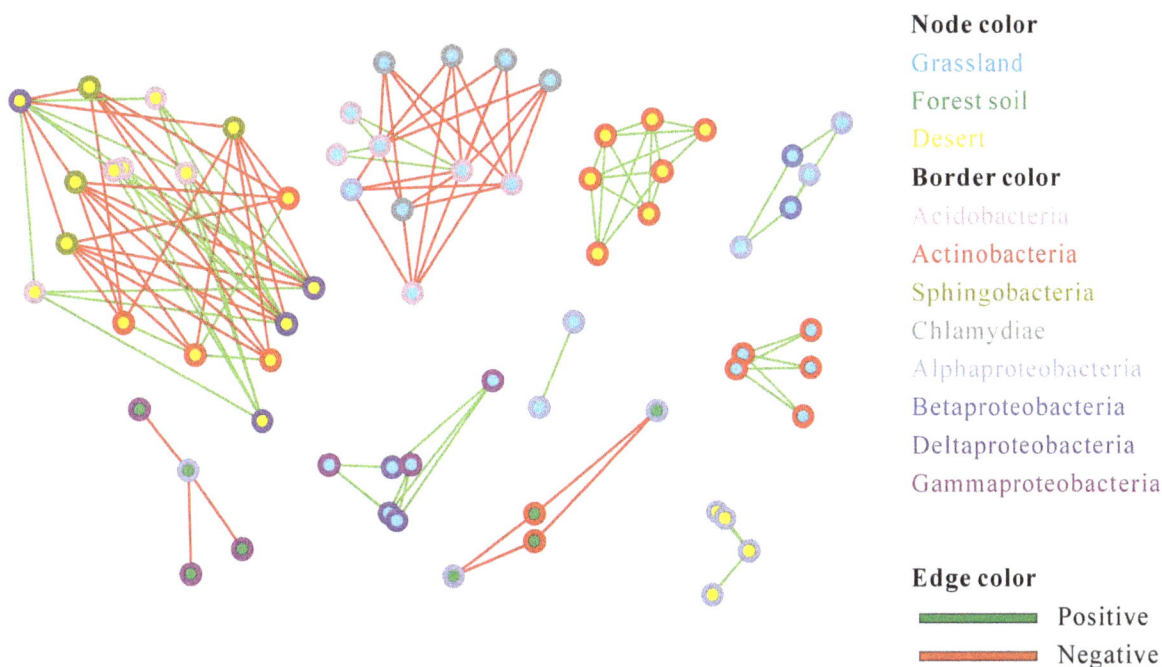

Node color
Grassland
Forest soil
Desert

Border color
Acidobacteria
Actinobacteria
Sphingobacteria
Chlamydiae
Alphaproteobacteria
Betaproteobacteria
Deltaproteobacteria
Gammaproteobacteria

Edge color
Positive
Negative

Figure 5. A global microbial interaction network of the soil microbial community. The network captured all significant associations (multiple corrected p-value <0.05) among the abundances of phylotypes at or above the genus level in the soil microbial community within and across the three soil sites. Phylotypes were illustrated by nodes (light blue for grasslands, blue for forest soils, and yellow for deserts) and edges denote significant correlations between phylotypes: positive correlation colored in green means co-occurrence whereas negative correlation in red means mutual exclusion. The border of nodes was colored according to taxonomic affiliations at the class level.

the highest LDA score among pairwise comparisons of all sites. According to the non-strict and strict statistical strategy, 46 taxonomic biomarkers were detected to be significantly differential in at least one of five soil habitats.

Table S3   The proportion of reads mapped to MetaPhlAn clade-specific marker genes and KEGG orthologous gene families.

Table S4   Estimated values for relative abundances of KEGG functional modules in the soil microbial community. Differentially abundant modules were detected by LEfSe and labeled by the soil

habitat with the highest LDA score among pairwise comparisons of all habitats.

Table S5   Estimated values represented by percentage of the coverage of KEGG functional modules in the soil microbial community. The presence/absence of modules was defined as follows: the median of coverage estimates of the samples per site > 0.9 stands for presence; the median <0.1 for absence.

Author Contributions

Conceived and designed the experiments: ZX SJS. Performed the experiments: ZX. Analyzed the data: ZX MAH SJ. Contributed reagents/materials/analysis tools: ZX LHH SJS. Wrote the paper: ZX MAH LHH SJ SJS.

References

1. Daniel R (2005) The metagenomics of soil. Nat Rev Microbiol 3: 470–478.
2. Vogel TM, Simonet P, Jansson JK, Hirsch PR, Tiedje JM, et al. (2009) TerraGenome: a consortium for the sequencing of a soil metagenome. Nat Rev Microbiol 7: 252.
3. Delmont TO, Robe P, Cecillon S, Clark IM, Constancias F, et al. (2011) Accessing the soil metagenome for studies of microbial diversity. Appl Environ Microbiol 77: 1315–1324.
4. Torsvik V, Øvreås L (2002) Microbial diversity and function in soil: from genes to ecosystems. Curr Opin Microbiol 5: 240–245.
5. Henne A, Daniel R, Schmitz RA, Gottschalk G (1999) Construction of environmental DNA libraries in *Escherichia coli* and screening for the presence of genes conferring utilization of 4-hydroxybutyrate. Appl Environ Microbiol 65: 3901–3907.
6. Rondon MR, August PR, Bettermann AD, Brady SF, Grossman TH, et al. (2000) Cloning the soil metagenome: a strategy for accessing the genetic and functional diversity of uncultured microorganisms. Appl Environ Microbiol 66: 2541–2547.
7. Fierer N, Lauber CL, Ramirez KS, Zaneveld J, Bradford MA, et al. (2012) Comparative metagenomic, phylogenetic and physiological analyses of soil microbial communities across nitrogen gradients. ISME J 6: 1007–1017.
8. Barberán A, Bates ST, Casamayor EO, Fierer N (2012) Using network analysis to explore co-occurrence patterns in soil microbial communities. ISME J. 6: 343–351.
9. De Angelis KM, Gladden JM, Allgaier M, Dhaeseleer P, Fortney JL, et al. (2010) Strategies for enhancing the effectiveness of metagenomic based enzyme discovery in lignocellulolytic microbial communities. Bioenerg Res 3: 146–158.
10. Delmont TO, Prestat E, Keegan KP, Faubladier M, Robe P, et al. (2012) Structure, fluctuation and magnitude of a natural grassland soil metagenome. ISME J 6: 1677–1687.
11. Stewart FJ, Sharma AK, Bryant JA, Eppley JM, DeLong EF. (2011) Community transcriptomics reveals universal patterns of protein sequence conservation in natural microbial communities. Genome Biol 12: R26.
12. Fierer N, Leff JW, Adams BJ, Nielsen UN, Bates ST, et al. (2012) Cross-biome metagenomic analyses of soil microbial communities and their functional attributes. Proc Natl Acad Sci U S A 109: 21390–21395.
13. Yergeau E, Sanschagrin S, Beaumier D, Greer CW (2012) Metagenomic analysis of the bioremediation of diesel-contaminated Canadian high arctic soils. PLoS One. 7: e30058.
14. Langille MG, Zaneveld J, Caporaso JG, McDonald D, Knights D, et al. (2013) Predictive functional profiling of microbial communities using 16S rRNA marker gene sequences. Nat Biotechnol 31: 814–821.
15. Meyer F, Paarmann D, DSouza M, Olson R, Glass EM, et al. (2008) The metagenomics RAST server - a public resource for the automatic phylogenetic and functional analysis of metagenomes. BMC Bioinformatics 9: 386.
16. Markowitz VM, Chen IM, Chu K, Szeto E, Palaniappan K, et al. (2012) IMG/M: the integrated metagenome data management and comparative analysis system. Nucleic Acids Res 40: D123–129.
17. Sun S, Chen J, Li W, Altintas I, Lin A, et al. (2011) Community cyberinfrastructure for Advanced Microbial Ecology Research and Analysis: the CAMERA resource. Nucleic Acids Res 39: D546–551.
18. Wu S, Zhu Z, Fu L, Niu B, Li W (2011) WebMGA: a customizable web server for fast metagenomic sequence analysis. BMC Genomics 12: 444.
19. Gevers D, Knight R, Petrosino JF, Huang K, McGuire AL, et al. (2012) The Human Microbiome Project: a community resource for the healthy human microbiome. PLoS Biol 10: e1001377.
20. Segata N, Boernigen D, Tickle TL, Morgan XC, Garrett WS, et al. (2013) Computational meta'omics for microbial community studies. Mol Syst Biol 9: 666.
21. Segata N, Izard J, Waldron L, Gevers D, Miropolsky L, et al. (2011) Metagenomic biomarker discovery and explanation. Genome Biol 12: R60.
22. Faust K, Sathirapongsasuti JF, Izard J, Segata N, Gevers D, et al. (2012) Microbial co-occurrence relationships in the human microbiome. PLoS Comput Biol 8: e1002606.
23. Andreote FD, Jiménez DJ, Chaves D, Dias AC, Luvizotto DM, et al. (2012) The microbiome of Brazilian mangrove sediments as revealed by metagenomics. PLoS One 7: e38600.
24. Segata N, Waldron L, Ballarini A, Narasimhan V, Jousson O, et al. (2012) Metagenomic microbial community profiling using unique clade-specific marker genes. Nat Methods 9: 811–814.
25. Altschul SF, Gish W, Miller W, Myers EW, Lipman DJ (1990) Basic local alignment search tool. J Mol Biol 215: 403–410.
26. Bray JR, Curtis JT (1957) An ordination of the upland forest communities of southern Wisconsin. Ecol Monographs, 27: 325–349.
27. Oksanen J (2011) Multivariate analysis of ecological communities in R: vegan tutorial. URL: http://cc.oulu.fi/~jarioksa/opetus/metodi/vegantutor.pdf.
28. R Development Core Team (2007) R: A language and environment for statistical computing [http://www.R-project.org].R Foundation for Statistical Computing, Vienna, Austria.
29. Abubucker S, Segata N, Goll J, Schubert AM, Izard J, et al. (2012) Metabolic reconstruction for metagenomic data and its application to the human microbiome. PLoS Comput Biol 8: e1002358.
30. Kanehisa M, Goto S, Furumichi M, Tanabe M, Hirakawa M. (2010) KEGG for representation and analysis of molecular networks involving diseases and drugs. Nucleic Acids Res 38: D355–360.
31. Edgar RC (2010) Search and clustering orders of magnitude faster than BLAST. Bioinformatics 26: 2460–2461.
32. Ye Y, Doak TG (2009) A parsimony approach to biological pathway reconstruction/inference for genomes and metagenomes. PLoS Comput Biol 5: e1000465.
33. Goecks J, Nekrutenko A, Taylor J, Galaxy Team (2010) Galaxy: a comprehensive approach for supporting accessible, reproducible, and transparent computational research in the life sciences. Genome Biol 11: R86.
34. Sarkar SK, Chang CK (1997) The Simes method for multiple hypothesis testing with positively dependent test statistics. J Am Stat Assoc 92: 1601–1608.
35. Benjamini Y, Hochberg Y (1995) Controlling the false discovery rate: a practical and powerful approach to multiple testing. J Royal Statis Soc B 57: 289–300.
36. Smoot ME, Ono K, Ruscheinski J, Wang PL, Ideker T. (2011) Cytoscape 2.8: new features for data integration and network visualization. Bioinformatics 27: 431–432.
37. Qin J, Li R, Raes J, Arumugam M, Burgdorf KS, et al. (2010) A human gut microbial gene catalogue established by metagenomic sequencing. Nature 464: 59–65.
38. Segata N, Haake SK, Mannon P, Lemon KP, Waldron L, et al. (2012) Composition of the adult digestive tract bacterial microbiome based on seven mouth surfaces, tonsils, throat and stool samples. Genome Biol 13: R42.
39. Janssen PH (2006) Identifying the dominant soil bacterial taxa in libraries of 16S rRNA and 16S rRNA genes. Appl Environ Microbiol 72: 1719–1728.
40. Dedysh SN (2009) Exploring methanotrophs diversity in acidic northern wetlands: molecular and cultivation-based studies. Microbiology 78: 655–669.
41. Dedysh SN (2011) Cultivating uncultured bacteria from northern wetlands: knowledge gained and remaining gaps. Front Microbiol 2: 184.
42. Ojima DS, Valentine DW, Mosier AR, Parton WJ, Schimel (1993) Effect of land use change on methane oxidation in temperate forest and grassland soils. Chemosphere 26: 675–685.
43. Jang I, Lee S, Hong JH, Kang H (2006) Methane oxidation rates in forest soils and their controlling variables, a review and a case study in Korea. Ecol Res 21: 849–854.
44. Imhoff JF (2005) Family II. *Ectothiorhodospiraceae*. In: Brenner DJ, Krieg NR, Staley JT, Garrity GM, editors. *Bergey's manual of systematic bacteriology*, 2nd. Springer, New York, NY. pp. 41–52.

45. Kuever J, Rainey FA, Widdel F (2005) Family II. *Desulfobulbaceae* fam. nov. In: Brenner DJ, Krieg NR, Staley JT, Garrity GM, editors. p.*Bergey's manual of systematic bacteriology*, 2nd. New York: Springer, p.988.

46. Lau E, Fisher MC, Steudler PA, Cavanaugh CM (2013) The methanol dehydrogenase gene, mxaF, as a functional and phylogenetic marker for proteobacterial methanotrophs in natural environments. PLoS One 8: e56993.

47. Mulkidjanian AY, Koonin EV, Makarova KS, Mekhedov SL, Sorokin A, et al. (2006) The cyanobacterial genome core and the origin of photosynthesis. Proc Natl Acad Sci U S A 103: 13126–13131.

48. Empadinhas N, da Costa MS (2011) Diversity, biological roles and biosynthetic pathways for sugar-glycerate containing compatible solutes in bacteria and archaea. Environ Microbiol 13: 2056–2077.

49. Kaneko T, Nakamura Y, Sato S, Minamisawa K, Uchiumi T, et al. (2002) Complete genomic sequence of nitrogen-fixing symbiotic bacterium *Bradyrhizobium japonicum* USDA110. DNA Res 9: 189–197.

50. Desrosiers DC, Sun YC, Zaidi AA, Eggers CH, Cox DL, et al. (2007) The general transition metal (Tro) and Zn2+ (Znu) transporters in Treponema pallidum: analysis of metal specificities and expression profiles. Mol Microbiol 65: 137–152.

51. Wolosiuk RA, Ballicora MA, Hagelin K (1993) The reductive pentose phosphate cycle for photosynthetic CO2 assimilation: enzyme modulation. FASEB J 7: 622–637.

52. Maloney GS, Kochevenko A, Tieman DM, Tohge T, Krieger U, et al. (2010) Characterization of the branched-chain amino acid aminotransferase enzyme family in tomato. Plant Physiol 153: 925–936.

53. Maeda H, Dudareva N (2012) The Shikimate pathway and aromatic amino acid biosynthesis in plants. Annu Rev Plant Biol 63: 73–105.

2

Spatial Distribution of Soil Organic Carbon and Its Influencing Factors in Desert Grasslands of the Hexi Corridor, Northwest China

Min Wang[1,2]*, Yongzhong Su[1], Xiao Yang[1]

1 Linze Inland River Basin Research Station, Chinese Ecosystem Network Research, Cold and Arid Regions Environmental and Engineering Research Institute, Chinese Academy of Sciences, Lanzhou, Gansu, China, **2** University of Chinese Academy of Sciences, Beijing, China

Abstract

Knowledge of the distribution patterns of soil organic carbon (SOC) and factors that influence these patterns is crucial for understanding the carbon cycle. The objectives of this study were to determine the spatial distribution pattern of soil organic carbon density (SOCD) and the controlling factors in arid desert grasslands of northwest China. The above- and belowground biomass and SOCD in 260 soil profiles from 52 sites over 2.7×10^4 km^2 were investigated. Combined with a satellite-based dataset of an enhanced vegetation index during 2011–2012 and climatic factors at different sites, the relationships between SOCD and biotic and abiotic factors were identified. The results indicated that the mean SOCD was 1.20 (SD:+/− 0.85), 1.73 (SD:+/− 1.20), and 2.69 (SD:+/− 1.91) kg m^{-2} at soil depths of 0–30 cm, 0–50 cm, and 0–100 cm, respectively, which was smaller than other estimates in temperate grassland, steppe, and desert-grassland ecosystems. The spatial distribution of SOCD gradually decreased from the southeast to the northwest, corresponding to the precipitation gradient. SOCD increased significantly with vegetation biomass, annual precipitation, soil moisture, clay and silt content, and decreased with mean annual temperature and sand content. The correlation between BGB and SOCD was closer than the correlation between AGB and SOCD. Variables could together explain about 69.8%, 74.4%, and 78.9% of total variation in SOCD at 0–30 cm, 0–50 cm, and 0–100 cm, respectively. In addition, we found that mean annual temperature is more important than other abiotic factors in determining SOCD in arid desert grasslands in our study area. The information obtained in this study provides a basis for accurately estimating SOC stocks and assessing carbon (C) sequestration potential in the desert grasslands of northwest China.

Editor: Ben Bond-Lamberty, DOE Pacific Northwest National Laboratory, United States of America

Funding: The Strategic Priority Research Program - Climate Change: Carbon Budget and Relevant Issues of the Chinese Academy of Sciences, grant number XDA05050406-3. (2) National Natural Science Foundation of China (91125022). The funders had no role in study design, data collection and analysis, decision to publish, or preparation of the manuscript.

Competing Interests: The authors have declared that no competing interests exist.

* E-mail: wmin85@126.com

Introduction

Soil plays a crucial role in the global carbon cycle by linking carbon transformation with the pedosphere, biosphere, and atmosphere. Therefore, minor changes in the soil carbon pool will greatly affect the alteration of atmospheric CO_2 concentration and have potential feedbacks to climate change [1–4]. SOC storage in temperate grasslands is heavily studied [5–14], while there is little research examining SOC storage in arid regions such as desert-grassland or desert-steppe. Arid regions cover about 47.2% of the earth's land area, with soils containing nearly 241 Pg of soil organic carbon, which is about 40 times more than what was added into the atmosphere through anthropogenic activities [15]. Additionally, soils in these regions are fragile and may experience degradation, desertification, wind erosion, and overgrazing. Small changes in soil conditions can modify the original balance of soil carbon cycle, increase the C loss from soil, and release more greenhouse gases into the atmosphere. Therefore, SOC storage in the desert-grassland ecosystem is a critical component of global C cycle and has a considerable effect on reducing the rate of enrichment of atmospheric CO_2.

In northwest China, desert-grasslands are widely distributed (near 6.5×10^7 hm^2); however, the SOC storage here has not been widely studied. Among the limited estimates of SOC storage in desert-grassland in northwest China [16–18], large differences were found, potentially due to different data sources or approaches. Data from the Second National Soil Survey were usually used for these previous estimates [16–18], but few soil profiles were sampled from the grasslands in northwest China, and these soil profiles lacked data on bulk density and gravel fractions [19]. Regarding different data approaches, previous studies usually calculated SOC stock using average SOC density (SOCD); however, this approach could be constrained by limited soil profiles and large soil heterogeneity. Accordingly, satellite-based approaches will be useful to scale up site-level observations to regional-scale estimates [19].

SOC storage in grasslands is closely correlated with biological, climatic, and edaphic factors. SOC storage exhibits a balance between C inputs from organic material, and C losses through decomposition and mineralization [20–23]. In particular, this balance depends on climatic conditions [18,24–26]. Precipitation and temperature determine the vegetation types, size of plant productivity, and the speed of microbial degradation of soil organic matter. In addition to climate, soil texture plays an important role, which results in an increasing clay content and decreasing C outputs through its stabilizing effect on SOC [5,26,27].

The Hexi Corridor, one characteristically arid area in northwest China, with annual precipitation ranging from about 200 mm in the east to less than 50 mm in the west, represents a desert-grassland ecosystem. Desert-grasslands here always contain transition zones between desert and grassland or between desert and oases, and play a crucial role in maintaining a stable ecological environment and productivity. The desert grassland ecosystem in the Hexi Corridor has unique features, such as limited precipitation, low vegetation cover, coarse soil particles, large gravel content, and highly intensified wind erosion. Accordingly, the vegetation composition varies more than other desert-grasslands. For example, desert steppe in Inner Mongolia is mainly composed of gramineous plants and shrubs with deep roots [28], whereas small shrubs or subshrubs with shallow roots dominate the desert grassland in the Hexi Corridor [29], which results in small biomass productivities and low soil C inputs. These characteristics could lead to greater differences in SOC storage compared with other typical grasslands, but few studies have focused on this desert grassland. The working hypothesis for our study was that the spatial distribution of SOC will show clear relationships with unique climate and vegetation biomass, as well as specific soil conditions. Throughout our field investigations, our research objectives were: (1) to identify SOC density and its spatial distribution characteristics; and (2) to analyze the influence of biotic, climatic, and edaphic factors on SOC density and its distribution in an arid desert grassland. Well water conditions and soil particle composition can promote plant growth and soil organic carbon fixation, and high temperature can accelerate the decomposition of soil organic carbon. For these reasons we hypothesized that SOCD would increase with the increase of moisture and soil clay content, and decrease with the rise of temperature.

Materials and Methods

Ethics Statements

The location of field studies is not privately-owned or protected in any way, so no specific permission was required. All field studies in the desert-grassland were undertaken with support from Linze Inland River Basin Research Station, Cold and Arid Regions Environmental and Engineering Research Institute, Chinese Academy of Sciences. The field studies did not involve endangered or protected species.

Study area

The study area is found in the central region of the Hexi Corridor in Gansu province, northwestern China (spanning from $101°42'36''$ to $97°45'36''E$ and $40°31'12''$ to $38°8'26''N$, elevation ranging from 1200 m to 1500 m); the study region has an area of $\sim 2.7 \times 10^4$ km^2. The mean annual precipitation varied from 250 mm to 50 mm from the southeast to the northwest, and 70–80% of the rainfall occurs between June and August. The annual mean temperature varies from 5 to 9°C. The main soil types are Calcic-Orthic Aridosols according to Chinese Soil Taxonomy, which is equivalent to the Aridosols and Entisols of the USDA soil taxonomy classification (Group of Chinese Soil Taxonomy, Institute of Soil Science, Chinese Academy of Sciences, 2001). Soil thickness ranges from 0.2 m to 1.5 m, and most soils contain a large amount of gravel (in 0.5–6 cm),

especially below the 30 cm soil layer. The vegetation population structure in desert grassland is relatively simple, and the main plant species is composed of small shrubs and sub-shrubs including *Asterothamnus centraliasiaticus*, *Reaumuria songarica*, *Salsola passerina*, *Sympegma regelii* and some ephemeral plant species such as *Suaeda glauca*, *Bassia dasyphylla* and *Artemisia scoparia*. The research areas have been subjected to grazing prohibition since 2000.

Sampling sites

A total of 52 sites were selected (Figure 1). Each site contains five plots (the area of a square plot is 1 m×1 m for herbaceous plants and small semi-shrubs or 5 m×5 m for shrubs). A total of 260 soil profiles were sampled (i.e., five profiles at each site) in August of 2011–2012. Based on the pre- investigation of vegetation community types in the study area,we set 52 sampling points. These sampling points basically cover the main vegetation communities of the area (*Reaumuria songarica* community, *Salsola passerine* community, *Sympegma regelii* community, and *Asterothamnus centraliasiaticus* community), and can represent the community characteristics, biomass and soil organic carbon contents and other information of desert grassland. On the other hand, we chose areas with no animal dung, no trace of vegetation were eaten, no trace of trampling, and within the barbed wire enclosure as sampling points. That can ensure selected sample sites without human grazing interference.

Soil sampling, analysis, and biomass survey

At each sampling plot, three soil pits were randomly excavated and mixed into a composite sample at seven depths of 0–5, 5–10, 10–20, 20–30, 30–50, 50–70, and 70–100 cm. Bulk density samples for each depth interval were obtained using a cutting ring (volume of 100 cm^3). Soil moisture (SM) was measured gravimetrically after 24 h desiccation at 105°C. Bulk density was also calculated as the ratio of the oven-dry soil weight to the cutting ring volume. Soil samples were air-dried, hand-picked to remove plant residues, visible soil organisms, and stones, and weighed. The air-dried samples were then sieved through a screen with 2 mm openings, and gravels (>2 mm) were weighed. The weight percent of gravel to soil was obtained. A proportion of samples that passed through the 2 mm sieve were finely ground to pass through a 0.10 mm sieve and analyzed for soil organic carbon

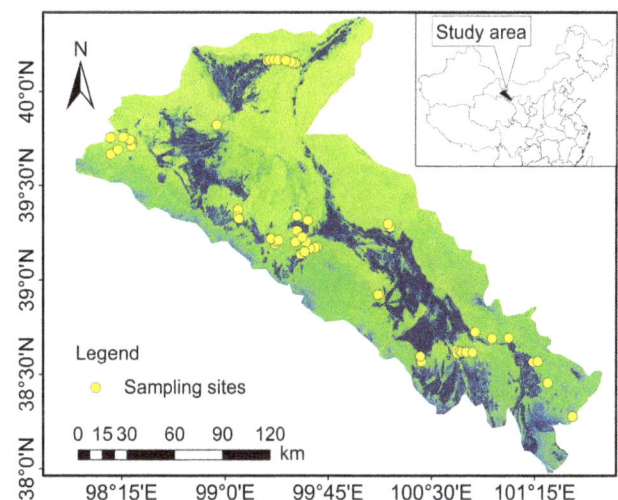

Figure 1. Spatial distribution of sampling sites in the Hexi Corridor.

by the $K_2Cr_2O_7$-H_2SO_4 oxidation method developed by Walkley-Black [30]. A subsample was then analyzed for soil texture by the wet sieve method [31].

Aboveground biomass (AGB) and belowground biomass (BGB) were harvested at 260 plots. At sites with herbaceous plants, which were always present at a low plant density (1 m×1 m), we excavated the total plants to obtain aboveground and below-ground biomass. At sites with shrubs (5 m×5 m), one or several of the dominant plant species in the plot were chosen according to the crown breadth proportion (large: length × width ≥ 50 cm×50 cm; medium: 20 cm×20 cm≤length × width <50 cm×50 cm, and small: length × width <20 cm×20 cm); the whole plants were then dug out in accordance with the crown breadth survey to estimate above and belowground biomass. In the laboratory, the roots samples were soaked in water and cleaned of residual soil using a 0.5 mm sieve. Biomass samples were oven-dried at 65°C to a constant weight and weighed to the nearest 0.01 g.

MODIS data and climate information

The MODIS-EVI data used in this study were obtained from the United States Geological Survey at a spatial resolution of 250 m×250 m and 16-day intervals for the period 2011 to 2012 (http://LPDAAC.usgs.gov). Monthly maximum EVI composites were generated using the Maximum Value Composition method proposed by Holben [32] from 2011 to 2012. The EVI data used were the average of monthly EVI during the growing season from July to August.

Climate data, such as mean annual air temperature (MAT) and annual precipitation (AP), were separated from the climate database of the China monthly ground weather dataset during 2011–2012 (http://cdc.cma.gov.cn). These data were spatially interpolated from the records of 25 climatic stations located throughout the Hexi Corridor.

SOC evaluation

SOC densities for each soil profile at 0–30 cm, 0–50 cm, and 0–100 cm depth intervals were calculated:

$$SOCD = \sum_{h=1}^{n} H_h \times BD_h \times SOC_h \times (1 - C_h)/100 \qquad (1)$$

where SOC density (SOCD) in kg m^{-2}, soil thickness (H_h) in cm, bulk density (BD_h) in g cm^{-3}, SOC (SOC_h) in g kg^{-1}, and volume percentage of the fraction >2 mm (C_h) at layer h were used.

To investigate the spatial distribution of SOCD, we established the relationship between SOCD and MODIS-EVI for three soil depth intervals (Table 1), which was based on the linear relationship between AGB-EVI (Figure 2A) and AGB-SOCD (Figure 2B, and equation for AGB-SOCD$_{0-50 cm}$ and AGB-SOCD$_{0-100 cm}$ was SOCD$_{0-50 cm}$ = 0.014AGB+0.091, R^2 = 0.41, SOCD$_{0-100 cm}$ = 0.023AGB+0.0039, R^2 = 0.44, respectively).

Using the regression equations (Table 1), each pixel of EVI was converted to SOCD. We then obtained the spatial distribution of SOCD for different soil layers (Figure 3A–C). The spatial distribution of SOCD was performed in ArcGIS, version 9.3 (ESRI, RedLands, California).

Statistical analysis

We used simple linear regression to analyze the relationship between dependent and independent variables (AGB-EVI, SOCD-AGB, SOCD-EVI, SOCD-BGB, and SOCD-TB,

Table 1. Relationships between soil organic carbon density and enhanced vegetation index at three soil depth intervals (0–30 cm, 0–50 cm, and 0–100 cm).

Soil depth	Equation	R^2	P	RMSE	SSE	F-statistic
0–30 cm	SOCD$_{0-30 cm}$ = 18.87 EVI - 1.05	0.62	<0.001	0.52	13.97	82.77
0–50 cm	SOCD$_{0-50 cm}$ = 25.64 EVI - 1.33	0.58	<0.001	0.70	30.76	69.98
0–100 cm	SOCD$_{0-30 cm}$ = 37.12 EVI - 1.75	0.48	<0.001	1.35	95.32	47.69

Notes: RMSE, root-mean-square error; SSE, sum of squares for error.

respectively). Additionally, to evaluate integrative effects of MAT, AP, SM, and soil texture on SOCD, a general linear model (GLM) was employed. Ordinary least squares regression was used to fit these models. Residuals from the linear models were examined for normality (by Shapiro-Wilk test), independence (by Durbin-Watson test), and linearity (by plotting the residuals after fitting linear regression between dependent variable and independent variable) and data met the assumption of normality, independence and linearity. Therefore, linear regressions and GLM models were effective and meaningful for these analyses. Statistical analyses were conducted with R version 3.0.2 package (http://www.R-project.org). We also analyzed the correlations between SOCD and environmental factors (MAT, AP, SM, Silt, Clay, and Sand) using nonlinear regression (by Levenberg-Marquardt and Universal Global Optimization analyses). We constructed exponential equations to describe relationships between SOCD and MAT, AP, SM, soil texture. Nonlinear regression analyses were performed using 1stOpt software, version 1.5 (First Optimization, 7D-Soft High Technology Inc., Xian, China).

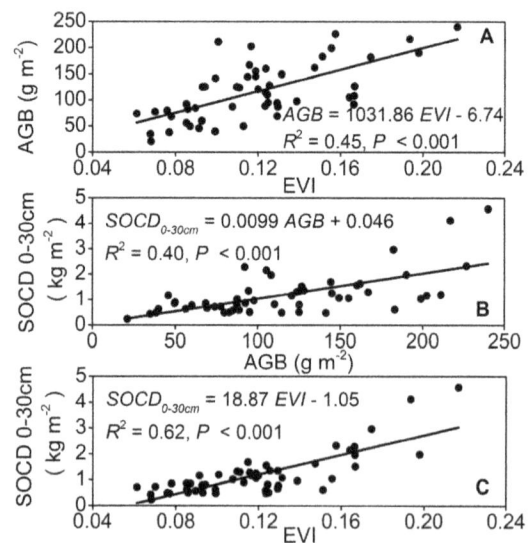

Figure 2. Relationships between above-ground biomass, enhanced vegetation index, and soil organic carbon density at depth interval of 0–30 cm. Notes: AGB, above-ground biomass; EVI, enhanced vegetation index; SOCD, soil organic carbon density.

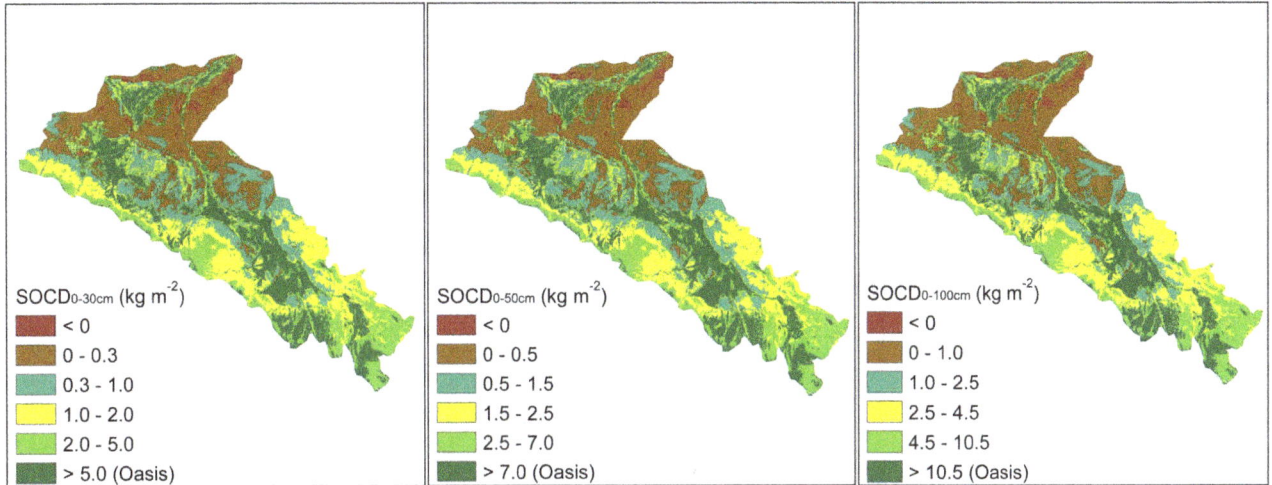

Figure 3. Spatial distributions of soil organic carbon density at different soil layers (0–30 cm, 0–50 cm, and 0–100 cm). *Notes:* SOCD, soil organic carbon density.

Results

SOC stocks and spatial distribution

The statistical description of SOCD across 52 sites at soil depths of 0–30 cm, 0–50 cm, and 0–100 cm is shown in Table 2. As shown, SOCD exhibited large variations among the three soil depths, ranging from 0.24–4.58 kg m^{-2} for 30 cm in depth, 0.35–6.32 kg m^{-2} for 50 cm, and 0.59–9.57 kg m^{-2} for 100 cm, respectively. The corresponding average SOC densities were 1.20, 1.73, and 2.69 kg m^{-2}, respectively. The SOC content in 0–30 cm interval accounted for almost 45% of total SOC in the top 1 m of soil.

The density of SOC decreased from the southeast to the northwest (Figure 3A–C), which corresponds to the precipitation gradient.

Effects of biomass and environmental factors on SOCD

The significant positive relationships between SOCD and AGB, BGB, and TB at different soil depths were characterized by linear functions (Figure 4, $P<0.001$). The R^2 values of regression functions between TB and SOCD were the highest of three biological variables. The R^2 values of regression functions between BGB and SOCD were higher than that between AGB and SOCD and thus showed the closer correlation between BGB and SOCD compared with the correlation between AGB and SOCD (Figure 4, A–C for BGB, D–F for AGB) and exhibited the crucial influence of BGB on SOC distribution (Figure 4G–I). The association of SOC content with BGB and TB was closest in the top soil and decreased at deeper intervals (Figure 4D and G). Nevertheless, the

relationships between AGB and SOCD exhibited only minimal differences at three soil depths (Figure 4A–C).

We established regression relationships between SOCD and various environmental factors, such as MAT, AP, SM, silt, clay, and sand content (Figure 5). In the top 0–30 cm interval, SOCD decreased markedly with MAT (Figure 5A) and sand content (Figure 5P). On the contrary, SOCD increased significantly with an increase SM (Figure 5G). Furthermore, SOCD was positively related with silt (Figure 5J) and clay content (Figure 5M) as well as AP and SM. The regression curves indicated that the closest relationship between SOCD and environmental variables in the top 30 cm soil was with MAT (Figure 5A, $R^2 = 0.79$, $P<0.001$), followed by SM (Figure 5G, $R^2 = 0.67$, $P<0.001$). Additionally, similar relationships between SOCD and environmental factors were also observed in other soil intervals (Figure 5B, E, H, K, N, and Q for 0–50 cm; Figure 5C, F, I, L, O, and R for 0–100 cm). The $\tilde{R}^2$ value of fitted curves for associations of SOCD with MAT, AP, SM, and sand content declined with soil depth, but an increasing trend was found for association of SOCD with silt and clay content (Figure 5).

According to the above-described relationship between SOCD and environmental factors, we chose five variables (MAT, SM, clay, silt, and sand content, and soil moisture as a measure of water availability) to establish a GLM model. The results suggested that environmental factors explained 69.81%, 74.41%, and 78.87% of the overall variation of SOCD at the soil intervals of 0–30 cm, 0–50 cm, and 0–100 cm, respectively (Table 3).

MAT was the most important parameter for SOCD at 0–30 cm and 0–50 cm (and accounted for 35.53% and 33.58% of

Table 2. Statistics of soil organic carbon density at three soil depth intervals (0–30 cm, 0–50 cm, and 0–100 cm).

SOCD	N	Mean (kg m^{-2})	Std.D. (kg m^{-2})	Min (kg m^{-2})	Median (kg m^{-2})	Max (kg m^{-2})
SOCD$_{0-30 cm}$	52	1.20	0.85	0.24	0.99	4.58
SOCD$_{0-50 cm}$	52	1.73	1.20	0.35	1.44	6.32
SOCD$_{0-100 cm}$	52	2.69	1.91	0.59	2.16	9.57

Notes: SOCD, soil organic carbon density; N, number of samples; Std.D., standard Deviation.

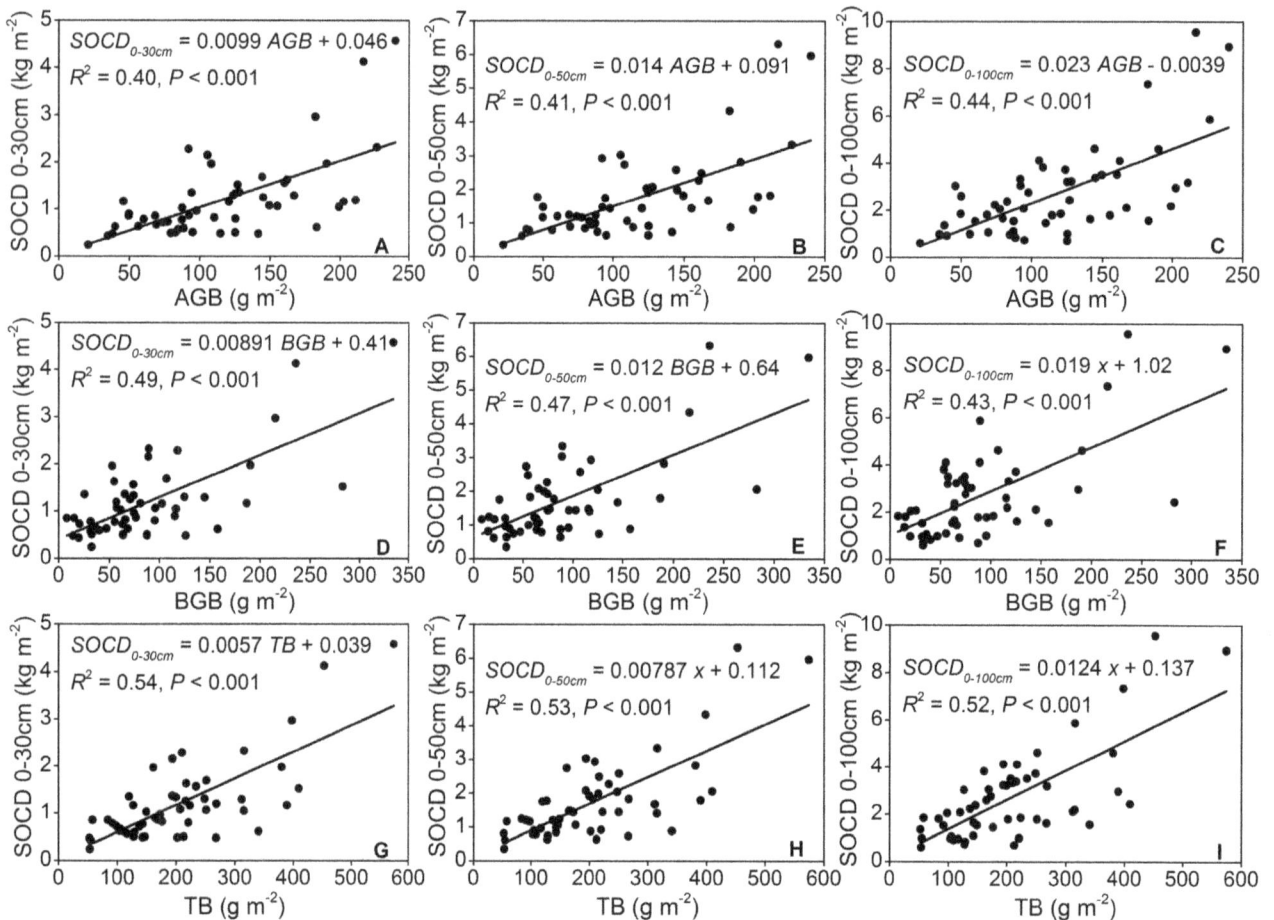

Figure 4. Relationships between soil organic carbon density and biomass at different depth intervals. *Notes:* AGB, above-ground biomass; BGB, below-ground biomass; TB, total biomass; SOCD, soil organic carbon density.

variation), whereas SM was the most important parameter for SOCD at 0–100 cm (where it accounted for 23.61% of variation). The proportion of variances explained by both factors decreased with an increase of soil depth. Soil texture variables (clay, silt, and sand content) explained 7.41%, 15.71%, and 35.95% of the variance at 0–30 cm, 0–50 cm, and 0–100 cm, respectively. In contrast to MAT and SM, the proportion of variance explained by soil texture markedly increased along with soil depth. The proportion of variance explained by sand content increased more rapidly than those explained by the other two variables, and instead silt content become the most important textural variable at soil depth of 0–100 cm (accounted for 13.60% of variation).

Discussion

SOC storage estimation

We summarized previous estimations on SOC storage on different vegetation types at global and regional scales in Table 4. In this study, the mean SOCD of 260 soil profiles at a soil depth of 0–100 cm in the desert grassland of the Hexi Corridor was 2.69 kg m^{-2}. Our results were generally lower than the global mean SOCD (10.8 kg m^{-2}), the average SOCD in China (7.8 kg m^{-2}), and other records based on vegetation types (temperate desert, steppe, and grassland). This difference is most likely due to differences in climate and soil conditions, which play a critical role in determining vegetation types and biomass productivity. The

drier local climate, together with a greater gravel content and thinner soil layer thickness compared to other temperate steppe, grassland, and desert-grassland regions lead to lower vegetation production and limited SOC inputs [29].

Relationship between SOCD and biomass

SOC concentrations are closely linked to biotic processes, such as biomass production, decomposition, and the placement of aboveground litter and root litter in and on the soil [33]. Through regression analyses, we detected a higher coefficient of R^2 for the linear functions between SOCD and BGB compared with the association between SOCD and AGB at the three soil depth intervals. This finding indicated that BGB was most likely the main resource of C inputs and a dominant biological factor on the determination of SOCD. Belowground roots can provide abundant and stable organic material into soil and enhance SOC density. However, due to intense wind erosion in our research area, large amounts of litter fall were blown away by wind, which resulted in only a small amount of aboveground organic matter entering the soil.

Both BGB and TB exhibited the highest correlation in the top 30 cm soil, and this correlation decreased at deeper intervals likely due to high organic matter inputs at surface soil [26,34]. Large gravel content at depths below 30 cm make it hard for roots to extend to deep soil layers, so most of root biomass is concentrated in the upper 0–30 cm soil interval (almost 97%) [29]. According to

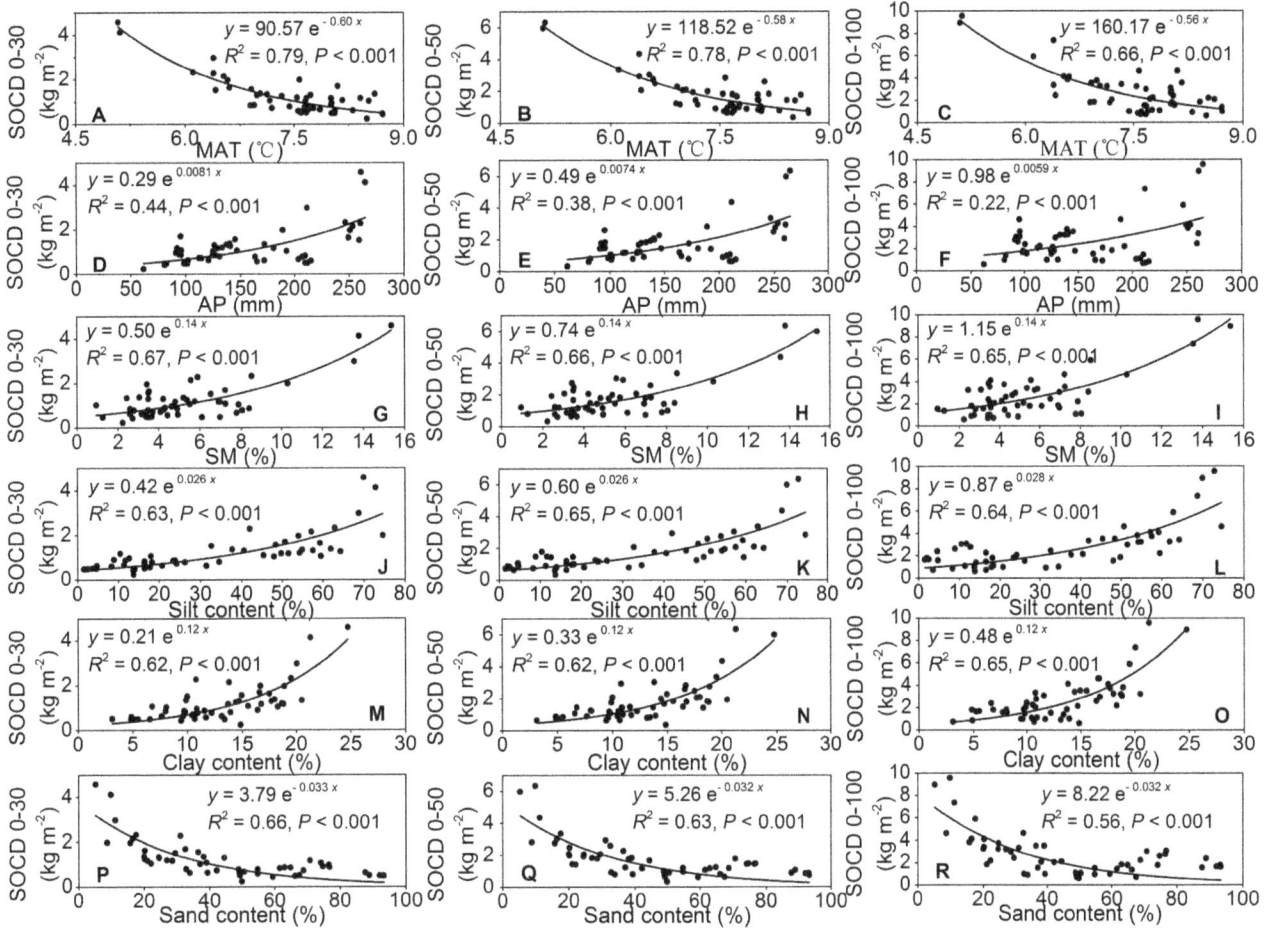

Figure 5. Relationships between soil organic carbon density and environmental factors at different depth intervals. *Notes*: SOCD, soil organic carbon density; MAT, mean annual temperature; AP, annual precipitation; SM, soil moisture.

Table 3. Integrative effects of mean annual temperature, soil moisture and soil texture (clay, silt, and sand) on soil organic carbon density at three soil depth intervals (0–30 cm, 0–50 cm, and 0–100 cm).

Source	0–30 cm				0–50 cm				0–100 cm			
	df	MS	SS%	P	df	MS	SS%	P	df	MS	SS%	P
MAT	1	4.45***	35.53	−0.47	1	8.70***	33.58	−0.66	1	15.13***	19.31	−0.87
SM	1	3.37***	26.87	0.10	1	6.51***	25.12	0.14	1	18.49***	23.61	0.24
Clay	1	0.33*	2.60	0.026	1	1.22**	4.69	0.051	1	7.04***	8.99	0.12
Silt	1	0.52*	4.14	0.013	1	1.89**	7.28	0.024	1	10.46***	13.36	0.057
Sand	1	0.08	0.67	0.004	1	0.97*	3.74	0.014	1	10.65***	13.60	0.047
Residuals	46	0.08	30.19		46	0.14	25.59		46	0.36***	21.13	
Intercept				3.22				3.78				2.25

Notes:
***$P<0.001$;
**$P<0.01$;
*$P<0.05$.
df, degree of freedom; MS, mean squares; SS%, proportion of variances explained by the variable; P, parameters of best-fitted GLM equations; MAT, mean annual temperature; SM, soil moisture.
The results were obtained from general linear model analysis.

Table 4. Comparisons of soil organic carbon density in desert grasslands in the Hexi Corridor with previous estimates.

Research types	Soil organic carbon density (kg m^{-2})				Reference
	0–30 (cm)	0–50 (cm)	0–100 (cm)	Actual depth (cm)	
Global mean	–	–	10.8	–	[23]
Global cool temperate desert	–	–	9.7	–	
Global cool temperate steppe	–	–	13.3	–	
Global temperate grassland	–	–	11.7	–	[26]
Global desert	–	–	6.2	–	
Average of China	3.7	–	7.8	–	[18]
Temperate grassland of Inner Mongolia	4	5.19	6.68		[28]
Desert steppe of Inner Mongolia	2.3	3.1	4.01	–	
				–	
Temperate typical steppe	–	–	12.3	–	[40]
Temperate deserted steppe	–	–	8.7	–	
Temperate desert	–	–	6.2	–	
Temperate steppe-desert	–	–	8	–	[10]
Temperate desert-steppe	–	–	8.7	–	
Temperate desert	–	–	6.2	–	
Desert	–	–	4.39	–	[41]
Desert steppe	–	–	7.09	–	
Temperate typical steppe	–	–	–	8	[16]
Temperate deserted steppe	–	–	–	2.8	
Temperate desert	–	–	–	2.2	
Desert grasslands in the Hexi Corridor	1.2	1.73	2.69	–	This study

Notes: "–" mean not measured.

the shallow distribution of roots, the majority of C inputs from roots are concentrated at soil depths of 0–30 cm; additionally, the topsoil is the first layer that directly receives C inputs from aboveground biomass.

Relationship between SOCD and climate and soil environmental factors

Under natural conditions, the distribution of SOC was controlled by climate, vegetation, parent material, and soil texture [3,5,23,26,35]. Our studies observed SOC distribution was positively associated with precipitation and clay content and negatively correlated with temperature and sand content and indicated that the strength of relationships between SOCD and environmental variables decreased with the increase of soil depth.

As shown in Figure 5, SOCD decreased markedly with MAT, likely due to the accelerated mineralization of soil organic matter with increasing temperature [3,24,26]. On the other hand, increasing temperature in arid regions results in a considerable decline in water use efficiency by increasing evapotranspiration can lead to low biomass production and low SOC density [15]. Additionally, the effect of MAT on SOCD in each soil layer was stronger than that for AP and edaphic factors. This result underlines the importance of MAT as a predicator for the SOC density in desert grasslands.

In arid ecosystems, precipitation and soil moisture constrain plant production and decomposition [26]. Our results revealed a substantial increase in the SOCD in desert grasslands with an increase in AP and SM. These results imply that water availability is a powerful parameter for assessment of SOCD. Water is the limiting factor for plant production in desert grasslands, as a small increase of water could significantly stimulate bio-productivity and thus contribute to the accumulation of SOCD [26,36]. Meanwhile, higher precipitation and soil moisture will affect SOC sequestration mainly through higher soil acidity and lower base saturation at the exchange sites, which would reduce the litter decomposition rate. In addition to the abovementioned results, the relationship between SOCD and SM was stronger than that between SOCD and AP. Precipitation in arid desert-grassland has difficulty entering the soil through infiltration due to low vegetation coverage, less litter content, and coarse soil texture, and consequently, soil moisture can represent the actual water content and play a more important role in model construction between SOCD and environmental factors.

In our study, we observed that the increased SOCD was positively correlated with the accumulation of silt and clay content but negatively correlated with sand content. Finely textured soils with appropriate clay and silt content can increase physical and hydrological protection of SOC by inhibiting decomposition through stabilizing SOC, and increasing residence time to decrease C leaching [3,5,26,37,38]. Moreover, an increase in clay and silt content could enhance the formation of aggregates, which have two advantages for improving SOC content: (1) by improving the soil quality and capacity to efficiently retain water [3,5,39] and (2) by mitigating wind erosion [27]. These two advantages could stimulate plant productivity and thus result in additional C inputs.

The GLM analysis suggested that climate factors were more important in determining the distribution of SOC than soil texture, but the effect decreased gradually with an increase in soil depth. In contrast, the impact of soil texture on SOC distribution in surface soils was not obvious but increased alongside soil depth, which indicates that soil texture plays a critical role in SOC distribution at deeper soil layers. Among all climatic and edaphic parameters, MAT had the highest contribution to explain the distribution of SOC in surface soils (0–30 cm and 0–50 cm). Considering the vast majority of SOC accumulates in the surface soil [26], MAT plays a decisive role in the spatial distribution of SOC in the arid desert- grassland ecosystem in the Hexi Corridor. That may be because biomass production is strictly limited by water in arid desert- grasslands, and the rise of temperature can promote the decomposition of SOC, and conversely promote the accumulation of SOC.

Author Contributions

Conceived and designed the experiments: MW. Performed the experiments: MW YS XY. Analyzed the data: MW YS. Contributed reagents/materials/analysis tools: MW. Wrote the paper: MW.

References

1. Davidson EA, Janssens IA (2006) Temperature sensitivity of soil carbon decomposition and feedbacks to climate change. Nature 440: 165–173.
2. Albaladejo J, Ortiz R, Garcia-Franco N, Navarro AR, Almagro M, et al. (2013) Land use and climate change impacts on soil organic carbon stocks in semi-arid Spain. Journal of Soils and Sediments 13: 265–277.
3. Schimel DS, Braswell BH, Holland EA, McKeown R, Ojima DS, et al. (1994) Climatic, edaphic, and biotic controls over storage and turnover of carbon in soils. Global biogeochemical cycles 8: 279–293.
4. Lal R (2004) Soil carbon sequestration impacts on global climate change and food security. Science 304: 1623–1627.
5. Yang YH, Fang JY, Tang YH, Ji CJ, Zheng CY, et al. (2008) Storage, patterns and controls of soil organic carbon in the Tibetan grasslands. Global Change Biology 14: 1592–1599.
6. Conant R, Paustian K (2002) Spatial variability of soil organic carbon in grasslands: implications for detecting change at different scales. Environmental Pollution 116: S127–S135.
7. Fan JW, Zhong HP, Harris W, Yu GR, Wang SQ, et al. (2008) Carbon storage in the grasslands of China based on field measurements of above-and below-ground biomass. Climatic Change 86: 375–396.
8. Wang GX, Qian J, Cheng GD, Lai YM (2002) Soil organic carbon pool of grassland soils on the Qinghai-Tibetan Plateau and its global implication. Science of the Total Environment 291: 207–217.
9. Han X, He N, Wu L, Wang Y (2008) Storage and Dynamics of Carbon and Nitrogen in Soil after Grazing Exclusion in Grasslands of Northern China. Journal of Environmental Quality 37: 663–668.
10. Ni J (2002) Carbon storage in grasslands of China. Journal of Arid Environments 50: 205–218.
11. Parton W, Scurlock J, Ojima D, Gilmanov T, Scholes R, et al. (1993) Observations and modeling of biomass and soil organic matter dynamics for the grassland biome worldwide. Global biogeochemical cycles 7: 785–809.
12. Piao SL, Fang JY, Zhou LM, Tan K, Tao S (2007) Changes in biomass carbon stocks in China's grasslands between 1982 and 1999. Global biogeochemical cycles 21 Doi: 10.1029/2005gb002634.
13. Qibin W, Linghao L, Xianhua L (1998) Spatial heterogeneity of soil organic carbon and total nitrogen in Xilin River basin grassland, Inner Mongolia. Acta Phytoecologica Sinica 22: 409–414.
14. Yang YH, Fang JY, Guo DL, Ji CJ, Ma WH (2010) Vertical patterns of soil carbon, nitrogen and carbon: nitrogen stoichiometry in Tibetan grasslands. Biogeosciences Discussions 7: 1–24.
15. Lal R (2004) Carbon sequestration in dryland ecosystems. Environ Manage 33: 528–544.
16. Wu HB, Guo ZT, Peng CH (2003) Distribution and storage of soil organic carbon in China. Global biogeochemical cycles 17 Doi: 10.1029/2001gb001844.
17. Wang SQ, Tian HQ, Liu JY, Pan SF (2003) Pattern and change of soil organic carbon storage in China: 1960s–1980s. Tellus B 55: 416–427.
18. Yang YH, Mohammat A, Feng JM, Zhou R, Fang JY (2007) Storage, patterns and environmental controls of soil organic carbon in China. Biogeochemistry 84: 131–141.
19. Fang JY, Yang YH, Ma WH, Mohammat A, Shen HH (2010) Ecosystem carbon stocks and their changes in China's grasslands. Science China Life sciences 53: 757–765.
20. Schlesinger WH, Andrews JA (2000) Soil respiration and the global carbon cycle. Biogeochemistry 48: 7–20.
21. Bastida F, Moreno JL, Hernandez T, Garcia C (2007) The long-term effects of the management of a forest soil on its carbon content, microbial biomass and activity under a semi-arid climate. Applied Soil Ecology 37: 53–62.
22. Johnston CA, Groffman P, Breshears DD, Cardon ZG, Currie W, et al. (2004) Carbon cycling in soil. Frontiers in Ecology and the Environment 2: 522–528.
23. Post WM, Emanuel WR, Zinke PJ, Stangenberger AG (1982) Soil carbon pools and world life zones. Nature 298: 156–159.
24. Burke IC, Yonker CM, Parton WJ, Cole CV, Flach K, et al. (1989) Texture, climate, and cultivation effects on soil organic matter content in US grassland soils. Soil Sci Soc Am J 53: 800–805.
25. Canadell JG, Kirschbaum MUF, Kurz WA, Sanz M-J, Schlamadinger B, et al. (2007) Factoring out natural and indirect human effects on terrestrial carbon sources and sinks. Environmental Science & Policy 10: 370–384.
26. Jobbágy EG, Jackson RB (2000) The vertical distribution of soil organic carbon and its relation to climate and vegetation. Ecological Applications 10: 423–436.
27. Su YZ, Wang XF, Yang R, Lee J (2010) Effects of sandy desertified land rehabilitation on soil carbon sequestration and aggregation in an arid region in China. Journal of Environmental Management 91: 2109–2116.
28. Ma WH (2006) Carbon Storage in the Temperate Grassland of Inner Mongolia. Beijing: Peking University.
29. Wang M, Su YZ, Yang R, Yang X (2013) Allocation patterns of above- and below-ground biomass in desert grassland in the middle reaches of Heihe River, Gansu Province, China. Chinese Journal of Plant Ecology 37: 209–219 (in Chinese).
30. Nelson DW, Sommers LE (1982) Total carbon, organic carbon, and organic matter. In: Methods of soil analysis Part 2 Chemical and microbiological properties: 539–579.
31. Chaudhari SK, Singh R, Kundu DK (2008) Rapid textural analysis for saline and alkaline soils with different physical and chemical properties. Soil Sci Soc Am J 72: 431–441.
32. Holben BN (1986) Characteristics of maximum-value composite images from temporal AVHRR data. International Journal of Remote Sensing 7: 1417–1434.
33. Lal R, Kimble J (2001) Soil erosion and carbon dynamics on grazing land. The potential of US grazing lands to sequester carbon and mitigate the greenhouse effect: Lewis Publishers, Boca Raton, Florida: 231–247.
34. Li Z, Zhao QG (2001) Organic carbon content and distribution in soils under different land uses in tropical and subtropical China. Plant and Soil 231: 175–185.
35. Wang SQ, Huang M, Shao XM, Mickler RA, Li KR, et al. (2004) Vertical distribution of soil organic carbon in China. Environmental Management 33: 200–209.
36. Callesen I, Liski J, Raulund-Rasmussen K, Olsson MT, Tau-Strand L, et al. (2003) Soil carbon stores in Nordic well-drained forest soils - relationships with climate and texture class. Global Change Biology 9: 358–370.
37. Paul EA (1984) Dynamics of organic matter in soils. Plant and Soil 76: 275–285.
38. Wynn JG, Bird MI, Vellen L, Grand-Clement E, Carter J, et al. (2006) Continental-scale measurement of the soil organic carbon pool with climatic, edaphic, and biotic controls. Global biogeochemical cycles 20 Doi: 10.1029/2005gb002576.
39. Schimel DS, Parton WJ (1986) Microclimatic controls of nitrogen mineralization and nitrification in shortgrass steppe soils. Plant and Soil 93: 347–357.
40. Zinke PJ, Stangenberger AG, Post WM, Emanuel WR, Olson JS (1984) Worldwide organic soil carbon and nitrogen data. Oak Ridge National Lab., TN (USA) .
41. Liu WJ, Chen SY, Qin X, Baumann F, Scholten T, et al. (2012) Storage, patterns, and control of soil organic carbon and nitrogen in the northeastern margin of the Qinghai–Tibetan Plateau. Environmental Research Letters 7 Doi: 10.1088/1748-9326/7/3/035401.

3

SoilGrids1km — Global Soil Information Based on Automated Mapping

Tomislav Hengl[1]*, **Jorge Mendes de Jesus**[1], **Robert A. MacMillan**[2], **Niels H. Batjes**[1],
Gerard B. M. Heuvelink[1,3], **Eloi Ribeiro**[1], **Alessandro Samuel-Rosa**[4], **Bas Kempen**[1], **Johan G. B. Leenaars**[1],
Markus G. Walsh[5], **Maria Ruiperez Gonzalez**[1]

1 ISRIC — World Soil Information, Wageningen, the Netherlands, 2 LandMapper Environmental Solutions Inc., Edmonton, Canada, 3 Wageningen University, Wageningen, the Netherlands, 4 Federal Rural University of Rio de Janeiro, Rio de Janeiro, Brazil, 5 The Earth Institute, Columbia University, New York, New York, United States of America, and Selian Agricultural Research Inst., Arusha, Tanzania

Abstract

Background: Soils are widely recognized as a non-renewable natural resource and as biophysical carbon sinks. As such, there is a growing requirement for global soil information. Although several global soil information systems already exist, these tend to suffer from inconsistencies and limited spatial detail.

Methodology/Principal Findings: We present SoilGrids1km — a global 3D soil information system at 1 km resolution — containing spatial predictions for a selection of soil properties (at six standard depths): soil organic carbon (g kg$-$1), soil pH, sand, silt and clay fractions (%), bulk density (kg m$-$3), cation-exchange capacity (cmol+/kg), coarse fragments (%), soil organic carbon stock (t ha$-$1), depth to bedrock (cm), World Reference Base soil groups, and USDA Soil Taxonomy suborders. Our predictions are based on global spatial prediction models which we fitted, per soil variable, using a compilation of major international soil profile databases (ca. 110,000 soil profiles), and a selection of ca. 75 global environmental covariates representing soil forming factors. Results of regression modeling indicate that the most useful covariates for modeling soils at the global scale are climatic and biomass indices (based on MODIS images), lithology, and taxonomic mapping units derived from conventional soil survey (Harmonized World Soil Database). Prediction accuracies assessed using 5–fold cross-validation were between 23–51%.

Conclusions/Significance: SoilGrids1km provide an initial set of examples of soil spatial data for input into global models at a resolution and consistency not previously available. Some of the main limitations of the current version of SoilGrids1km are: (1) weak relationships between soil properties/classes and explanatory variables due to scale mismatches, (2) difficulty to obtain covariates that capture soil forming factors, (3) low sampling density and spatial clustering of soil profile locations. However, as the SoilGrids system is highly automated and flexible, increasingly accurate predictions can be generated as new input data become available.

Editor: Ben Bond-Lamberty, DOE Pacific Northwest National Laboratory, United States of America

Funding: ISRIC is a non-profit organization primarily funded by the Dutch government. The authors are especially thankful for support from the Africa Soil Information Service (AfSIS) project, funded by the Bill and Melinda Gates foundation and the Alliance for a Green Revolution in Africa (AGRA). The funders had no role in study design, data collection and analysis, decision to publish, or preparation of the manuscript.

Competing Interests: RAM is owner and retired principle of LandMapper Environmental Solutions Inc. There are no patents, products in development or marketed products to declare.

* Email: tom.hengl@wur.nl

Introduction

There is increasing recognition of the urgent need to improve the quality, quantity and spatial detail of information about soils to respond to challenges presented by growing pressures on soils to support a large variety of critical functions [1–4]. Arrouays et al. [3] argue that existing soils information is not well suited to addressing vital questions related to mapping, monitoring or modelling soil processes that are driven by fluxes or changes in soils of water, nutrients, carbon, solutes or energy. Conventional models of soil variation describe variation in the horizontal dimension using polygons comprising classes of named soils [5]. In the vertical dimension, variation is described in terms of classes of horizons or layers that vary in their properties, thickness and depth. These conceptual models of discrete variation of classes of soil in horizontal and vertical directions are not well suited for use in many of the (global) simulation models and decision making systems currently used to describe and interpret soil functions and processes, such as supporting crop growth modelling, modelling hydrological and climatological processes, soil carbon dynamics or erosion [2,5]. Most modern spatial models that require information about soils as an input need accurate numerical information about continuous variation in soil properties. Models also require input data layers that are complete, consistent and as correct and current as possible. These requirements are not well met by current sources of soils information, especially sources of global extent.

Soil is probably one of the least well described thematic layers at the global scale, and existing global soil maps are often of undocumented or unknown accuracy [5]. At the moment, only coarse scale soil maps of the world are available at an effective resolution of about ~20 km [10]. The most commonly used global soil maps include [2,5]: Harmonized World Soil Database (HWSD) [11], USGS-produced soil property maps (http://soils.usda.gov/use/worldsoils/mapindex/) and ISRIC-WISE based soil property maps [12].

While widely used and cited, these various coarse resolution soil maps tend to suffer from artefacts due to use of different soil mapping concepts between countries and regions, from variation in the underlying soil mapping scale (usually between 1:0.5 M to 1:5 M) and from differences in reliability of source data within and between continents [2,5]. They can also not easily be updated with new information and often lack any measure of uncertainty, which is assumed to be significant. In summary, currently available global soil maps are not comparable in level of detail, spatial accuracy and usability with other global environmental layers such as global land cover and climatic products (Figure 1).

In this paper, we present and describe SoilGrids1km — a global 3D soil information system at 1 resolution — as a first response to the need for a new, consistent and coherent, global soil information. SoilGrids1km was produced using the Global Soil Information Facilities (GSIF), which was recently developed at ISRIC as a framework and platform to support widespread, open collaboration in the assembly, collation and production of global soil information.

Materials and Methods

Global Soil Information Facilities

ISRIC — World Soil Information has a mandate to serve the international community with information about the world's soil resources to help addressing major global issues. Over the last four years, in collaboration with a growing number of international partners and with a direct support from the Bill and Melinda Gates Foundation (AfSIS project; http://africasoils.net), ISRIC has been developing a cyberinfrastructure called Global Soil Information Facilities (GSIF).

GSIF has a particular emphasis on supporting the assembly and collation of geo-registered soil profile descriptions with associated analytical data, and on supporting the production of new maps of 3D continuous soil properties and soil classes at global to regional scales. GSIF consists of several components: data portals for assembling and hosting soil profile data and covariate data, software for global soil data analysis and mapping, and facilities for documenting data and methods and for automating workflows.

One of these components is "SoilGrids" — an automated system for global soil mapping. SoilGrids is an implementation of model-based geostatistics [13,14] for the purpose of predicting soil properties (in 2D or 3D) and soil classes for a global soil mask (see further Figure 3c) using automated mapping. Automated mapping is the computer- aided generation of maps from point observations and covariate layers, with minimal human intervention, so that map updating is easy. In the context of geostatistical mapping, automated mapping implies that model fitting, prediction and visualization are run using fully automated and reproducible workflows [14,15]. The current implementation of SoilGrids

focuses on producing predictions at 1 km spatial resolution and for a selection of soil properties and classes of interest to modelers and to international organizations such as FAO, Intergovernmental Panel on Climate Change (IPCC), the Consultative Group on International Agricultural Research (CGIAR) and similar.

We have imagined GSIF as a crowd-sourcing system, largely inspired by systems such as OpenStreetMap, Geo-wiki [16] and the R Open Source environment for statistical computing [17]. In this context, GSIF follows the "Agile" approach to software/IT development [18] meaning that we support rapid development, integration of soil field data, output validation, and rapid publishing of results. A new development cycle with new outputs (in principle of improved accuracy) is implemented in succession within an automated processing framework until the desired target specifications have been reached.

Input data for SoilGrids1km

The main input data sources for SoilGrids1km are global compilations of publicly available (shared) soil profile data and environmental layers at 1 km resolution; both are freely accessible via portals (http://worldsoilprofiles and http://www.worldgrids.org). The main sources of soil profile data used to produce the first version of SoilGrids1km are: the USA National Cooperative Soil Survey Soil Characterization database (http://ncsslabdatamart.sc.egov.usda.gov/) and profiles from the USA National Soil Information System (http://soils.usda.gov/technical/nasis/), LUCAS Topsoil Survey database [19], Africa Soil Profiles database [20], Mexican National soil profile database [21], Brazilian national soil profile database [22], Chinese soil profile database [23], and the soil profile archive from the Canadian Soil Information System [24]. Other significant sources of profile data used are: ISRIC-WISE [25], SOTER [26], SPADE [27], and Russian soil reference profiles [28].

The compilation of points shown in Figure 2 is possibly the largest compilation of soil ground-truth data in the world. It can be compared, for example, to a compilation of meteorological station data used to generate the WorldClim dataset [29]. A large part of the soil profile data used to generate SoilGrids1km can be accessed via the WorldSoilProfiles.org data portal, however some data sets such as LUCAS [19] have strict data use policies and can only be obtained from the original data provider.

As covariates for SoilGrids1km we used a selection of GIS layers (75): mainly MODIS images, but also climate surfaces [29], Global Lithological Map (GLiM) [30], HWSD mapping units [11], and SRTM DEM-derived surfaces. These layers (apart for the GLiM) are all available via the WorldGrids.org data portal. The actual number of covariates used during the analyses is different for each soil variable as these are iteratively selected for each soil attribute, based on their statistical significance to help predict the specific attribute.

Before model fitting, the original covariates were converted to principal components ($n = 95$) to reduce data overlap and help remove noise and artefacts [7]. Number of components is larger than the number of original covariates because covariates such as lithology and land form classes are converted to indicators before the principal component analysis.

Soil mask map

We make no spatial predictions for global land cover categories that represent non-active soil areas, such as: artificial surfaces and associated areas (>50% of pixel covered with urban areas), bare rock areas, water bodies [31], shifting sands, permanent snow and ice. The global mask map of soils with vegetation cover and world deserts is shown in Figure 3c.

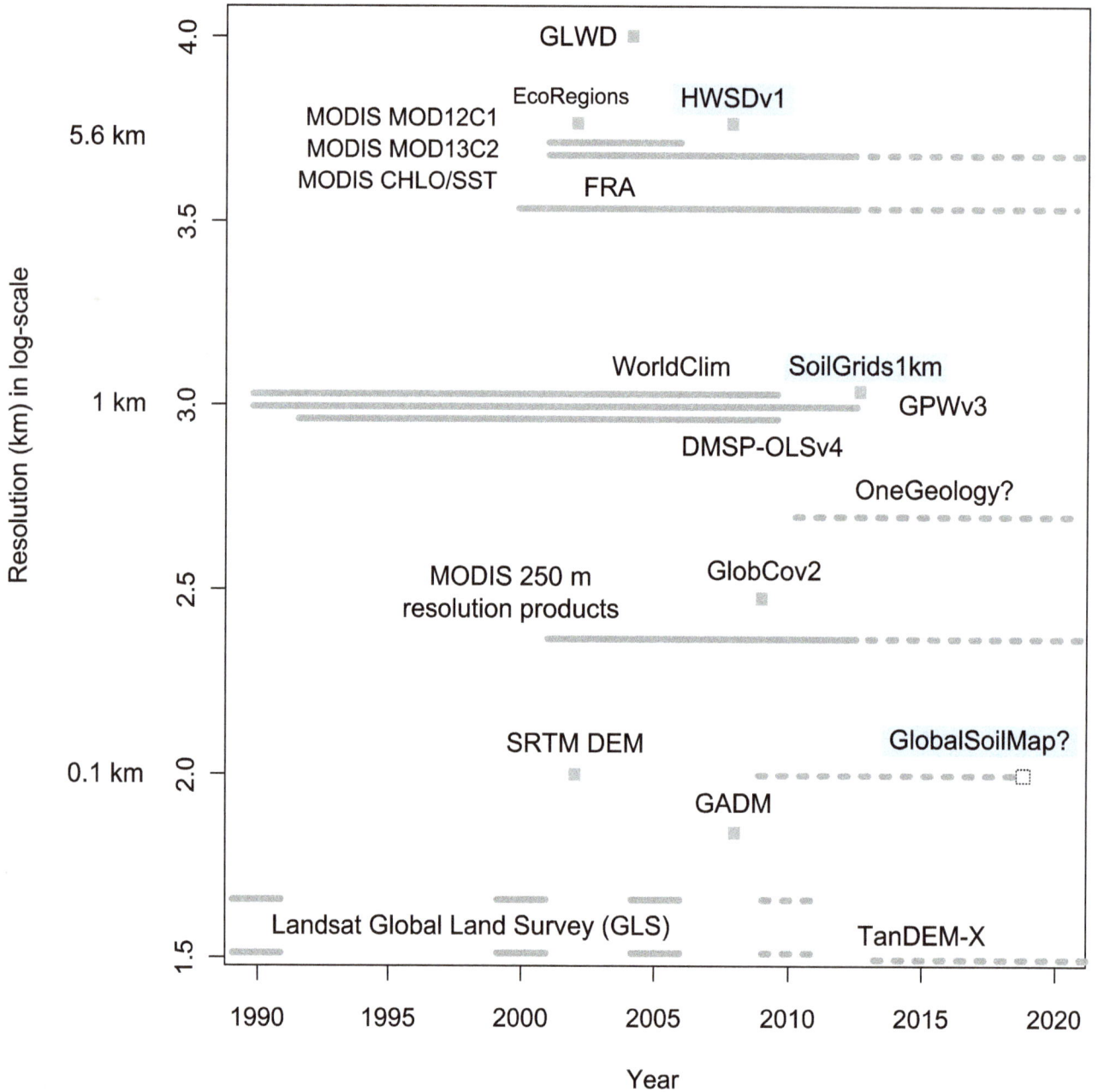

Figure 1. Spatial resolution and temporal coverage/publication time of some widely used global environmental data layers (global soil layers have been highlighted): GLWD — Global Lakes and Wetlands Database, HWSD — Harmonized World Soil Database, MOD12C1 — MODIS Land Cover Type Yearly L3, MOD13C2 — Vegetation Indices Monthly L3, CHLO/SST — MODIS Aqua Level-3 annual Chlorophyll/mid-IR Sea Surface Temperature, FRA — Forest Resources Assessment, GPW — Gridded Population of the World, DMSP-OLS — Nighttime Lights Time Series, GlobCov — Land Cover classes based on the MERIS FR images, GADM — Global Administrative Areas, TanDEM-X — Germany's topographic radar mission. Key agenda setters in the terms of production and dissemination of remote sensing and thematic environmental layers at the beginning of the 21st century include: NASA's MODIS (Moderate-resolution Imaging Spectroradiometer) and Landsat products — in terms of thematic content and usability [6–8], and Germany's TanDEM-X new global 12 m resolution DEM with ±2 m vertical accuracy [9]. Based on information retrieved on February 15th 2014. was produced using the Global Soil Information Facilities (GSIF), which was recently developed at ISRIC as a framework and platform to support widespread, open collaboration in the assembly, collation and production of global soil information.

The soil mask map was derived using the long term MODIS LAI images (MOD15A2), MODIS land cover product (MOD12Q1) [6], and global water mask [31] products. We distinguish three classes in the soil mask:

1. soils with vegetation cover — pixels with MODIS LAI>0 for at least one month in the last 12+ years (2000–2011),

2. urban areas — equal to the MODIS land cover product "*Urban and built-up*" class,

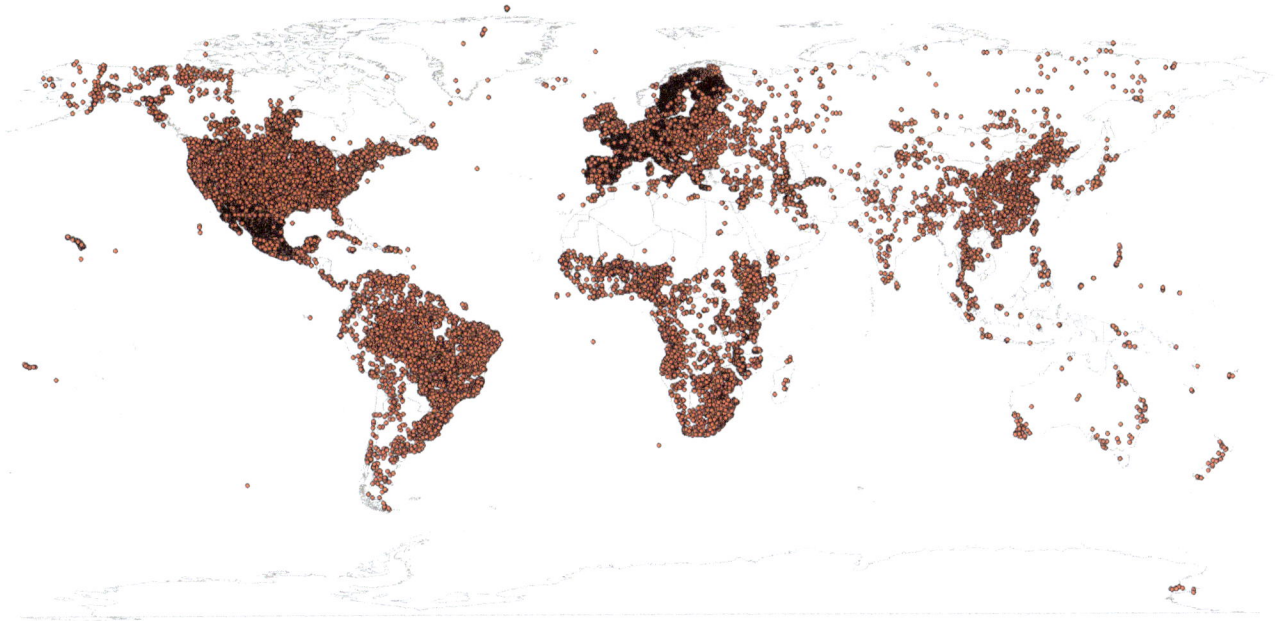

Figure 2. World distribution of soil profiles used to generate the SoilGrids1km product (about 110,000 points). Courtesy of various national and international agencies (see: Acknowledgments).

3. bare soil areas — areas without any biological activity but classified as "*Barren or sparsely vegetated*" in the MODIS land cover product.

Spatial prediction models

Two groups of spatial prediction models were implemented:

1. 2D or 3D regression and/or regression-kriging [32,33] combined with splines for numerical properties as implemented in the GSIF package for R. Here, the regression part is fitted using either:

   - Multiple linear regression [34] (for predicting pH, sand, silt and clay percentages and bulk density),
   - General Linear Models (GLM's) with log-link function [35,36] (for predicting organic carbon content and CEC),
   - Zero-inflated models [37] (for predicting coarse fragments and depth to bedrock; Figure 4),

2. Multinomial logistic regression (as implemented in the nnet package for R) for predicting distribution of soil classes [36].

As a general framework for mapping soil properties and classes we use the regression-kriging method commonly used in geostatistical mapping of soil properties [32,33,38]. We extend the existing 2D regression-kriging method to 3D space i.e. to predict values at voxels (Figure 4 right). In addition, we combine regression with splines, so that relationships between the soil property and covariates as well as soil-depth are modelled simultaneously:

$$\hat{z}(\mathbf{s}_0, d_0) = \sum_{j=0}^{p} \hat{\beta}_j \cdot X_j(\mathbf{s}_0, d_0) + \hat{\mathbf{g}}(d_0) + \sum_{i=1}^{n} \lambda_i(\mathbf{s}_0, d_0) \cdot e(\mathbf{s}_i, d_i) \quad (1)$$

where $\hat{z}$ is the predicted soil property, $\mathbf{s}_i$ are geographical coordinates, d_i is depth expressed in meters below land surface. Note that $\hat{\beta}_j \cdot X_j$ and $\hat{\mathbf{g}}(d_0)$ are the trend part of the model, where $X_j(\mathbf{s}_0, d_0)$ are covariates at the target location $\mathbf{s}_0$ and depth d_0, $\hat{\mathbf{g}}(d_0)$ is the predicted vertical trend, modelled by a spline function, and $e(\mathbf{s}_i, d_i)$ are residuals interpolated using 3D kriging using kriging weights $\lambda_i(\mathbf{s}_0, d_0)$. Because all covariates in the current version of SoilGrids1km are in fact 2D (i.e. values available at surface or for top-soil only), we copy the values of covariates for all depths in the regression matrix, which is a simplification. With the increasing availability of gamma radiometrics and similar, we anticipate that also 3D covariates will be used more in the near future with values differing per depth, although many covariates (e.g. elevation) will always remain 2D by definition.

3D regression and/or regression-kriging can be considered novel approaches to modeling soil variation. For comparison, the GlobalSoilMap project (http://globalsoilmap.net) proposes that soil-depth spline functions and spatial prediction functions should be fitted separately [3,40]. This spatial prediction system can be considered 2.5D because 2D models need to be fitted for each standard depth, i.e. each depth is modelled using a separate model that includes different combinations of covariates and in which data from predictions at one depth do not influence predictions at another. In the case of 3D modelling, a single model (Eq.1) is used for predicting in both X,Y and d for any property or class of interest, and fitting of the regression equation and residuals occurs at the same time as part of a single step. Another advantage of using a full 3D spatial prediction system, in comparison to the 2.5D, is also that it allows for producing spatial predictions and confidence intervals at any 3D location and not only at standard depths.

For each soil property, we have evaluated which version of the model in Eq.(1) would be most applicable. For example, initial tests showed that, for some soil properties e.g. soil organic carbon content and bulk density, the soil-depth relationship ($\hat{\mathbf{g}}(d_0)$) can often be better modelled using a log-log relationship. Consider for example:

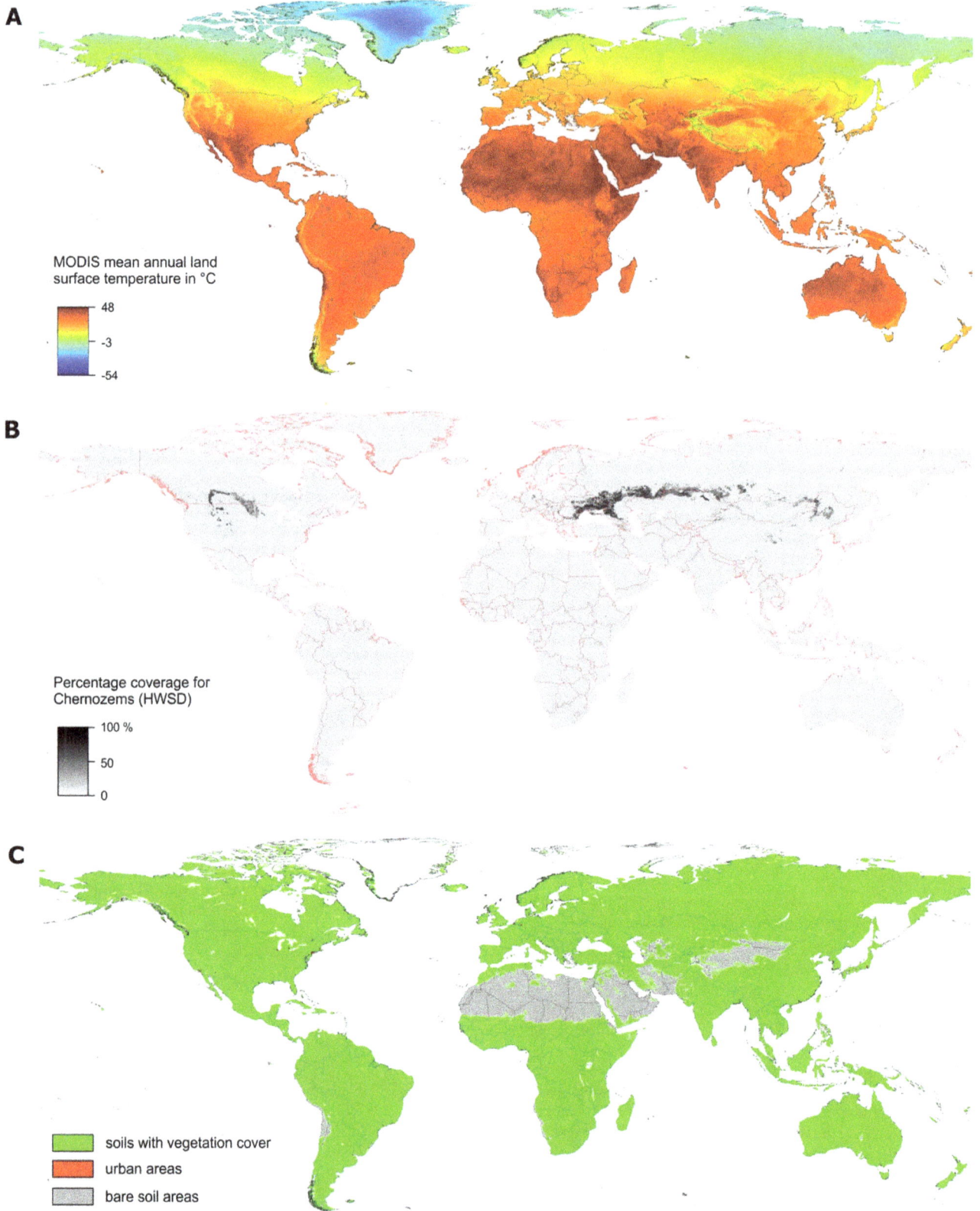

Figure 3. Examples of input layers used to generate SoilGrids1km: (a) long-term day-time MODIS land surface temperature, (b) percent cover Chernozems (based on the HWSD data set), and (c) global soil mask map. The spatial prediction domain of SoilGrids1km are the areas with vegetation cover and urban areas, while bare soil areas have been masked out. See text for more explanation.

PEDON

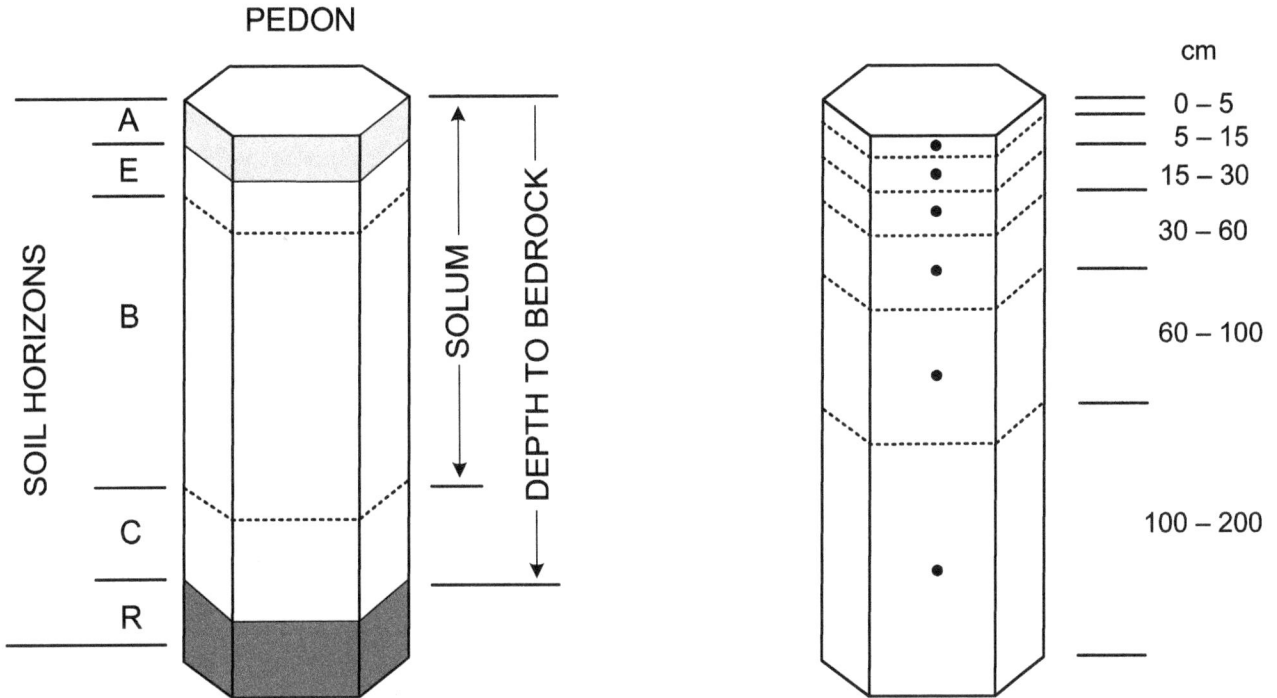

Figure 4. Standard stratification and designation of a soil profile: (left) soil horizons, solum thickness and depth to bedrock ('R' layer), and (right) six standard depths used in the *GlobalSoilMap* project [3].

$$\widehat{\mathrm{ORC}}(d) = \exp(\tau_0 + \tau_1 \cdot \log(d)) \qquad (2)$$

where $\widehat{\mathrm{ORC}}(d)$ is the predicted soil organic carbon content at depth d and τ_1 is the rate of decrease with depth. The model fitted using the global compilation of soil profiles (Figure 5b) has $\tau_0 = 4.1517$ (standard error 0.005326) and $\tau_1 = -0.60934$ (standard error 0.00145). This model explains 36% of the variation in the log-transformed ORC, which is a significant portion. This illustrates that any global soil property model can significantly profit from including depth into the statistical modelling. For other soil properties that do not show a monotonic vertical trend, higher order splines implemented via the ns function in the package splines [35] have been used to account for complex, non-linear relationships.

Further, soil covariate layers (X_j) used to produce SoilGrids1km were selected to represent the CLORPT model originally presented by Jenny [38,41]:

$$S = f(\mathrm{cl, o, r, p, t}) \qquad (3)$$

where S stands for soil (properties and classes), *cl* for climate, *o* for organisms (including humans), *r* is relief, *p* is parent material or geology and *t* is time. Most of the cl,o,r,p,t covariates are now publicly available and can be obtained at low cost thanks to NASA's/USGS Earth Observation projects such as MODIS and SRTM. We have also included soil class information (WRB reference groups) extracted from the HWSD (Figure 3b). These are basically traditional soil polygon delineations, comparable to other categorical covariates e.g. land cover classes or geological units.

The 3D regression function used for modelling changes of the of soil organic carbon content in 3D was thus (in R syntax):

```
formulaString = (ORCDRC + 1) ~ PC1 + PC2 + ... + PC95
+ ns(altitude, df = 2) glm (formula = formulaString,
family = gaussian(link = log),
data = rmatrix)
```

where ORCDRC is the organic carbon content, PC1 to PC95 are the principal components derived from some 75 covariate layers representing Jenny's soil forming factors, altitude is depth in meters from the soil surface, rmatrix is the regression matrix with values of target variable and predictors, ns is the natural spline function and df = 2 sets the number of allowed breakpoints (in this case two breakpoints to allow for curvilinear relationship). Soil classes are useful *'carriers of soil information'* [42], hence for SoilGrids1km we also provide global predictions for standard soil classes classified according to the two most widely used international soil classification systems:

- FAO's World Reference Base (WRB) — with focus on mapping soil groups e.g. Chernozem, Luvisols, Gleysols and similar. The current system [43] defines 32 reference soil groups.
- United States Department of Agriculture (USDA) Soil Taxonomy — with focus on mapping the soil suborders. The current system [44] defines 67 soil suborders (subdivision of 12 orders: Alfisols, Andisols, Aridisols, Entisols, Gelisols, Histosols, Inceptisols, Mollisols, Oxisols, Spodosols, Ultisols and Vertisols).

Models for predicting WRB soil groups and USDA soil orders were fitted using the nnet package (fits multinomial log-linear models via neural networks) using the default settings of 100

A

B

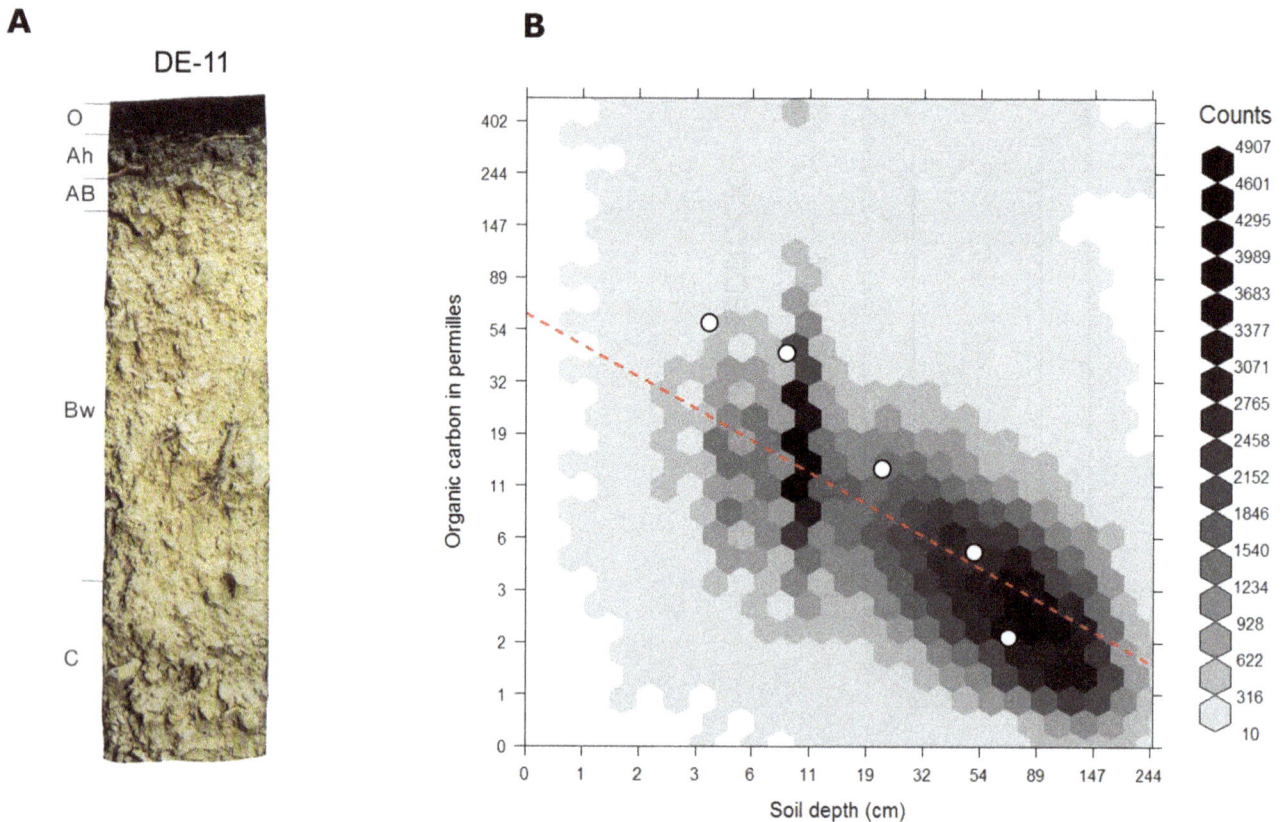

Figure 5. Individual soil profile from the ISRIC soil monolith collection (a) and globally fitted regression model for predicting soil organic carbon using depth only (b). The individual profile horizons are described by Mokma and Buurman [39]. Adjusted R-square for the model on the right is 0.363. Open circles show measured values for the profile on the left.

maximum iterations [36]. Soil classes are modeled as 2D variables i.e. the model does not include depth component, e.g.:

$$formulaString = TAXGWRB \sim PC1 + PC2 + ... + PC95$$

$$nnet :: multinom(formula = formulaString,$$

$$data = rmatrix, MaxNWts = 7000)$$

where TAXGWRB is the field observed WRB soil group, nnet:: multinom is the function to fit a multinomial logistic regression and MaxNWts sets the maximum allowable number of weights high enough for such a large regression data (regression model with ca. 100 covariates).

Note that all predictions in the initial version of SoilGrids1km were made using regression modelling alone. 3D kriging on a sphere at almost one billion locations (130 million pixels times 6 depths) was beyond our technical capacities in 2013/2014. Efforts to use full 3D regression-kriging to produce the first version of SoilGrids1km were abandoned in response to two main issues. Firstly, the computational load to undertake global kriging was too demanding for the processing resources and time we initially had at our disposal. We are working to both increase our processing power and to make the global kriging algorithms more efficient so we can run them globally for subsequent versions of SoilGrids1km. Secondly, there are very large areas of the world (e.g. Russia, northern Canada) that presently have almost no point profile data. These areas lack a sufficient number and density of point observations to successfully compute residuals, which can then

be kriged (otherwise kriging leads to serious artifacts). Since we were unable to produce residuals for large parts of the world, we decided not to try to krige residuals globally at first, at least until we obtain enough new point data to support computing and kriging residuals for all major portions of the globe. A full implementation of the 3D regression-kriging model built for SoilGrids has been run successfully at the continental level in Africa but, for the present (February 2014), we have not been able to apply full 3D regression-kriging globally. As soon as these technical limitations are solved, future versions of SoilGrids1km will likely also include a 3D kriging component.

Quality control

Resulting spatial predictions in SoilGrids1km are evaluated using two groups of methods:

- Cross-validation: We used 5–fold cross-validation to estimate the average mapping accuracy for each target variable. For continuous soil properties, we evaluate the amount of variation explained by the models [45]; and for soil classes we evaluate the map purity (i.e. proportion of observations correctly classified) and kappa statistic.

- Visual checking and overlay analysis: Because there is a large amount of spatial data, we have requested users to visually explore maps and look for artefacts and inconsistencies. Inconsistencies and artefacts in maps can be continuously reported through a Global Soil Information mailing list.

To derive amount of variation explained by the models for numeric variables we first derive Root Mean Square Error [46]:

$$RMSE = \sqrt{\frac{1}{l} \cdot \sum_{i=1}^{l} [\hat{z}(\mathbf{s}_i) - z(\mathbf{s}_i)]^2} \qquad (4)$$

where l is the number of validation points. Amount of the variation explained by the model is then:

$$\Sigma_{\%} = \left[1 - \frac{SSE}{SSTO}\right] = \left[1 - \frac{RMSE^2}{\sigma_z^2}\right][0-100\%] \qquad (5)$$

where SSE is the sum of squares for residuals at cross-validation points (i.e. $RMSE^2 \cdot n$), and $SSTO$ is the total sum of squares.

Derivation of secondary soil properties: soil organic carbon stock

The SoilGrids1km output maps can be further used for estimation of secondary soil properties which are typically not measured directly in the field and need to be derived from primary soil properties. For instance, consider estimation of the global carbon stock (in t ha^{-1}). This secondary soil property can be derived from a number of primary soil properties [47]:

$$\text{OCS } [\text{kg m}^{-2}] = \frac{\text{ORC}}{1000} \text{ [kg kg}^{-1}] \cdot \frac{\text{HOT}}{100} \text{ [m]}$$
$$\cdot \text{BLD } [\text{kg m}^{-3}] \cdot \frac{100 - \text{CRF } [\%]}{100} \qquad (6)$$

where OCS is soil organic carbon stock, ORC is soil organic carbon mass fraction in permilles, HOT is horizon thickness in cm, BLD is soil bulk density in kg m^{-3} and CRF is volumetric fraction of coarse fragments (>2 mm) in percent (see also Figure 6).

The propagated error of the soil organic carbon stock (Eq.6) can be estimated using the Taylor series method [48]:

$$\sigma_{\text{OCS}} = \frac{1}{10,000,000} \cdot \text{HOT} \cdot$$
$$\sqrt{\text{BLD}^2 \cdot (100 - \text{CRF})^2 \cdot \sigma_{\text{ORC}}^2 + \sigma_{\text{BLD}}^2 \cdot (100 - \text{CRF})^2 \cdot \text{ORC}^2 + \text{BLD}^2 \cdot \sigma_{\text{CRF}}^2 \cdot \text{ORC}^2} \qquad (7)$$

where σ_{ORC}, σ_{BLD} and σ_{CRF} are standard deviations of the predicted soil organic carbon content, bulk density and coarse fragments, respectively. Note that we first predict OCS values for all depths/horizons, then aggregate values for the whole profile (0–2 m). We further use a map of predicted depth to bedrock to remove all predictions outside the effective soil depth (areas where soil is shallower than 2 m). A more robust way to estimate the propagated uncertainty of deriving OCS would be to use geostatistical simulations (e.g. derive standard error from a large number of realizations 100) that incorporate spatial and vertical correlations. Because we are dealing with massive data sets, running geostatistical simulations for millions of pixels was not yet considered as an option.

Software implementation

SoilGrids1km predictions are generated via the GSIF package for R, which makes use of a large number of other basic and contributed packages — gstat, raster, rgdal and other R packages for spatial analysis [49]. GSIF package for R contains most of the functions required to produce SoilGrids, and will remain the main

platform in the future to obtain global model parameters and access SoilGrids through an API.

As previously mentioned, the target resolution of SoilGrids1km is relatively coarse, nevertheless, the computational intensity and memory required to produce SoilGrids1km is high: one run of SoilGrids1km takes about 12–16 hours on a 12–core HP Z420 workstation with 64 GiB RAM running on a Windows 7 64-bit system. Note also that since we produce predictions at six depths and uncertainty for each depth, the quantity of GeoTIFF maps produced is in the order of $250 \times 912 \text{MiB} \approx 250$ GiB. To deal with processing such large data sets we used a combination of tiling and parallel processing, as implemented via the snowfall package for R [50], to maximize the CPU usage and minimize the time required to produce predictions.

The spatial prediction process consists of four main steps:

1. preparation of gridded covariates (principal component analysis),
2. preparation of point data,
3. model fitting and
4. spatial prediction and construction of GeoTiffs.

From the steps listed above, spatial prediction take the longest computing time, which is often in the order of 20 or more hours using the computer specification listed above. As a rule of thumb, we look for mapping frameworks that can generate outputs within 48 hrs. If the whole process from model fitting to prediction and export of maps to GeoTiffs consumes >48 hrs of computing, we consider the system to be impractical for routine operational use.

Results

Model fitting

The results of model fitting (Table 1) indicate that the distribution of soil organic carbon content is mainly controlled by climatic conditions, i.e. monthly temperatures and rainfall [51], while the distribution of texture fractions (sand, silt and clay) is mainly controlled by topography and lithology. These key predictors agree with expectations based on existing knowledge. The regression models account for between ca. 20–50% of observed variability in the target variables (Table 1). Detailed model parameters can be obtained from the SoilGrids1km homepage at http://soilgrids.org.

Figure 7 illustrates two examples of spatial predictions for soil organic carbon content and pH. As mentioned previously, soil organic carbon clearly decreases with depth (see also the soil-depth curves shown in Figure 8). Areas mapped as having elevated values of organic carbon are typically associated with cooler and wetter climate regimes and boreal-tundra type vegetation [51–54]. Note that several soil variables have skewed distributions hence also the output predictions are skewed, so that we use log-transformed legends to maximize contrast in the map (Figure 7).

Figure 8 shows predicted values for organic carbon and pH (mean value and confidence intervals) for the same location shown in Figure 5. The prediction intervals are rather wide (see also Figure 11), which is connected to the fact that the models explain only 23–51% of the variation. However, it is important to note that these are global maps of predictions made using relatively coarse resolution covariates. We assume that is unlikely that any effort to map the distribution of soils at a resolution of 1 km could explain a much larger proportion of the total variation in soil properties, as much of this variation occurs over distances less than 1 km [55].

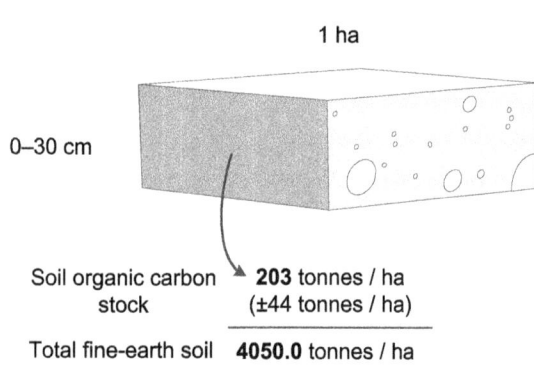

1 ha

0–30 cm

Soil organic carbon stock **203** tonnes / ha (±44 tonnes / ha)

Total fine-earth soil **4050.0** tonnes / ha

Bulk density (BLD): 1500 kg / m³ (s.d. = ±100)

Organic carbon (ORC): 50‰ (s.d. = ±10)

Coarse fragments (CRF): 10% (s.d. = ±5)

Total volume of the block (HOT): 30 cm (· 1 ha)

Soil organic carbon stock (OCS): 203 tonnes / ha (±44)

$$\textbf{OCS} = ORC/1000 \cdot BLD \cdot (100\text{-}CRF)/100 \cdot HOT/100$$
$$= 1/10{,}000{,}000 \cdot ORC \cdot BLD \cdot (100\text{-}CRF) \cdot HOT$$
$$= 1/10{,}000{,}000 \cdot 50 \cdot 1500 \text{ kg / m}^3 \cdot (100\text{-}10) \cdot 30 \text{ cm}$$
$$= 20.25 \text{ kg / m}^2 = 203 \text{ tonnes / ha}$$

$$\textbf{OCS.sd} = 1/10{,}000{,}000 \cdot HOT \cdot \sqrt{ BLD^2 \cdot (100 - CRF)^2 \cdot ORC.sd^2 + } $$
$$+ BLD.sd^2 \cdot (100 - CRF)^2 \cdot ORC^2 + BLD^2 \cdot CRF.sd^2 \cdot ORC^2)$$
$$= 4.4 \text{ kg / m}^2 = 44.1 \text{ tonnes / ha}$$

Figure 6. Soil organic carbon stock calculus scheme. Example of how total soil organic carbon stock (OCS) and its propagated error can be estimated for a given volume of soil using organic carbon content (ORC), bulk density (BLD), thickness of horizon (HOT), and percentage of coarse fragments (CRF). See text for more detail.

Also note that SoilGrids1km predictions are not capable of representing abrupt changes in values through depth e.g. due to buried horizons, textural heterogeneity or similar. Because we have used linear or close to linear models (plus smoothing splines) to predict values of targeted soil properties and not e.g. regression-trees, these models have smoothed out a significant amount of the variability in the point data, so that it is not realistic to expect abrupt changes in soil properties; at least not vertically (as illustrated previously in Figure 8).

Figure 9 (with a zoom in on Italy) shows that the SoilGrids1km predictions exhibit an order of magnitude greater spatial detail than previous global soil information products e.g. HWSD. This is mainly because a large stack of fine resolution remote sensing based covariate layers have been used to generate SoilGrids1km,

and many of these have shown to be significantly correlated with soil properties and classes. Spatial classification accuracy for mapped soil classes, when evaluated using kappa statistics (Table 1), shows a somewhat better match between what was observed on the ground for the USDA classification system (ground-truth classification available for 16,212 profiles) than for the WRB system (classification available for 37,015 profiles).

For many WRB classes our models predicted occurrences in areas that are inconsistent with a strict definition of geographic areas where these classes can occur. The most difficult to map seem to be WRB classes such as Andosols, Solonchaks, Calcisols and Cryosols. These classes are strictly defined (e.g. Andosols are connected with volcanic activities and specific geology) and we need to explore ways to prepare covariates that will prevent

Table 1. Mapping performance of SoilGrids1km — amount of variation explained (from 100%) or purity/kappa for categorical variables — for eight targeted soil properties and two soil classes distributed via SoilGrids1km.

Variable name	Type	GSIF code	Units	Range (observed)	Amount of var. explained
Soil organic carbon (dry combustion)	3D	ORCDRC	g kg−1	0–450	22.9%
pH index (H2O solution)	3D	PHIH5X	10−1	2.1–11.0	50.5%
Sand content (gravimetric)	3D	SNDPPT	kg kg−1	1–94	23.5%
Silt content (gravimetric)	3D	SLTPPT	kg kg−1	2–74	34.9%
Clay content (gravimetric)	3D	CLYPPT	kg kg−1	2–68	24.4%
Coarse fragments (volumetric)	3D	GRAVOL	cm3 cm−3	0–89	-
Bulk density (fine earth fraction)	3D	BLDVOL	kg m−3	250–2870	31.8%
Cation-exchange capacity (fine earth fraction)	3D	CEC	cmol+/kg	0–234	29.4%
Depth to bedrock	2D	DBR	cm	0–240	-
Soil group (WRB taxonomy)	2D	TAXGWRB	-	-	28.1% (kappa)
Soil suborder (USDA taxonomy)	2D	TAXOKST	-	-	40.3% (kappa)

WRB = "World Reference Base"; USDA = "United States Department of Agriculture".
Amount of variation explained by the models (Eq.5) i.e. kappa statistics for soil types was determined using 5-fold cross-validation.

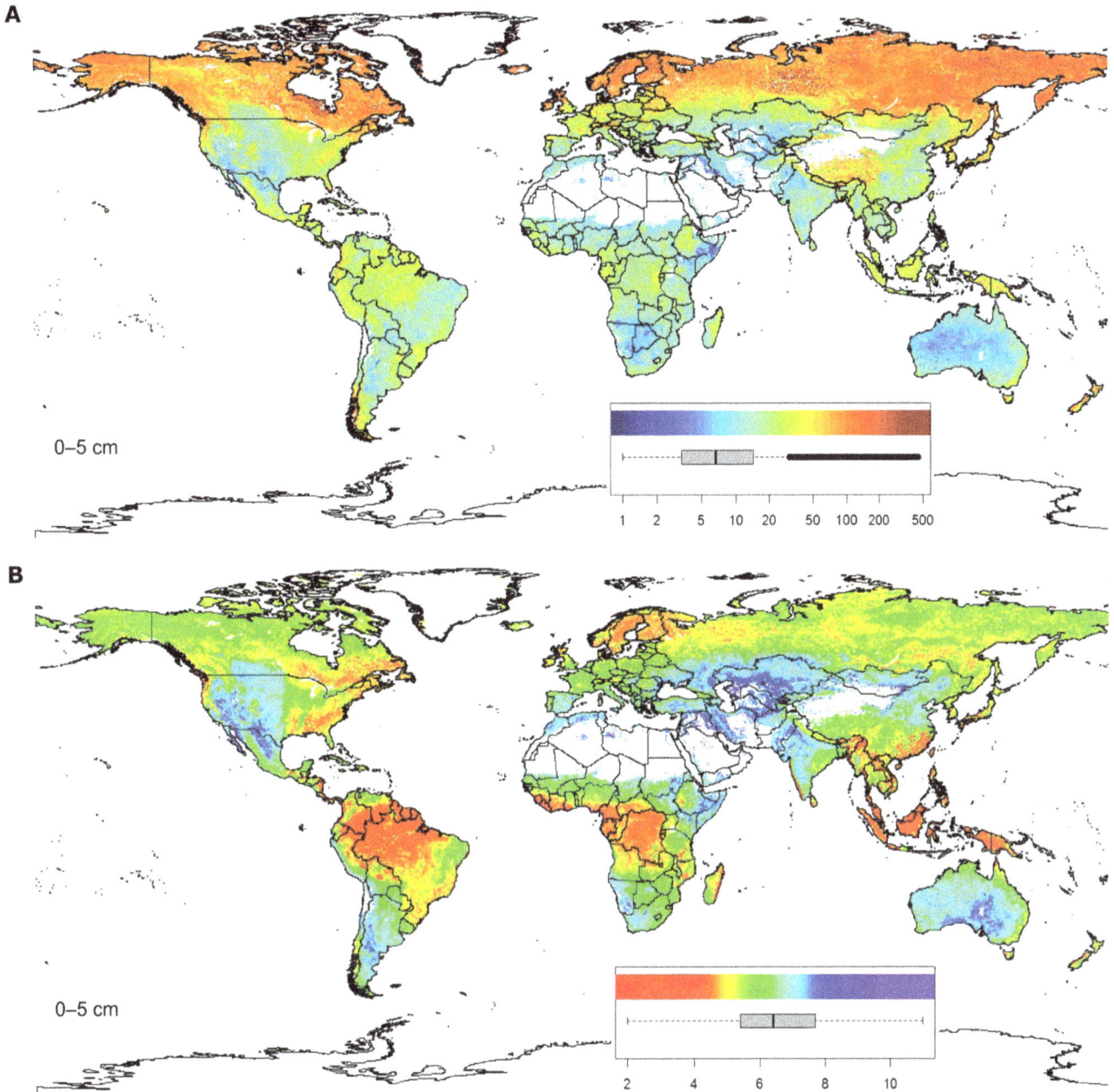

Figure 7. Example of SoilGrids1km layers: (A) soil organic carbon content in permille, and (B) soil pH for the topsoil (0–5 centimetres). Boxplots show the sampled distribution of the soil property based on the present compilation of global soil profile data.

prediction of those classes in areas where, by definition, they should not occur. Likewise, USDA suborders are based on soil moisture and climate regimes, for which we did not currently have global covariate maps, and consequently strictly defined classes such as Xerolls (Mollisols in Mediterranean climate; xeric moisture regime) were predicted in Brazil, which probably does not match the definition of the class.

Multinomial logistic regression is a purely data-driven method, so that the overall mapping performance highly depends on representation of environmental conditions by soil samples. All classes that are poorly represented in the environmental space, due to under-sampling, are understandably difficult to map accurately using a purely data- driven model [56]. Nevertheless, the final results of automated extraction of soil classes using multinomial

logistic regression are promising, especially for mapping the USDA classes. The mapping accuracy could probably be improved by adding more classification-related covariates and more field observations of soil taxonomy, hopefully through crowd-sourcing, in areas where the accuracy is critically low.

Figure 10 shows derived total soil organic carbon stock based on Eq.(6). According to this map, the total (baseline) amount of soil organic carbon (up to 2 m depth; excluding deserts, bare rock areas and ice caps) is about 330 t ha^{-1} on average. The highest concentrations of soil organic carbon are in areas of cooler climate and high rainfall, i.e. northern parts of Canada and Russia seem to be pools for most of the world's soil organic carbon. This largely agrees with results by Hugelius et al. [53] and Scharlemann et al. [57].

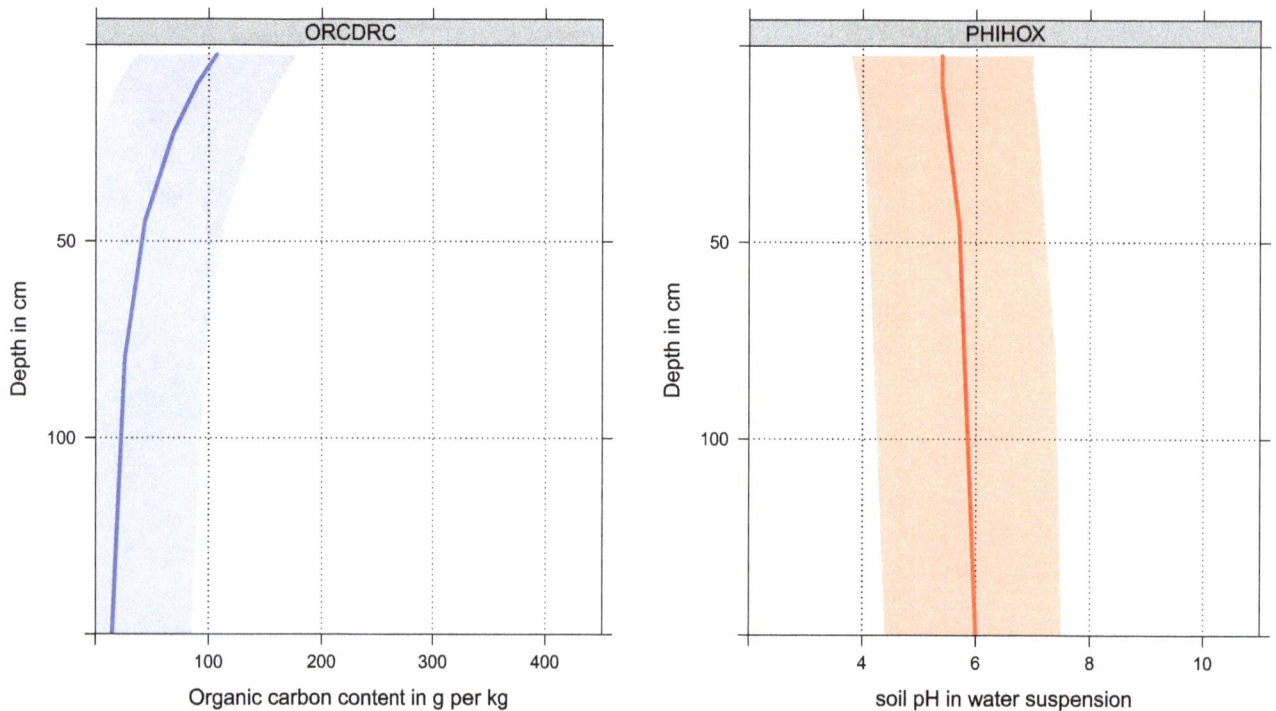

Figure 8. SoilGrids1km-derived soil-depth curves for the profile shown in Figure 5. Location of the profile: 6.3831°E, 50.479167°N. The shaded background indicates the 90% prediction interval for each depth. ORCDRC = soil organic carbon content in permilles; PHIHOX = soil pH in water suspension. See also Table 1.

The map shown in Figure 10 can be used to supplement maps of total aboveground biomass (see e.g. Ruesch and Gibbs [58] and Scharlemann et al. [57]). Our results also confirm that, overall, the amount of organic carbon below ground is greater than held in biomass above ground [51].

Quality issues

The results of cross-validation are shown in Table 1. The cross-validation results, as expected, largely reflect the model fitting success — properties that can be modeled successfully can also be mapped with higher accuracy. The soil properties that were most difficult to map are soil texture fractions, CEC and WRB soil groups. Although the accuracies of the predictions rarely exceed 50% of the total variation, all statistical models are significant showing clear spatial patterns (see e.g. Figure 7). Low cross-validation percentages are common in soil mapping [38,55], these numbers were not unexpected. Nevertheless, these can be considered promising initial results considering the complexity of harmonization of input point data (see further discussion).

Based on the feedback we received to date from users visiting the project homepage at http://soilgrids. org, the main limitations of SoilGrids1km are:

1. problems arising from poor relationships between covariates and dependent variables e.g. covariates can only explain part of the variability, which could possibly be improved by using more sophisticated statistical models;

2. problems arising from high spatial clustering of sampling locations (see Figure 2; observations are too sparse to improve on the regression using a kriging step);

3. problems associated with using partially-harmonized soil profile data;

4. problems arising from use of HWSD soil mapping units that are of too coarse scale and often not completely harmonized so that the country borders are still visible (obvious artefact);

5. limitations in the usability of SoilGrids1km for spatial planning at county or farm scale due to coarse resolution of the maps;

6. inability to consider and model significant sources of variability e.g. temporal variability due to changes in land use and/or land cover [59];

7. limitations arising from insufficient use of higher quality and finer resolution conventional soil maps prepared at national to regional scales.

Discussion

SoilGrids1km were released on December 5th 2013 (World Soil Day) at the FAO Rome, as a proposed contribution of the Netherlands to the Global Soil Partnership [60]. The system, at the moment, includes predicted values for (Table 1): soil organic carbon (g kg^{-1}), soil pH, sand, silt and clay fractions (%), bulk density (kg m^{-3}), cation-exchange capacity (cmol+/kg) of the fine earth fraction, coarse fragments (%), soil organic carbon stock (t ha^{-1}), depth to bedrock (in cm; see Figure 4), World Reference Base soil groups [43], and USDA Soil Taxonomy suborders [44]. We focused on generating spatial predictions at six standard depths (0–5 cm, 5–15 cm, 15–30 cm, 30–60 cm, 60–100 cm and 100–200 cm), for which spatially distributed estimates of upper and lower level 90% prediction intervals are presented. As such, we follow the corresponding specifications of the GlobalSoilMap project [3].

Initial predictions of soil classes were made at higher (more general) taxonomic levels for both WRB (soil groups) and Soil

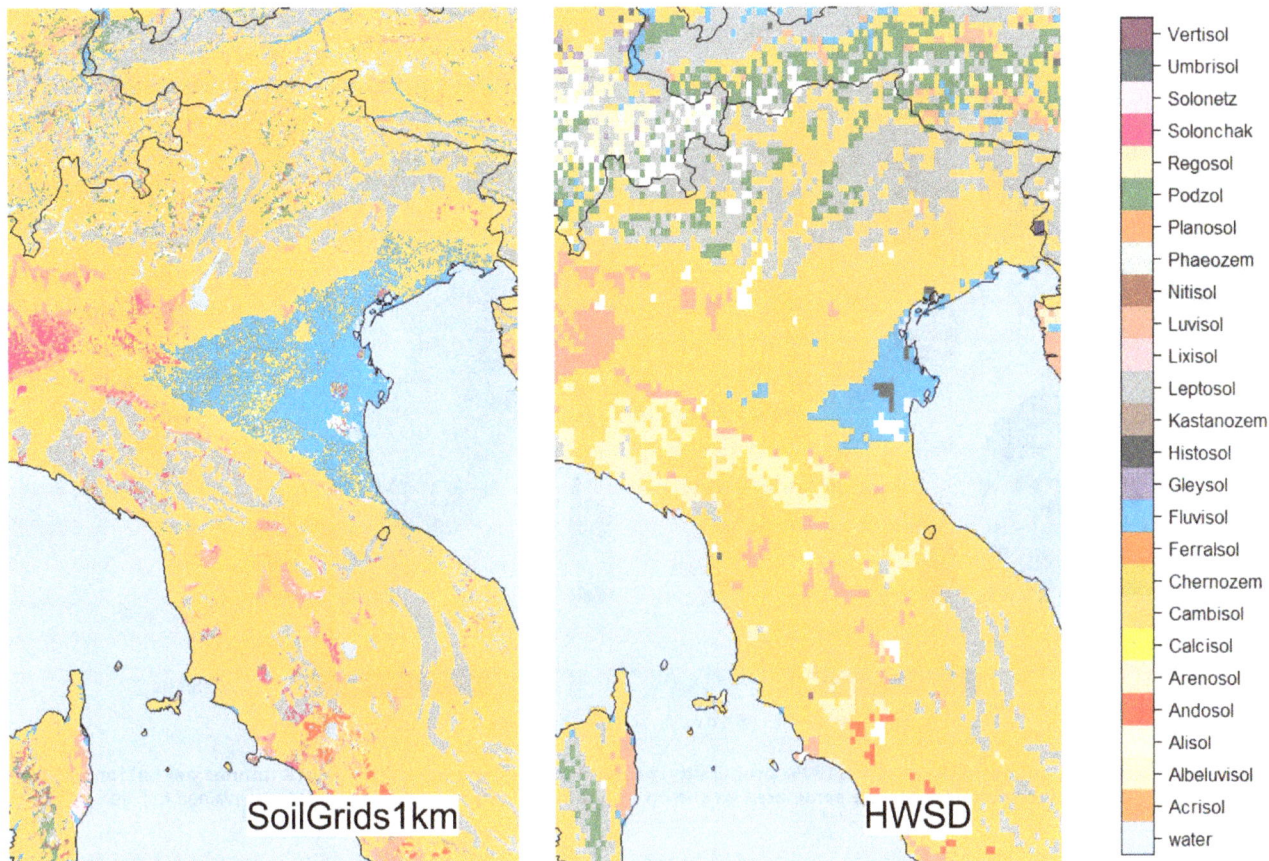

Figure 9. Spatial predictions of WRB soil groups for SoilGrids1km (left) and HWSD data set representing conventional soil maps (right). A zoom in on North of Italy. White pixels indicate missing values.

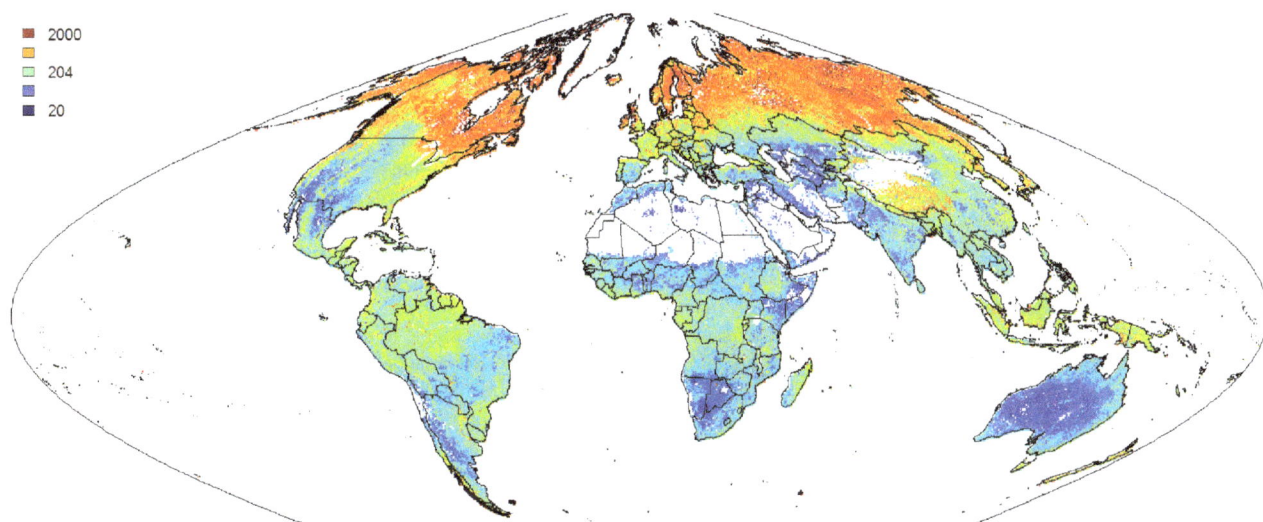

Figure 10. Predicted global distribution of the soil organic carbon stock in tonnes per ha for 0–200 centimetres. Total soil organic carbon stock (here displayed on a log-scale) was estimated as a sum of soil organic carbon stocks for six standard depths and adjusted for the depth to bedrock. Projected in the Sinusoidal equal area projection to give a realistic presentation of areas. Vast deserts (e.g. Sahara or Gobi) can be assumed to contain close to zero organic carbon stock. See also Figure 11.

0–30 cm

lower   upper

30–60 cm

lower   upper

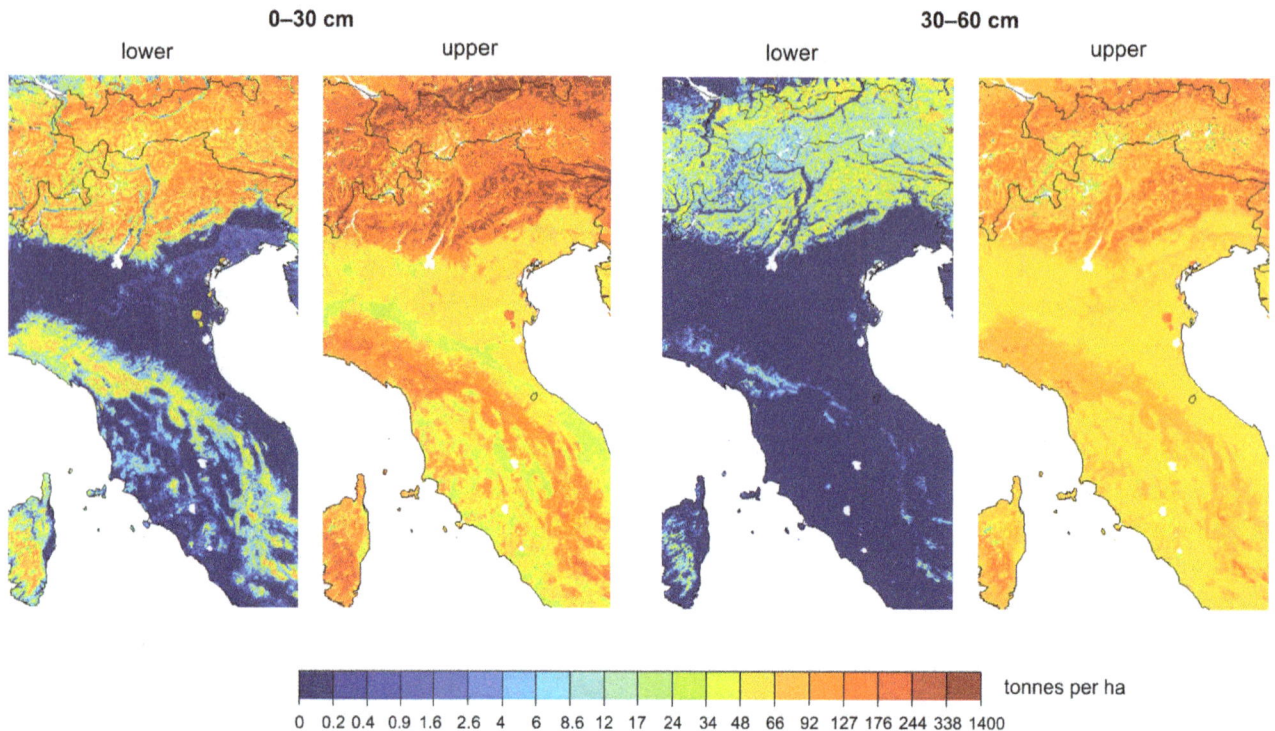

tonnes per ha

0   0.2 0.4 0.9 1.6 2.6   4   6   8.6 12   17   24   34   48   66   92 127 176 244 338 1400

Figure 11. Lower and upper confidence limits (90% probability) of estimated soil organic carbon stock (tonnes per ha) for standard depths 0–30 and 30–60 centimeters for the same area as shown in Figure 9. Derived using the procedure explained in Figure 6.

Taxonomy (suborders). This was done because the available point profile data sets do not provide a sufficient number of locations representative of all of the lower levels of classification in each system. Without a sufficient number of examples for all lower classes, distributed fully across all of the feature space within which each class can occur, it is not possible to successfully predict many of the lower classes defined for either system. Once we have more point observations that encompass the full range of lower level classes across the entire environmental and geographic spectrum of their distribution, we will be able to predict at a more detailed taxonomic level for both classification systems.

The main purpose of SoilGrids1km is to provide initial, fully worked, examples of how complete and consistent global maps of soil properties, and soil classes, can be produced using currently available legacy soil profile data, freely available gridded maps of global covariates and an on-line automated soil mapping system (GSIF). Additionally, we want to use these initial example maps to implement and demonstrate procedures and systems for supporting free and unrestricted access to what we consider to be the best possible current, globally-complete, estimates of soil properties and soil classes. It is hoped that the production, distribution and use of these new, initial, global soil maps will stimulate additional efforts to both improve these maps and to launch new efforts to collect and use new soils information in new soil mapping and monitoring projects. We especially aim at supporting countries in Africa, and large parts of Asia and Latin America, that often have limited infrastructures to produce soil information at fine resolution [2,5]. We think that there is a great potential in using the existing field observations and Open Source software to map spatial and spatio-temporal patterns, i.e. without doing any major financial investments.

A number of legitimate concerns exist relative to the initial SoilGrids1km outputs. Probably the most immediate and signif-

icant concern has to do with the accuracy and usability of the initial predictions of soil property and class values. We acknowledge that the accuracy of these initial predictions rarely exceeds 50% of the total variation and, for many properties, is often closer to 20–30% (Table 1). The results of cross-validation are informative but need to be taken with caution because most of the soil profiles (Figure 2) were not collected using probability sampling, so that the cross-validation results possibly carry the same sampling bias as the original data [61]. Also note that the accuracy of mapping WRB groups is likely lower than the accuracy of mapping USDA soil suborders because over 40% of the soil profiles that were used for the WRB classification were actually classes translated from national systems. Translation i.e. harmonization of international soil records probably introduces additional noise that cannot be solved by regression modelling.

We argue that it is unreasonable to expect any global map of variation in soil properties to explain much more than 50% of the total observed variation. It is well known that a significant proportion of spatial variation in soil properties occurs over relatively short distances of meters to tens of meters [55,56]. It is therefore unreasonable to expect that a map of global variation in soil properties, portrayed at a spatial resolution of 1 km, will be able to capture and portray the 50% or more of total variation that occurs at resolutions shorter than 1000 m. Our hope and plan is to gradually improve the accuracy of the predictions by addressing these issues and concerns one by one, in a systematic way (Figure 13). This should be done primarily by working with national and regional soil data agencies, i.e. by adding additional covariates at increasingly finer spatial resolutions and by adding more field/point data from areas that are under-represented.

Although millions of soil profile records have undoubtedly been collected throughout the world, they are often unequally distributed (Figure 2). Likewise, many soil profiles funded by

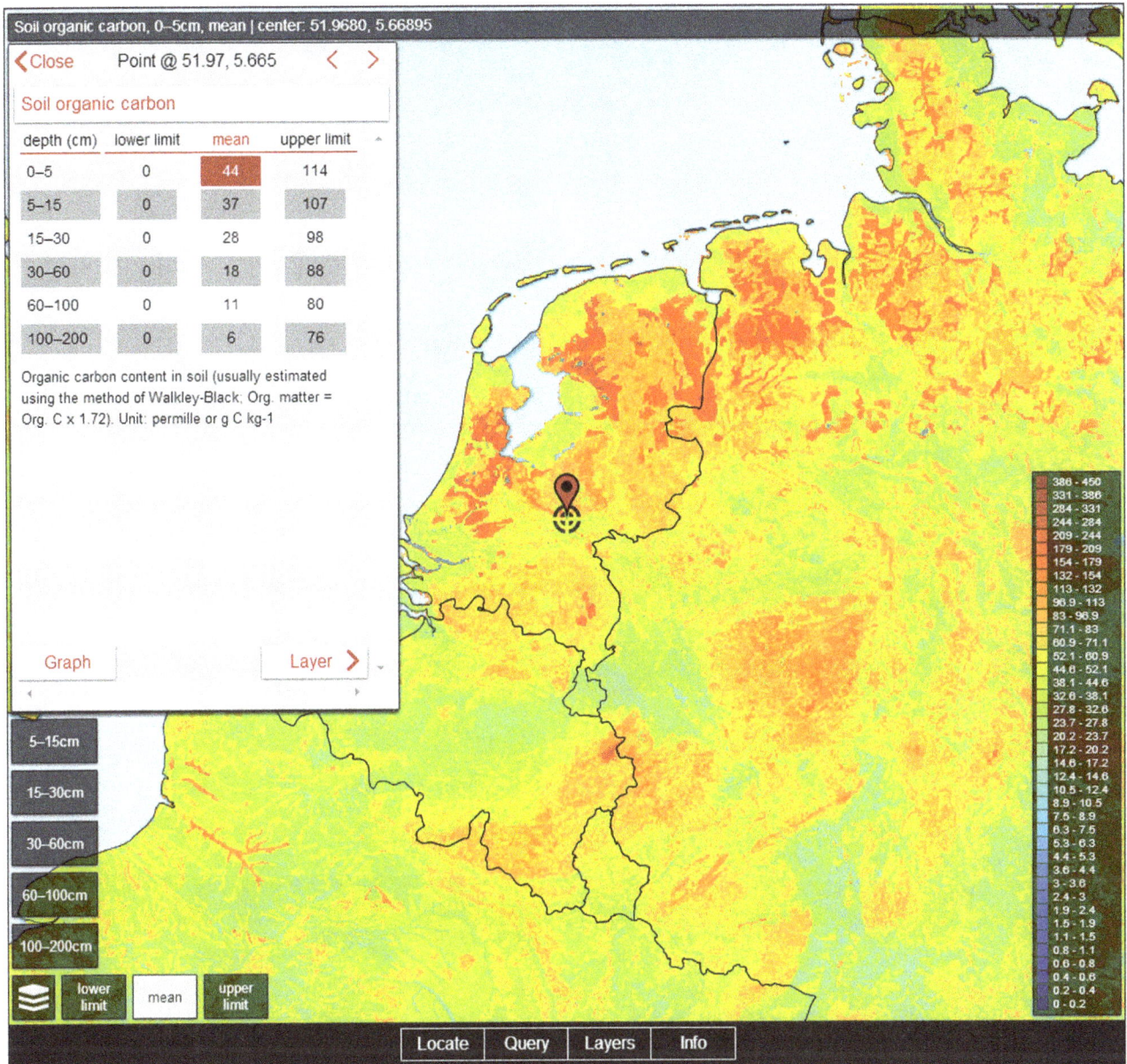

Figure 12. Accessing SoilGrids1km from the SoilInfo app for mobile devices. SoilInfo app is available for download via http://soilinfo.isric.org.

public money are not publicly available or are available in paper format only. Due to unbalanced representation and spatial clustering, predictions in the current version of SoilGrids1km are largely controlled by point data sets available for the USA and Europe. Most of these are from agricultural soils, which inflicts additional bias. Our predictions are therefore likely to exhibit lower accuracy for poorly represented areas such as most of the former Russian Federation, the northern Circumpolar Region, semi-arid and arid areas.

We have also purposely excluded all areas that show no evidence of historical vegetative cover. Our predictions are hence not globally complete. This is a definite drawback for use in global modelling and we acknowledge a need to use either expert judgment or data from other mapping sources to provide alternative predictions for areas with missing values. Again, for

deserts and bare rock areas it is perfectly valid to assume a 0 value for soil organic carbon, but it is not as straightforward to estimate soil pH for shifting sand areas for example. For the present, we argue that it is inappropriate to try to make predictions for areas that completely lack vegetative cover e.g. shifting sands of Sahara. These areas have very few to zero point profile observations which can be used to calibrate statistical prediction models. In addition, even if they did have a sufficient number of point profile measurements, the environments of extreme climatic conditions are so different from vegetated ones so that any prediction model is likely to be very different from ones we develop for vegetated areas. We recommend that SoilGrids1km users who require values for the complete land mask fill in the gaps by using expert knowledge or best regional estimates as available from conventional soil mapping (e.g. HWSD, ISRIC-WISE).

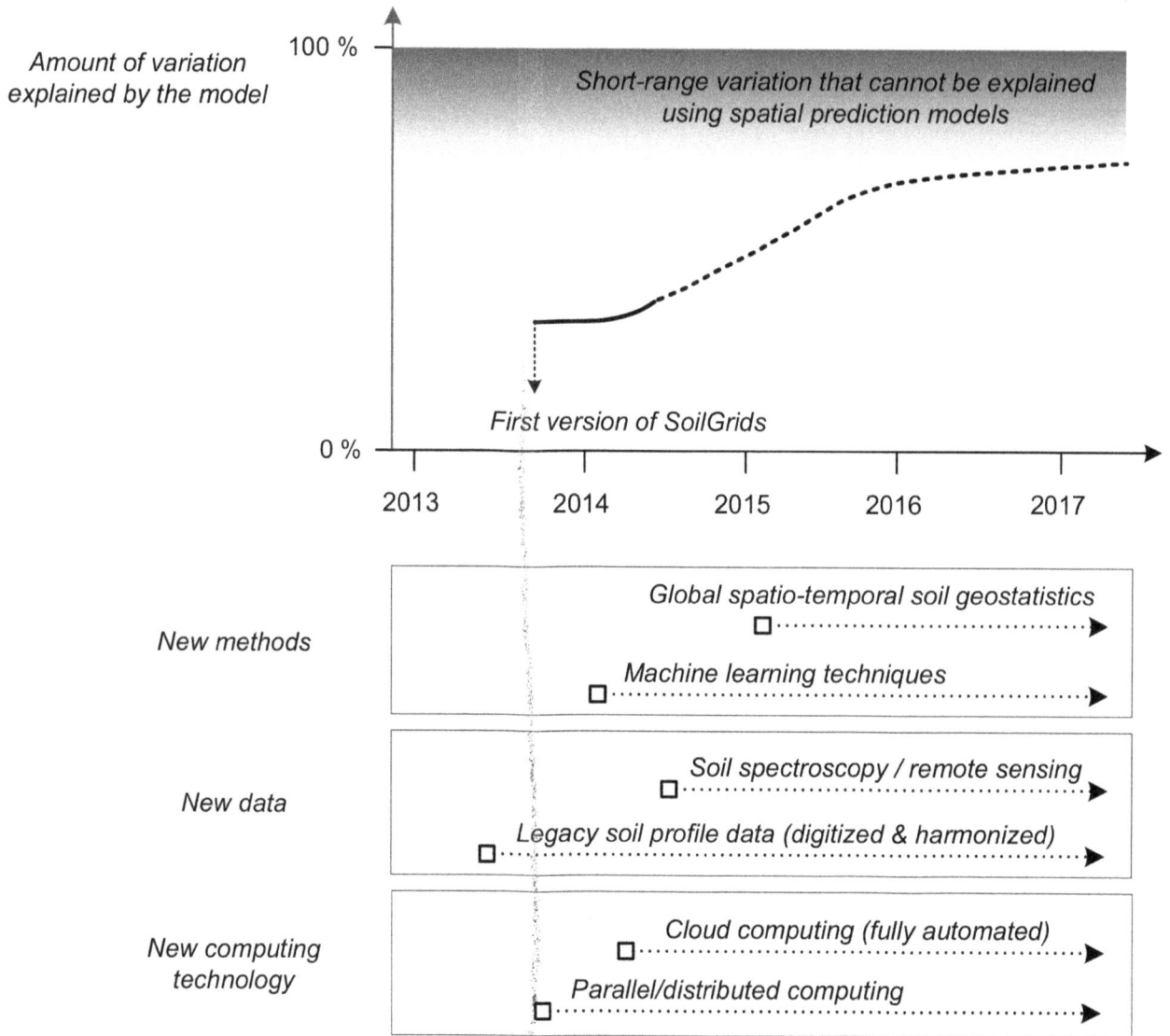

Figure 13. Projected evolution of SoilGrids in the years to come. We anticipate that the main drivers of success of SoilGrids will be use of machine learning methods for model fitting, development of spatio-temporal geostatistical models, use of new sources of field and remote sensing data and use of faster and more powerful computing capacities. Amount of variation explained by these models will eventually reach a *'natural limit'* (short-range variation that cannot be explained using spatial prediction models), until there is a technological jump in soil remote sensing technology e.g. ground penetrating scanners.

It is worth emphasizing that we designed GSIF as a flexible framework with respect to the choice of depths, dimensions (2D or 3D spatial predictions), spatial support size, soil properties and classes and prediction models. Outputs from GSIF are reproducible as a result of use of scripting. Consequently, all maps can be easily updated as new inputs (point and covariate data) become available. We used the GSIF system to generate SoilGrids1km maps for the standard depths defined by the GlobalSoilMap project, but basically one could use the same system for any depth and also for any new property. GSIF is therefore scalable and can be used to produce spatial predictions for virtually any soil property, at any depth and at any spatial or temporal resolution. This, of course, assumes the existence of a sufficient number of point soil observations of appropriate quality and of sufficient covariate layers at sufficiently fine spatial resolution to support modelling at a given spatial resolution.

All methods and models fitted for the purpose of producing SoilGrids1km are available via an Open Source platform (GSIF package for R) and could be adapted for both regional and local mapping. As with input data, the models used to make predictions in GSIF can be improved or replaced in subsequent iterations once better performing models are identified. Prediction models that could be considered in the future include those based on hierarchical Bayes models, regression trees, Random Forests and other machine learning techniques. Regression- trees and similar models could help model better abrupt changes in values vertically, and Random Forests could help emphasize relative importance of specific covariates. The actual modelling approach used to produce any set of predictions will be reviewed continuously to identify and apply the approach that produces the most correct, consistent and usable outputs.

Because the SoilGrids1km maps can be easily updated (or changed) the process used to produce the map (i.e. SoilGrids system) becomes more important than the map itself. Previously, the map product was seen as more important than the process used to produce it, because any map had to be considered as valid and useful for an extended period, as it took so long, and cost so much, to revise or update the map. Under the GSIF model, the final (or most current) map is no longer the most important output and any system that only provides a final map is considered deficient. We hence argue that it is more important to provide access to all data and models needed to produce (and reproduce) the map than to simply provide the final map itself.

In the future, we hope that GSIF will be used by an increasing number and variety of interested parties, including national and regional soil mapping agencies, commercial consulting agencies, advocacy groups and non-governmental organizations. We envisage GSIF as a platform for cooperation, collaboration, innovation and sharing. It will become so if interested parties decide to participate and contribute as committed partners. The number of soil profiles freely shared by the soil science community is constantly growing and national agencies and other data providers are encouraged to contribute their point data to help improve the prediction accuracy locally for specific countries/ regions, for the benefit of the global user community and in support of the global UN conventions.

SoilGrids1km are available for download under a Creative Commons non-Commercial license via http://soilgrids.org. Soil-Grids1km are also accessible via a Representational State Transfer (http://rest. soilgrids.org) service and via a mobile phone app "SoilInfo App" (http://soilinfo-app.org; Figure 12).

Acknowledgments

SoilGrids were developed as a part of the Global Soil Information Facilities (GSIF) initiative — tools for collating and serving global soil data developed jointly by ISRIC — World Soil Information and collaborators. ISRIC is a non-profit organization primarily funded by the Dutch government. SoilGrids1km were possible due to generous contributions of point and covariate data by various national and international agencies and organizations to which we are sincerely grateful: USDA National Resources Conservation Service, European Commission's Joint Research Centre, Africa Soil Information Service project, Mexican Instituto Nacional de Estadística y Geografía/CONABIO, University of São Paulo, Agriculture and Agri-Food Canada, Institute of Soil Science, Chinese Academy of Sciences, Tehran University (National Database of Iran), and others. We are especially thankful for support from the Africa Soil Information Service (AfSIS) project, funded by the Bill and Melinda Gates foundation and the Alliance for a Green Revolution in Africa (AGRA), and to the Food and Agriculture Organization (FAO) of the United Nations for supporting the international soil standards and initiatives developed at ISRIC (SOTER, GSIF). The funders had no role in study design, data collection and analysis, decision to publish, or preparation of the manuscript. We gratefully acknowledge the past and present leadership of ISRIC (Prem Bindraban, Peter de Ruiter and Hein van Holsteijn) for recognizing the potential value of GSIF and SoilGrids1km and for consistently and actively supporting development both within ISRIC and in interactions with external agencies and partner organizations. We also gratefully acknowledge the significant contributions of Dr. H.I. Reuter, formerly with ISRIC, presently statistical officer at EUROSTAT, in developing the initial concepts and design for GSIF, in helping to conceptualize and assemble the global maps of covariates and in implementing some of the early algorithms.

Author Contributions

Conceived and designed the experiments: TH. Contributed to the writing of the manuscript: TH BM GH BK MW. Prepared input profile and covariate data: NB JL MG AR. Web mapping services and user interfaces: JJ ER. Cross-validation and quality control: AR GH BK NB RAM. Design of the system, programming and preparation of input data: TH JJ.

References

1. Sanchez PA, Ahamed S, Carré F, Hartemink AE, Hempel J, et al (2009) Digital Soil Map of the World. Science 325: 680–681.

2. Omuto C, Nachtergaele F, Vargas Rojas R (2012) State of the Art Report on Global and Regional Soil Information: Where are we? Where to go? Global Soil Partnership technical report. Rome: FAO, 66 pp.

3. Arrouays D, Grundy MG, Hartemink AE, Hempel JW, Heuvelink GB, et al. (2014) Chapter Three — GlobalSoilMap: Toward a Fine-Resolution Global Grid of Soil Properties. In: Sparks DL, editor, Soil carbon, Academic Press, volume 125 of Advances in Agronomy. pp. 93–134.

4. Viscarra Rossel RA, Webster R, Bui EN, Baldock JA (2014) Baseline map of organic carbon in Australian soil to support national carbon accounting and monitoring under climate change. Global Change Biology 20(9): 2953–2970.

5. Grunwald S, Thompson JA, Boettinger JL (2011) Digital Soil Mapping and Modeling at Continental Scales: Finding Solutions for Global Issues. Soil Science Society of America Journal 75: 1201–1213.

6. Savtchenko A, Ouzounov D, Ahmad S, Acker J, Leptoukh G, et al. (2004) Terra and Aqua MODIS products available from NASA GES DAAC. Advances in Space Research 34: 710–714.

7. Scharlemann JPW, Benz D, Hay SI, Purse BV, Tatem AJ, et al. (2008) Global data for ecology and epidemiology: a novel algorithm for temporal Fourier processing MODIS data. PLoS One 3: e1408.

8. Hansen MC, Potapov PV, Moore R, Hancher M, Turubanova SA, et al. (2013) High-resolution global maps of 21st-century forest cover change. Science 342: 850–853.

9. González JH, Bachmann M, Krieger G, Fiedler H (2010) Development of the TanDEM-X calibration concept: analysis of systematic errors. Geoscience and Remote Sensing, IEEE Transactions on 48: 716–726.

10. Hartemink AE, Krasilnikov P, Bockheim J (2013) Soil maps of the world. Geoderma 207/208: 256–267.

11. FAO/IIASA/ISRIC/ISS-CAS/JRC (2012) Harmonized World Soil Database (version 1.2). Rome: FAO.

12. Batjes NH (2012) ISRIC-WISE derived soil properties on a 5 by 5 arc-minutes global grid (ver. 1.2). Report 2012/01. Wageningen: ISRIC — World Soil Information, 57 pp.

13. Diggle PJ, Ribeiro Jr PJ (2007) Model-based Geostatistics. Springer Series in Statistics. Springer, 288 pp.

14. Kilibarda M, Hengl T, Heuvelink GB, Graeler B, Pebesma E, et al. (2014) Spatio-temporal interpolation of daily temperatures for global land areas at 1 km resolution. Journal of Geophysical Research: Atmospheres, 119(5): 2294–2313.

15. Pebesma E, Cornford D, Dubois G, Heuvelink GB, Hristopulos D, et al. (2011) INTAMAP: The design and implementation of an interoperable automated interpolation web service. Computers & Geosciences 37: 343–352.

16. Fritz S, McCallum I, Schill C, Perger C, See L, et al. (2012) Geo-Wiki: An online platform for improving global land cover. Environmental Modelling & Software 31: 110–123.

17. R Development Core Team (2009) R: A language and environment for statistical computing. Vienna, Austria: R Foundation for Statistical Computing, 409 pp. ISBN 3-900051-07-0.

18. Shore J (2007) The Art of Agile Development. Theory in practice. O'Reilly Media, 440 pp.

19. Tóth G, Jones A, Montanarella L, editors (2013) LUCAS Topsoil Survey. Methodology, data and results. JRC Technical Reports EUR 26102. Luxembourg: Publications Office of the European Union.

20. Leenaars J (2012) Africa Soil Profiles Database, Version 1.0. A compilation of geo-referenced and standardized legacy soil profile data for Sub Saharan Africa (with dataset). Wageningen, the Netherlands: Africa Soil Information Service (AfSIS) project and ISRIC — World Soil Information, 45 pp. Available: http://www.isric.org. ISRIC report 2012/03.

21. Instituto Nacional de Estadística y Geografía (INEGI) (2000) Conjunto de Datos de Perfiles de Suelos, Escala 1: 250 000 Serie II. (Continuo Nacional). Aguascalientes, Ags. México: INEGI. INEGI2000.

22. Cooper M, Mendes LMS, Silva WLC, Sparovek G (2005) A national soil profile database for brazil available to international scientists. Soil Science Society of America Journal 69: 649–652.

23. Shangguan W, Dai Y, Liu B, Zhu A, Duan Q, et al. (2013) A China data set of soil properties for land surface modeling. Journal of Advances in Modeling Earth Systems 5: 212–224.

24. MacDonald KB, Valentine KWG (1992) CanSIS/NSDB. A general description. Ottawa: Centre for Land and Biological Resources Research, Research Branch, Agriculture Canada.

25. Batjes NH (2009) Harmonized soil profile data for applications at global and continental scales: Updates to the WISE database. Soil Use and Management 25: 124–127.

26. Van Engelen V, Dijkshoorn J, editors (2012) Global and National Soils and Terrain Digital Databases (SOTER), Procedures Manual, version 2.0. ISRIC Report 2012/04. Wageningen, the Netherlands: ISRIC — World Soil Information, 192 pp.

27. Hollis JM, Jones RJA, Marshall CJ, Holden A, Van de Veen JR, et al. (2006) SPADE-2: The soil profile analytical database for Europe, version 1.0. Luxembourg: Office for official publications of the European Communities. EUR22127EN.

28. Stolbovoi V, McCallum I (2002) Land Resources of Russia (CD-ROM). Vienna: IIASA and RAS.

29. Hijmans RJ, Cameron SE, Parra JL, Jones PG, Jarvis A (2005) Very high resolution interpolated climate surfaces for global land areas. International Journal of Climatology 25: 1965–1978.

30. Hartmann J, Moosdorf N (2012) The new global lithological map database GLiM: A representation of rock properties at the Earth surface. Geochemistry, Geophysics, Geosystems 13: n/a–n/a.

31. Carroll M, Townshend J, DiMiceli C, Noojipady P, Sohlberg RA (2009) A new global raster water mask at 250 m resolution. International Journal of Digital Earth 2: 291–308.

32. Odeh I, McBratney AB, Chittleborough D (1995) Further results on prediction of soil properties from terrain attributes: heterotopic cokriging and regression-kriging. Geoderma 67: 215–226.

33. Hengl T, Heuvelink G, Rossiter DG (2007) About regression-kriging: from equations to case studies. Computers & Geosciences 33: 1301–1315.

34. Kutner M, Neter J, Nachtsheim C, Li W (2005) Applied Linear Statistical Models. Operations and decision sciences series. McGraw-Hill Irwin.

35. Hastie TJ (1992) Statistical Models in S, Wadsworth & Brooks/Cole, Chapter: Generalized additive models. pp. 249–307.

36. Venables WN, Ripley BD (2002) Modern applied statistics with S. New York: Springer-Verlag, 4th edition, 481 pp.

37. Agarwal DK, Gelfand AE, Citron-Pousty S (2002) Zero-inflated models with application to spatial count data. Environmental and Ecological statistics 9: 341–355.

38. McBratney AB, Minasny B, MacMillan RA, Carré F (2011) Digital soil mapping. In: Li H, Sumner M, editors, Handbook of Soil Science, CRC Press, volume 37. pp. 1–45.

39. Mokma DL, Buurman P (1982) Podzols and podzolization in temperate regions. ISM monograph. Wageningen: International Soil Museum, 126 pp.

40. Minasny B, McBratney AB (2010) Methodologies for Global Soil Mapping. In: Boettinger JL, Howell DW, Moore AC, Hartemink AE, Kienast-Brown S, editors, Digital Soil Mapping: Bridging Research, Environmental Application, and Operation, Springer, volume 2 of Progress in Soil Science, chapter 34. pp. 429–454.

41. Jenny H (1994) Factors of soil formation: a system of quantitative pedology. Dover books on Earth sciences. Dover Publications.

42. Bouma J, Batjes NH, Groot JJR (1998) Exploring land quality effects on world food supply. Geoderma 86: 43–59.

43. IUSS Working Group WRB (2006) World reference base for soil resources 2006: a framework for international classification, correlation and communication. World soil resources reports No. 103. Rome: Food and Agriculture Organization of the United Nations.

44. US Department of Agriculture (2010) Keys to Soil Taxonomy. U.S. Government Printing Office, 11th edition.

45. Hengl T, Nikolić M, MacMillan RA (2013) Mapping efficiency and information content. International Journal of Applied Earth Observation and Geoinformation 22: 127–138.

46. Goovaerts P (2001) Geostatistical modelling of uncertainty in soil science. Geoderma 103: 3–26.

47. Nelson DW, Sommers L (1982) Total carbon, organic carbon, and organic matter. In: Page A, Miller R, Keeney D, editors, Methods of soil analysis, Part 2, Madison, WI: ASA and SSSA, Agron. Monogr. 9. 2nd edition, pp. 539–579.

48. Heuvelink G (1998) Error propagation in environmental modelling with GIS. London, UK: Taylor & Francis.

49. Bivand R, Pebesma E, Rubio V (2013) Applied Spatial Data Analysis with R. Use R Series. Heidelberg: Springer, 2nd edition, 401 pp.

50. Knaus J, Porzelius C, Binder H, Schwarzer G (2009) Easier parallel computing in R with snowfall and sfCluster. The R Journal 1: 54–59.

51. Jobbágy EG, Jackson RB (2000) The vertical distribution of soil organic carbon and its relation to climate and vegetation. Ecological applications 10: 423–436.

52. Todd-Brown KEO, Randerson JT, Post WM, Hoffman FM, Tarnocai C, et al. (2013) Causes of variation in soil carbon simulations from CMIP5 Earth system models and comparison with observations. Biogeosciences 10: 1717–1736.

53. Hugelius G, Strauss J, Zubrzycki S, Harden JW, Schuur E, et al. (2014) Improved estimates show large circumpolar stocks of permafrost carbon while quantifying substantial uncertainty ranges and identifying remaining data gaps. Biogeosciences Discussions 11: 4771–4822.

54. Minasny B, McBratney AB, Malone BP, Lacoste M, Walter C (2014) Quantitatively Predicting Soil Carbon Across Landscapes. In: Hartemink AE, McSweeney K, editors, Soil Carbon, Springer International Publishing, Progress in Soil Science. pp. 45–57.

55. Heuvelink GBM, Webster R (2001) Modelling soil variation: past, present, and future. Geoderma 100: 269–301.

56. Antonić O, Pernar N, Jelaska SD (2003) Spatial distribution of main forest soil groups in Croatia as a function of basic pedogenetic factors. Ecological modelling 170: 363–371.

57. Scharlemann JPW, Tanner EVJ, Hiederer R, Kapos V (2014) Global soil carbon: understanding and managing the largest terrestrial carbon pool. Carbon Management 5: 81–91.

58. Ruesch A, Gibbs HK (2008) New IPCC Tier-1 global biomass carbon map for the year 2000. Oak Ridge National Laboratory, Tennessee: Carbon Dioxide Information Analysis Center.

59. Verburg PH, Neumann K, Nol L (2011) Challenges in using land use and land cover data for global change studies. Global Change Biology 17: 974–989.

60. Montanarella L, Vargas R (2012) Global governance of soil resources as a necessary condition for sustainable development. Current Opinion in Environmental Sustainability 4: 559–564.

61. Brus D, Kempen B, Heuvelink G (2011) Sampling for validation of digital soil maps. European Journal of Soil Science 62: 394–407.

4

Spatial Variability and Stocks of Soil Organic Carbon in the Gobi Desert of Northwestern China

Pingping Zhang[1], Ming'an Shao[1,2]*

1 State Key Laboratory of Soil Erosion and Dryland Farming on the Loess Plateau, Northwest A & F University, Yangling, China, **2** Key Laboratory of Ecosystem Network Observation and Modeling, Institute of Geographical Science and Natural Resources, Chinese Academy of Sciences, Beijing, China

Abstract

Soil organic carbon (SOC) plays an important role in improving soil properties and the C global cycle. Limited attention, though, has been given to assessing the spatial patterns and stocks of SOC in desert ecosystems. In this study, we quantitatively evaluated the spatial variability of SOC and its influencing factors and estimated SOC storage in a region (40 km^2) of the Gobi desert. SOC exhibited a log-normal depth distribution with means of 1.6, 1.5, 1.4, and 1.4 g kg^{-1} for the 0–10, 10–20, 20–30, and 30–40 cm layers, respectively, and was moderately variable according to the coefficients of variation (37–42%). Variability of SOC increased as the sampling area expanded and could be well parameterized as a power function of the sampling area. Significant correlations were detected between SOC and soil physical properties, i.e. stone, sand, silt, and clay contents and soil bulk density. The relatively coarse fractions, i.e. sand, silt, and stone contents, had the largest effects on SOC variability. Experimental semivariograms of SOC were best fitted by exponential models. Nugget-to-sill ratios indicated a strong spatial dependence for SOC concentrations at all depths in the study area. The surface layer (0–10 cm) had the largest spatial dependency compared with the other layers. The mapping revealed a decreasing trend of SOC concentrations from south to north across this region of the Gobi desert, with higher levels close to an oasis and lower levels surrounded by mountains and near the desert. SOC density to depths of 20 and 40 cm for this 40 km^2 area was estimated at 0.42 and 0.68 kg C m^{-2}, respectively. This study provides an important contribution to understanding the role of the Gobi desert in the global carbon cycle.

Editor: Ben Bond-Lamberty, DOE Pacific Northwest National Laboratory, United States of America

Funding: This work was supported by grants from the Major Program of the National Natural Science Foundation of China No. 91025018. The funders had no role in study design, data collection and analysis, decision to publish, or preparation of the manuscript.

Competing Interests: The authors have declared that no competing interests exist.

* E-mail: mashao@ms.iswc.ac.cn

Introduction

Soil organic carbon (SOC) has an important influence on the physical, chemical, and biological properties of soil and is critical for improving soil fertility and quality, increasing the water holding capacity of soil, reducing soil erosion, and enhancing crop productivity [1,2]. With climate change and environmental issues dominating global concerns, SOC has received increasing attention worldwide because of its important role in the global C cycle and its potential feedback on the global warming [3–6]. As one of the largest and most dynamic component in the global C cycle, the SOC stock is at least two times the amount of C stored in the vegetation and atmosphere [7]. Thus, a small loss of SOC pool due to changes in fertilization, cropping system, farming practices, and soil erosion could significantly increase the atmospheric CO_2 [8–11]. On the other hand, soils can increase the existing SOC pool by sequestration of C from the atmosphere [12–15], the processes of which are an active area of study. Reliable assessment of the spatial patterns and stocks of SOC at one timeline as a baseline is essential for understanding the potential of soils to sequester C, for quantifying the SOC sink or source capacity of soils in changing environments, and for developing the strategies necessary to mitigate the effects of global warming [16,17].

In recent years, extensive work has been conducted toward estimating the SOC stocks and distribution patterns at the global,

continental, country, and regional scales [11,18–24]. For example, the global SOC stock has been estimated to be about 2400 Pg C in the top 2 m [4]. However, these estimations are highly uncertain because of the gaps in spatial coverage for many regions that causes difficulties to develop a harmonized SOC baseline [22–24]. In addition, the selection of the type of SOC database, the land use and/or soil map, the mapping resolution, reference depth, bulk density or other information can also have a great effect on the final SOC stock estimation [25]. Similarly, due to inconsistent estimation methods and limited data, the SOC stock estimations in China are also uncertain and has varied greatly, from 50 to180 Pg, and SOC density from 54.6 to 190.5 t C/ha [22]. The accuracy of these large-scale SOC stock estimations largely depends on the data availability from site-based or small-scale measurements [6,24]. To reduce the uncertainty of SOC stocks estimation and better understand the role of SOC in the global C cycle, reliable baseline datasets providing information on SOC stocks in all types of sites and ecosystems are necessary.

Desertification is one of the most severe types of land degradation in arid and semiarid areas of the world [26]. Due to the harsh natural conditions and the fragile ecological environment, desert ecosystems are more sensitive to climate change, leading to the emission of CO_2 to the atmosphere and a reduction in the pool of SOC [27,28]. In contrast, it is possible to increase SOC concentrations in desert soils through the adoption of restorative measures such as the establishment of plants [14,29,30] and the prohibition of grazing [31]. [32] indicated that the control of desertification could globally sequester 0.9–1.9 Pg C yr^{-1} over a period of 25–50 years.

China is also seriously threatened by desertification [33,34]. [27] estimated that desertified land in China potentially covers 158 Mha, comprising 81 Mha of slight, 61 Mha of moderate, and 35 Mha of severe desertification. The widely distributed desertified lands in China thus likely have a considerable effect on the regional terrestrial C balance and the feedbacks that affect climate change [35]. Although some studies have been conducted on assessment of SOC concentrations/stocks in desert area, many questions still remain open. So far, most studies on SOC stock estimates from desert ecosystems have been carried out in sandy desert [36–39], and only few from the Gobi desert are available [40]. Due to the difficulties and associated costs of soil sampling, most studies estimating SOC stocks rely on information taken from a relatively small number of representative sampling points or profiles [41]. This decreases the estimation accuracy of SOC stocks in desert ecosystems and limits our capability to evaluate the C budget, to assess its contribution to the increasing global concentrations of CO_2, and to propose measures to increase the sequestration of organic C in soils. Therefore, to better understand the SOC pool in desert environments, it is necessary to conduct more intensive site and local scale estimates of the variance in SOC, and the possible spatial controls on this variance structure.

Gansu province is one of the main desertified areas induced by wind and one of the source regions of sandstorms in northern China. In this paper, a typical fenced region of the Gobi desert was chosen as a study case. The objectives of this study were: (1) to estimate SOC concentrations and determine the spatial distribution of SOC, (2) to analyse the relationships between SOC concentrations and environmental factors, and (3) to estimate SOC density and storage in the study area.

Materials and Methods

Ethics statement

The study area belong to Linze Inland River Basin Research Station (39°21′ N and 100°07′E,1389 m), a department of the Cold and Arid Regions Environmental and Engineering Research Institute, Chinese Academy of Sciences. The study was approved by the Cold and Arid Regions Environmental and Engineering Research Institute, Chinese Academy of Sciences.

Study area

The study area, occupying approximately 40 km^2 (5 km×8 km), is located in the Gobi desert in the middle of Gansu province (the central reaches of the Heihe River Basin) of Northwestern China, between latitudes 39°24′ and 39°28′N and longitudes 100°08′ and 100°11′E (Fig. 1a,b). The region is a relatively flat alluvial plain (elevation ranging from 1390 to 1470 m) bordered by a young oasis to the southwest, the remnants of the Qilian Mountains to the north, and an extension of the Badain Jaran Desert to the southeast (Fig. 1c). The area is characterized by low and seasonal variability in rainfall and is classified as a typical temperate desert. The mean annual precipitation and air temperature are 117 mm and 7.6°C, respectively. Rainfall in brief summer showers contributes 65% of the annual total precipitation. The mean annual pan-evaporation is approximately 2390 mm, twenty times greater than the annual precipitation. The average annual wind speed is 3.2 m s^{-1}, with the resultant wind coming from the northwest, and the dominant windy days and wind storms occur between

March and May [42]. The zonal soil is classified as gray-brown desert soil, derived from gravelly diluvial-alluvial materials of the denuded monadnock [43]. Stones are present in a significant proportion of the surface and sub-soil horizons. The aboveground plant cover is discontinuous and can be described as patches of sub-shrubs surrounded by bare areas. The study area has been fenced and is protected from grazing for the purpose of revegetation and reclamation. The dominant plant species are *Nitraria sphaerocarpa* Maxim. and *Reaumuria soongorica* (Pall.) Maxim., and the accompanying plant species are mainly *Kalidium gracile* Fenzl., *Allium mongolicum* Rgl., *Bassia dasyphylla* (Fisch. and Mey.) Kuntze. and *Halogeton arachnoideus* Moq.

Soil sampling and laboratory analysis

Soil samples were collected from a total of 187 locations on a regular grid of 500 m×500 m throughout the study area from August to September in 2011. A portable GPS receiver (Garmin GPSmap 62 s) was used to locate the sampling site, as displayed in Figure 1c. At each location, soil was collected from four depths (0–10, 10–20, 20–30, and 30–40 cm) at five randomly selected sampling points within a radius of approximately 20 cm. The five sub-samples were then combined to produce one representative sample. In total, 748 soil samples were collected. The samples were all air dried, weighed, and sieved to 2 mm to separate the coarse (>2 mm) and fine (<2 mm) fractions. The former was reweighed to determine the stone content. The latter was separated into two parts: one was subject to further particle-size analyses by laser diffraction using a Mastersizer 2000 (Malvern Instruments, Malvern, England), and the other was ground to pass through a 0.25-mm mesh for SOC concentration analysis. The SOC concentration (g kg^{-1}) was measured using the potassium dichromate-wet combustion method [44]. Undisturbed soil samples cores (100 cm^3) were also collected for determining soil bulk density (BD) in each layer. To reduce the influences of stones on BD measurement, we averaged five replicate measurements for each location and layer.

Re-sampling method

Estimating SOC variability at different scales is important for effective soil survey sampling design and SOC change prediction [2]. In this study, to detect the change tendency of SOC variability with the size of the sampling area, a series of sampling point allocations with different areas were done through re-sampling using all sampling points (n = 187) in the study area (5 km×8 km) [45,46]. For convenience, the west-east and south-north distances of the re-sampling area were set integral multiples of 1 km to generate five re-sampling options for the west-east distance and eight options for the south-north distance. The random combination of these two sets of options thus produced 40 potential re-sampling methods. Some of these methods, though, were the same within the study area (e.g., the re-sampling methods of 2 km×6 km, 3 km×4 km, and 4 km×3 km yielded the same area), so finally a total of 24 kinds of re-sampling areas were obtained (Table 1). SOC variability at a certain area was computed by averaging the CV values of all the possible re-sampling scenarios with the same area.

Calculation of SOC density and stock

The SOC density of a single layer was estimated based on Eq. (1):

$$SOCD_i = \frac{SOC_i \times BD_i \times d_i \times (1 - CF_i/100)}{100} \qquad (1)$$

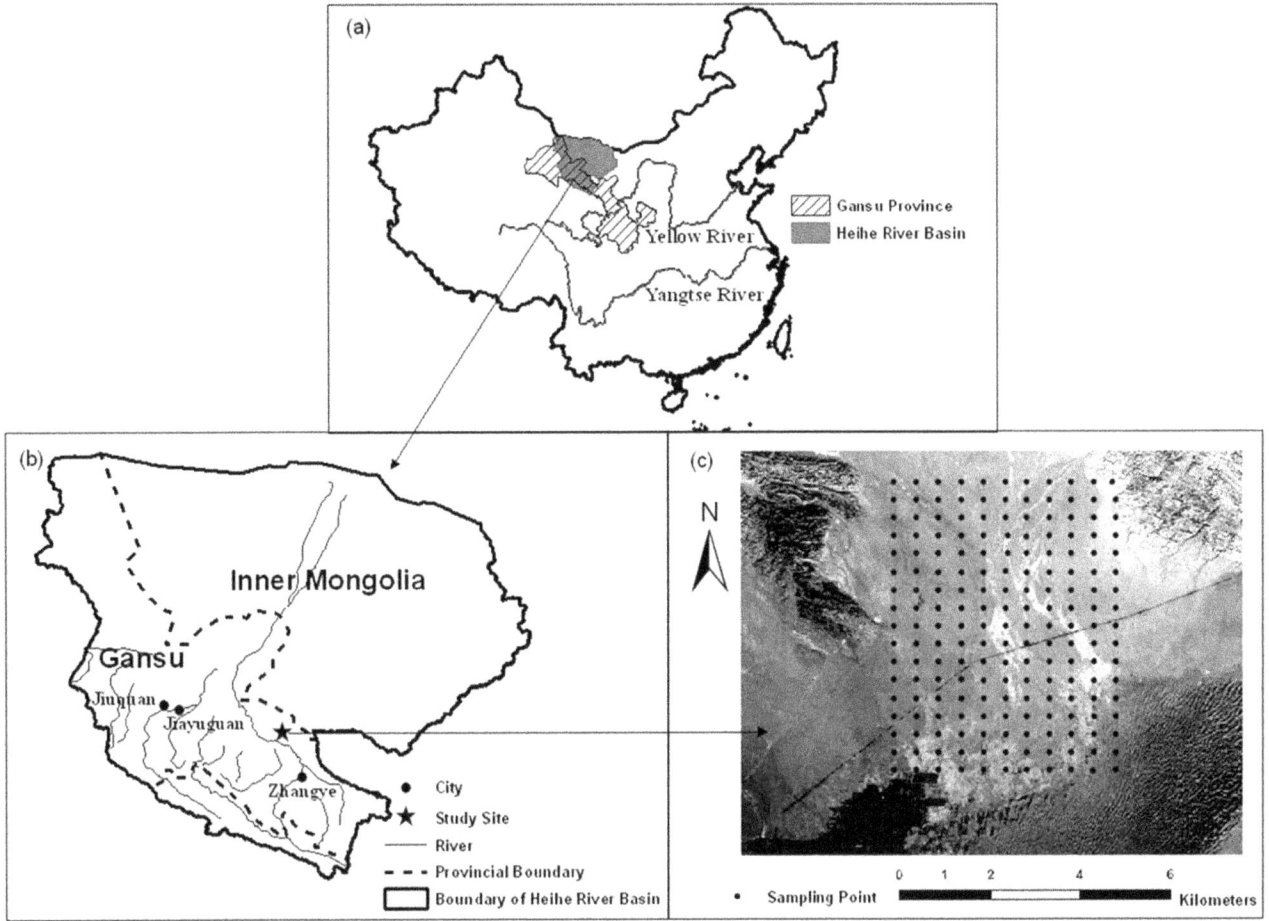

Figure 1. The location of Gansu Province and the Heihe River Basin, China (a), the study site in the Heihe River Basin (b), and the soil-sampling points in the study area (c).

Table 1. Details of the re-sampling areas and sampling methods.

Re-sampling area (km^2)	Sampling method	Re-sampling area (km^2)	Sampling method
1	1×1*	15	3×5; 5×3
2	1×2; 2×1	16	2×8; 4×4
3	1×3; 3×1	18	3×6
4	1×4; 4×1; 2×2	20	4×5; 5×4
5	1×5; 5×1	21	3×7
6	1×6; 2×3; 3×2	24	3×8; 4×6
7	1×7	25	5×5
8	1×8; 2×4; 4×2	28	4×7
9	3×8	30	5×6
10	2×5; 5×2	32	4×8
12	2×6; 3×4; 4×3	35	5×7
14	2×7	40	5×8

*The digit before the multiplication sign represents the west-east sampling distance (km), and the digit after it represents the south-north sampling distance (km).

Table 2. Selected soil physical properties at different soil depths.

Variables	Statistical parameter	Soil depth			
		0–10 cm	10–20 cm	20–30 cm	30–40 cm
Stones	Mean (%)	12.1 a	13.3 ab	15.2 b	69.1 c
	CV (%)	100.5	103.4	101.3	24.5
Sand	Mean (%)	70.9 a	72.4 a	71.1 a	26.0 a
	CV (%)	20.1	19.7	21	22.1
Silt	Mean (%)	7.5 a	6.5 a	6.5 a	2.4 a
	CV (%)	85.5	97.5	107.5	111.9
Clay	Mean (%)	9.5 a	7.8 ab	7.2 b	2.6 b
	CV (%)	84.4	102.4	111.3	123
BD	Mean (g cm^{-3})	1.4 a	1.4 a	1.4 a	0.5 a
	CV (%)	8.8	10.4	9.7	10.1

BD, bulk density; CV, coefficient of variation; Stones, >2 mm; Sand, 2–0.2 mm; Silt, 0.2–0.002 mm; Clay, <0.002 mm. Mean values followed by the same letter are not significantly different ($\alpha = 0.05$) at different soil depths using the LSD method.

where $SOCD_i$ and SOC_i are SOC density (kg C m^{-2}) and concentration (g kg^{-1}) of the ith layer, BD_i is the bulk density of the ith layer (g cm^{-3}), d_i is the depth of the ith layer (cm), and CF_i is the fraction (%) of coarse fragments >2 mm in the ith layer.

For an individual profile with a depth D (cm), SOC densities were then calculated by summarising the SOC density of each soil layer i:

$$SOCD_D = \sum_{i=1}^{n} SOCD_i \qquad (2)$$

The total SOC storage is calculated as:

$$SOC_{stock} = \sum_{j=1}^{m} S_j \times SOCD_{D,j} \qquad (3)$$

where S_j is the total area (m^2) of a given land-use type j and $SOCD_{D,j}$ is the average SOC density of the jth land-use type (kg m^{-2}).

Statistical analysis

A descriptive statistical analysis was first used to illustrate the central trend and the overall variation of the variables. This analysis included descriptions of the minimum, maximum, mean, median, skewness, Kurtosis, standard deviation (SD), and coefficients of variation (CVs). A one-sample Kolmogorov-Smirnov (K-S) test was used to examine the normality of the data, and natural logarithmic transformations were performed where necessary to meet the normality requirement of geostatistical analysis. The means of different layers were compared by a one-way analysis of variance (ANOVA). Correlation analysis and stepwise linear regression analysis were performed to understand relationships between SOC concentrations and the environmental factors. All statistical analyses used the programme SPSS v. 16.0 (SPSS Inc., Chicago, IL, USA).

Geostatistical methods such as semivariogram calculation, cross-validation, kriging, and mapping have been widely applied in the study of SOC spatial distribution [47,48]. Semivariograms were used to determine the degree of spatial dependence. Before semivariogram calculation, a preliminary semivariogram surface analysis was performed to detect any zonal effect or trend in direction [49]. The experimental semivariogram, $\gamma(h)$, is half the

Table 3. Summary of statistical parameters for soil organic carbon (SOC) concentrations at different soil depths.

Variables	SOC (g kg^{-1})			
	0–10 cm	10–20 cm	20–30 cm	30–40 cm
Maximum	4.5	3.9	4.3	3.6
Minimum	0.4	0.5	0.5	0.5
Mean	1.6 a	1.5 ab	1.4 b	1.4 b
SD	0.6	0.6	0.6	0.6
CV	36.8	38.5	41.0	42.1
Skewness	1.83	1.26	1.64	1.15
Kurtosis	5.95	2.40	4.66	1.04
P of K-S test	0.001	0.012	0.009	0.006

SD, standard deviation; CV, coefficient of variation. Mean values followed by the same letter are not significantly different ($\alpha = 0.05$) at different soil depths using the LSD method.

expected squared difference between paired data separated by a distance h and is expressed as:

$$\gamma(h) = \frac{1}{2N(h)} \sum_{i=1}^{N(h)} [Z(x_i) - Z(x_i + h)]^2 \qquad (4)$$

where $Z(x_i + h)$ and $Z(x_i)$ are observations at positions $x_i + h$ and x_i, respectively, and $N(h)$ denotes the number of data pairs separated by the lag distance, h. A semivariogram model contains three important parameters which interpret the spatial structure of soil properties: nugget (C_0), sill ($C+C_0$), and range (A). Nugget represents the undetectable measurement error, inherent variability or the variation within the minimum sampling distance. Sill is the upper limit of the semivariogram model, representing the total variation. The separation distance at which the sill is reached is the range of spatial dependence. Samples separated by distances smaller than the range are spatially related, whereas samples separated by larger distances are not spatially related. The nugget ratio (C_0/C_0+C) can be regarded as a criterion for classifying the spatial dependence of soil properties. A variable is considered to have strong, moderate, or weak spatial dependence if the ratio is less than 0.25; between 0.25 and 0.75; and over 0.75, respectively [50].

In this study, all semivariograms were estimated with fixed distance intervals of 500 m, and the maximum distance was set to 4700 m. Because 4700 m is less than half the maximum distance between sampling sites, it coincides with the requirement of geostatistical analysis [51]. A total of nine sets of class intervals were generated. There are several commonly used semivariogram models (such as spherical, exponential, Gaussian, and linear models). The best model was based on two criteria: small residual sum of squares (RSS) and high coefficient of determination (R^2).

The best-fit parameters were subsequently estimated using weighted least squares regression method.

After selecting the best-fit semivariogram models, ordinary kriging was used as an interpolation method to predict values for SOC concentrations. The prediction was calculated as the linear sum:

$$Z^*(x_0) = \sum_{i=1}^{n} \lambda_i Z(x_i) \qquad (5)$$

where $Z^*(x_0)$ is the value to be predicted at the location x_0, $Z(x_i)$ is the known value at the sampling location x_i, n is the number of locations within the search neighbourhood used for the prediction, and λ_i is the kriging weight assigned to $Z(x_i)$.

The predicted map quality of the SOC was tested by cross-validation with replacement. Three indices were calculated to assess the effectiveness of kriging: mean error (ME), root mean square error (RMSE), and root mean square standardised error (RMSSE) [52].

The geostatistical analysis was performed with GS+ v. 7.0 (Gamma Design Software, Plainwell, Michigan, USA), and contour maps through ordinary kriging were produced with GIS software ArcView v. 3.3 and its extension module of Spatial Analysis v. 2 (ESRI Inc., Redlands, California, USA).

Results and Discussion

Soil physical properties

Due to the lack of mineral weathering and to long-term wind and water erosion, this natural desert exhibited a high content of stones, with an average of 27.42% stones in the top 40 cm of soil. The stone content increased slowly from 12.2% in the 0–10 cm layer to 15.2% in the 20–30 cm layer, and then abruptly increased

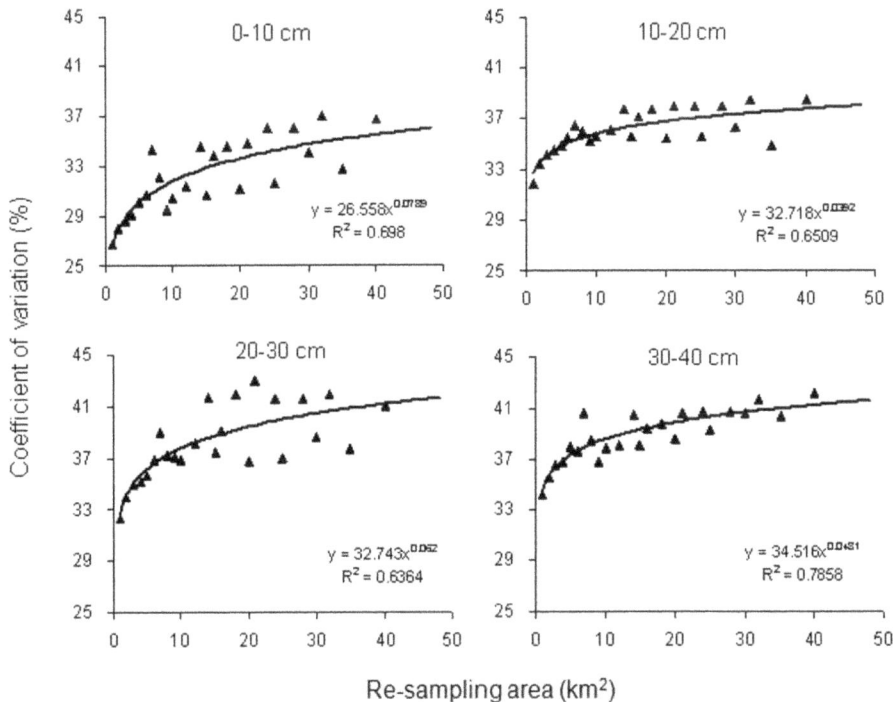

Figure 2. Coefficient of variation of soil organic carbon (SOC) concentrations at different re-sampling areas.

Table 4. Correlation analysis between soil organic carbon (SOC) concentrations and soil physical properties at different soil depths.

Variables	SOC (g kg^{-1})			
	0–10 cm	10–20 cm	20–30 cm	30–40 cm
Stones (%)	−0.196**	−0.201**	−0.201**	−0.448**
Sand (%)	−0.679**	−0.605**	−0.572**	−0.564**
Silt (%)	0.676**	0.606**	0.572**	0.567**
Clay (%)	0.663**	0.593**	0.561**	0.552**
BD (g cm^{-3})	−0.413**	−0.387**	−0.432**	−0.398**

BD, bulk density; ** denotes significance of correlation at $P<0.01$.

to 69.1% in the 30–40 cm layer (Table 2). All soil layers, except for 30–40 cm, contained more than 70% sand and much less silt and clay. Surface layer (0–10 cm) tended to have a higher content of clay ($P<0.01$), perhaps due to erosion. Clay contents tend to increase at shallower depths as the amount of soil erosion increases [53]. Beyond that, other soil properties (sand, silt, and BD) remained constant throughout the entire soil profile. Compared with the lower variability of sand (CVs of 19.6–22.1%), stone (CVs of 24.5–100.5%), silt (CVs of 85.5–111.9%) and clay (CVs of 84.4–123.0%) showed moderate or strong variability. The CV of the BD at each depth was ≤10.5%, indicating that BD was not very variable throughout the study area. The CVs indicated that the variabilities of all constituent contents, except for stone content, generally increased with soil depth.

SOC concentration and its variability

The calculation of variation function generally should be in accord with normal distribution, otherwise it may cause the proportional effect, raising the sill or nugget values [54]. As shown in Table 3, the raw SOC exhibited a positively skewed distribution in this study area. Thus, we used the logarithmic transformation to reduce the data skewness. The Ln-transformed data for all four layers passed the Kolmogorov-Smirnov test at a significance level higher than 0.05 (not shown) and consequently could be used in the analysis of the geostatistical variation function.

SOC concentrations were generally variable, ranging between 0.4–4.5 g kg^{-1}, 0.5–3.9 g kg^{-1}, 0.5–4.3 g kg^{-1}, and 0.5–3.6 g kg^{-1} for the four descending layers, respectively (Table 3). According to the soil-nutrient classification standards from the

Table 5. Stepwise multiple linear regression of soil organic carbon (SOC) concentrations with selected soil variables at different soil depths.

Soil depth	Independent variables	Coefficient	Explained Variance (%)
0–10 cm	Constant	3.589 **	
	Sand	−2.617 **	39.78
	Stones	0.006 **	7.30
	Adjusted R^2	0.470	
	MSE	0.417	
10–20 cm	Constant	0.989 **	
	Silt	5.200 **	31.18
	Stones	0.006 **	6.52
	Adjusted R^2	0.377	
	MSE	0.444	
20–30 cm	Constant	0.950 **	
	Silt	4.599 **	27.20
	Stones	0.006 **	6.80
	Adjusted R^2	0.340	
	MSE	0.466	
30–40 cm	Constant	1.613 **	
	Silt	3.206 **	24.20
	Stones	−0.007 **	10.00
	Adjusted R^2	0.342	
	MSE	0.477	

MSE, mean squared error; R^2, coefficient of determination; ** Denotes significance of correlation at $P<0.01$.

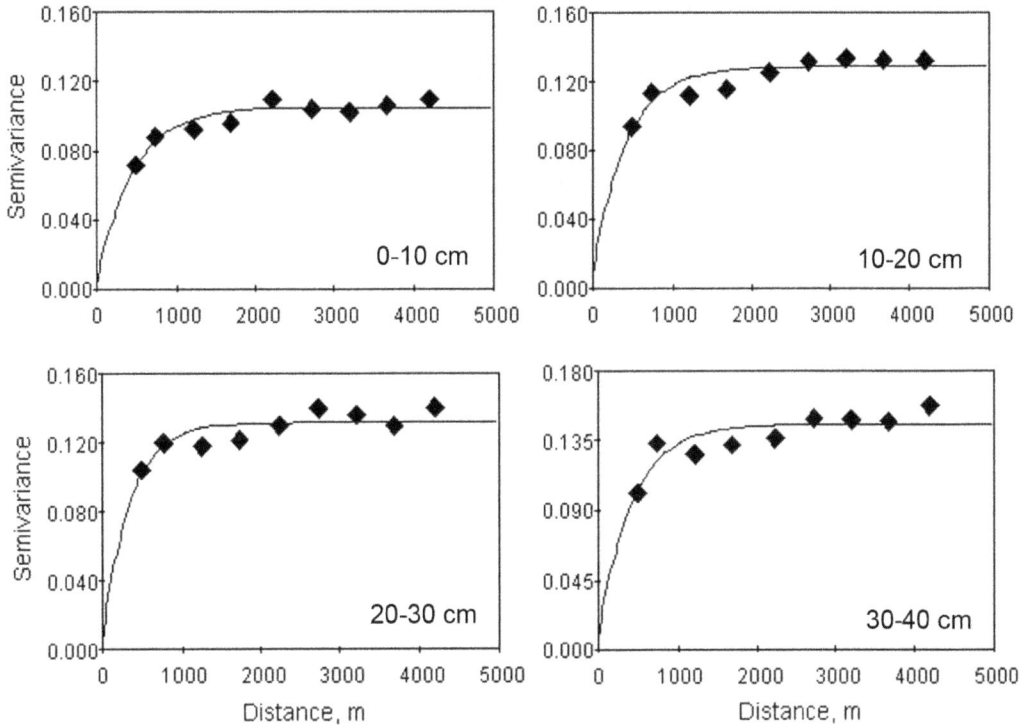

Figure 3. Isotropic semivariograms of soil organic carbon (SOC) concentrations at different soil depths.

Second National Soil Survey in China [55], the SOC level in this area was very low; only seven of the 748 soil samples were above the lowest classification standard (3.5 g kg^{-1}). The mean were 1.6, 1.5, 1.4, and 1.4 g kg^{-1} for the 0–10, 10–20, 20–30, and 30–40 cm layers, respectively, and progressively decreased with soil depth. The surface layer (0–10 cm) had the highest SOC concentrations, due to inputs of organic material that accumulated from organic litter and root residues [25,56], and higher soil aeration enabled higher soil enzyme activities in the surface layer than in the deeper layers [29]. Some researchers have also reported that SOC content decreased with depth in other natural ecosystems [6,10], but this trend is unlikely at sites with substantial human intervention such as orchards and tree nurseries [6]. In these sites, the topsoil may not have SOC contents much higher than the underlying layers, as a result of the management practices, such as heavy tillage, that could have enhanced SOC runoff and soil respiration [6].

The CVs of SOC concentrations varied from 36.8% to 42.1%, which are considered as moderate variation [57]. A moderate variability of SOC has also been reported in other studies at

multiple scales [11,25,41,58]. [11,25,58] reported a decreasing trend of SOC variability with increasing soil depth. Our study, however, found the opposite trend. This can probably be explained by the different land-use types and vegetation characteristics. In their studies, croplands, grasslands, and forestland are the main land-use types. SOC is thus greatly affected by human activities, such as grazing and deforestation, and the agricultural managements of plowing, fertilisation, harvesting, and crop rotation which can have a greater impact on the surface soil layer than on the deeper layers [2,6,31,59]. Moreover, the high vegetation coverage in these land-use types indicates a high biomass of plant roots and litter, and an active soil microbial and enzyme activities, which mainly occur in the surface soil layer [29,56]. By contrast, the natural desert region in our study is rarely disturbed by human activities and the vegetation growth is limited by inferior soil and water conditions. The surface layer is more susceptible to processes, such as sedimentation and erosion, which can homogenise the distribution of SOC [60]. Therefore, SOC in the deeper layer tend to have greater spatial variability. Additionally, the preferential transport of SOC via cracks during

Table 6. Parameters of the semivariogram models estimated for soil organic carbon (SOC) concentrations at different soil depths.

Soil depth	Model	Nugget C_0	Still (C_0+C)	C_0/(C_0+C)	Range (m)	R^2
0–10 cm	exponential	0.0019	0.1048	0.018	1347	0.884
10–20 cm	exponential	0.0069	0.1288	0.054	1251	0.770
20–30 cm	exponential	0.0047	0.1324	0.035	1047	0.681
30–40 cm	exponential	0.0069	0.1458	0.047	1254	0.724

R^2, coefficient of determination.

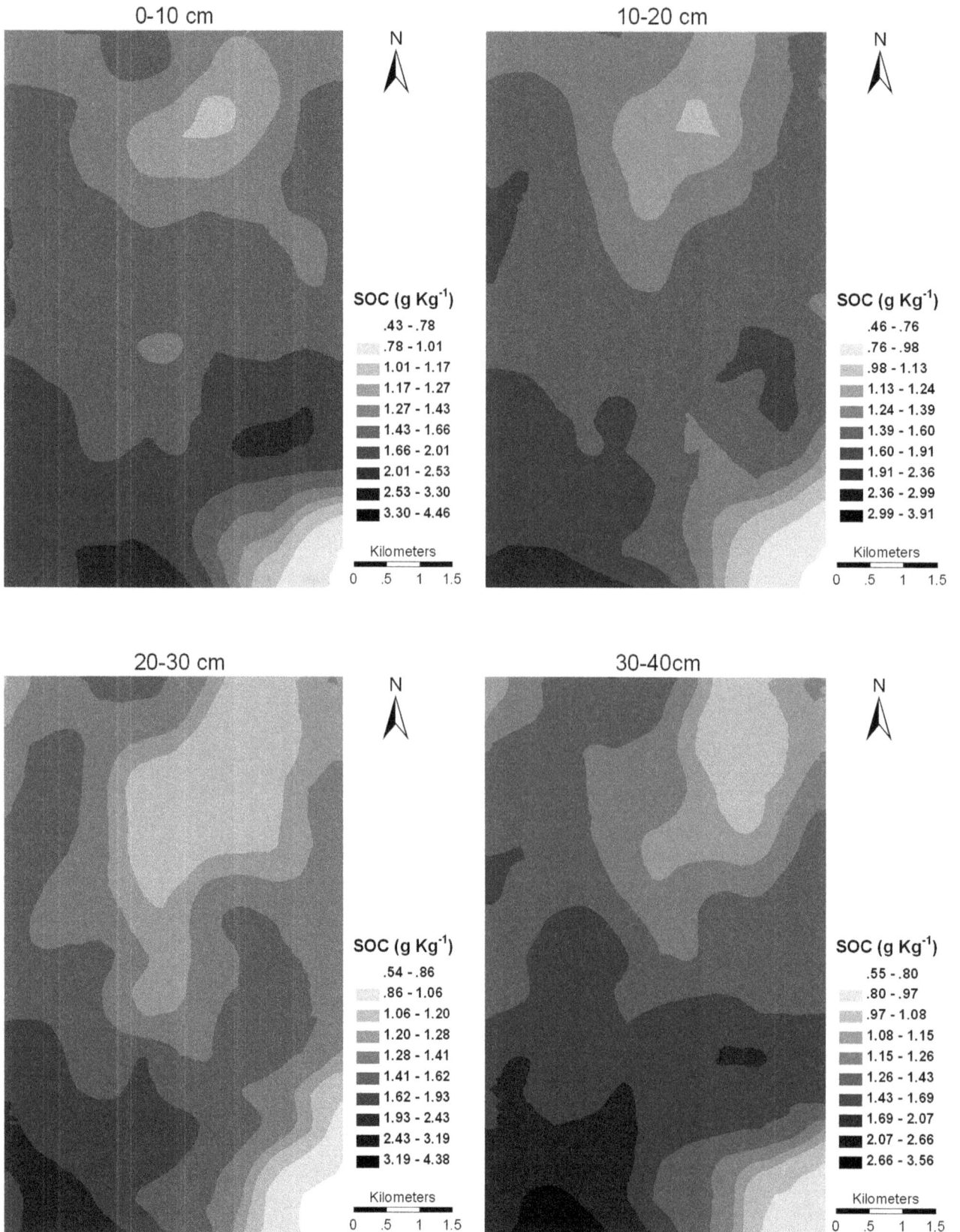

Figure 4. Distribution of soil organic carbon (SOC) concentrations across the study area.

Table 7. Verification of interpolation reliability for soil organic carbon (SOC) concentrations at different soil depths.

Soil depth	ME	RMSE	RMSSE
0–10 cm	−0.00821	0.535	1.105
10–20 cm	−0.00083	0.544	1.056
20–30 cm	0.00173	0.504	1.017
30–40 cm	−0.00143	0.570	1.104

ME, mean error; RMSE, root mean square error; RMSSE, root mean square standardized error.

dry periods could further increase the heterogeneity of SOC in the deeper soil horizons [10].

Response of SOC variability to the expansion of area

Fig. 2 demonstrated that CV of SOC concentration increased with the expansion of area and could be well parameterized as a power function of the sampling area for all the four soil layers. The factors influencing SOC concentration variability are scale dependent. For example, parent material, precipitation and geological history are of major importance to affect SOC at large scales. However, microtopography (such as the run-off gullies) and vegetation may be the dominant factors of SOC variability at small scales [61]. As sampling area increases, the origin of SOC variations above may get increasingly complex and heterogeneous, contributing to greater variability [2]. Moreover, the value of the fractal power parameter in the surface layer (0.0789) was much larger than that in the other three layers (0.0392–0.0620), indicating SOC concentration variability in the surface layer was more sensitive to the expansion of area. If increasing the same area, the SOC variability will increase more in the surface layer than that in the deeper layers.

The function between CV of SOC concentration and sampling area can also be used to estimate the variability of SOC concentration at a desired area and, consequently, the number of required samples (NRS). In order to obtain the mean value of SOC concentration with an accuracy level of Δ, at a confidence level of $1-\alpha$, the sample size should reach the requirement of $NSR = \lambda_\alpha^2 (CV\%/\Delta)^2$ (λ_α is the value of the Student's t-distribution at the confidence level of $1-\alpha$) [48]. Take the surface layer (0–10 cm) for example, NRS is found to be equal to 110, 227 and 326 for an area of 1, 100 and 1000 km^2, respectively, by assuming an accuracy level of 5% and a confidence level of 95%.

Pearson correlation and stepwise linear regression analyses

Under the extremely arid climate, the vegetational cover in this region of the Gobi desert is very low, and the chemical and biotic influences on soil development are relatively minor. Climate, soil type, and terrain can be disregarded as variables considering the small size and flatness of the study area. The physical properties of the soil were thus most likely responsible for the SOC variability in our study area.

Table 4 shows the correlation analyses between the soil physical properties and SOC concentrations. The four soil layers had similar patterns. SOC concentrations were positively correlated with both silt and clay contents and were negatively correlated with stone and sand contents and BD. The positive effects of clay and silt contents on SOC concentrations are likely due to the ability of clay and silt particles to adsorb organic matter. Finer particles are better than larger particles in protecting bound

organic matter for longer times [62]. Moreover, the distribution of soil particle size and BD can indirectly impact SOC dynamics by affecting the physical structure, drainage, and aeration of soils [63].

The stepwise linear regression analysis was further performed to delineate the effect of different factors on SOC and to find the best predictive variables for SOC. A summary of these linear models is shown in Table 5. For the surface layer (0–10 cm), the regression model explained 47.0% of the overall SOC variability in which most of the variability was attributable to sand (39.7%) and stone contents (7.3%). For the other three layers, however, silt and stone contents together accounted for approximately 35% of the total variance of SOC concentrations. These results indicated that the relatively coarse fractions (stones, sand, and silt) were more important for the explanation of variability in SOC concentrations in this study area

Semivariogram and parameters

Spatial structure was not significantly associated with direction in the study area. Only isotropic semivariograms were thus plotted for SOC concentrations by using the model best fitted by the least squares regression method. Exponential models were theoretically optimal for all four soil layers.

The semivariogram models and best-fitted model parameters of SOC concentration at different soil depths are given in Fig. 3 and Table 6. The semivariogram of the SOC concentrations indicated a slightly smaller nugget effect (C_0) in the surface soil layer (0–10 cm) than in the other three layers, implying that the deeper soil layers had higher undetectable experimental error, short-range variability, and random and inherent variability of SOC concentration than the surface soil layer [3]. The sill values, representing total variation, showed an increasing trend from the surface soil layer to the deepest layer, which further validated the results obtained by conventional statistical methods. The nugget ratio (C_0/C_0+C) ranged from 0.018 to 0.054, indicating a strong spatial dependence for SOC concentrations for all four soil layers in our study [50]. Strong spatial dependency of soil properties can usually be attributed to intrinsic factors. In this natural desert, fenced enclosures for natural restoration have been implemented for many years and are rarely disturbed by human activities and grazing, so we suggest that the variability of SOC concentrations in this region may be highly dependent on the mineralogical composition of the parental material and on the weathering processes that have led to its formation [64–66].

The range also changed significantly with soil depth, which is likely due to the control of the distribution pattern of SOC by different soil processes. The range for the 0–10 cm layer was the largest, indicating a larger spatial autocorrelation in the surface soil layer than in the deeper soil layers. The surface soil layer is most sensitive to erosion, which could homogenise the distribution of SOC and increase the spatial autocorrelation distance [60]. [67]

Table 8. Comparison of SOC concentrations and SOC stocks in other desert areas in China.

Region	Environment	Mean annual precipitation (mm)	Area (km²)	Reference depth (cm)	SOC concentration (g kg⁻¹)	SOC stock (kg C m⁻²)	Reference
Erdos, Inner Mongolia	Temperate shrub desert	170 (161–209)	Profile measurement	0–300	0.42		[75]
Aershan, Inner Mongolia	Temperate desert	102 (61–121)	Profile measurement	0–300	0.25		[75]
Alxa league, Inner Mongolia	Desert steppe	134	Point measurement (sites with different grazing degrees)	0–40	1.73–2.05		[74]
Horqin Sand Land, Inner Mongolia	Desert grassland	350–450	Point measurement (sites with different desertification types and degrees)	0–30	0.37–4.35	0.19–1.76	[38]
Alxa Left County, Inner Mongolia	Desert shrubland	60–160	2.5	0–30		0.21–2.80	[39]
Horqin region, Inner Mongolia	Sand Land (sand dunes)	360	Point measurement (grazed and restored sites)	0–20	1.31–2.03		[37]
Horqin region, Inner Mongolia	Sand Land (mobile dunes)	360	Point measurement (sites with different restoration processes of dune vegetation)	0–20	0.68–1.29		[77]
Hunshandake, eastern Inner Mongolia	Sandy Land	300 (165–572)	21 400	variable	0.44–4.37		[76]
Central Gansu province	Gobi desert	117	40	0–40	1.45 (0.43–4.46)	0.68	This study

also showed that wind erosion significantly changed the spatial distribution patterns of SOC over two or three windy seasons. With the increasing of soil depth, the influence of erosion on SOC distribution was weakened, and the SOC tended to be much more independent and was characterized by a stochastic pattern. As shown in Table 6, the range decreased from 1347 to 1047 m from 0–10 cm to 20–30 cm. The range in the 30–40 cm layer, however, increased slightly, reflecting the influence of parental materials on the spatial structure of SOC in deeper layers. In deeper soil layers, the SOC was probably inherited from the soil parent materials because the soils are young with little weathering or anthropogenic impacts [65]. Since the parental materials are distributed quite uniformly across the study area [66], it leads to a better distribution of spatial continuity. Because the range was larger than our sampling interval (500 m), our sampling system was sufficiently robust to detect spatial relationships on the scale of the landscape.

Kriging of spatial variation of SOC concentrations

The semivariogram models were used as input to ordinary kriging, and the resulting distribution maps are shown in Fig. 4. The interpolation cross-validation was carried out to test the effectiveness of the prediction maps, and the associated prediction errors for each map are shown in Table 7. ME determined the degree of bias in the estimates and should be close to zero. RMSE quantified the average differences between prediction and observation and should be as small as possible. If the model accurately described the data, RMSSE should be close to unity. Based on the above criteria, the predicted maps of SOC concentration for the study area were reliable.

The character of SOC distribution is clearly similar for each layer, and the entire study area is characterised by low concentrations of SOC and high variation. SOC concentrations generally decreased gradually from south to north. The southwest part of the study area, close to the oasis, tended to have higher concentrations of SOC. The shelter belts in front of the oasis functioned as natural barriers to reduce wind velocity and intercept the fine material, which tends to be rich in SOC [68]. The increased fine fractions also improved the soil properties, encouraging the colonisation of some annuals, e.g. *B. dasyphylla* and *H. arachnoideus*. Their rapid growth and death provided an important influx of SOC. In contrast, relatively low concentrations of SOC were mainly distributed in the north and the southwest. The northern part of the site was surrounded by mountains and was characterised by coarse soils and little vegetations, facilitating the drifting of fine particles and a noticeable decline in SOC. On the other hand, the southwestern region was linked with the Badain Juran desert and the longtime encroachment of drifted sand from the desert produced a more sandy texture in this area.

SOC density and stocks

SOC density is indispensable for the assessment of SOC stocks and is the required measurement of account for the Clean Development Mechanism of the United Nations Framework Convention on Climate Change [69]. SOC density was 0.22, 0.20, 0.19, and 0.07 kg C m⁻² for the 0–10, 10–20, 20–30, and 30–40 cm layers, respectively, and dropped sharply with increasing depth due to the high stone content. The low value in the 30–40 cm layer indicates that the SOC density would be even lower below 40 cm because of the increasing number of stones. We can thus reasonably conclude that SOC in this region of the Gobi desert was mainly stored in the upper 30 cm of soil. The overall SOC stocks in the upper 20 cm was 0.42 kg C m⁻² and to a depth of 40 cm 0.68 kg C m⁻². When compared to the reported values for other regions in China [11,25,70,71,72], due to natural drought, large number of stones, and intense soil erosion, SOC stocks in the Gobi desert region was very low. However, since the Gobi desert (568,980 km^2) accounts for about 5.9% of China's total territory [73] and is sensitive to climate change [17], it is likely to have a considerable effect on the terrestrial C balance in China.

Table 8 summarises the results of SOC studies (including SOC concentrations and densities) in other desert regions in China. The SOC concentrations and densities in our study area were generally in the same range as those in other regions, except that SOC concentrations were a little higher than those in the Erdos and Aershan regions. Different reference soil depths, different sampling methods, and the higher patchiness of our study area may account for our higher SOC concentrations. Even though the soils in our study area had higher fractions of stones, patches with more fertile soil still allowed the growth of vegetation, which thus increased the SOC concentrations. Compared with some of the other regions mentioned in Table 8, SOC in our study site is less affected by grazing, which can give rise to a considerable decrease in ground coverage and primary productivity, and thus accelerate soil erosion by wind and result in loss of SOC [74]. However, the improvement of soil quality through the adoption of grazing prohibition can increase SOC concentrations [36]. Moreover, higher SOC concentrations in our study site can also be attributed to the reduced mineralization rate of SOC, because of the lack of water in this region [66]. Comparisons among the studies listed in Table 8, however, remain limited due to differences in sampling methods of SOC measurement and reference soil depths. Identifying the dynamics of SOC in changing desert environments is difficult. We should thus develop more site inventories of SOC in desert environments or use a comparable approach to better understand the potential changes of SOC in desert environments, which will lay the groundwork for developing more effective strategies to combat soil desertification and reduce the risk of desertification in the future.

Conclusions

We have provided estimates of spatial SOC concentrations and stocks for this region of the Gobi desert that are more accurate than previous estimates. Classical statistics indicated that SOC concentrations decreased with increasing soil depth and were moderately variable in the study area. The deepest soil layer (30–40 cm) had the highest amount of variation in SOC concentrations. Significant correlations were detected between SOC and selected physical properties of the soil, especially the stone, sand, and silt contents. The composition of the parental material (such as the distribution of soil particle size) and the weathering (such as erosion and sedimentation) that led to its formation may be responsible for the strong spatial dependence of SOC. This dependence implies that SOC in desert ecosystem is sensitive to climate change and thus represents an important dynamic pool of C in the global C cycle. The kriging interpolated maps indicated a decreasing trend of SOC concentrations from south to north across the study area, which was apparently related to the location of the study area. This study contributes to our understanding of the role of Gobi desert ecosystem in the global C cycle and incorporation of small-scale spatial variations of SOC into large-scale spatiotemporal models.

Acknowledgments

Special thanks go to the staff of the Linze Inland River Basin Research Station of the Institute of Cold and Arid Regions Environmental and Engineering Research of CAS.

Author Contributions

Conceived and designed the experiments: PPZ MAS. Performed the experiments: PPZ. Analyzed the data: PPZ. Wrote the paper: PPZ.

References

1. Rossi J, Govaerts A, De Vos B, Verbist B, Vervoort A, et al. (2009) Spatial structures of soil organic carbon in tropical forests-A case study of Southeastern Tanzania. Catena 77: 19–27.
2. Wang DD, Shi XZ, Lu XX, Wang HJ, Yu DS, et al. (2010) Response of soil organic carbon spatial variability to the expansion of scale in the uplands of Northeast China. Geoderma 154: 302–310.
3. Schlesinger WH, Raikes JA, Hartley AE, Cross AF (1996) On the spatial pattern of soil nutrients in desert ecosystems. Ecology 77: 364–374.
4. Amundson R (2001) The carbon budget in soils. Annual Review of Earth and Planetary Sciences 29: 535–562.
5. Davidson EA, Janssens IA (2006) Temperature sensitivity of soil carbon decomposition and feedbacks to climate change. Nature 440: 165–173.
6. Su ZY, Xiong YM, Zhu JY, Ye YC, Ye M (2006) Soil organic carbon content and distribution in a small landscape of Dongguan, South China. Pedosphere 16: 10–17.
7. IPCC (Intergovernmental Panel on Climate Change) (2000) Land use, land use change and forestry, a special report of the IPCC. In: Watson RT, Noble IR, Bolin B, Ravondranath NH, Verardo DJ, Dokken DJ, editors. Cambridge: Cambridge University Press. pp. 1–51.
8. Li H, Parent LE, Karam A, Tremblay C (2004) Potential of Sphagnum peat for improving soil organic matter, water holding capacity, bulk density and potato yield in a sandy soil. Plant and soil 265: 355–365.
9. Zhang C, McGrath D (2004) Geostatistical and GIS analyses on soil organic carbon concentrations in grassland of southeastern Ireland from two different periods. Geoderma 119: 261–275.
10. Don A, Schumacher J, Scherer-Lorenzen M, Scholten T, Schulze ED (2007) Spatial and vertical variation of soil carbon at two grassland sites-implications for measuring soil carbon stocks. Geoderma 141: 272–282.
11. Liu ZP, Shao MA, Wang YQ (2011) Effect of environmental factors on regional soil organic carbon stocks across the Loess Plateau region, China. Agriculture, Ecosystems and Environment 142: 184–194.
12. Batjes NH (1999) Management options for reducing CO_2-concentrations in the atmosphere by increasing carbon sequestration in the soil. Dutch National

Research Programme on Global Air Pollution and Climate Change. Wageningen: ISRIC. Technical Paper 30, Report 410-200-031.
13. Marland G, Schlamadinger B (1999) The Kyoto Protocol could make a difference for the optimal forest-based CO_2 mitigation strategy: some results from GORCAM. Environmental Science and Policy 2: 111–124.
14. Schlesinger WH (1999) Carbon sequestration in soils. Science 284: 2095.
15. Lal R (2004) Carbon sequestration in dryland ecosystems. Environmental management 33: 528–544.
16. Venteris E, McCarty G, Ritchie J, Gish T (2004) Influence of management history and landscape variables on soil organic carbon and soil redistribution. Soil science 169: 787–795.
17. Hoffmann U, Yair A, Hikel H, Kuhn NJ (2012) Soil organic carbon in the rocky desert of northern Negev (Israel). Journal of Soils and Sediments 12:811–825.
18. Eswaran H, Van Den Berg E, Reich P (1993) Organic carbon in soils of the world. Soil Science Society of America Journal 57: 192–194.
19. Batjes NH (1996) Total carbon and nitrogen in the soils of the world. European Journal of Soil Science 47: 151–163.
20. Batjes N, Dijkshoorn J (1999) Carbon and nitrogen stocks in the soils of the Amazon Region. Geoderma 89: 273–286.
21. Arrouays D, Deslais W, Badeau V (2001) The carbon content of topsoil and its geographical distribution in France. Soil use and Management 17: 7–11.
22. Yu D, Shi X, Wang H, Sun W, Chen J, et al. (2007) Regional patterns of soil organic carbon stocks in China. Journal of Environmental Management 85: 680–689.
23. Grimm R, Behrens T, Märker M, Elsenbeer H (2008) Soil organic carbon concentrations and stocks on Barro Colorado Island-digital soil mapping using Random Forests analysis. Geoderma 146: 102–113.
24. Baritz R, Seufert G, Montanarella L, Van Ranst E (2010) Carbon concentrations and stocks in forest soils of Europe. Forest Ecology and Management 260: 262–277.
25. Wang YF, Fu B, Lü Y, Song C, Luan Y (2010) Local-scale spatial variability of soil organic carbon and its stock in the hilly area of the Loess Plateau, China. Quaternary Research 73: 70–76.

26. Murdock LW, Frye WW (1983) Erosion: its effect on soil properties, productivity and profit. publication AGR-102. College of Agriculture, University of Kentucky.

27. Duan ZH, Xiao HL, Dong ZB, He XD, Wang G (2001) Estimate of total CO_2 output from desertified sandy land in China. Atmospheric Environment 35: 5915–5921.

28. Li X, Wang Y, Liu L, Luo G, Li Y, et al. (2013) Effect of land use history and pattern on soil carbon storage in arid region of Central Asia. PloS one 8: e68372.

29. Cao CY, Jiang SY, Ying Z, Zhang FX, Han XS (2011) Spatial variability of soil nutrients and microbiological properties after the establishment of leguminous shrub Caragana microphylla Lam. plantation on sand dune in the Horqin Sandy Land of Northeast China. Ecological Engineering 37: 1467–1475.

30. Deng L, Shangguan ZP, Sweeney S (2013) Changes in soil carbon and nitrogen following land abandonment of farmland on the Loess Plateau, China. PloS one 8: e71923.

31. Zhou ZY, Li FR, Chen SK, Zhang HR, Li G (2011) Dynamics of vegetation and soil carbon and nitrogen accumulation over 26 years under controlled grazing in a desert shrubland. Plant and Soil 341: 257–268.

32. Lal R (2001) Potential of desertification control to sequester carbon and mitigate the greenhouse effect. Climatic Change 51: 35–72.

33. Zhu ZD, Chen GT (1994) Sandy desertification in China. Science Press, Beijing, pp. 20–33.

34. Feng Q, Endo K, Guodong C (2002) Soil carbon in desertified land in relation to site characteristics. Geoderma 106: 21–43.

35. Qi F, Guoduong C, Masao M (2001) The carbon cycle of sandy lands in China and its global significance. Climatic Change 48: 535–549.

36. Zhou RL, Li YQ, Zhao HL, Drake S (2008) Desertification effects on C and N content of sandy soils under grassland in Horqin, northern China. Geoderma 145: 370–375.

37. Zuo XA, Zhao HL, Zhao XY, Zhang TH, Guo YR, et al. (2008) Spatial pattern and heterogeneity of soil properties in sand dunes under grazing and restoration in Horqin Sandy Land, Northern China. Soil and Tillage Research 99: 202–212.

38. Zhao HL, He YH, Zhou RL, Su YZ, Li YQ, et al. (2009) Effects of desertification on soil organic C and N content in sandy farmland and grassland of Inner Mongolia. Catena 77: 187–191.

39. Zhou ZY, Li FR, Chen SK, Zhang HR, Li G (2011) Dynamics of vegetation and soil carbon and nitrogen accumulation over 26 years under controlled grazing in a desert shrubland. Plant and soil 341: 257–268.

40. Lioubimtseva E, Simon B, Faure H, Faure-Denard L, Adams J (1998) Impacts of climatic change on carbon storage in the Sahara–Gobi desert belt since the Last Glacial Maximum. Global and Planetary Change 16: 95–105.

41. Grüneberg E, Schöning I, Kalko EK, Weisser WW (2010) Regional organic carbon stock variability: A comparison between depth increments and soil horizons. Geoderma 155: 426–433.

42. Li QY, He ZB, Zhao WZ, Li QS (2004) Spatial pattern of Nitraria spaerocarpa population and dynamics in different habitats. Journal of Desert Research 24: 484–488 (in Chinese).

43. FAO/UNESCO (1988) 'Soil map of the world, revised legend.' (FAO/UNESCO: Rome)

44. Nelson DW, Sommers LE (Eds.) (1982) Total carbon, organic carbon and organic matter. In: Page AL, Miller RH, Keeney DR, editors. Methods of Soil 591 Analysis. Part 2. Agronomy Monograph, 2nded. ASA and SSSA, Madison, WI. pp 534–580.

45. Hu W, Shao MA, Wang QJ (2005) Scale-dependency of spatial variability of soil moisture on a degraded slope-land on the Loss Plateau. Transactions of the CSAE 21: 11–16 (in Chinese).

46. Zhang FS, Liu ZX, Zhang Y, Miao YG, Qu W, et al. (2009) Scaling effect on spatial variability of soil organic matter in crop land. Journal of the Graduate School of the Chinese Academy of Sciences 26: 350–356 (in Chinese).

47. Bond-Lamberty B, Brown KM, Goranson C, Gower ST (2006) Spatial dynamics of soil moisture and temperature in a black spruce boreal chronosequence. Canadian Journal of Forest Research-Revue Canadienne De Recherche Forestiere 36: 2794–2802.

48. Loescher HW, Ayres E, Duffy P, Luo H, Brunke M (2014) Spatial variation in soil properties among North American ecosystems and guidelines for sampling designs. Plos one 9: e83216.

49. Trangmar BB, Yost RS, Uehara G (1985) Application of geostatistics to spatial studies of soil properties. Advances in agronomy 38: 45–94.

50. Cambardella C, Moorman T, Parkin T, Karlen D, Novak J, et al. (1994) Field-scale variability of soil properties in central Iowa soils. Soil Science Society of America Journal 58: 1501–1511.

51. Wei JB, Xiao DN, Zhang XY, Li XZ, Li XY (2006) Spatial variability of soil organic carbon in relation to environmental factors of a typical small watershed in the black soil region, northeast China. Environmental Monitoring and Assessment 121: 597–613.

52. Marchetti A, Piccini C, Francaviglia R, Mabit L (2012) Spatial distribution of soil organic matter using geostatistics: A key indicator to assess soil degradation status in central Italy. Pedosphere 22: 230–242.

53. Arriaga FJ, Lowery B (2005) Spatial distribution of carbon over an eroded landscape in southwest Wisconsin. Soil and Tillage Research 81: 155–162.

54. Wang SQ, Zhou CH (1999) Estimating soil carbon reservoir of terrestrial ecosystem in China. Geographical Research 18: 349–356.

55. Office for the Second National Soil Survey of China (1993) Second National Soil Survey of China. Beijing: China Agricultural Press.

56. Liu ZP, Shao MA, Wang YQ (2012) Large-scale spatial variability and distribution of soil organic carbon across the entire Loess Plateau, China. Soil Research 50: 114–124.

57. Lei ZD, Yang SX, Xu ZR. (1985) Research spatial variability of soil properties. Journal of Hydraulic Engineering 9: 10–21 (in Chinese).

58. Fang X, Xue ZJ, Li BC, An SS (2012) Soil organic carbon distribution in relation to land use and its storage in a small watershed of the Loess Plateau, China. Catena, 88: 6–13.

59. Zibilske LM, Bradford JM, Smart JR (2002) Conservation tillage induced changes in organic carbon, total nitrogen and available phosphorus in a semi-arid alkaline subtropical soil. Soil and Tillage Research 66: 153–163.

60. Li J, Okin GS, Alvarez L, Epstein H (2008) Effects of wind erosion on the spatial heterogeneity of soil nutrients in two desert grassland communities. Biogeochemistry 88: 73–88.

61. Wang J, Fu B, Qiu Y, Chen L (2001) Soil nutrients in relation to land use and landscape position in the semi-arid small catchment on the loess plateau in China. Journal of Arid Environments 48: 537–550.

62. Six J, Conant RT, Paul EA, Paustian K (2002) Stabilization mechanisms of soil organic matter: Implications for C-saturation of soils. Plant and Soil 241: 155–176.

63. Saxton KE, Rawls WJ (2006) Soil water characteristic estimates by texture and organic matter for hydrologic solutions. Soil Science Society of America Journal 70: 1569–1578.

64. Tack FMG, Verloo MG, Vanmechelen L, Van Ranst E (1997) Baseline concentration levels of trace elements as a function of clay and organic carbon contents in soils in Flanders (Belgium). Science of the Total Environment 201: 113–123.

65. Zhang XP, Deng W, Yang XM (2002) The background concentrations of 13 soil trace elements and their relationships to parent materials and vegetation in Xizang (Tibet), China. Journal of Asian Earth Sciences 21: 167–174.

66. Su YZ, Yang R (2008) Background concentrations of elements in surface soils and their changes as affected by agriculture use in the desert-oasis ecotone in the middle of Heihe River Basin, North-west China. Journal of Geochemical Exploration 98: 57–64.

67. Li J, Okin GS, Alvarez L, Epstein H (2007) Quantitative effects of vegetation cover on wind erosion and soil nutrient loss in a desert grassland of southern New Mexico, USA. Biogeochemistry 85: 317–332.

68. Yan H, Wang S, Wang C, Zhang G, Patel N (2005) Losses of soil organic carbon under wind erosion in China. Global Change Biology 11: 828–840.

69. United Nations Framework Convention on Climate Change (2007) Tool for the Demonstration and Assessment of Additionality in A/R CDM Project Activities, Annex 17, CDM Executive Board 35, Bonn, Germany. Available: http://cdm.unfccc.int/methodologies/ARmethodologies/tools /ar-am-tool-01-v2.pdf

70. Li Z, Jiang X, Pan XZ, Zhao QG (2001) Organic carbon storage in soils of tropical and subtropical China. Water, Air, and Soil Pollution 129: 45–60.

71. Fan Y, Liu S, Zhang S, Deng L (2006) Background organic carbon storage of topsoil and whole profile of soils from Tibet District and their spatial distribution. Acta Ecologica Sinica 20: 2834–2846 (in Chinese).

72. Zhang Y, Zhao YC, Shi XZ, Lu XX, Yu DS, et al. (2008) Variation of soil organic carbon estimates in mountain regions: a case study form Southwest China. Geoderma 146: 449–456.

73. Feng YZ, Yang GH (2003) Countermeasures of sustainable development of land resources in the Northwest region. Gansu Science and Technology Press, Lanzhou, pp. 102.

74. Pei SF, Fu H, Wan CG (2008) Changes in soil properties and vegetation following exclosure and grazing in degraded Alxa desert steppe of Inner Mongolia, China. Agriculture, ecosystems and environment 124: 33–39.

75. Wang Y, Li Y, Ye X, Chu Y, Wang X (2010) Profile storage of organic/inorganic carbon in soil: From forest to desert. Science of the Total Environment 408: 1925–1931.

76. Yang X, Zhu B, Wang X, Li C, Zhou Z, et al. (2008) Late Quaternary environmental changes and organic carbon density in the Hunshandake Sandy Land, eastern Inner Mongolia, China. Global and Planetary Change 61, 70–78.

77. Zuo XA, Zhao XY, Zhao HL, Zhang TH, Guo YR, et al. (2009) Spatial heterogeneity of soil properties and vegetation-soil relationships following vegetation restoration of mobile dunes in Horqin Sandy Land, Northern China. Plant and soil 318, 153–167.

5

Negative Effects of an Exotic Grass Invasion on Small-Mammal Communities

Eric D. Freeman[1]*, **Tiffanny R. Sharp**[1], **Randy T. Larsen**[1,2], **Robert N. Knight**[3], **Steven J. Slater**[4], **Brock R. McMillan**[1]

1 Brigham Young University, Department of Plant and Wildlife Sciences, Provo, Utah, United States of America, 2 Monte L. Bean Life Science Museum, Provo, Utah, United States of America, 3 United States Army Dugway Proving Ground, Environmental Programs, Dugway, Utah, United States of America, 4 HawkWatch International, Conservation Director, Salt Lake City, Utah, United States of America

Abstract

Exotic invasive species can directly and indirectly influence natural ecological communities. Cheatgrass (*Bromus tectorum*) is non-native to the western United States and has invaded large areas of the Great Basin. Changes to the structure and composition of plant communities invaded by cheatgrass likely have effects at higher trophic levels. As a keystone guild in North American deserts, granivorous small mammals drive and maintain plant diversity. Our objective was to assess potential effects of invasion by cheatgrass on small-mammal communities. We sampled small-mammal and plant communities at 70 sites (Great Basin, Utah). We assessed abundance and diversity of the small-mammal community, diversity of the plant community, and the percentage of cheatgrass cover and shrub species. Abundance and diversity of the small-mammal community decreased with increasing abundance of cheatgrass. Similarly, cover of cheatgrass remained a significant predictor of small-mammal abundance even after accounting for the loss of the shrub layer and plant diversity, suggesting that there are direct and indirect effects of cheatgrass. The change in the small-mammal communities associated with invasion of cheatgrass likely has effects through higher and lower trophic levels and has the potential to cause major changes in ecosystem structure and function.

Editor: Fei-Hai Yu, Beijing Forestry University, China

Funding: Support was provided by US Army Dugway Proving Grounds Grant # R0202329. Robert N. Knight of Dugway Proving Ground, a coauthor of this manuscript, was key in gaining access to military lands, facilitating interaction between authors and the military, took part in concept development, and participated in writing the manuscript.

Competing Interests: The authors have declared that no competing interests exist.

* Email: edfreeman1@gmail.com

Introduction

Exotic invasive species can directly and indirectly influence natural ecological communities by modifying structure [1], decreasing diversity [2], and altering ecosystem function [3]. Specifically, exotic plant species alter the hydrology of ecosystems, increase soil erosion, decrease native plant diversity, and alter fire cycles [4,5]. Examples in North America include salt cedar (*Tamarix ramosissima*), kudzu (*Pueraria lobata*), leafy spurge (*Euphorbia esula*), red brome (*Bromus rubens*), and cheatgrass (*Bromus tectorum*). Alternatively, some exotic species can be benign or have net positive effects on systems where they are introduced or become established [6]. As exotic species continue to successfully invade and transform ecosystems, understanding the effects of these invasions on all trophic levels will be important for mitigating negative impacts, maintaining ecosystem integrity, preserving biodiversity, and predicting the consequences of further invasions [7,8].

Cheatgrass invades and impacts communities worldwide [9], but has had particular success in the Great Basin Desert(s) of the western United States. In this ecosystem, cheatgrass has altered the fire cycle, outcompeted native vegetation, and altered ecosystem dynamics [10,11,12]. Remotely sensed data indicate that at least 40,000 square kilometers of the Great Basin (nearly 10%) are dominated by monocultures of cheatgrass [13]. Additionally, cheatgrass is a major understory component across a larger area and 200,000 additional square kilometers are vulnerable to invasion [14]. This invasion is noteworthy and concerning when considering that as little as 200 years ago, cheatgrass was isolated to a single population on the East coast and that the earliest records of cheatgrass in the Great Basin were from around 1900 [12,15].

The invasion of this robust annual and other *Bromus* spp. is associated with increased frequency of fire and decreased plant diversity across much of the western US [16]. As cheatgrass invades a desert system, it fills the inter-plant spaces that normally separate native plant species. Where fires were once confined to relatively small areas because of limited connectivity, cheatgrass allows fire to carry over larger areas, impacting the native plant community more widely and frequently. In addition to increasing pressures on native plants through fire, cheatgrass often out-competes native plants through higher rates of root growth, seedling germination, and adult survival [17,18].

Dramatic changes to the composition and structure of plant communities subjected to invasion often have cascading effects at higher trophic levels [19,20,21]. For example, invasive plant species in western North American grasslands alter predator-prey interactions (and subsequent survival of native species at higher trophic levels) by changing the structure of the physical environment and the abundance and diversity of native plants [22]. Specifically, a bottom-up effect occurs as cheatgrass invasion changes the availability of native plant resources used for forage and cover by a variety of taxa (e.g., small mammals) [23,24]. Desert small mammals are primarily granivorous and prefer native seed over that of cheatgrass, which is nutritionally inferior [25]. Additionally, different species of small mammals forage in different microhabitats [26,27], the variety of which is reduced when cheatgrass invades, fills inter-plant spaces, alters natural communities, and increases the frequency of fire [24]. Changes to habitat and food resources that result from cheatgrass invasion likely affect small-mammal communities and subsequently, higher trophic levels (e.g., canids and raptors).

As a keystone guild in North American deserts, granivorous small mammals drive and maintain plant diversity [28,29]. This top-down effect occurs as granivorous small mammals modify the availability of reproductive propagules via seed gathering, caching, and consumption behaviors, which vary among species [26]. Seed caching behaviors of small mammals make them an important vector of dispersal for plants [30]. Because different species of small mammal have different consumption and caching behaviors, a change in the rodent community likely affects seed survival, dispersal, and plant recruitment. The potential impact of cheatgrass invasion on small-mammal communities may also have indirect effects on native plant diversity.

Our objective was to assess the effects of invasion by cheatgrass on small-mammal communities. To make this assessment, we determined abundance and diversity of small mammals at sites with varying levels of invasion by cheatgrass across the Great Basin Desert in northwest Utah. We predicted that as the percentage of cheatgrass cover increased: 1) the overall abundance of small mammals would decrease, but the responses of individual functional groups and species would vary, and 2) the species diversity (or indices of richness and evenness) of the small-mammal community would decrease. Because small-mammal assemblages often have significant effects on other trophic levels (e.g., primary producers and predators) and ecosystem processes (e.g., seed dispersal/consumption and soil disturbance), understanding the impact of invasion by cheatgrass on small-mammal communities is pertinent to the conservation of ecosystem structure and function.

Methods

Study site location and selection

During the summers of 2011 and 2012, we sampled small-mammal and plant communities at 70 sites across the Great Basin Desert in Box Elder, Tooele, and Juab Counties, Utah (Figure 1). These sites were located between 41°43′ N – 39°40′ N (North-South) and 113°57′ W – 112°39′ W (East-West). Plant communities were dominated by sagebrush (*Artemesia* spp.), saltbush (*Atriplex* spp.), greasewood (*Sarcobatus vermiculatus*), or cheatgrass, if lacking a shrub layer. We established a 90-m by 90-m trapping grid at each site, which were separated by at least 500 m.

We used remotely sensed MODIS and SWReGAP vegetation data and field observations made in 2011 of cheatgrass cover to select sites for sampling. We identified areas that differed in extent of cheatgrass invasion (low, medium or high) to ensure that we sampled sites across a continuum of cheatgrass cover. Within these areas, we established sites at computer-generated random points with an distribution across the levels of invasion. When generating

random points, we only queried areas that were relatively close to roads (to ensure access) and requested a clumped distribution (separated by <50 km) for sampling of nearby sites within each week. This ensured that long-distance transportation between sites was limited so that traps were checked in a timely manner each morning. We selected and sampled 70 sites in a stratified random manner to avoid confounding level of invasion with time (small mammals may be less active at some times of the year).

Plant survey

At each site sampled for small mammals, we established three 50-m transects that ran through the trapping grid, paralleling the 2nd, 4th, and 6th lines of traps. We used a point-intercept method and sampled at 50-cm intervals along each transect. For each sample, we recorded if cheatgrass was present. The percentage of cheatgrass cover was estimated as the number of points with cheatgrass present divided by the total number of points at a site (300). Similarly, we recorded other plant species (or bare ground) present at each point and estimated the percentages of herbaceous plant and shrub cover in a similar manner. We used shrub cover as a proxy for fire (an indirect effect of cheatgrass) because burned sites in Great Basin shrub communities (which includes all of the sites we sampled) generally have reduced or no shrub cover, while shrubs are a major part of the plant community at unburned sites [12].

Small mammal survey

We sampled small mammals at each site for three consecutive nights during the summers of 2011 and 2012. To sample small mammals, we established a 7×7 trapping grid at each site with 15 m spacing between trap stations and placed one trap at each station (49 traps; 90 m×90 m trapping grid). We used 7.6 cm×7.6 cm×30.5 cm collapsible Sherman live traps baited with commercially available birdseed. We checked traps each morning and closed them until evening if daytime temperatures were expected to exceed 22°C. We added cotton batting to the traps if nighttime lows were projected below 4.5°C. We took these precautions to decrease the likelihood of temperature-induced stress and to reduce incidental mortality. We identified small mammals captured to species and collected basic live-trap data (e.g., sex, age, mass, anatomical measurements, and reproductive condition). We temporarily marked all animals by shaving a small patch of fur and released them at the capture site. All capture and handling methods were approved by the Institutional Animal Care and Use Committee (IACUC) of Brigham Young University (Protocol Numbers 110306 and 120601). Additionally, we acquired a Certificate of Registration (# 1COLL8652) from the Utah Division of Wildlife Resources (permission to trap) and obtained permission to access lands whenever needed. Our study did not involve endangered or threatened species.

Data analysis

We used linear regression analyses to assess relationships between factors potentially impacted by cheatgrass invasion (diversity of the plant and small-mammal community, the abundance of small mammals, and the percentage of shrub cover) and the percentage of cheatgrass cover. Given the structure and type (count) of our small mammal data, we used negative binomial distributions for the error structure in these analyses [31]. In other cases (e.g., for diversity indices and the percentage of shrub cover), we examined distributions and utilized a normal error structure (glm function in R). We calculated the minimum number of small mammals known to be alive (MNA; the total number of unique individuals captured) for each species at each site. We then

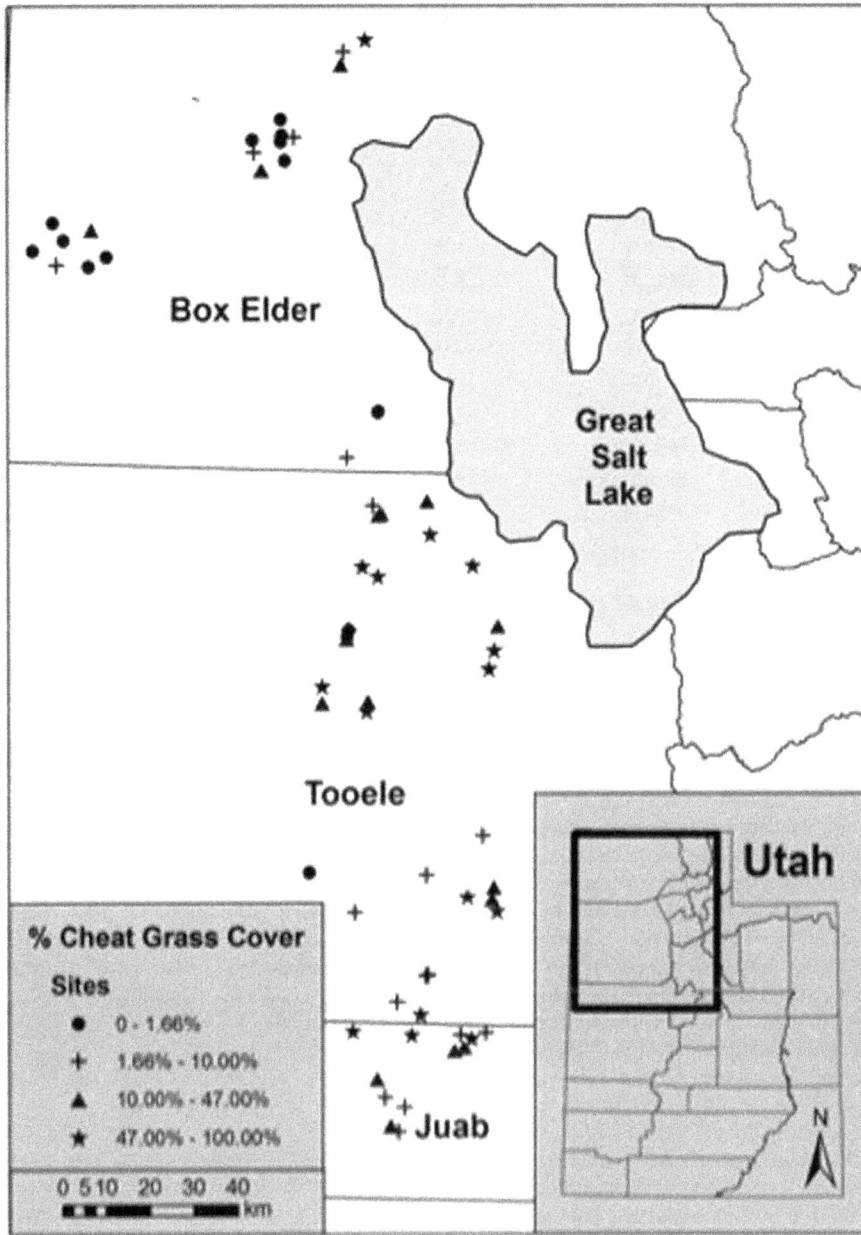

Figure 1. Study Area. Sites sampled for small mammals and the percentage of cheatgrass (*Bromus tectorum*) cover in the Great Basin Desert, Utah during the summers of 2011 and 2012. We divided sites into quartiles based on the percentage of cheatgrass cover.

examined the overall abundances of small mammals relative to the percentage of cheatgrass cover at each site using simple linear regression. We used the glm.nb function within the MASS package of program R [32] for these analyses [33]. Each regression model included sampling year to account for annual differences in small-mammal communities.

To better understand any ecological relationship illustrated in the overall regression, we also evaluated trends in abundance when the data were partitioned by family. We captured individuals from the Families Cricetidae, Sciuridae, and Heteromyidae. Specific representatives that we captured from these family groups fall into 3 distinct functional groups: nocturnal generalists (Cricetidae), diurnal generalists (Sciuridae), and nocturnal specialists (Heteromyidae). We quantified the variation in the abundance

of each functional group explained by the percentage of cheatgrass cover using additional linear regression analyses (again using a negative binomial distribution for count data).

We also regressed species-specific abundances of small mammals as a function of the percentage of cheatgrass cover. These analyses excluded species that were found at less than 5% of sites. After analyzing the species-specific abundance data, we re-ran the overall abundance regression, but excluded small-mammal species that showed significant declines in the individual analyses. The purpose of this additional analysis was to determine if the declining but non-significant trends exhibited by many species was a significant decline when pooled. Similarly, we wanted to determine if the decline in overall small-mammal abundances was only significant because a few species were experiencing

Table 1. Abundance of Small Mammals.

Family *Species*	Cheatgrass Cover (%)			
	Low (0–1.66)	Medium-Low (1.67–10)	Medium-High (10–47)	High (47–100)
Heteromyidae				
Dipodomys ordii	3.00±1.19	1.47±0.75	2.38±0.66	2.71±1.25
Dipodomys microps	1.59±0.58	0.82±0.30	1.31±0.44	0.12±0.12
Chaetodipus formosus	0.35±0.35	0.82±0.49	1.13±0.62	0.18±0.13
Perognathus parvus	-	0.12±0.12	0.44±0.26	0.35±0.35
Perognathus longimembris	-	0.24±0.14	-	-
Microdipodops megacephalus	0.06±0.06	-	-	-
Cricetidae				
Peromyscus maniculatus	4.29±0.91	4.29±1.35	3.31±0.93	2.29±0.75
Onychomys leucogaster	0.06±0.06	0.18±0.10	0.19±0.14	-
Neotoma lepida	0.24±0.14	-	0.06±0.06	-
Microtus montanus	-	-	0.13±0.13	-
Sciuridae				
Ammospermophilus leucurus	1.24±0.57	0.76±0.41	0.38±0.22	0.06±0.06
Neotamias minimus	-	0.12±0.12	-	-
Total	10.82±1.80	8.82±1.41	9.31±1.19	5.71±1.40

Note. – Mean abundance of small mammals (minimum number known alive) captured by species ± SE in each quartile of cheatgrass cover (%) for 66 sites in the Great Basin Desert, Utah (data collected in 2011–2012).

serious declines while most populations were stable. Because each of these analyses included count data, we continued to assume a negative binomial distribution for the error structure.

To assess the diversity of the small-mammal community at each site we used the exponential form of the Shannon-Wiener index (a measure more sensitive to species richness), and the reciprocal form of Simpson's index (a measure more sensitive to species evenness) [34]. Similarly, to allow us to account for the loss of plant diversity concurrent with cheatgrass invasion when modeling small-mammal abundances, we calculated these indices for the plant community. We used the Shannon and inverse Simpson functions within the vegan package in Program R to calculate these indices [35,36]. We examined the indices relative to the percentage of cheatgrass cover at each site using linear regression in program R [33]. To meet normality assumptions of regression, we square-root transformed these indices for the small-mammal community. Additionally, regression models included sampling year to account for annual differences in capture rates.

Lastly, we used linear regression analyses (with a negative binomial distribution) to assess whether changes in the small-mammal community are due to a direct or indirect effect of cheatgrass invasion. We used regression equations that modeled abundances of small mammals as a function of capture year, the percentage of shrub cover, the diversity of the plant community, and the percentage of cheatgrass cover. We verified that these variables were suitable for inclusion in the same model by examining collinearity between variables. This allowed us to determine if cheatgrass cover impacted abundances after accounting for the decrease in shrub cover (resulting from increased frequency of fire associated with invasion of cheatgrass) and the decrease in the diversity of the plant community that occurred with increased cheatgrass. If cheatgrass cover did not significantly explain abundance after accounting for shrub cover and diversity of the plant community, our analyses would indicate that changes in abundance of small mammals were primarily indirect effects of

invasion (e.g., fire, loss of overhead cover, or loss of plant diversity). However, if the percentage of cheatgrass cover remained a significant variable explaining variation in abundance of small mammals after accounting for these parameters, our analyses would indicate a direct negative impact of cheatgrass invasion on small mammals (e.g., inherent effects such as decreased food availability or decreased ability to move across the landscape through matted grass stems). We set alpha to be equal to 0.05 for all analyses.

Results

We detected 113 plant species across the 70 sites we sampled. The percentage of cheatgrass cover ranged from 0–94% at these sites. As the percentage of cheatgrass cover increased, shrub cover (a variable strongly influenced by fire in Great Basin Desert shrub communities) significantly declined (estimate = −0.201, SE = 0.051, t_{67} = −3.983, P<0.001). This change is indicative of the negative effects of increased fire frequency (caused by cheatgrass) on abundance of shrubs (i.e., at high densities of cheatgrass the shrub layer is often absent or diminished). Both the exponential Shannon-Wiener (estimate = −4.683, SE = 0.8440, t_{67} = −5.549, P<0.001) and the reciprocal Simpson's (estimate = −0.365, SE = 0.070, t_{67} = −5.187, P<0.001) indices (diversity of the plant community) also decreased with increased cover of cheatgrass.

We captured 580 unique small mammals during 10,437 trap nights. Individuals captured included representatives from 12 species, 10 genera, and 3 families (Table 1). In decreasing order of abundance, we captured: deer mouse (*Peromyscus maniculatus*), Ord's kangaroo rat (*Dipodomys ordii*), chisel-toothed kangaroo rat (*Dipodomys microps*), white-tailed antelope squirrel (*Ammospermophilus leucurus*), long-tailed pocket mouse (*Chaetodipus formosus*), Great Basin pocket mouse (*Perognathus parvus*), northern grasshopper mouse (*Onychomys leucogaster*), desert woodrat

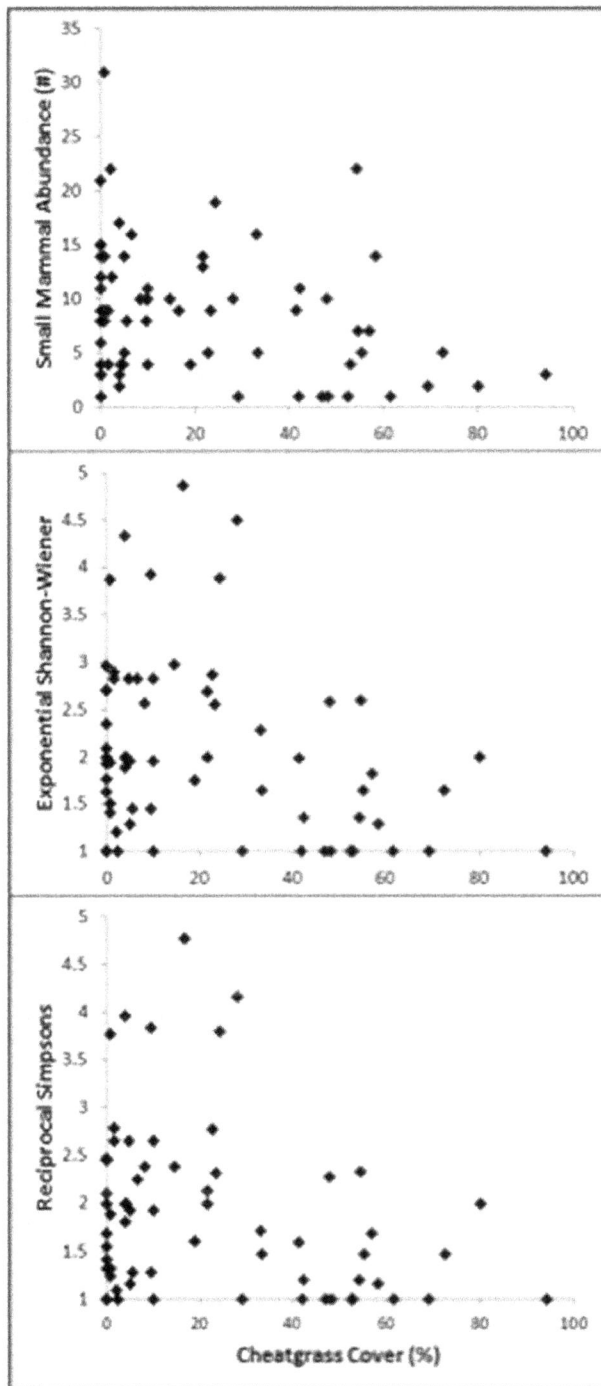

Figure 2. Cheatgrass and the Small-Mammal Community. Abundance (top), Exponential Shannon-Wiener index (middle), and Reciprocal Simpson's index (bottom) for small-mammal assemblages at 66 sites with varying cheatgrass (*Bromus tectorum*) cover in the Great Basin Desert, Utah, 2011–2012.

(*Neotoma lepida*), little pocket mouse (*Perognathus longimembris*), montane vole (*Microtus montanus*), least chipmunk (*Tamias minimus*), and dark kangaroo mouse (*Microdipodops megacephalus*; Table 1). We did not capture any small mammals at four sites (2 sites with relatively low percentage of cheatgrass cover and 2 with relatively high percentage of cheatgrass cover); therefore,

these sites were excluded from subsequent analyses because indices of diversity could not be calculated. Because these sites were evenly split between low and high percentage of cheatgrass cover their removal was unlikely to have significant effects on our analysis. Of the remaining 66 sites, the number of individuals captured at each site ranged from 1 to 31.

The overall abundances of small mammals had a negative slope when regressed against the percentage of cheatgrass cover (estimate = −1.007, SE = 0.348, z_{67} = −2.893, P = 0.004; Figure 2). When the small-mammal assemblages were partitioned into functional groups, only one group, Sciuridae, had a significant decrease in abundance with increased cover of cheatgrass (estimate = −4.968, SE = 1.915, z_{67} = −2.595, P = 0.009); neither Cricetid (estimate = −0.870, SE = .537, z_{67} = −1.621, P = 0.105) nor Heteromyid (estimate = −0.994, SE = 0.600, z_{67} = −1.656, P = 0.098) abundances decreased significantly (although there were negative trends for both groups). When the data were examined at the species level, 2 of 8 species decreased significantly in abundance with increased cheatgrass: Chisel-toothed kangaroo rat and white-tailed antelope squirrel (Table 2). After removing the chisel-toothed kangaroo rat and white-tailed antelope squirrel from the small-mammal community, there remained a negative trend in abundance with increased cheatgrass cover (estimate = −0.658, SE = 0.365, z_{67} = −1.801, P = 0.072), suggesting that the decreasing trend in small mammals was not solely due to declines by these two species.

The diversity of the small-mammal community decreased with increased cover of cheatgrass. Both the square root-transformed exponential Shannon-Wiener (estimate = −0.345, SE = 0.152, t_{63} = −2.263, P = 0.027) and the reciprocal Simpson's index (estimate = −0.312, SE = 0.146, t_{63} = −2.139, P = 0.036) decreased with increased cover of cheatgrass (Figure 2). Additionally, the negative relationship between cheatgrass and small-mammal abundance remained significant even after accounting for declines in shrub cover (likely associated with fire) and plant diversity (estimate = −1.464, SE = 0.482, z_{64} = −3.041, P = 0.002), indicating that there is a direct negative effect of cheatgrass invasion on small-mammal communities in the Great Basin.

Discussion

Changes in the percentage of cheatgrass cover were associated with changes in the small-mammal community of the Great Basin Desert. These changes likely resulted from a decrease in habitat suitability when cheatgrass invaded [19]. Overall abundance, richness, and evenness of the small-mammal community decreased with increasing cover of cheatgrass. Additionally, there were significant declines in abundance (with increasing cheatgrass cover) of two small-mammal species and one functional group when analyzed individually. Similarly, the combined abundances of all species, excluding the two species (chisel-toothed kangaroo rat and white-tailed antelope squirrel) with a significant decline, was negatively associated with the percentage of cheatgrass cover. This finding indicates that decreases in overall community abundances were not entirely driven by the two species exhibiting statistically significant declines. In other words, there is likely a biologically significant decrease in the abundance of other species that our analysis did not detect due to sample size limitations. Additional samples of some species may have illustrated patterns more clearly, but the stratified random nature of our selection process prevented the purposeful selection of additional sites likely to be inhabited by certain species.

Changes in small-mammal abundance (overall and species-specific) and diversity often result from decreased niche availability

Table 2. Regression Coefficients.

Species	estimate	SE	z_{67}[a]	p[b]
Chaetodipus formosus	−1.242	0.817	−1.521	0.128
Neotoma lepida	−4.854	4.287	−1.132	0.258
Peromyscus maniculatus	−0.786	0.570	−1.379	0.168
Onychomys leucogaster	−2.209	2.218	−0.996	0.319
Perognathus parvus	1.821	1.644	1.107	0.268
Dipodomys microps	−3.077	1.118	−2.752	0.006*
Dipodomys ordii	−0.531	0.925	−0.574	0.566
Ammospermophilus leucurus	−4.759	1.942	−2.451	0.014*

Note. – Linear regression coefficients and standard errors (SE) for each species when small-mammal abundance is calculated as a function of the percentage of cheatgrass cover for sites in the Great Basin Desert, Utah (data collected 2011–2012). The asterisk denotes significance at the $P<0.05$ level.
[a]test statistic.
[b]p-value.

[19,37]. Niche selection enables otherwise similar small-mammal species to coexist and several partitioning theories may explain changes in abundance and diversity of small mammals [26]. These theories include the spatial partitioning of resources (e.g., Ord's kangaroo rats prefer to forage in open areas, while pocket mice generally forage under shrubbery or other cover) and the partitioning of food resources by seed size [26,38]. Cheatgrass decreases both open space and shrub cover in a system, potentially negatively affecting these species [24]. Similarly, the decrease in shade availability and increase in restrictions on mobility that are concurrent with invasion by cheatgrass may impact small-mammal communities [39,40]. The observed changes in the Great Basin likely indicate that species respond differently to cheatgrass invasion and that cheatgrass invasion has reduced the number of different niches available to partition.

While changes in microhabitat availability may explain some of the trends in small-mammal populations, changes in the food supply may explain others. For example, small mammals prefer native seed over that of cheatgrass, which is relatively low in calories and protein [25]. Foliage is also an important food source for small mammals and is altered by cheatgrass invasion. For example, deer mice are known to increase consumption of foliage when there is a lack of precipitation, presumably because it is a source of moisture [41]. Additionally, there is a correlation between ingestion of green vegetation and reproductive activity in small mammals [42]. Because cheatgrass is green for only a short time when moisture is available [10], the lack of available green vegetation throughout the summer where cheatgrass is a dominant plant may impact survival or reproduction in small-mammal populations, contributing to the reduced abundance and diversity that we witnessed.

These potential mechanisms of small mammal decline may be linked directly or indirectly to cheatgrass invasion. Direct effects likely include a reduction in quality forage and increased obstruction to mobility [25,40]. Examples of indirect effects of cheatgrass include changes in the native herbaceous plant community, fire, and the associated reduction in shrub species [24]. The percentage of cheatgrass cover remained a significant correlate of small-mammal abundance even after accounting for changing shrub cover and decreasing diversity of the plant community. This regression accounts for likely indirect effects, indicating that cheatgrass was likely directly affecting small-mammal abundance (in addition to indirect effects). Although the

fire history at our sites was unknown, shrub cover was a plausible substitute [24].

The described changes in the small-mammal community may have effects at both higher and lower trophic levels. For example, several desert canid species are dependent on small mammals for the majority of their diet and changes in the small-mammal community may impact these populations (e.g., kit fox populations have declined following declines in small-mammal abundance) [43]. Additionally, many avian predators (raptors) in the Great Basin depend on small mammals for the majority of their diet [44,45] and changes in small-mammal abundance have been associated with changes in population size of some raptor species [46]. As a primary prey source, the significant decline in small-mammal abundance that occurred with increasing cover of cheatgrass may have negative effects on both diurnal and nocturnal species of raptors.

Removal or modification of top-down pressures can also have extreme effects on desert plant communities. Experimental removal of kangaroo rats from plots in the Chihuahuan Desert revealed that grasses released from pressures exerted by kangaroo rats had increased leaf and tiller growth and inflorescence production [47]. Other plots exhibited a three-fold increase in annual and perennial grass density where kangaroo rats were experimentally removed [28]. This change was dominated by a 20-fold increase in a single perennial grass species. In addition to direct effects of cheatgrass invasion on plant communities, the decline in abundance of small mammals – and one species of kangaroo rat in particular –associated with invasion likely has additional top-down influences on native vegetation.

Our analysis indicates that cheatgrass invasion is associated with changes in the small-mammal community. Ostoja and Schupp [19] reported similar findings but wondered whether their localized results were applicable to the Great Basin as a whole (they believed that they were). Our results build on that previously available as we collected data from sites across a broad spatial scale and spectrum of cheatgrass invasion. Additionally, we made several measurements of the plant community at each site, allowing our analyses to account for variation in the plant community and the presence of fire. Our results indicated that cheatgrass decreased the diversity and abundance of small mammals at invaded sites across a large portion of the Great Basin. Moreover, changes in the small-mammal community are likely the result of both direct (e.g., decreases in niche/food availability or increased mobility restrictions) and indirect effects (e.g., decreased shrub cover and increased fire frequency) associated with cheatgrass invasion. Negative effects of invasion on the small-mammal community will reduce food availability for higher trophic levels and remove a top-down pressure in this system, likely modifying plant population dynamics and resulting in a system regulated by bottom-up forces (cheatgrass). As the invasion of cheatgrass continues to spread across the western United States, implications for both plant and animal biodiversity and ecosystem function are severe.

Acknowledgments

We recognize the contributions of Hawkwatch International, Dugway Proving Grounds, and several field technicians and thank them for their contributions to this project.

Author Contributions

Conceived and designed the experiments: EDF RTL TRS RNK SJS BRM. Performed the experiments: EDF BRM. Analyzed the data: EDF RTL TRS BRM. Contributed reagents/materials/analysis tools: RNK SJS. Wrote the paper: EDF RTL TRS BRM RNK SJS. Input on manuscript: EDF RTL TRS RNK SJS BRM.

References

1. Cuddington K, Hastings A (2004) Invasive engineers. Ecological Modelling 178: 335–347.
2. Hejda M, Pysek P, Jarosek V (2009) Impact of invasive plants on the species richness, diversity and composition of invaded communities. Journal of Ecology 97: 393–403.
3. Weidenhamer JD, Callaway RM (2010) Direct and indirect effects of invasive plants on soil chemistry and ecosystem function. Journal of chemical ecology 36: 59–69.
4. Brooks ML, D'Antonio CM, Richardson DM, Grace JB, Keeley JE, et al. (2004) Effects of invasive alien plants on fire regimes. Bioscience 54: 677–688.
5. Dukes JS, Mooney HA (2004) Disruption of ecosystem processes in western North America by invasive species. Revista chilena de historia natural 77: 411–437.
6. D'Antonio C, Meyerson LA (2002) Exotic plant species as problems and solutions in ecological restoration: a synthesis. Restoration Ecology 10: 703–713.
7. Hiebert RD (1997) Prioritizing invasive plants and planning for management. Assessment and management of plant invasions. pp. 195–212.
8. Walker LR, Smith SD (1997) Impacts of invasive plants on community and ecosystem properties. Assessment and management of plant invasions. pp. 69–86.
9. Novak SJ, Mack RN (2001) Tracing plant introduction and spread: genetic evidence from Bromus tectorum (cheatgrass). Bioscience 51: 114–122.
10. Young JA, Clements CD (2009) Cheatgrass: fire and forage on the range: University of Nevada Press.
11. Melgoza G, Nowak RS, Tausch RJ (1990) Soil water exploitation after fire: competition between Bromus tectorum (cheatgrass) and two native species. Oecologia 83: 7–13.
12. Knapp PA (1996) Cheatgrass (Bromus tectorum L) dominance in the Great Basin Desert: History, persistence, and influences to human activities. Global Environmental Change 6: 37–52.
13. Balch JK, Bradley BA, D'Antonio CM, Gomez-Dans J (2013) Introduced annual grass increases regional fire activity across the arid western USA (1980-2009). Global Change Biology 19: 173–183.
14. Zouhar K (2003) Bromus tectorum. Fire Effects Information System Fort Collins, CO: USDA Forest Service, Rocky Mountain Research Station Fire Sciences Laboratory http://www fs fed us/database/feis/Author: Michele A James Scanned Images: Dave Egan Reviewers: Dave Brewer, Dave Egan, Chris McClone, and Judy Springer Series Editor: Dave Egan.
15. Mack RN (2011) Fifty years of waging war on cheatgrass: research advances, while meaningful control languishes. Fifty Years of Invasion Ecology: The Legacy of Charles Elton. West Sussex: Blackwell Publishing Ltd. pp. 253–265.
16. D'Antonio CM, Vitousek PM (1992) Biological invasions by exotic grasses, the grass/fire cycle, and global change. Annual Review of Ecology and Systematics 23: 63–87.
17. Young JA, Evans RA (1973) Downy brome: intruder in the plant succession of big sagebrush communities in the Great Basin. Journal of Range Management 26: 410–415.
18. Humphrey LD, Schupp EW (2004) Competition as a barrier to establishment of a native perennial grass (Elymus elymoides) in alien annual grass (Bromus tectorum) communities. Journal of Arid Environments 58: 405–422.
19. Ostoja SM, Schupp EW (2009) Conversion of sagebrush shrublands to exotic annual grasslands negatively impacts small mammal communities. Diversity and Distributions 15: 863–870.
20. Crooks JA (2002) Characterizing ecosystem level consequences of biological invasions: the role of ecosystem engineers. Oikos 97: 153–166.
21. Slater SJ, Frye Christensen KW, Knight RN, MacDuff R, Keller K (2012) Great Basin Avian Species-at-risk and Invasive Species Management through Multi-Agency Monitoring and Coordination Final Report. Department of Defense, Legacy Resources Management Program (Project #10-102).
22. Pearson DE (2009) Invasive plant architecture alters trophic interactions by changing predator abundance and behavior. Oecologia 159: 549–558.
23. Stewart G, Hull A (1949) Cheatgrass (Bromus tectorum L.)—an ecologic intruder in southern Idaho. Ecology 30: 58–74.
24. Whisenant SG (1990) Changing fire frequencies on Idaho's Snake River Plains: ecological and management implications. General Technical Report-Intermountain Research Station, USDA Forest Service: 4–10.
25. Kelrick M, MacMahon J, Parmenter R, Sisson D (1986) Native seed preferences of shrub-steppe rodents, birds and ants: the relationships of seed attributes and seed use. Oecologia 68: 327–337.
26. Brown JH, Lieberman GA (1973) Resource utilization and coexistence of seed-eating desert rodents in sand dune habitats. Ecology 54: 788–797.
27. Price MV (1978) The role of microhabitat in structuring desert rodent communities. Ecology 59: 910–921.
28. Brown JH, Heske EJ (1990) Control of a desert-grassland transition by a keystone rodent guild. Science(Washington) 250: 1705–1707.
29. Guo Q, Thompson DB, Valone TJ, Brown JH (1995) The effects of vertebrate granivores and folivores on plant community structure in the Chihuahuan Desert. Oikos 73: 251–259.
30. Vander Wall SB (1992) The role of animals in dispersing a" wind-dispersed" pine. Ecology 73: 614–621.
31. White GC, Bennetts RE (1996) Analysis of frequency count data using the negative binomial distribution. Ecology 77: 2549–2557.
32. Ripley R, Hornik K, Gebhardt A, Firth D (2011) MASS: support functions and datasets for Venables and Ripley's MASS. R package version: 7.3–16.
33. Team RDC (2014) R: A Language and Environment for Statistical Computing. Vienna, Austria: R Foundation for Statistical Computing.
34. Hill MO (1973) Diversity and evenness: a unifying notation and its consequences. Ecology 54: 427–432.
35. Oksanen J, Blanchet FG, Kindt R, Legendre P, Minchin P, et al. (2011) vegan: Community Ecology Package. R package version 2.0-2. http://CRAN R-project org/package= vegan.
36. Kindt R, Coe R (2005) Tree diversity analysis: A manual and software for common statistical methods for ecological and biodiversity studies: World Agroforestry Centre Eastern and Central Africa Program.
37. Gano K, Rickard W (1982) Small mammals of a bitterbrush-cheatgrass community. Northwest Science 56.
38. Price MV, Brown JH (1983) Patterns of morphology and resource use in North American desert rodent communities. Great Basin Naturalist Memoirs 7: 117–134.
39. Parmenter RR, MacMahon JA (1983) Factors determining the abundance and distribution of rodents in a shrub-steppe ecosystem: the role of shrubs. Oecologia 59: 145–156.
40. Rieder J, Newbold T, Ostoja S (2010) Structural changes in vegetation coincident with annual grass invasion negatively impacts sprint velocity of small vertebrates. Biological Invasions 12: 2429–2439.
41. Sieg C, Uresk D, Hansen R (1986) Seasonal diets of deer mice on bentonite mine spoils and sagebrush grasslands in southeastern Montana. Northwest Science 60: 81–89.
42. Reichman O, van de Graaff KM (1975) Association between ingestion of green vegetation and desert rodent reproduction. Journal of Mammalogy 56: 503–506.
43. White P, White CAV, Ralls K (1996) Functional and numerical responses of kit foxes to a short-term decline in mammalian prey. Journal of Mammalogy 77: 370–376.
44. Fitch HS, Swenson F, Tillotson DF (1946) Behavior and food habits of the Red-tailed Hawk. Condor 48: 205–237.
45. Blair CL, Schitoskey F Jr (1982) Breeding biology and diet of the Ferruginous Hawk in South Dakota. The Wilson Bulletin 94: 46–54.
46. Schmutz JK, Hungle DJ (1989) Populations of ferruginous and Swainson's hawks increase in synchrony with ground squirrels. Canadian Journal of Zoology 67: 2596–2601.
47. Kerley G, Whitford W (2009) Can kangaroo rat graminivory contribute to the persistence of desertified shrublands? Journal of Arid Environments 73: 651–657.

6

Transcriptomic Analysis of a Tertiary Relict Plant, Extreme Xerophyte *Reaumuria soongorica* to Identify Genes Related to Drought Adaptation

Yong Shi[1][9], Xia Yan[2][9], Pengshan Zhao[1], Hengxia Yin[1], Xin Zhao[1], Honglang Xiao[2], Xinrong Li[1,3], Guoxiong Chen[1], Xiao-Fei Ma[1]*

1 Key Laboratory of Stress Physiology and Ecology in Cold and Arid Regions, Department of Ecology and Agriculture Research, Cold and Arid Regions Environmental and Engineering Research Institute, Chinese Academy of Sciences, Lanzhou, People's Republic of China, 2 Key Laboratory of Eco-hydrology and of Inland River Basin, Cold and Arid Regions Environmental and Engineering Research Institute, Chinese Academy of Sciences, Lanzhou, People's Republic of China, 3 Shapotou Desert Research and Experiment Station, Cold and Arid Regions Environmental and Engineering Research Institute, Chinese Academy of Sciences, Lanzhou, People's Republic of China

Abstract

Background: *Reaumuria soongorica* is an extreme xerophyte shrub widely distributed in the desert regions including sand dune, Gobi and marginal loess of central Asia which plays a crucial role to sustain and restore fragile desert ecosystems. However, due to the lacking of the genomic sequences, studies on *R. soongorica* had mainly limited in physiological responses to drought stress. Here, a deep transcriptomic sequencing of *R. soongorica* will facilitate molecular functional studies and pave the path to understand drought adaptation for a desert plant.

Methodology/Principal Findings: A total of 53,193,660 clean paired-end reads was generated from the Illumina HiSeq[TM] 2000 platform. By assembly with Trinity, we got 173,700 contigs and 77,647 unigenes with mean length of 677 bp and N50 of 1109 bp. Over 55% (43,054) unigenes were successfully annotated based on sequence similarity against public databases as well as Rfam and Pfam database. Local BLAST and Kyoto Encyclopedia of Genes and Genomes (KEGG) maps were used to further exhausting seek for candidate genes related to drought adaptation and a set of 123 putative candidate genes were identified. Moreover, all the C_4 photosynthesis genes existed and were active in *R. soongorica*, which has been regarded as a typical C_3 plant.

Conclusion/Significance: The assembled unigenes in present work provide abundant genomic information for the functional assignments in an extreme xerophyte *R. soongorica*, and will help us exploit the genetic basis of how desert plants adapt to drought environment in the near future.

Editor: Zhanjiang Liu, Auburn University, United States of America

Funding: This work was supported by Key Project of Chinese National Programs for Fundamental Research and Development (973 Program, 2013CB4229906), by Grant 29Y127E71 from the "One Hundred Talents" Project of the Chinese Academy of Sciences, and by National Natural Science Foundation of China (No. 91125025). The funders had no role in study design, data collection and analysis, decision to publish, or preparation of the manuscript.

Competing Interests: The authors have declared that no competing interests exist.

* E-mail: maxiaofei@lzb.ac.cn

[9] These authors contributed equally to this work.

Introduction

Understanding the genetic basis of how organisms adapt to climate change is one of the most challenging tasks [1,2]. Although many attempts have exploited the adaptation mechanism in the species with known genome sequences, the molecular basis of adaptation in the non-genomic species is still poorly understood [1,2], especially in the plants from arid regions which contain plenty of potential genetic resources for ecology engineering. As pioneer and constructive species in all kinds of desert ecosystems, *Reaumuria* plants play important roles to sustain fragile desert ecosystems by keeping the vital process of the transport of energy and substances [3–5], and preventing from wind erosion, sand drifting and the further desertification of these regions [3,6–8]. These plant species were widely used as fine pioneer plants in the restoration of degraded ecosystems with natural rainfall [9] and in the sustainable development of arid regions due to their extreme tolerance to saline-alkaline conditions [10–12].

Reaumuria (Tamaricaceae) plants are perennial xeric shrubs, and all the 12 species classified in this genus were distributed in the arid regions from North Africa, Asia, and South Europe, among which four species including *R. soongorica*, *R. alternifolia*, *R. kaschgarica*, and *R. trigyna* were found in China (www.eflora.org, Figure 1). *R. soongorica* (2n = 22 with 778 Mb genome size [13]) is one of the constructive and dominant species in kinds of desert ecosystems in central Asia, such as Taklamakan, Gurbantunggut, Kumtag, Badain Jaran, Qaidam, South Russia, South Mongolia and Tenger deserts, and Mu Us, Ulan Buh and Horqin sandy lands, and marginal Loess [14,15] (Figure 1). Desertification of these regions is getting worse due to accelerating global climate change and human activity [14,16]. *R. soongorica* has undergone desertification of Asia which initiated at least 22 million years ago according to the palaeomagnetic measurements and fossil evidence [17]. During the process of adaptation to desertification, *R. soongorica* has evolved specific traits including extremely thick cuticle, hollow stomata, specialized leaf shape, deep root system, and effective physiological mechanisms such as reduced transpiration rate, increased water use efficiency, and maintaining the stem vigor to survive desiccation by leaf abscission [7,18–21]. Much effort has been made in *R. soongorica* to elucidate the mechanism of drought adaptation during last decade, however, due to paucity of genomic information, most of the previous studies have limited to its physiological characteristics [21–27]. Little work had focused on the genetic diversity based on neutral markers (RAPD [28], ISSR [29,30] and cpDNA [31]). However, all these studies failed to dissolve the adaptive evolution of *R. soongorica*. Therefore, transcriptome sequences are in urgent to supply sufficient functional genomic information to address systemically the genetic mechanisms of drought adaptation of *R. soongorica*.

In this study, the *R. soongorica* transcriptome was sequenced by the Illumina paired-end sequencing technology (Illumina HiSeq^{TM} 2000 platform). A total of 4.8 gigabases raw data was assembled into 173,700 contigs and further constructed into 77,647 unigenes (mean length = 677 bp, N50 = 1109 bp). Moreover, 123 unigenes were identified to be potentially involved in drought adaptation. To our surprise, all the C$_4$ photosynthesis genes were existed and active in *R. soongorica* which has been regarded as a typical C$_3$ plant [32]. The *R. soongorica* transcriptomic information provides a prime reference point for the subsequent exploitation of this important genetic resource and will facilitate to unravel the mechanism of adaptation to extreme arid environment.

Results

Sequencing and *de novo* Assembly

To obtain a global overview of the *R. soongorica* transcriptome, a pooled cDNA library representing the inflorescences, leaves, and seedlings was constructed, and then sequenced on the Illumina HiSeq^{TM} 2000 platform. A total of 4.8 gigabases dataset was generated from 53,193,660 clean paired-end reads with length of 90 bps and Q20 over 96% (Table 1). This suggested that the sequencing output and quality were good enough for further analysis.

A total of 173,700 contigs with average length of 321 bp and an N50 length of 532 bp were assembled by Trinity program [33].

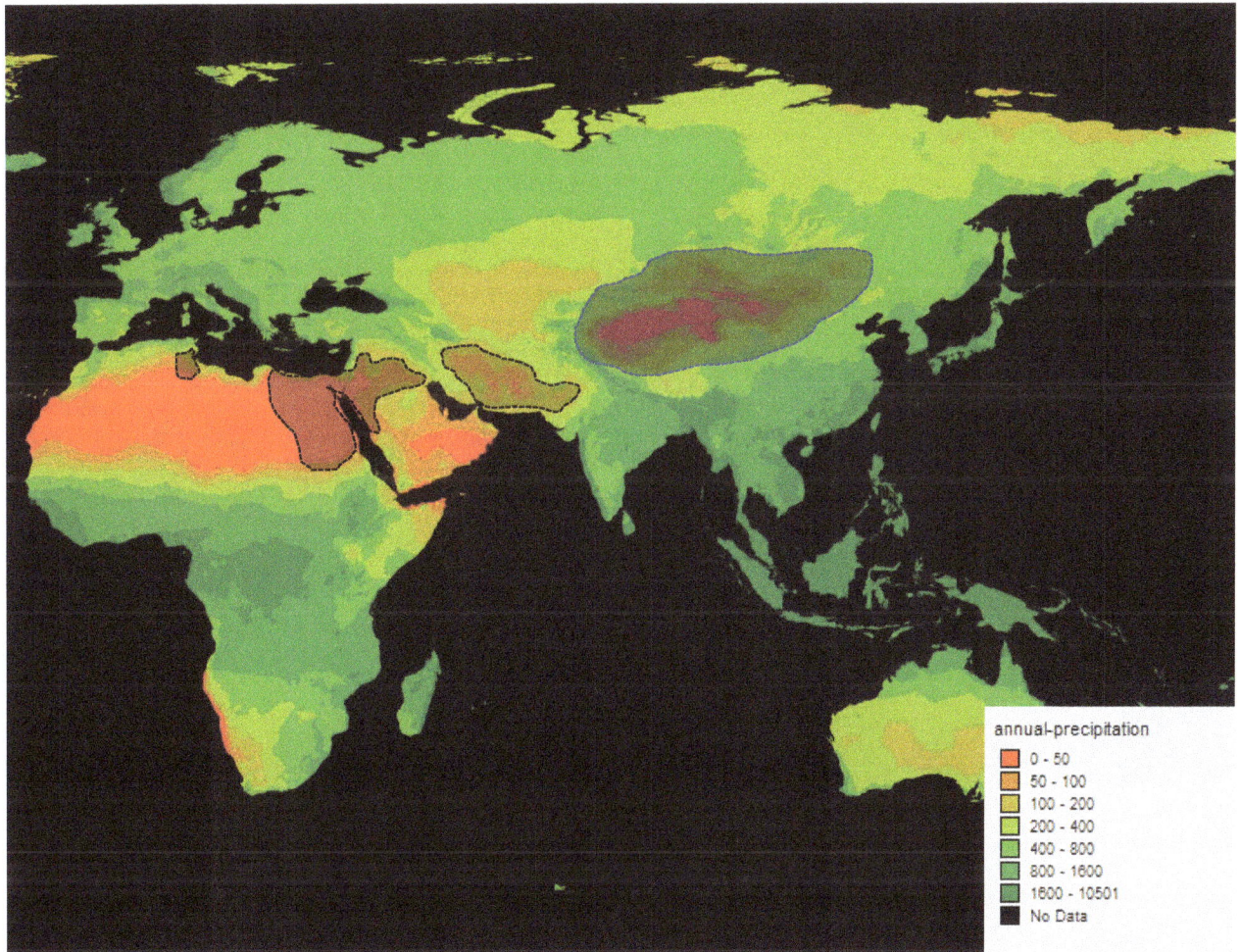

Figure 1. Distribution of *R. soongorica* in northern China. The blue shadow indicates the *Reaumuria* distribution in China and neighboring districts. The black shadow indicates the *Reaumuria* distribution in North Africa and Central Asia.

There were 103,541 contigs (59.61%) ranged from 100 to 200 bp, 67,087 contigs (38.62%) ranged from 201 bp to 2000 bp, and 3,072 contigs (1.77%) longer than 2 kb (Figure S1A). After further clustering and assembly, a total of 77,647 unigenes was generated with average length of 677 bp (200 to 7,546 bp) and N50 length of 1109 bp (Figure S1B). About 40% (30,680) unigenes were longer than 500 bp, among which 5.66% (4,392) unigenes were more than 2000 bp.

Sequence Annotation

Several complementary approaches were utilized to validate and annotate the assembled unigenes. The unigenes were first aligned to the National Center for Biotechnology Information (NCBI) non-redundant protein (Nr) database, non-redundant nucleotide sequence (Nt) database, and the Swiss-Prot protein database with E-values less than 1e-5. Among the 77,647 unigenes, 41,582 (53.55%), 28,197 (36.31%) and 25,297 (32.58%) unigenes were significantly matched to the known genes in the Nr, Nt and Swiss-Prot protein databases, respectively (Table S1 and Table 2). The E-value distribution of the top hits in the Nr database showed that 46.01% of the sequences were mapped to the known genes in plants with best hits (E-value<1e-50, mean identity is 69.54%), and approximately 16% of unigenes can hit deposited sequences with similarity over 80% (Figure 2A and 2B). About 82% of annotated unigenes can be assigned with a best score to the sequences from the top seven species, i.e., *Vitis vinifera* (41.65%), *Ricinus communis* (13.46%), *Populus trichocarpa* (11.39%), *Glycine max* (8.12%), *Medicago truncatula* (3.29%), *Arabidopsis thaliana* (2.08%) and *A. lyrata* subsp. *lyrata* (1.78%) (Figure 2C). Interestingly, the phylogenetic relationship based on the internal transcribed spacer (ITS) also showed *R. soongorica* in among the other rosids species firstly diverged from *V. vinifera*, though the bootstrap value of their relationship is below 50% (Figure 2D).

GO and COG classification. To identify functional categories among the 77,647 unigenes, all the best BLAST hits were input into the Gene Ontology (GO) Software Blast2GO for GO

Table 1. Summary of *de novo* sequence assembly for *Reaumuria soongorica*.

Item	Number
Total Raw Reads	57,745,560
Total Clean Reads	53,193,660
Total Clean Nucleotides (nt)	4,787,429,400
Q20 percentage (%)	96.32
N percentage (%)	0.01
GC percentage (%)	45.52
Number of contigs	173,700
Shortest contig (bp)	100
Longest contig (bp)	7561
N50 of contigs (bp)	532
Mean length of contig (bp)	321
Number of unigenes	77,647
Shortest unigene (bp)	200
Longest unigene (bp)	7546
N50 of unigenes (bp)	1109
Mean length of unigene (bp)	677

functional enrichment analysis by performing Fisher's exact test [34,35]. In total, 38.21% of unigenes (29,666) could be assigned to gene ontology classes with 202,607 functional terms (Figure 3, Table S2). Interestingly, cellular process ($p = 0.007$), metabolic process ($p = 0.007$) and response to stimulus ($p = 0.009$) are strong significantly overrepresented in the 29 biological process GO groups.

For further functional prediction and classification, all of the 77,647 unigenes were aligned to the Clusters of Orthologous Groups of proteins (COG) category. Overall, 19.30% of unigenes (14,987) were assigned into 25 COG categories with 28,537 COG functional terms (Figure 4 and Table S3). The categories including transcription (2,455, 8.60%), carbohydrate transport and metabolism (1,795, 6.29%), signal transduction mechanisms (1,788, 6.27%) and secondary metabolites biosynthesis, transport and catabolism (860, 3.01%) were identified, which might be related to response for drought stress in plants.

KEGG pathway mapping. To further gain insights into the biological functions and interactions of our unigenes, a pathway-based analysis was performed in the light of the Kyoto Encyclopedia of Genes and Genomes (KEGG) Pathway database which based on the roles in biochemical pathways. There were 23,569 (30.35%) out of the 77,647 unigenes were mapped to 128 KEGG pathways. Among them, 5,056 unigenes (21.45%) were related to metabolic pathways (Ko01100, no maps in KEGG database), 2,460 (10.87%) to biosynthesis of secondary metabolites (Ko01100, no maps in KEGG database). The highest representation with KEGG map was Plant-pathogen interaction (Ko04626, 1,156 unigenes, 4.90%), followed by Plant hormone signal transduction (Ko04075, 1,113 unigenes, 4.72%) and RNA transport (Ko03013, 1,002 unigenes, 4.25%) (Table S4). To be mentioned, 57 core enzymes was detected in the biosynthetic pathway of flavonoid (Ko00941, 228 unigenes) which involved in secondary metabolism under abiotic stresses in plants [36] (Figure S2). Also, all of the core components were found in the circadian rhythm pathway (Ko04712, 192 unigenes), which is crucial for timing of flower and budset in response to the seasonal rhythm of temperature and light length [37,38] (File S1). *R. soongorica* has been regarded as a C_3 plant based on the photosynthesis characteristics [39], but all the core genes of C_4 carbon fixation were found in our transcriptome, surprisingly (Ko00710, 155 unigenes, File S2). Absisic acid (ABA) is a crucial hormone involved in many stress responses [40]. The key enzymes in its biosynthetic and catabolic pathways (Ko00906) and receptor genes (Ko04075) were discovered as well (Table S5 and File S3).

Rfam and Pfam analysis. From Nr, Nt, Swiss-Prot, GO, COG, and KEGG databases, more than half of the unigenes (42,839, 55.17%, mean length was 975 bp, Table 2 and Figure 5A) were annotated, while 34,808 (44.83%) unigenes (mean length was 334 bp, Figure 5A) failed to be annotated. As shown in Figure 5B, the annotated transcripts are significant more abundant in the pool (ANOVA p value = 2.2e-16).

To further exploit the potential function of the un-annotated unigenes, Rfam and Pfam analysis were conducted sequentially. Among the un-annotated unigenes, 1,904 showed significant hits (E-value<0.01, Identity ≥82.30%) with the deposits in the Rfam database, but only 42 were trustable (E-value<1e-5, Identity >84.47%, BLAST coverage >50%, File S4). Furthermore, 173 out of 34,808 un-annotated unigenes contained 92 kinds of Pfam domains (E-value<1e-5), among which reverse transcriptase (RVT_3) and zinc-binding in reverse transcriptase (zf-RVT) domains were highly represented (26 and 19 times, respectively; File S4).

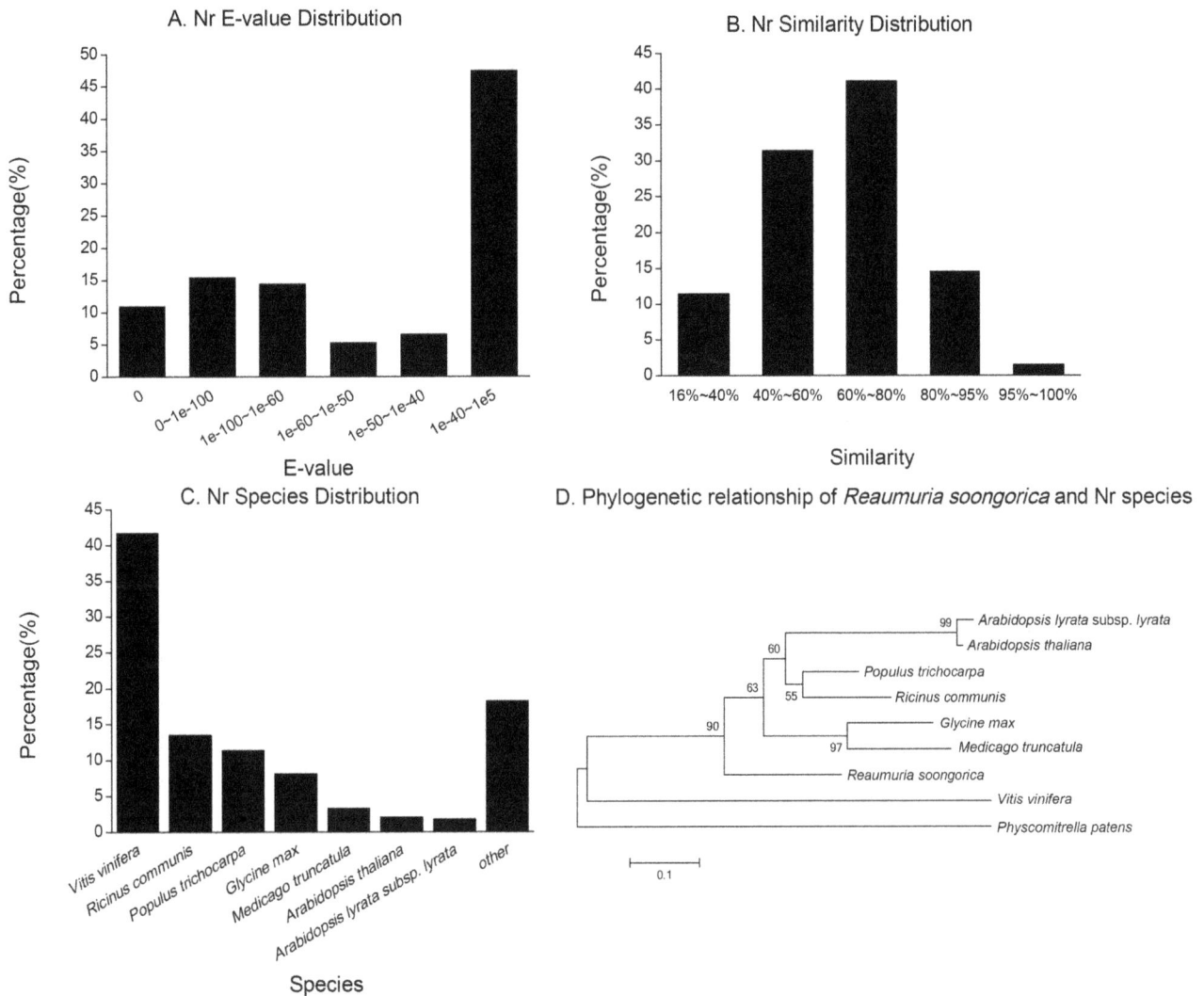

Figure 2. Characteristics of similarity search of unigenes against Nr databases. (A) The E-value distribution of Nr annotation results. (B) The similarity distribution of Nr annotation results. (C) The species distribution of Nr annotation results. (D) The phylogenetic relationship of *Reaumuria soongorica* and Nr species. Phylogenetic tree constructed by the Neighbor-Joining method for the internal transcribed spacer (ITS) sequences. *Physcomitrella patens* was included as an out-group. Neighbor-Joining was consensus tree used 1000 bootstrap replicates. The number represents the percentage of bootstrap values.

Functional Genes Related to Drought Adaptation

To screen functional genes related to drought adaptation through three main strategies (drought escape, drought avoidance and drought tolerance; reviewed in [41,42]), candidate genes from *Arabidopsis* were local BLASTed against 77,647 unigenes in *R. soongorica*. A total of 123 unigenes homologous to 113 *Arabidopsis* candidate genes potentially involved in drought adaptation were identified in our transcriptomic dataset (Table S5). Among them, 46 putative flowering time unigenes were identified to be involved in drought escape (E-value<1e-5) which is a common cost-effective strategy to avoid drought stress in natural populations

Table 2. Summary of sequence annotation for *Reaumuria soongorica*.

Database	Nr	Nt	Swiss-Prot	GO	COG	KEGG	Rfam	Pfam
Hit numbers (percentage)	41582 (53.55%)	28197 (36.31%)	25297 (32.58%)	29666 (38.20%)	14987 (19.30%)	23569 (30.35%)	42 (0.05%)	173 (0.22%)
Annotated numbers (percentage)	42839 (55.17%)						215 (0.28%)	
Total annotated numbers (percentage)	43054 (55.45%)							

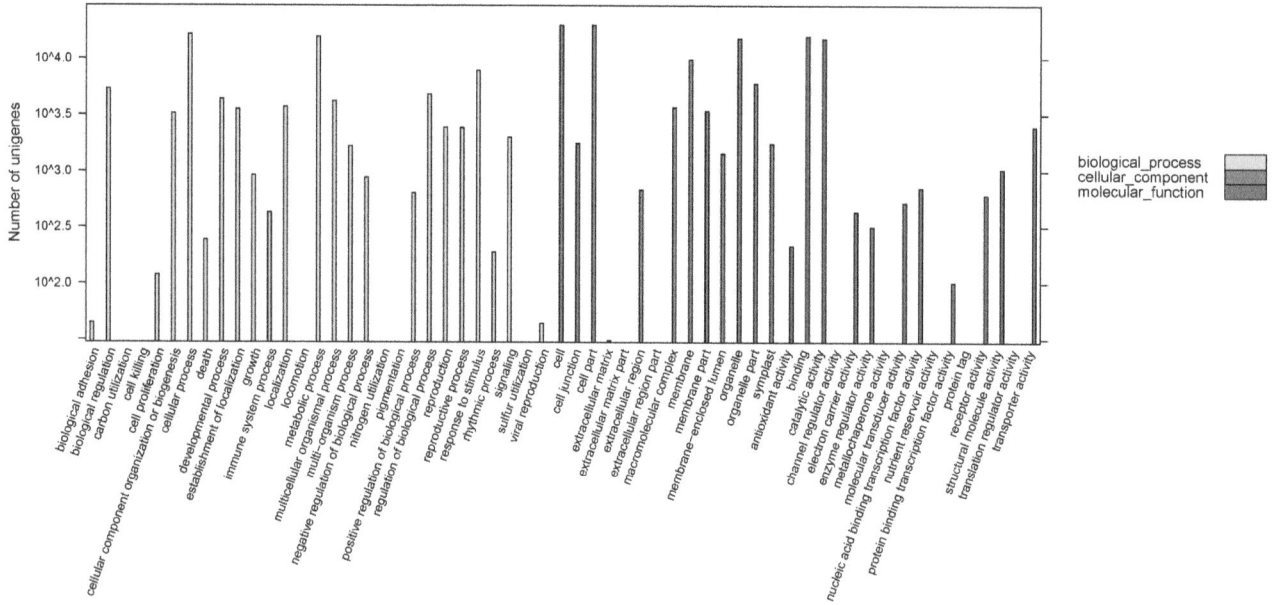

Figure 3. Gene Ontology classification of assembled unigenes. Total 29,666 unigenes were categorized into three main categories: Biological process (81,460, 40.21%), Cellular component (83,650, 41.29%) and Molecular function (37,497, 18.51%).

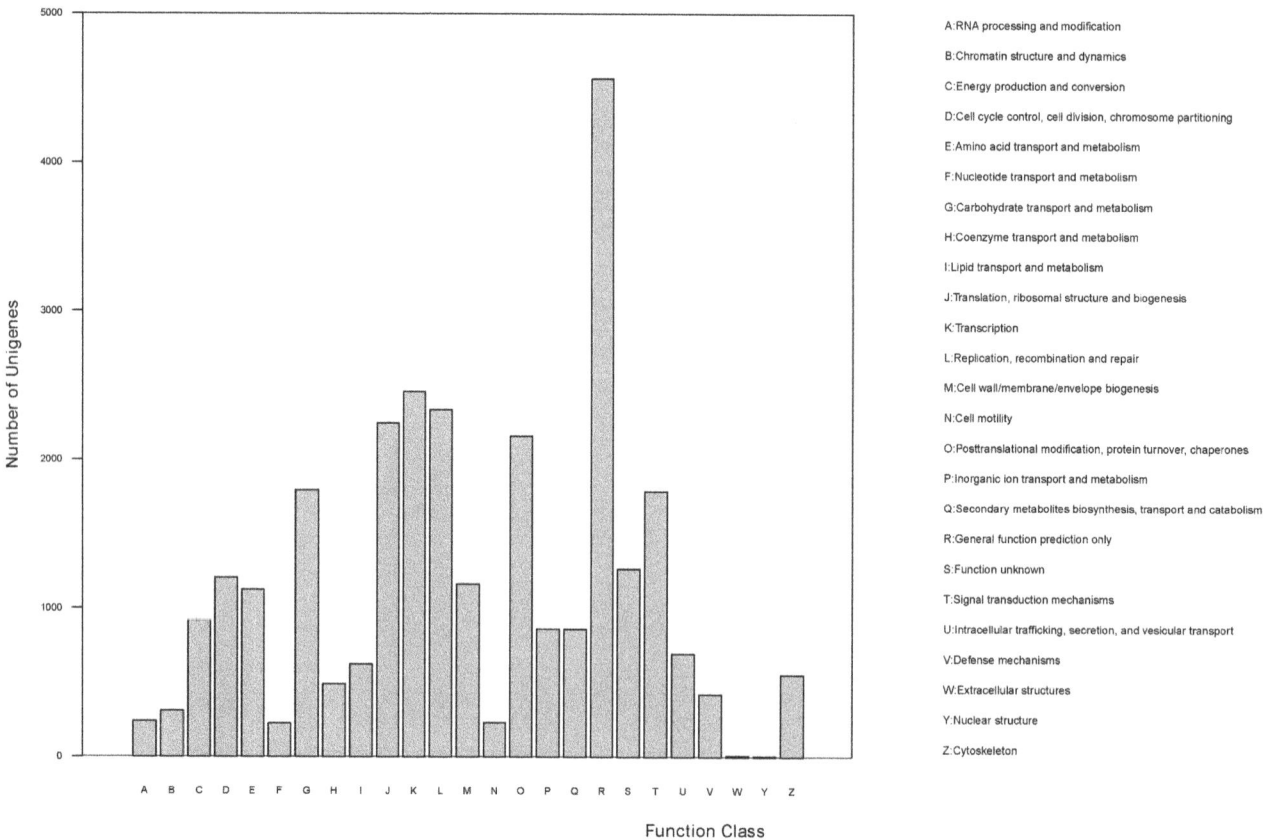

Figure 4. COG function classification. Total 14,987 unigenes were assigned to 25 COG classifications. The largest cluster was for general function prediction only (4,564, 15.99%), followed by transcription (2,455, 8.60%), replication, recombination and repair (2,335, 8.18%), translation, ribosomal structure and biogenesis (2,243, 7.86%), posttranslational modification, protein turnover, and chaperones (2,157, 7.56%), carbohydrate transport and metabolism (1,795, 6.29%) and signal transduction mechanisms (1,788, 6.27%).

A

B

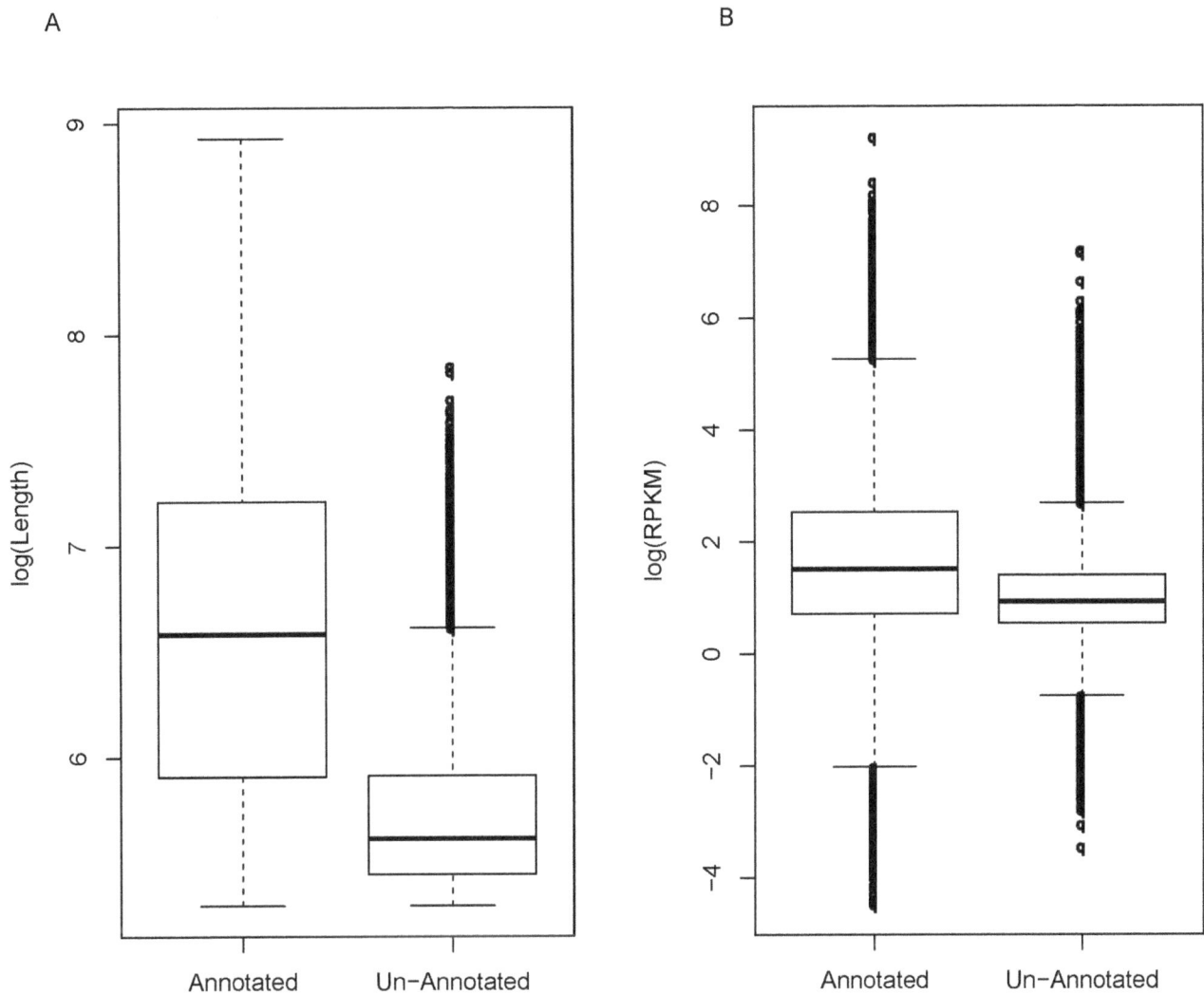

Figure 5. Boxplot distribution of unigenes in annotated and un-annotated unigenes. (A) Length distribution inferred with log(Length). (B) Expression level inferred by log(RPKM) value. Both show significant in ANOVA tests $p = 2.2e-16$.

[41,42]. A total of 40 unigenes were found to be potentially involved in drought avoidance (E-value<1e-5), which regulate the biogenesis and development of the cuticle (eight unigenes), stomata (six unigenes), and trichomes and root system (26 unigenes). For drought tolerance strategy, the ABA-dependent and -independent pathways have been extensively studied. Totally, 32 unigenes were involved in ABA-dependent pathways, including biosynthetic and catabolic genes, ABA receptors, while eight unigenes were involved in ABA-independent pathway. These genes will be helpful for exploiting the genetic mechanism of how *R. soongorica* adapts to drought natural environment in northern China.

Discussion

In this study, a large amount of *R. soongorica* transcriptomic unigenes (77,647) were sequenced with Illumina HiSeq[TM] 2000 platform (Table 1). The N50 length of unigenes was 1,109 bp and the average length was 677 bp; these results were comparable to the recently published plant trancriptomic analysis, such as *Hevea brasiliensis* (N50 = 436 bp, average length = 485 bp [43]) and *Dendrocalamus latiflorus* Munro (N50 = 1,132 bp, average length = 736 bp [44]). Up to date, Trinity is one of most powerful

packages prevailed in *de novo* assembly of short reads, especially in dealing with alternative splicesomes and paralogs [33]. A total of 24,271 (31.26%) unigenes were generated as clusters which might correspond to putative alternative spliced transcripts and/or paralogous transcripts ("CL"-unigene). The number of clusters with only two types of "CL"-unigenes (i.e. CL1.Contig1_A and CL1.Contig2_A) was notably high in our dataset (5,662 clusters, Figure S3). A considerable number of clusters with more than ten types of "CL"-unigenes were also found in our dataset (Figure S3). For example, thirteen "CL2023.Contigs" showed high identities (E-value<1e-41, Identity>78%) with *Arabidopsis cryptochrome 1* (*CRY1*) (Table S5, File S5), which plays a crucial role in sensing seasonal change of blue light and UV-A to initiate flower properly [45,46]. The thirteen contigs can be divided into three types based on sequence divergence with identities less than 90% (File S5). Together with the overrepresented numbers of the clusters with more than two types of "CL"-unigenes, these results suggest that a large amount of paralogous transcripts were presented in our dataset, indicating at least a whole genomic duplicate might happen in the evolutionary history of *R. soongorica*. In addition, three alternative splicing sites were found in the alignment of cluster "CL11.contigs" which contained 32 contigs (File S6), showing that splice-isoforms were also produced by the Trinity assembly. Of course, imperfect assembly of short reads by the Trinity cannot be ruled out.

More than half of the unigenes (42,839, 55.17%) were successfully assigned as annotated genes by BLASTing against with public databases Nr, Nt, Swiss-Prot, GO, COG and KEGG, given the absence of genomic information of *R. soongorica* (Table 2). Notably, the percentage of annotated is the lowest in among the previous studies with the same sequencing strategy during the last year (58.24 to 78.9%, [44,47,48]). One possibility of lacking annotation is due to the technical limitation, such as sequencing depth and read length [49], which was common in the all studies with *de novo* transcriptome analysis [44,47,48]. We did find the un-annotated sequences were averagely much shorter than the annotated unigenes (334 bp vs 975 bp, Figure 5A). However, there was still a considerable percentage of unigenes (7.96%, 2,770 of 34,808) with length over 500 bp and reads per kilobase per million reads (RPKM) over three failed to hit any homologs in the known plant species (Figure 5). In addition, there were 173 unigenes contained at least one Pfam domain (E-value<1e-5, File S4), among which two reverse transcriptase domains RVT_3 and zf-RVT were highly represented with 26 and 19 times, respectively. These results suggested that a considerable portion of genes in *R. soongorica* might originate from novel retrotransposon mechanisms [50,51], which not found in any known genomes yet, and the high frequency of reverse transcriptase genes could be an indispensable part of *R. soongorica* genome (778 Mb, 2n = 22 [13]).

So far, *V. vinifera* was the highest related species with known genome to *R. soongorica* (Figure 2C), but fewer than half of annotated unigenes in *R. soongorica* hit protein sequences in *V. vinifera* (Figure 2C). This is consistent with that these two species were classified into different orders, Caryophyllales (*Tamaraceae, Reaumuria*) and Vitales (*Vitaceae, Vitis*) [52,53], and with their low bootstrap value support in the ITS phylogenetic tree (Figure 2D). Therefore, the vast un-annotated unigenes (44.83%, 34,808 of 77,647) could only be novel genes compared with the known genomes and specific in the genome of the relict Tertiary plant *R. soongorica*. Functional and expressional studies by Digital Gene Expression analysis and real-time PCR are needed to further corroborate this hypothesis in detail.

After deep sequencing and exhaustive annotation, this endeavor provided a large amount of unigenes that will facilitate to exploit genetic resources in the functional studies and to identify candidate genes responsible for drought adaptation in *R. soongorica*. Candidate genes out of the 77,647 unigenes in *R. soongorica* involved in three drought adaptation strategies (drought avoidance, drought escape and drought tolerance; reviewed in [41,42]) were analyzed (Table S5). At least eight unigenes were possibly involved in the formation of the thick cuticle on *R. soongorica* leaf surface (8.3 μm [18]), which plays an important role in regulating the exchange of gases and water in plants and can enhance tolerance to drought [54]. To be mentioned, three unigenes were found as homologs of *HvABCG31/Eibi1*, an ATP-binding cassette subfamily G full transporter (also found in KEGG, ko02010), which is essential for the cutin formation and the preservation leaf water in wild barley [55]. Stomata, trichomes and root hairs are crucial for water usefulness and maintenance under drought environment (reviewed in [54,56]), and the molecular mechanisms of the differentiation of these tissues have been extensively studied in *Arabidopsis* [57–59]. *FAMA* is required for the terminal differentiation of guard cells [60]. *Glabra 1* (*GL1*) is an important regulator of trichome initiation in *Arabidopsis* [61]. The development of root

hairs and trichomes is regulated by *GL2, GL3, Enhancer of Glabra 3* (*EGL3*) and *Transparent Testa Glabra 1* (*TTG1*) with similar molecular mechanisms in *Arabidopsis* [62]. Here, a set of unigenes potentially involved in the biogenesis and development of these structures were identified (Table S5), which will facilitate to disentangle the formation of the specific traits such as hollow stomata and deep root system [18] and to further understand the molecular mechanism of drought avoidance strategy in desert plants.

Drought stress can promote plants flowering earlier [63–65]. In our field survey, flowering time of *R. soongorica* obviously varied between the populations along a precipitation gradient. The plants in Haishiwan, Gansu province (annual precipitation 600 mm) were still flowering in November, while the plants in Shashichang, Gansu province (precipitation 180 mm) had finished flowering and started to disperse seeds in September (data not shown). We propose that the genes control flowering time might have undergone strong natural selection in *R. soongorica* populations from extreme arid regions. Finally, including all the core components in circadian rhythm in plants KEGG pathway (Ko04712, File S1), 29 out of 51 flowering time genes in *Arabidopsis* can hit their homologous unigenes in *R. soongorica* (Table S5). Further investigating the correlation between the genetic diversity of these candidate genes and the variation of flowering time along drought gradient could shed light on how *R. soongorica* adapts to natural arid environment by drought escape strategy.

The molecular response to drought stress has been studied intensively in model plants, and at least two pathways are suggested to be involved into drought tolerance: ABA-dependent and -independent pathways (reviewed in [66,67]). In this study, most of the key genes related to ABA biosynthesis, catabolic and receptor complex were identified by KEGG annotation and local BLAST (File S3 and Table S5), such as rate-limiting enzyme *9-cis-epoxycarotenoid dioxygenases* (*NCED*) in the biosynthetic pathway [68], receptor complex components *pyrabactin resistance/PYR1-like/regulatory components of ABA receptor* (*PYR/PYL/RCAR*) [69], *protein phosphatase 2Cs* (*PP2Cs*) to sensor ABA signal [70], and important kinases and transcription factors (i.e. *SNF1-related protein kinase 2s* (*SnRK2s*) [71], *ABRE-binding factor/ABA-responsive elements* (*ABF/AREBs*) [72]). All of these key unigenes may function in ABA signal transduction in response to drought stress as in model plants. Furthermore, several transcription factors like *Dehydration Responsive Element Binding/C-repeat Binding Factors* (*DREB/CBFs*) which are involved in ABA-independent pathways were identified by local BLASTing (Table S5). These genes will help to uncover the molecular basis of physiological responses to drought stress in *R. soongorica*.

The C_4 pathway has been acknowledged to be more adaptive than the C_3 pathway in response to abiotic stresses, such as high temperature, radiation and drought [73,74]. The C_3 and C_4 photosynthesis can occur simultaneously in different leaves within the same plant [75], and the transition between the C_3 and C_4 pathways can also occur in some C_3 plants under some conditions (i.e. *Eleocharis vivipara* [76], *Flaveria brownii* [77]). Nevertheless, the origin of C_4 pathway and the transition between the C_3 and C_4 pathways remained elusive because of the absence of genetic evidence. *R. soongorica* is a C_3 plant according to its physiological characteristics [32]. In this study, all of the genes encode key enzymes in C_4 carbon fixation pathway were presented in the transcriptomic dataset from the annotation of KEGG (File S2). The lengths of the C_4 genes were not statistically different with the C_3 genes (ANOVA $p = 0.83$), but the C_3 genes were significantly abundant than C_4 genes ($p = 2.7e-05$, Figure 6). This is partially concordant to the previous studies which characterized *R.*

A

B

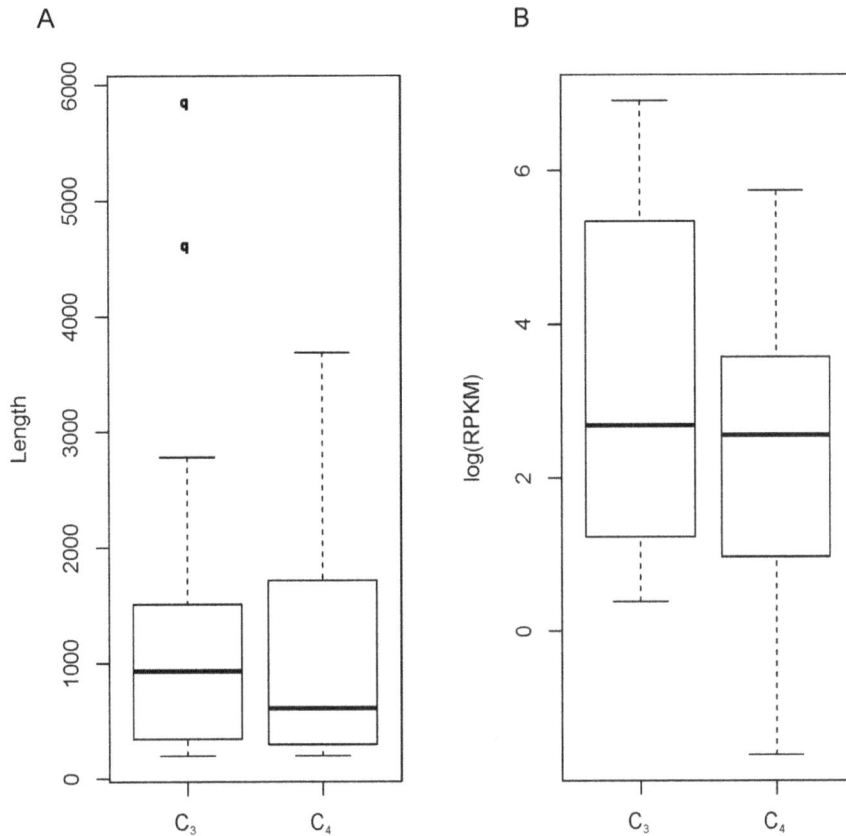

Figure 6. Boxplot distribution of C$_3$ and C$_4$ unigenes. (A) Length distribution inferred with log(Length). (B) Expression level inferred by log(RPKM) value.

soongorica as a C$_3$ plant [32,78,79]. To our knowledge, the present and active of the C$_4$ genes in *R. soongorica* might be the first transcriptomic evidence to support that a Tamaricaceae plant could also orchestrate the C$_3$ and C$_4$ pathways in response to environmental changes [76,80,81].

Conclusion and Perspectives

Desert plants have attracted more and more attentions, because they contain plenty of potential genetic resources to adapt to the extremely harsh conditions in their habitats. In the recent years, transcriptome sequencing became a most powerful and efficient approach to uncover genomic information in non-model organisms [82,83], few studies (but see [84]) focused on exploitation of the molecular basis of drought adaptation of desert plants. In present study, 77,647 unigenes were generated from a Tertiary relic species *R. soongorica* with the Illumina HiSeq^{TM} 2000 platform and more than half of unigenes has been annotated. At least 123 candidate genes related to drought adaptation were identified by the KEGG annotation and local BLAST, and population genetic study on these candidate genes will help us to better understand the adaptive evolutionary mechanism of *R. soongorica*. Expression and function analysis of the un-annotated unigenes will be also employed to unravel the specific drought adaptation mechanism in *R. soongorica*. These endeavors will advance our knowledge of a dominant plant species coping with the global climate change in the fragile desert ecosystems.

Materials and Methods

Ethics Statement

R. soongorica is a species widely distributed in Shashichang, Jingtai County, Gansu province and other arid regions, and it is not included into any list of endangered or protected species. Before collecting the samples, an oral permission was got from the local management of forestry after applying with introduction letters of CAREERI (Cold and Arid Regions Environmental and Engineering Research Institute, Chinese Academy of Sciences).

Plant Materials

Leaves and inflorescences of *R. soongorica* (Figure S4A) were collected in wild field, Shashichang, Jingtai County, Gansu province, northwest of China ($37°21'41''N,104°8'11''E$), where the average annual precipitation is 180 mm from 1950 to 2000 (http://www.worldclim.org/). Tissues were immediately frozen in liquid nitrogen for later extraction of total RNA. *R. soongorica* seeds from the same place were planted on damp filter papers and incubated at $4°C$ for 4 days before being placed at $23°C$ under long-day (16 h light/8 h dark) condition with a photosynthetic photon flux density of 150 μmol m^{-2} s^{-1}. Two weeks after germination, seedlings were harvested for RNA isolation (Figure S4B).

RNA Extraction, Library Preparation and RNA-seq

Total RNA was extracted with E.Z.N.A® Plant RNA Kit (Omega Bio-tek, Doraville, GA, USA). The concentration and quality of each RNA sample were determined by NanoDrop

2000^{TM} micro-volume spectrophotometer (Thermo Scientific, Waltham, MA, USA) and gel electrophoresis. One sample of total RNA was extracted from mature plant tissues including flowers and leaves. Another sample was from whole seedlings including roots, hypocotyls, and cotyledons. The two RNA samples were pooled equally to construct the cDNA library with a final concentration of 610.4 ng/μl.

The poly (A) mRNA was enriched by magnetic Oligo (dT) beads, and then was interrupted into 200–700 nt fragments. Using these short fragments as templates, the first cDNA strand was synthesized by random hexamers primers, followed by the second-stand cDNA synthesis using DNA polymerase I (New England BioLabs) and RNase H (Invitrogen). The short DNA fragments were purified with a QiaQuick PCR extraction kit (QIAGEN Inc., Valencia, CA, USA) for following end repairing and tailing A. Then the DNA fragments were ligated to sequencing adaptors, and the DNA fragments with required length were purified by agarose gel electrophoresis and gathered by PCR amplification. Finally, a paired-end library with insert sizes of 200–700 bp was sequenced using Illumina HiSeq^{TM} 2000 with the average read length of 90 bp. The raw data are available in the ArrayExpress database, E-MTAB-1543 (http://www.ebi.ac.uk/arrayexpress/experiments/E-MTAB-1543/).

De novo Assembly and Expression Profiling

The clean reads were obtained after filtering adapter sequences, reads with 5% ambiguous sequences 'N', and low-quality reads (reads with a base quality less than Q20), with a custom PERL script. Then, Trinity, a package consisting of Inchworm, Chrysalis and Butterfly, was used to perform the de novo assembly of high-quality clean reads [33]. The command-line parameters used were "–seqType fq –min_contig_length 200–CPU 5–bflyHeapSpace-Max 4G –JM 20G". Short reads with overlapping sequences were firstly assembled by Inchworm to form the longest contigs without gaps. These contigs were pooled to build into de Brujin graphs by Chrysalis. According to the paired end information, Butterfly reconciled the de Bruijn graphs and output longer sequences without overlooking the possibility of splice forms and paralogous transcripts. Such sequences which cannot be extended on either end were defined as unigenes. After clustering, the unigenes were divided into clusters (prefix is "CL") and singletons (prefix is "Unigene").

RPKM of each unigene was normalized by ERANGE3.1 software to determine the unigene expression profiles [85].

Functional Annotation

All sequences were annotated by aligning with public protein and nucleotide databases, such as the NCBI Nr, Nt database, the Swiss-Prot protein database, the KEGG database, and the COG with an E-value cutoff of 1e-5. Based on the alignment results, the further annotation analysis with GO terms, which described biological processes, molecular functions and cellular components, was performed by Blast2GO software [35,86]. The distribution of the gene functions was plotted by WEGO [87].

Then non-coding RNA was retrieved from the Rfam database (http://rfam.sanger.ac.uk/). The sequences excluded non-conding RNAs were BLASTed against Pfam-A database (http://pfam.sanger.ac.uk/) for further seeking functional domains using a hidden Markov model (HMM) algorithm (version 26.0) [88].

Identification of Putative Candidate Genes Involved in Drought Adaptation

Generally, plants take three main strategies (drought escape, drought avoidance and drought tolerance) to adapt to drought conditions (reviewed in [41,42]). To uncover the potential candidate genes related to the three strategies in *R. soongorica*, genes in model plant *Arabidopsis* involved in flowering time (drought escape), epidermal development such as stomata, cuticle waxes, trichomes and root hairs (drought avoidance), and ABA and drought stress signals (drought tolerance), were selected to screen the potential orthologs from the unigene dataset by Local BLASTN with E-value cutoff of 1e-5 (ftp://ftp.ncbi.nlm.nih.gov/blast/executables/LATEST-BLAST/).

Supporting Information

Figure S1 Length distributions of contigs and unigenes. (A) Summary distribution of the lengths of the 173,700 assembled contigs (>100 bp, mean length = 321 bp, N50 = 532 bp). (B) Summary distribution of the lengths of the 77,647 assembled unigenes (>200 bp, mean length = 677 bp, N50 = 1109 bp).

Figure S2 The flavonoid pathway from KEGG annotation.

Figure S3 The distribution of "CL"-unigenes for *Reaumuria soongorica*. The x-axis represents the number of contigs a "CL"-unigene composed.

Figure S4 The picture of *Reaumuria soongorica*. (A) Adult *R. soongorica*. (B) Two-week old seedlings of *R. soongorica*.

Table S1 List of unigenes with significant BLASTN matches against Nr, Nt, Swiss-Prot and KEGG databases.

Table S2 The summary of GO annotation for *Reaumuria soongorica*.

Table S3 The summary of COG annotation for *Reaumuria soongorica*.

Table S4 The summary of KEGG annotation for *Reaumuria soongorica*.

Table S5 List of candidate genes involved in drought escape, drought avoidance and drought tolerance in *Reaumuria soongorica* (Unigenes appears multiple times were counted once in article).

File S1 The unigenes involved in circadian rhythm of *Reaumuria soongorica* from KEGG annotation.

File S2 The unigenes involved in carbon fixation in photosynthetic organisms of *Reaumurica soongorica* from KEGG annotation.

File S3 The unigenes involved in ABA biosynthetic and catabolic pathways and ABA receptors in *Reaumuria soongorica* from KEGG annotaton.

File S4 The Rfam and Pfam annotations of *Reaumuria soongorica* unigenes.

File S5 Multiple sequence alignment of *AtCRY1* and CL2023.Contigs by CLUSTALW.

File S6 Multiple sequence alignment of CL11.Contigs by CLUSTALW.

Acknowledgments

We thank the technical support for Illumina sequencing and initial data analysis from Beijing Genome Institute at Shenzhen, P. R. China, and appreciate the constructive and helpful suggestions from Xiao-Ru Wang and two anonymous reviewers.

Author Contributions

Conceived and designed the experiments: X-FM XY PZ. Performed the experiments: YS XY PZ X-FM. Analyzed the data: YS XY PZ X-FM. Contributed reagents/materials/analysis tools: XZ HY. Wrote the paper: YS XY PZ X-FM GC. Experimental help: XZ HY. Permission for use of laboratory: GC XZ XL HX.

References

1. Manel S, Joost S, Epperson BK, Holderegger R, Storfer A, et al. (2010) Perspectives on the use of landscape genetics to detect genetic adaptive variation in the field. Mol Ecol 19: 3760–3772.
2. Stapley J, Reger J, Feulner PGD, Smadja C, Galindo J, et al. (2010) Adaptation genomics: the next generation. Trends Ecol Evol 25: 705–712.
3. Liu JL, Li FR, Liu CA, Liu QJ (2012) Influences of shrub vegetation on distribution and diversity of a ground beetle community in a Gobi desert ecosystem. Biodivers Conserv 21: 2601–2619.
4. Li XR, Zhang P, Su YG, Jia RL (2012) Carbon fixation by biological soil crusts following revegetation of sand dunes in arid desert regions of China: A four-year field study. CATENA 97: 119–126.
5. Saul-Tcherkas V, Steinberger Y (2011) Soil microbial diversity in the vicinity of a Negev Desert shrub-*Reaumuria negevensis*. Microb Ecol 61: 64–81.
6. Slemnev NN, Gunin PD, Kazantseva TI (1994) On the problem of natural restitution of dominant plants in the ecosystems of the desert zone of Mongolia. Rastit Resur 30: 1–15.
7. Liu YB, Zhang TG, Li XR, Wang G (2007) Protective mechanism of desiccation tolerance in *Reaumuria soongorica*: Leaf abscission and sucrose accumulation in the stem. Sci China Ser C Life Sci 50: 15–21.
8. Xu L, Wang L, Yue M, Zhao GF, Wang YL (2003) Analysis of grey relatedness between the modular sturcture of *Reaumuria soongorica* population in the desert of Fukang,Xinjiang and the environmental factors. Acta Phytoecologica Sinica 27: 742–748.
9. Bai YF, Wu JG, Xing Q, Pan QM, Huang JH, et al. (2008) Primary production and rain use efficiency across a precipitation gradient on the Mongolia plateau. Ecology 89: 2140–2153.
10. Zhou Y, Pei ZQ, Su JQ, Zhang JL, Zheng YR, et al. (2012) Comparing soil organic carbon dynamics in perennial grasses and shrubs in a saline-alkaline arid region, northwestern China. PLoS One 7: e42927.
11. Ramadan T (1998) Ecophysiology of salt excretion in the xero-halophyte *Reaumuria hirtella*. New Phytol 139: 273–281.
12. Gorai M, Neffati M (2007) Germination responses of *Reaumuria vermiculata* to salinity and temperature. Ann Appl Biol 151: 53–59.
13. Wang XH, Zhang T, Wen ZN, Xiao HL, Yang ZJ, et al. (2011) The chromosome number, karyotype and genome size of the desert plant diploid *Reaumuria soongorica* (Pall.) Maxim. Plant Cell Rep 30: 955–964.
14. Yang XP, Rost KT, Lehmkuhl F, Zhu ZD, Dodson J (2004) The evolution of dry lands in northern China and in the Republic of Mongolia since the Last Glacial Maximum. Quaternary Int 118–119: 69–85.
15. Zhu ZD, Wu Z, Liu S, Di X (1980) An outline of Chinese deserts. Beijing: Science Press.
16. Yang XP, Scuderi L, Paillou P, Liu ZT, Li HW, et al. (2011) Quaternary environmental changes in the drylands of China - A critical review. Quaternary Sci Rev 30: 3219–3233.
17. Guo ZT, Ruddiman WF, Hao QZ, Wu HB, Qiao YS, et al. (2002) Onset of Asian desertification by 22 Myr ago inferred from loess deposits in China. Nature 416: 159–163.
18. Liu JQ, Qiu MX, Pu JC, Lu ZM (1982) The typical extreme xerophyte-*Reaumuria soongorica* in the desert of China. Act bot sinica 24: 485–488.
19. Chong PF, Li Y, Su SP, Gao M, Qiu ZJ (2010) Photosynthetic characteristics and their effect factors of *Reaumuria soongorica* on three geographical populations. Acta Ecol Sin 30: 914–922.
20. Finkelstein R, Gampala SS, Lynch TJ, Thomas TL, Rock CD (2005) Redundant and distinct functions of the ABA response loci ABA-insensitive(ABI)5 and ABRE-binding factor(ABF)3. Plant Mol Biol 59: 253–267.
21. Bai J, Gong CM, Chen K, Kang HM, Wang G (2009) Examination of antioxidative system's responses in the different phases of drought stress and during recovery in desert plant *Reaumuria soongorica* (Pall.) Maxim. J Plant Biol 52: 417–425.
22. Bai J, Gong CM, Wang G, Kang HM (2010) Antioxidative characteristics of *Reaumuria soongorica* under drought stress. Acta Bot Boreali-Occident Sin 30: 2444–2450.
23. Bai J, Xu DH, Kang HM, Chen K, Wang G (2008) Photoprotective function of photorespiration in *Reaumuria soongorica* during different levels of drought stress in natural high irradiance. Photosynthetica 46: 232–237.
24. Chong PF, Su SP, Li Y (2011) Comprehensive evaluation of drought resistance of *Reaumuria soongorica* from four geographical populations. Acta Prataculturae Sinica 20: 26–33.
25. Lv MT, Yang JY, Yang M, Zhang ZR, Ma X (2010) Effect of different drought stress conditions on germination of *Reaumuria soongorica* seeds. Chinese Journal of Grassland 32: 58–63.
26. Yan QD, Su PX, Gao S (2012) Response of photosynthetic characteristics of C₃ desert plant *Reaumuria soongorica* and C₄ desert plant Salsola passerina to different drought degrees. Journal of Desert Research 32: 364–371.
27. Zeng YJ, Wang YR, Zhuang GH, Yang ZS (2004) Seed germination responses of *Reaumuria soongorica* and *Zygophyllum xanthoxylum* to drought stress and sowing depth. Acta Ecol Sin 24: 1629–1634.
28. Xu L, Wang YL, Wang XM, Zhang LJ, Yue M, et al. (2003) Genetic structure of *Reaumuria soongorica* population in Fukang Desert, Xinjiang and its relationship with ecological factors. Acta Bot Sin 45: 787–794.
29. Li XL, Chen J, Wang G (2008) Spatial autocorrelation analysis of ISSR genetic variation of *Reaumuria soongorica* population in northwest of China. Journal of Desert Research 28: 468–472.
30. Qian ZQ, Xu L, Wang YL, Yang J, Zhao GF (2008) Ecological genetics of *Reaumuria soongorica* (Pall.) Maxim. population in the oasis-desert ecotone in Fukang, Xinjiang, and its implications for molecular evolution. Biochem Syst Ecol 36: 593–601.
31. Li ZH, Chen J, Zhao GF, Guo YP, Kou YX, et al. (2012) Response of a desert shrub to past geological and climatic change: A phylogeographic study of *Reaumuria soongorica* (Tamaricaceae) in western China. J Syst Evol 50: 351–361.
32. Ma JY, Chen T, Qiang WY, Wang G (2005) Correlations between foliar stable carbon isotope composition and environmental factors in desert plant *Reaumuria soongorica* (Pall.) Maxim. J Integr Plant Biol 47: 1065–1073.
33. Grabherr MG, Haas BJ, Yassour M, Levin JZ, Thompson DA, et al. (2011) Full-length transcriptome assembly from RNA-Seq data without a reference genome. Nat Biotechnol 29: 644–652.
34. Ashburner M, Ball CA, Blake JA, Botstein D, Butler H, et al. (2000) Gene Ontology: tool for the unification of biology. Nat Genet 25: 25–29.
35. Conesa A, Gotz S, Garcia-Gomez JM, Terol J, Talon M, et al. (2005) Blast2GO: a universal tool for annotation, visualization and analysis in functional genomics research. Bioinformatics 21: 3674–3676.
36. Winkel-Shirley B (2002) Biosynthesis of flavonoids and effects of stress. Curr Opin Plant Biol 5: 218–223.
37. Mouradov A, Cremer F, Coupland G (2002) Control of flowering time interacting pathways as a basis for diversity. The Plant Cell Online 14: S111–S130.
38. Böhlenius H, Huang T, Charbonnel-Campaa L, Brunner AM, Jansson S, et al. (2006) CO/FT regulatory module controls timing of flowering and seasonal growth cessation in trees. Science 312: 1040–1043.
39. Xu DH, Su PX, Zhang RY, Li HL, Zhao L, et al. (2010) Photosynthetic parameters and carbon reserves of a resurrection plant *Reaumuria soongorica* during dehydration and rehydration. Plant Growth Regul 60: 183–190.
40. Leung J, Giraudat J (1998) Abscisic acid signal transduction. Annu Rev Plant Physiol Plant Mol Biol 49: 199–222.
41. Chaves MM, Maroco JP, Pereira JS (2003) Understanding plant responses to drought - from genes to the whole plant. Funct Plant Biol 30: 239–264.
42. Mckay JK, Richards J, Mitchell-Olds T (2003) Genetics of drought adaptation in *Arabidopsis* thaliana: I. Pleiotropy contributes to genetic correlations among ecological traits. Mol Ecol 12: 1137–1151.
43. Xia ZH, Xu HM, Zhai JL, Li DJ, Luo HL, et al. (2011) RNA-Seq analysis and *de novo* transcriptome assembly of *Hevea brasiliensis*. Plant Mol Biol 77: 299–308.
44. Liu MY, Qiao GR, Jiang J, Yang H, Xie LH, et al. (2012) Transcriptome sequencing and *de novo* analysis for Ma Bamboo (*Dendrocalamus latiflorus* Munro) using the Illumina platform. PLoS One 7: e46766.

45. El-Assal SED, Alonso-Blanco C, Peeters AJM, Raz V, Koornneef M (2001) A QTL for flowering time in *Arabidopsis* reveals a novel allele of CRY2. Nat Genet 29: 435–440.

46. Mas P, Devlin PF, Panda S, Kay SA (2000) Functional interaction of phytochrome B and cryptochrome 2. Nature 408: 207–211.

47. Huang LL, Yang X, Sun P, Tong W, Hu SQ (2012) The first Illumina-based *de novo* transcriptome sequencing and analysis of Safflower Flowers. PLoS One 7: e38653.

48. Xu DL, Long H, Liang JJ, Zhang J, Chen X, et al. (2012) *De novo* assembly and characterization of the root transcriptome of *Aegilops variabilis* during an interaction with the cereal cyst nematode. BMC Genomics 13: 133.

49. Novaes E, Drost DR, Farmerie WG, Pappas GJ, Jr., Grattapaglia D, et al. (2008) High-throughput gene and SNP discovery in *Eucalyptus grandis*, an uncharacterized genome. BMC Genomics 9: 312.

50. Flavell AJ (1995) Retroelements, reverse transcriptase and evolution. Comp Biochem Physiol B Biochem Mol Biol 110: 3–15.

51. Xiong Y, Eickbush TH (1990) Origin and evolution of retroelements based upon their reverse-transcriptase sequences. EMBO J 9: 3353–3362.

52. Soltis DE, Soltis PS, Chase MW, Mort ME, Albach DC, et al. (2000) Angiosperm phylogeny inferred from 18S rDNA, rbcL, and atpB sequences. Bot J Linn Soc 133: 381–461.

53. The Angiosperm Phylogeny Group (2003) An update of the Angiosperm Phylogeny Group classification for the orders and families of flowering plants: APG II. Bot J Linn Soc 141: 399–436.

54. Yang J, Isabel Ordiz M, Jaworski JG, Beachy RN (2011) Induced accumulation of cuticular waxes enhances drought tolerance in *Arabidopsis* by changes in development of stomata. Plant Physiol Biochem 49: 1448–1455.

55. Chen GX, Komatsuda T, Ma JF, Nawrath C, Pourkheirandish M, et al. (2011) An ATP-binding cassette subfamily G full transporter is essential for the retention of leaf water in both wild barley and rice. Proc Natl Acad Sci U S A 108: 12354–12359.

56. Ishida T, Kurata T, Okada K, Wada T (2008) A genetic regulatory network in the development of trichomes and root hairs. Annu Rev Plant Biol 59: 365–386.

57. Pillitteri LJ, Sloan DB, Bogenschutz NL, Torii KU (2006) Termination of asymmetric cell division and differentiation of stomata. Nature 445: 501–505.

58. Schellmann S, Hulskamp M (2005) Epidermal differentiation: trichomes in *Arabidopsis* as a model system. Int J Dev Biol 49: 579–584.

59. Bruex A, Kainkaryam RM, Wieckowski Y, Kang YH, Bernhardt C, et al. (2012) A gene regulatory network for root epidermis cell differentiation in *Arabidopsis*. PLoS Genet 8: e1002446.

60. Ohashi-Ito K, Bergmann DC (2006) *Arabidopsis* FAMA controls the final proliferation/differentiation switch during stomatal development. Plant Cell 18: 2493–2505.

61. Schnittger A, Jurgens G, Hulskamp M (1998) Tissue layer and organ specificity of trichome formation are regulated by GLABRA1 and TRIPTYCHON in *Arabidopsis*. Development 125: 2283–2289.

62. Tominaga-Wada R, Ishida T, Wada T (2011) New Insights into the Mechanism of Development of *Arabidopsis* Root Hairs and Trichomes. Int Rev Cel Mol Bio 286: 67–106.

63. Fox GA (1990) Drought and the evolution of flowering time in desert annuals. Am J Bot 77: 1508–1518.

64. Franks SJ, Sim S, Weis AE (2007) Rapid evolution of flowering time by an annual plant in response to a climate fluctuation. Proc Natl Acad Sci U S A 104: 1278–1282.

65. Nevo E, Fu YB, Pavlicek T, Khalifa S, Tavasi M, et al. (2012) Evolution of wild cereals during 28 years of global warming in Israel. Proc Natl Acad Sci U S A 109: 3412–3415.

66. Shinozaki K, Yamaguchi-Shinozaki K (2007) Gene networks involved in drought stress response and tolerance. J Exp Bot 58: 221–227.

67. Xiong LM, Schumaker KS, Zhu JK (2002) Cell signaling during cold, drought, and salt stress. Plant Cell 14: S165–S183.

68. Nambara E, Marion-Poll A (2005) Abscisic acid biosynthesis and catabolism. Annu Rev Plant Biol 56: 165–185.

69. Gonzalez-Guzman M, Pizzio GA, Antoni R, Vera-Sirera F, Merilo E, et al. (2012) *Arabidopsis* PYR/PYL/RCAR receptors play a major role in Quantitative regulation of stomatal aperture and transcriptional response to abscisic acid. The Plant Cell Online 24: 2483–2496.

70. Ma Y, Szostkiewicz I, Korte A, Moes D, Yang Y, et al. (2009) Regulators of PP2C phosphatase activity function as abscisic acid sensors. Science 324: 1064–1068.

71. Fujita Y, Nakashima K, Yoshida T, Katagiri T, Kidokoro S, et al. (2009) Three SnRK2 protein kinases are the main positive regulators of abscisic acid signaling in response to water stress in *Arabidopsis*. Plant Cell Physiol 50: 2123–2132.

72. Kim SY (2006) The role of ABF family bZIP class transcription factors in stress response. Physiol Plant 126: 519–527.

73. Taiz L, Zeiger E (1991) Plant Physiology. Redwood City: The Benjamin/Cammings Publishing Company, Inc.

74. Raven PH, Evert RF, Eichhorn SE (1992) Biology of Plants. New York: Worth Publishers, Inc.

75. Raghavendra AS, Rajendrudu G, Das VSR (1978) Simultaneous occurrence of C_3 and C_4 photosyntheses in relation to leaf position in *Mollugo nudicaulis*. Nature 273: 143–144.

76. Ueno O (1998) Induction of Kranz anatomy and C_4-like biochemical characteristics in a submerged amphibious plant by abscisic acid. Plant Cell 10: 571–583.

77. Cheng SH, Demoore B, Wu JR, Edwards GE, Ku MSB (1989) Photosynthetic plasticity in *Flaveria Brownii* growth irradiance and the expression of C_4 photosynthesis. Plant Physiol 89: 1129–1135.

78. Sage RF, Christin PA, Edwards EJ (2011) The C_4 plant lineages of planet Earth. J Exp Bot 62: 3155–3169.

79. Sage RF, Sage TL, Kocacinar F (2012) Photorespiration and the evolution of C_4 photosynthesis. Annu Rev Plant Biol 63: 17.11–17.29.

80. Brautigam A, Kajala K, Wullenweber J, Sommer M, Gagneul D, et al. (2011) An mRNA blueprint for C_4 photosynthesis derived from comparative transcriptomics of closely related C_3 and C_4 species. Plant Physiol 155: 142–156.

81. Williams BP, Aubry S, Hibberd JM (2012) Molecular evolution of genes recruited into C_4 photosynthesis. Trends Plant Sci 17: 213–220.

82. Wang Z, Gerstein M, Snyder M (2009) RNA-Seq: a revolutionary tool for transcriptomics. Nat Rev Genet 10: 57–63.

83. Ekblom R, Galindo J (2011) Applications of next generation sequencing in molecular ecology of non-model organisms. Heredity (Edinb) 107: 1–15.

84. Zhou YD, Gao FD, Liu RM, Feng JD, Li HD (2012) *De novo* sequencing and analysis of root transcriptome using 454 pyrosequencing to discover putative genes associated with drought tolerance in *Ammopiptanthus mongolicus*. BMC Genomics 13: 266.

85. Mortazavi A, Williams BA, Mccue K, Schaeffer L, Wold B (2008) Mapping and quantifying mammalian transcriptomes by RNA-Seq. Nature Methods 5: 621–628.

86. Conesa A, Gotz S (2008) Blast2GO: A comprehensive suite for functional analysis in plant genomics. Int J Plant Genomics 2008: 619832.

87. Ye J, Fang L, Zheng H, Zhang Y, Chen J, et al. (2006) WEGO: a web tool for plotting GO annotations. Nucleic Acids Res 34: W293–W297.

88. Punta M, Coggill PC, Eberhardt RY, Mistry J, Tate J, et al. (2012) The Pfam protein families database. Nucleic Acids Res 40: D290–D301.

7

Detecting Leaf Pulvinar Movements on NDVI Time Series of Desert Trees: A New Approach for Water Stress Detection

Roberto O. Chávez[1]*, **Jan G. P. W. Clevers**[1], **Jan Verbesselt**[1], **Paulette I. Naulin**[2], **Martin Herold**[1]

1 Laboratory of Geo-Information Science and Remote Sensing, Wageningen University, Wageningen, The Netherlands, **2** Laboratorio de Biología de Plantas, Departamento Silvicultura y Conservación de la Naturaleza, Universidad de Chile, Santiago, Chile

Abstract

Heliotropic leaf movement or leaf 'solar tracking' occurs for a wide variety of plants, including many desert species and some crops. This has an important effect on the canopy spectral reflectance as measured from satellites. For this reason, monitoring systems based on spectral vegetation indices, such as the normalized difference vegetation index (NDVI), should account for heliotropic movements when evaluating the health condition of such species. In the hyper-arid Atacama Desert, Northern Chile, we studied seasonal and diurnal variations of MODIS and Landsat NDVI time series of plantation stands of the endemic species *Prosopis tamarugo* Phil., subject to different levels of groundwater depletion. As solar irradiation increased during the day and also during the summer, the paraheliotropic leaves of Tamarugo moved to an erectophile position (parallel to the sun rays) making the NDVI signal to drop. This way, Tamarugo stands with no water stress showed a positive NDVI difference between morning and midday ($\Delta NDVI_{mo-mi}$) and between winter and summer ($\Delta NDVI_{W-S}$). In this paper, we showed that the $\Delta NDVI_{mo-mi}$ of Tamarugo stands can be detected using MODIS Terra and Aqua images, and the $\Delta NDVI_{W-S}$ using Landsat or MODIS Terra images. Because pulvinar movement is triggered by changes in cell turgor, the effects of water stress caused by groundwater depletion can be assessed and monitored using $\Delta NDVI_{mo-mi}$ and $\Delta NDVI_{W-S}$. For an 11-year time series without rainfall events, Landsat $\Delta NDVI_{W-S}$ of Tamarugo stands showed a positive linear relationship with cumulative groundwater depletion. We conclude that both $\Delta NDVI_{mo-mi}$ and $\Delta NDVI_{W-S}$ have potential to detect early water stress of paraheliotropic vegetation.

Editor: Guy J-P. Schumann, NASA Jet Propulsion Laboratory, United States of America

Funding: This work has been funded by the CONICYT (Chile)-Wageningen University scholarship (Res. Ex. N. 281/2009). The funders had no role in study design, data collection and analysis, decision to publish, or preparation of the manuscript.

Competing Interests: The authors have declared that no competing interests exist.

* Email: roberto.chavez@wur.nl

Introduction

Heliotropism or 'solar tracking' is the ability of many desert plant species and crops to move leaves and flowers as a response to changes in the position of the sun throughout the day [1]. There are two types of heliotropic movements: diaheliotropic movements in which leaves adjust the leaf lamina to face direct solar irradiation, and paraheliotropic movements, in which leaves adjust to avoid facing incoming radiation by contraction of pulvinar structures located at the base of leaves [1–3]. Paraheliotropic movements are triggered by directional solar irradiation and allow partial regulation of the incident irradiance on the leaves. For desert species, the regulation of the solar irradiation intensity on the leaves is an important adaptation to avoid photosynthesis saturation (photoinhibition) and to enhance the water use efficiency [4–6].

Diurnal paraheliotropic movements have a direct impact on the canopy reflectance properties of vegetation [7–9]. Therefore, spectral vegetation indices derived from remote sensing data, like the normalized difference vegetation index (NDVI), can significantly vary during the day and during the year due to leaf movement as solar irradiation changes. Nevertheless, no studies have quantified the effects of solar tracking by plants on the NDVI

signal recorded from satellites. Diurnal leaf movements of Tamarugo plants (*Prosopis tamarugo* Phil.) were first described by Chávez et al. [9] under laboratory conditions, and later by Chávez et al. [8] for adult trees in the field. These diurnal leaf movements corresponded to paraheliotropic movements since the leaves moved to an erectophyle leaf distribution (facing away from the sun) around midday when solar irradiation was maximum. The paper of Chávez et al. [8] showed that leaf pulvinar movements caused diurnal changes of Tamarugo's canopy spectral reflectance and NDVI signal, which was negatively correlated to diurnal solar irradiation values.

In the present study, we hypothesize that the effects of Tamarugo's diurnal leaf pulvinar movements on the NDVI can also be recorded by remote sensors from space, since the acquisition time of the different sensors differ. A high solar irradiation at midday is assumed to cause a lower NDVI than in the morning as indicated in Figure 1a. In this context the MODIS (Moderate Resolution Imaging Spectroradiometer) sensor seems to be especially suitable to capture this difference in NDVI between morning (low solar irradiation, high NDVI) and midday (high solar irradiation, low NDVI), since the MODIS sensor on board of the Terra satellite acquires data for the study site at 10 a.m. (local

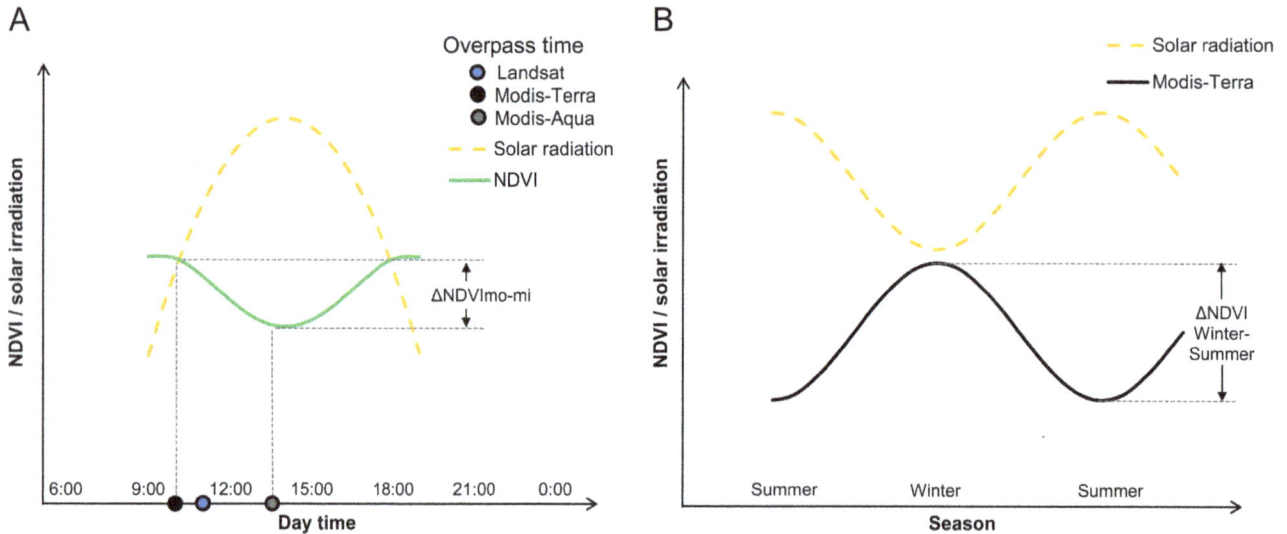

Figure 1. Conceptual diagram of the effect of leaf pulvinar movement on the NDVI signal. (A) NDVI difference between morning and midday ($\Delta NDVI_{mo\text{-}mi}$) occuring as solar irradiation changes during the day, and (B) NDVI difference between winter and summer ($\Delta NDVI_{W\text{-}S}$) occuring as solar irradiation varies between seasons. The time at which the Landsat (5–7), MODIS-Terra, and MODIS-Aqua satellites acquire data is displayed to illustrate the impact of pulvinar movements on the NDVI retrieved from these platforms.

time) and the MODIS sensor on board of the Aqua satellite acquires data at 1.30 p.m. (local time). Thus, the NDVI difference between morning and midday ($\Delta NDVI_{mo\text{-}mi}$) can be calculated as the difference between the NDVI MODIS Terra and the NDVI MODIS Aqua.

Considering the negative correlation between diurnal NDVI measurements and solar irradiation reported by Chávez et al. [8], we expect also seasonal NDVI variations associated with seasonal changes in solar irradiation with peaks in winter when the solar irradiation is the lowest. We hypothesize that this effect can also be recorded by sensors from space as indicated in Figure 1b. In this case, the Landsat TM (Thematic Mapper) and ETM (Enhanced Thematic Mapper) catalogue seems to be very suitable for studying NDVI seasonal variations since it offers one of the longest existing time series of systematically recorded satellite data worldwide [10]. Besides, the Landsat catalogue is considered the most relevant satellite dataset for ecological applications and environmental monitoring [11,12]. MODIS data might also be used to study seasonal effects of the pulvinar movements on the NDVI signal, providing images with a coarser spatial resolution (250 meters vs the 30 meters of Landsat), but with a higher temporal resolution (daily) enabling near real time vegetation monitoring [13]. However, the MODIS time series is considerably shorter than the Landsat time series (only since 2000).

Tamarugo is an endemic tree of the hyper-arid Atacama Desert, Northern Chile, a location considered among the most extreme environments for life [14,15]. The Tamarugo forest, locally known as Pampa del Tamarugal, sustains a biodiversity of about 40 species of plants and animals, some of them endemic for this particular ecosystem [16–19]. Precipitation events are very rare and the only source of water supply for vegetation is the groundwater (GW), from which Tamarugo is completely dependent. However, not only Tamarugo trees are demanding water: the main economic activity in Atacama is mining, which is also demanding water for human consumption and for many industrial processes. This has led to an overexploitation of the GW sources and a progressive depletion of the GW over the whole Pampa del Tamarugal [20].

The natural Tamarugo forest was almost extinct in the 19th century and during the 1970's an enormous reforestation effort was carried out by the Chilean government and 13,000 hectares of Tamarugo were planted in the Pampa del Tamarugal basin [21]. Currently, the Pampa del Tamarugal is under threat due to GW overexploitation. Chilean policy makers, scientists and private companies have debated intensively about defining environmentally safe GW extractions. To achieve this, good indicators of the Tamarugo water condition are needed and remote sensing, and specifically the NDVI, has proved to be useful for assessing Tamarugo's water condition [8,9]. Nevertheless, time series of NDVI have not been directly related to GW depletion yet and to do so, the effect of the leaf pulvinar movements must be considered to understand a) the natural NDVI dynamic in the absence of water stress, and b) how this dynamic may be altered by GW depletion. In this paper, we use MODIS and Landsat NDVI time series to study both the natural and the altered NDVI dynamics of Tamarugo stands located in the Pampa del Tamarugal basin. Furthermore, we explore other biological (phenology) and environmental factors (precipitation) with potential effects on the NDVI signal.

Material and methods

2.1 Species description

Tamarugo is a phreatophytic desert tree that is highly specialized to survive the hyper-arid conditions of the Atacama Desert. This species belongs to the Leguminoseae family, Mimosaceae subfamily and it can reach up to 25 meters height, 20–30 meters crown size and 2 meters stem diameter [22,23]. The branches are arched and twigs flexuous with composite leaves, often bipinnate with 6–15 pairs of folioles (Figure 2b, c, f) [24]. The Tamarugo petioles have a distinctive structure of motor cells in the pulvinus, responsible for the leaf paraheliotropic movements (Figure 2d, e, f). Differential turgor changes of the pulvinus cells make the leaves to stand up and orientate the leaf lamina parallel to the incoming sun rays. The composite leaves of Tamarugo have three levels of pulvinar structures: the first at the base of the

Figure 2. Pulvinar structures of Prosopis tamarugo leaves. (A) Tamarugo trees, (B) leaf angle randomly distributed during the morning when the solar radiation is low, (C) leaf angle in erectophyle position to avoid facing high solar irradiation at midday, (D) transversal section of a closed pulvinus (empty of water) during the morning, (E) transversal section of an open pulvinus (filled with water), which allows leaves to stand up and reach the erectophyle position, and (F) detail of the base of a Tamrugo pinna showing the three levels of pulvinar structures (at the base of the bipinna, of each pinna and each foliole).

bipinna, the second at the base of each pinna, and the third at the base of each of the folioles. This pulvinar mechanism at the three levels allows the Tamarugo canopy to adjust its internal structure to avoid facing excessive solar irradiation. Tamarugos are phreatophytic species [25,26] presenting a dual root system consisting of a deep taping root and a dense superficial root mat [27]. This dual system would allow Tamarugos to move water from the deep groundwater table to the superficial root mat layer during the night to ensure water supply during the growing season when the water demand at the capillary fringe increases [26].

2.2 Study area

The study area is located in the Atacama Desert (Northern Chile), specifically in the southern part of the Pampa del Tamarugal basin, where most of the remaining Tamarugo population is concentrated (Figure 3). The Tamarugo forest is practically the only ecosystem of the Absolute Desert eco-region [16], and it is characterized by almost null precipitation, high day-night temperature oscillation, and high potential evapotranspiration [28,29]. Most of the plantation stands (Pintados and Bellavista) are in the southern part of the basin and within the study area. Just little natural patches of Tamarugo remain in the

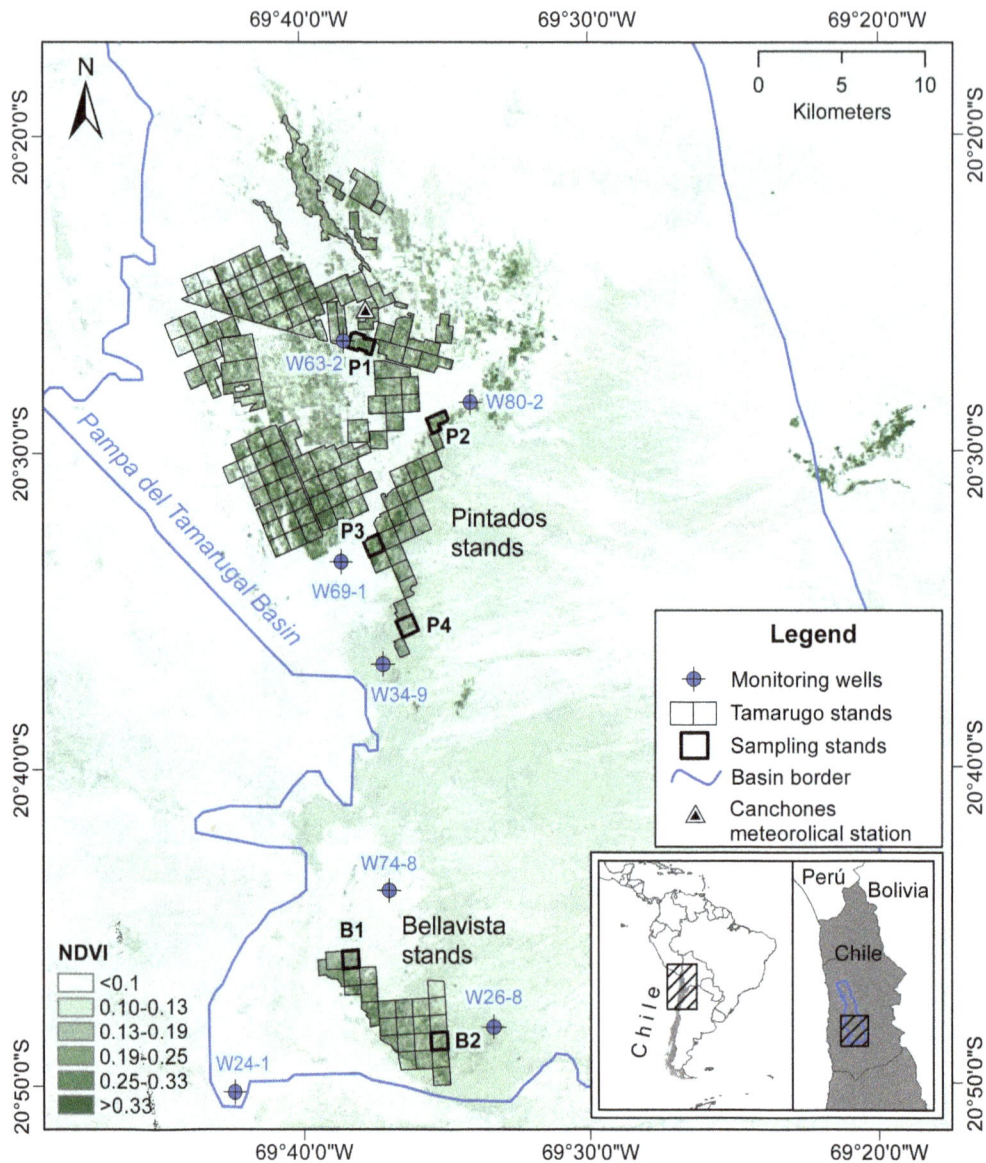

Figure 3. Landsat NDVI image showing the location of the Tamarugo stands (Winter 2007).

northern portion of the Pintados plantation (Figure 3). Although the oldest plantation stands were established as early as 1936, most of the existing plantation stands were planted between 1968 and 1972 [30]. The plantation scheme consisted of squared 1×1 kilometres stands and trees separated 10×10 meters. Besides Tamarugo plantations, there are plantations of other *Prosopis* species, sometimes mixed with Tamarugo. Only pure Tamarugo plantation stands and some natural forest patches were considered in this study and they can be identified in Figure 3 as the green areas highlighted in black.

2.3 Landsat and MODIS NDVI time series

We used all available Landsat 5 TM and Landsat 7 ETM data (referred from here onwards in the text as 'Landsat' data) as well as MODIS-Terra and MODIS-Aqua data of the study area covering the period 1989–2012. We selected this time frame since this is the period of time with available GW depth records for most of the monitoring wells located in the study area (Figure 3). For the

Landsat NDVI time series we used cloud free L1T images of 30 meters pixel resolution (471 scenes) corresponding to path 1 and row 34 and pre-processed using the Landsat Ecosystem Disturbance Adaptive Processing System (LEDAPS) to obtain surface reflectance values for all spectral bands [31]. Finally, we used the surface reflectance values of red and NIR to compute the NDVI for each date as follows: NDVI = (NIR-Red)/(NIR+Red). For the MODIS-Terra and MODIS-Aqua NDVI time series we used the MODIS 16-day composites at 250 meters pixel resolution (MOD13Q1 and MYD13Q1 data products). MODIS pixel reliability showed that 85% of the observations can be used with confidence (reliability = 0) and 15% were considered useful (reliability = 1) of which MODIS vegetation index quality indicated average aerosol quantity. MODIS pixels with reliability 0 and 1 were considered in this study and showed consistent values for the NDVI time series of all forest stands. Both MODIS and Landsat data were downloaded from the USGS Earth Explorer website. Complementary, we used a panchromatic WorldView2 image of

Table 1. Plantation stands close to monitoring wells in the Pampa del Tamarugal basin.

Stand	Plantation year	Canopy coverage (%)	Closest monitoring well (DGA code)	Distance to well (km)	Groundwater depth (m)			
					1989	1997	2007	2012
B1	1968–1969	17	017000-74-8	4.7	10.99*	11.19*	11.64*	11.90*
			017000-26-8	9.5				
			017000-24-1	10.1				
B2	1968–1969	21	017000-74-8	9.3	14.75*	14.87*	15.45*	15.82*
			017000-26-8	3.4				
			017000-24-1	12.8				
P1	unknown	22	017000-63-2	0.4	7.97	9.03	9.93	10.26
P2	1972	25	017000-80-2	1.6	9.90	11.91	12.97	13.11
P3	1972	27	017000-69-1	1.7	5.70	6.26	6.87	6.89
P4	1972	11	017000-34-9	1.9	7.93	8.65	9.54	9.55

(*)Groundwater depth and depletion estimated using inverse distance weighted interpolation of 3 neighbouring wells.

0.6 meters pixel resolution to quantify the tree coverage of each plantation stand. This was carried out by using object-based image classification and the eCognition software following the procedure used by Chávez et al. [8].

2.4 Groundwater and climatic data

GW records were obtained from the monitoring network of the Dirección General de Aguas (DGA), the Chilean Water Service. From this network, seven wells were close to the Tamarugo stands and had enough records to establish a direct relationship between the groundwater table and the forest status (Figure 3). We averaged the (three to twelve) records of each year to obtain annual values of groundwater depth for the seven monitoring wells. Figure 3 shows the location of the monitoring wells used in this study as well as the forest stands located close to each well. Basic data of each monitoring well are provided in Table 1. This way we obtained representative groundwater data for six Tamarugo stands for the period 1989–2012. In the case of stands B1 and B2, the groundwater depth was estimated using an inverse distance weighted interpolation of records from three wells (see Table 1).

Although groundwater is the main water source of the Tamarugo forest, sporadic precipitation may occur in the Atacama Desert, having a positive impact on the water status of the trees and the NDVI signal. For this reason, we included in our analysis precipitation records from the DGA meteorological station Huara en Fuerte Baquedano (20°07'51"S, 69°44'59"W) located about 30 km north from the study area and at a similar altitude (1,100 m). Solar irradiation records were obtained from the Canchones Experimental Station of the Universidad Arturo Prat (Chile), located next to the Tamarugo stand P1 in the northern part of the study area (Figure 3).

2.5 Data analysis

2.5.1 NDVI signal in the absence of water stress (natural dynamic). Finding Tamarugo vegetation without nearby GW depletion in the Atacama Desert was a difficult task. We identified a Tamarugo forest stand (B1) and a time frame (2005-2008) with almost null GW depletion and no precipitation events in the southern part of the study area (see Figure 3, Bellavista stand). We assumed the NDVI time series of this three year period was not

strongly influenced by the growth of trees. The Landsat, MODIS-Terra and MODIS-Aqua NDVI time series for the stand B1 were calculated using the median value of the pixels inside the 1×1 km stand. This aggregation enabled direct comparison of Landsat and MODIS NDVI time series. In the case of the Landsat time series we excluded pixels with NDVI values lower than 0.13, which were considered as no forest pixels. This threshold was set by considering the NDVI values observed outside the plantation stands, which correspond to completely bare areas (Figure 3). We first analysed the time series without any level of temporal aggregation, and then we aggregated the values to monthly averages in order to study the relationship between the NDVI and the monthly mean solar irradiation. For the latter purpose we used simple linear regression.

2.5.2 NDVI signal under water stress. After studying the natural dynamic of the NDVI time series for the three satellite sensors, we analysed the relationship between the average annual records of GW depletion and different metrics derived from the NDVI signal. To achieve this we used simple linear regression between the cumulative GW depletion and the NDVI derived metrics of the period between 1997 and 2007 with no precipitation. For the Landsat NDVI time series, these metrics were: annual NDVI average ($NDVI_{av}$), NDVI in winter ($NDVI_W$), and the NDVI difference between winter and summer ($\Delta NDVI_{W-S}$). For MODIS NDVI time series, these metrics were the $\Delta NDVI_{W-S}$ and the NDVI difference between morning and midday ($\Delta NDVI_{mo-mi}$). We calculated $\Delta NDVI_{W-S}$ and $\Delta NDVI_{mo-mi}$ for each year as follows:

1) $MODIS\ \Delta NDVI_{W-S} = MODIS\ Terra\ NDVI_W - MODIS\ Terra\ NDVI_S$,

2) $Landsat\ \Delta NDVI_{W-S} = Landsat\ NDVI_W - Landsat\ NDVI_S$, and

3) $\Delta NDVI_{mo-mi} = MODIS\ Terra\ NDVI_W - MODIS\ Aqua\ NDVI_W$.

Where:

$NDVI_W$
$= average\ of\ all\ NDVI\ scenes\ of\ May,\ June,\ July(winter)$

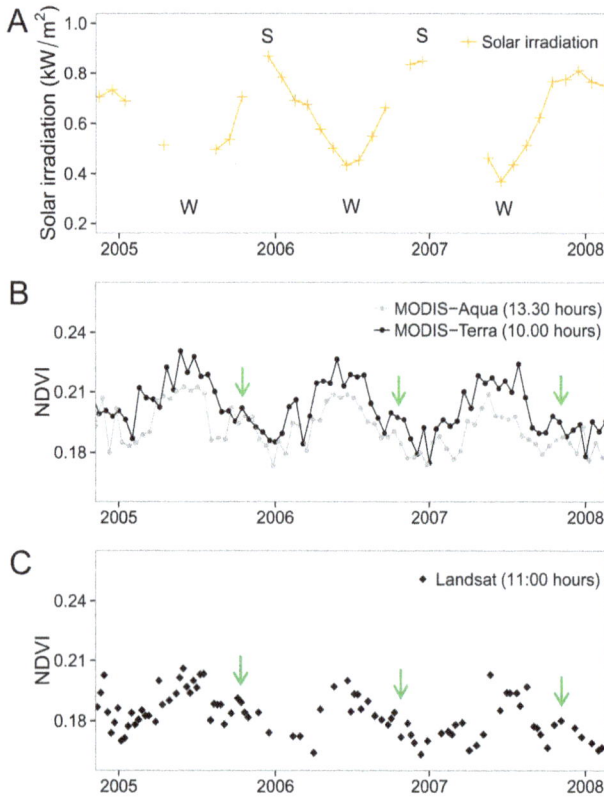

Figure 4. Time series of solar irradiation and NDVI for the B1 site (low groundwater depletion). (A) Solar irradiation, (B) MODIS 16 days composite NDVI, and (C) Landsat NDVI of the B1 site. Arrows indicate the peak of Tamarugo's vegetative period. S = summer, W = winter.

$NDVI_S$

$= average\ of\ all\ NDVI\ scenes\ of\ November,\ December,\ January(summer)$

In the case of the Landsat NDVI time series, we considered a minimum of three scenes for the summer and winter period to obtain a representative value of the respective season. For all Landsat and MODIS NDVI scenes we used the median value of the pixels inside the 1 × 1 km stands.

Results

3.1 Leaf pulvinar movement and the NDVI natural dynamic

In the absence of GW extraction or precipitation events, the NDVI signal of the Tamarugo stand B1 presented a strong seasonal variation for the period 2005–2008, mainly explained by the seasonal variation of the monthly average solar irradiation (Figure 4) influencing the pulvinar movement of paraheliotropic plants (Figure 1). The R^2 for the linear relationship between NDVI and solar irradiation was 0.66 for the Landsat NDVI time series, 0.65 for the MODIS-Terra NDVI, and 0.41 for the MODIS-Aqua NDVI. Partial foliage loss during the period May-September and the peak of the vegetative period occurring around October seemed to have only a marginal effect on the NDVI time series, noticeable as a small drop followed by a peak around

October (green arrows in Figure 4). Overall, the seasonal variation is the main feature of the annual NDVI signal, and therefore the $\Delta NDVI_{W-S}$ may be used to detect the leaf pulvinar movement occurring in the Tamarugo canopy under natural conditions.

Besides the seasonal variation, the MODIS-Terra and MODIS-Aqua NDVI time series allowed to identify the $\Delta NDVI_{mo-mi}$ reported by Chávez et al. [8] on single Tamarugo trees, for instance for the Tamarugo stand B1 (Figure 4b). Although the $\Delta NDVI_{mo-mi}$ was clearly noticeable during winter, it was close to zero in summer. This was expected since both the morning and midday solar irradiation in the Atacama Desert are much higher in summer than in winter. For example, the average solar irradiation of June 2007 (winter) was 0.21 kW/m^2 at 10.00 hours and 0.62 kW/m^2 at 13.30 hours while the average of December 2006 (summer) was 0.87 kW/m^2 at 10.00 hours and 0.94 kW/m^2 at 13.30 hours. As a result, in winter the leaves will only have an erectophile position at midday, but in summer this occurs already half way the morning (yielding a small $\Delta NDVI_{mo-mi}$). Based on what we observed in Figure 4 for a Tamarugo stand without water stress we can expect that it has a positive $\Delta NDVI_{mo-mi}$ in winter as well as a positive $\Delta NDVI_{W-S}$. Both NDVI derived metrics can be quantified and mapped using Landsat and MODIS images as shown in Figure 5.

Figure 5 displays the NDVI values at pixel level of all Bellavista plantation stands (including the stand B1) in the winter of 2007 (first row), the summer of 2006–2007 (second row), and the $\Delta NDVI_{W-S}$ of 2007 (third row) obtained from Landsat images (first column), MODIS-Terra images (second column), and MODIS-Aqua images (third column). The fourth column corresponds to the $\Delta NDVI_{mo-mi}$ in winter (Figure 5d) and summer (Figure 5h) based on Terra (morning) and Aqua (midday). This figure confirms that the $\Delta NDVI_{mo-mi}$ in winter and the $\Delta NDVI_{W-S}$ of 2007 was positive for the forested area. On the other hand, the $\Delta NDVI_{mo-mi}$ in summer was zero or close to zero. In a similar way, and as a consequence of the diurnal pulvinar movements, the $\Delta NDVI_{W-S}$ was higher when using MODIS-Terra images than when using MODIS-Aqua images. Thus, the most promising indicators of pulvinar movement seemed to be the $\Delta NDVI_{mo-mi}$ in winter and the $\Delta NDVI_{W-S}$ in the morning (MODIS-Terra). When using the NDVI as a potential indicator of Tamarugo's water status, the signal in winter was stronger. The canopy coverage can also play an important role in the strength of the NDVI signal and its effect has to be considered when using these NDVI derived metrics for monitoring purposes. We will discuss this issue further in the next section where more Tamarugo stands, with different canopy coverage, were analysed.

3.2 Groundwater depletion: the NDVI signal under water stress

Figure 6 displays the annual time series of GW depth and the Landsat $NDVI_W$ and $\Delta NDVI_{W-S}$ for the six Tamarugo stands analysed in this study. The precipitation events are indicated with arrows. Only four precipitation events were recorded in the 24 years period analysed: 3.0 mm in 1996, 1.8 mm in 2008, 7.9 mm in 2011, and 2.2 mm in 2012, three of them during the last five years. The Landsat $NDVI_W$ signal reacted to the precipitation event of 1996 by showing a short recovering phase (about one year) and quickly returned to the general decreasing trend. For the precipitation events in the last years this effect was difficult to observe since they occurred close to each other in time. These precipitation events did not have any impact on the groundwater table, so we assumed this water was only available for the trees in the superficial soil layers. Although these precipitation events contributed little water to the basin, we assume the moisture added

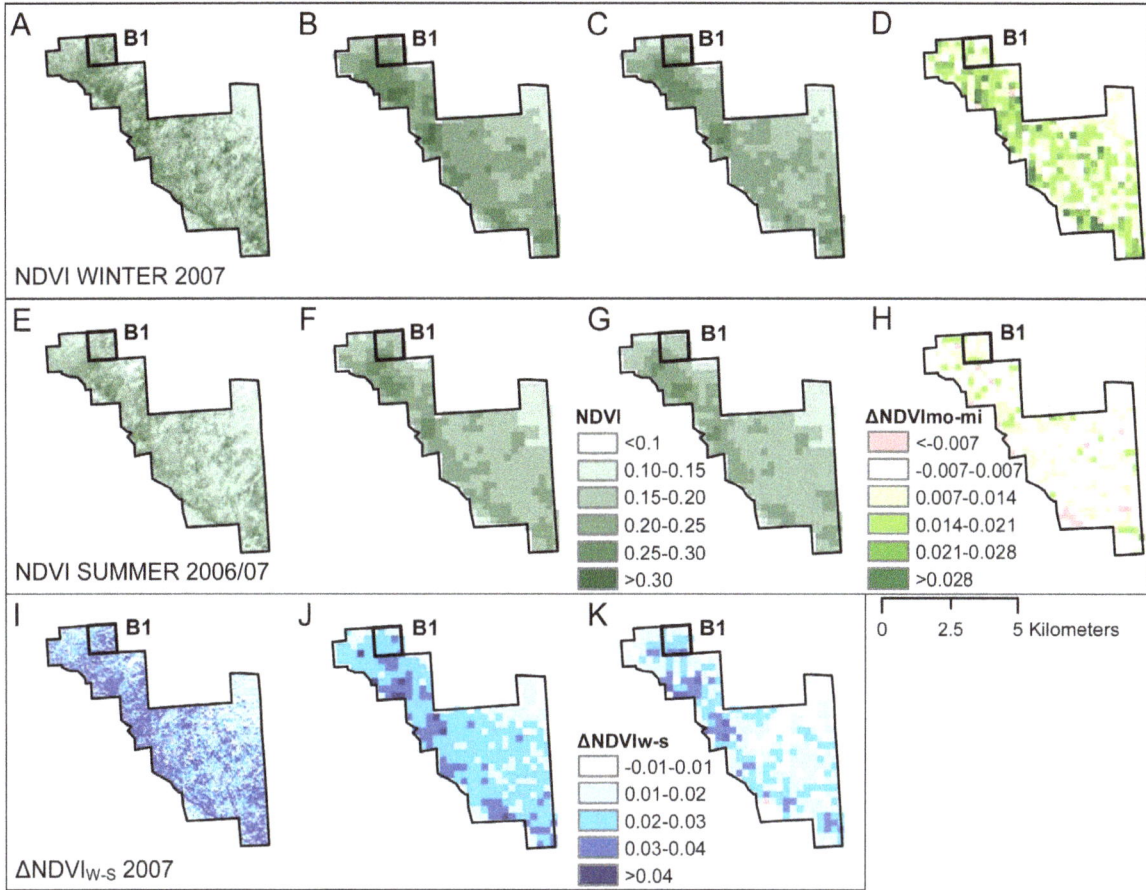

Figure 5. ΔNDVI morning-midday and ΔNDVI winter-summer of the Bellavista plantation in 2007. Winter 2007: (A) Landsat NDVI, (B) MODIS-Terra NDVI (morning), (C) MODIS-Aqua NDVI (midday), (D) ΔNDVI$_{mo-mi}$ = B–C; **Summer 2006–07**: (E) Landsat NDVI, (F) MODIS-Terra NDVI (morning), (G) MODIS-Aqua NDVI (midday), and (H) ΔNDVI$_{mo-mi}$ = F–G. Graphs I, J and K display the **ΔNDVI$_{W-S}$ 2007**, where (I) Landsat ΔNDVI$_{W-S}$ = A–E, (J) MODIS-Terra ΔNDVI$_{W-S}$ = B–F, and (K) the MODIS-Terra ΔNDVI$_{W-S}$ = C–G.

to the superficial root mat of Tamarugo trees temporally had a positive impact on the growth of the trees. Apart from the precipitation events, the analysed NDVI metrics seemed to follow the GW depth trend for all stands. To quantify this relationship, we calculated the R^2 for the linear regression between each of the NDVI metrics and the cumulative GW depletion for the period without precipitation (1997–2007). We also included in this analysis the Landsat annual NDVI$_{av}$ values to check whether the NDVI$_W$ was a better indicator than the simple annual NDVI average. The results are given in Table 2.

The Bellavista Tamarugo stands (B1 and B2) are located in the southern part of the basin and far from the area where the pumping wells are concentrated, which is towards the north and east of the Pintados stands (Figure 3). For this reason, the GW depletion in the stands B1 and B2 was less in comparison to the stands of the Pintados sector (P stands), especially in the case of P2. The stand B1 showed the lowest cumulative depletion (0.45 m) for the period 1997–2007 as well as the lowest R^2 (<0.1) for the relationship between Landsat NDVI$_{av}$ and GW depletion. Furthermore, the R^2 of the GW depletion - Landsat NDVI$_W$ relationship was also the lowest, but higher than the GW depletion - Landsat NDVI$_{av}$ relationship. In fact, this was the case for almost all stands. Thus, the Landsat NDVI$_W$ was more sensitive to changes in GW depth than the Landsat NDVI$_{av}$. This was also the case when comparing Landsat NDVI$_W$ with Landsat ΔNDVI$_{W-S}$.

Only for the stand B1, the R^2 of the Landsat ΔNDVI$_{W-S}$ - GW depletion relationship was higher than for the NDVI$_W$ - GW depletion relationship.

The rest of the stands showed GW depletions between 0.58 and 1.06 meters between 1997 and 2007 and R^2 values for the Landsat NDVI$_W$ - GW depletion relationship higher than 0.75 except for the stand P2 with an R^2 of 0.29. The stand P2 is located close to the pumping area, and therefore the GW depletion could have been influenced by short-term changes of the pumping rate. If the intra-annual GW values fluctuated too rapidly, the depletion may not have had an effect on the NDVI signal. However, this is difficult to detect in annually averaged records. Overall the Landsat NDVI$_W$ was the most sensitive NDVI derived metric to the 11-year changes in GW depletion.

In the case of the MODIS NDVI derived metrics, the R^2 values presented in Table 2 were difficult to interpret since the time series without precipitation events was very short (2003–2007). We found R^2 values as high as 0.70 when using the MODIS ΔNDVI$_{W-S}$ (stand B2) and the MODIS ΔNDVI$_{mo-mi}$ (stand P2), but also <0.1 (stands P1 and P4) for ΔNDVI$_{W-S}$ or ΔNDVI$_{mo-mi}$ (stand P3).

3.3 Mapping water stress using Landsat ΔNDVI$_{W-S}$

The NDVI$_W$ and ΔNDVI$_{W-S}$ showed good potential to assess the effect of GW depletion on the water status of Tamarugo trees.

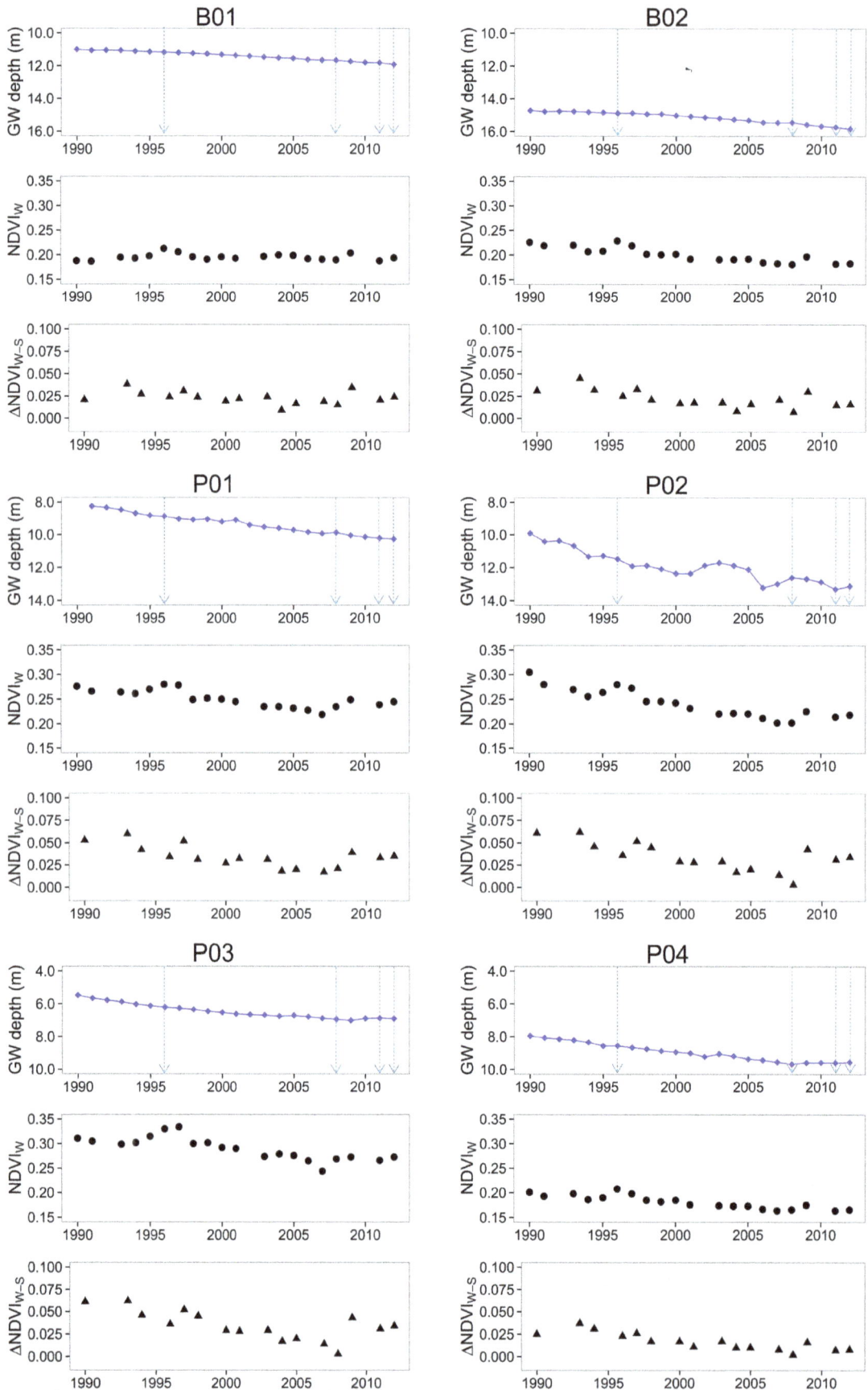

Figure 6. Time series of groundwater depth, Landsat NDVI$_W$, and Landsat ΔNDVI$_{W-S}$ for six Tamarugo plantation stands. Blue arrows indicate precipitation events.

Table 2. R^2 of the linear model of cumulative groundwater depletion v/s $NDVI_{av}$, $NDVI_W$, $\Delta NDVI_{W-S}$ and $\Delta NDVI_{mo-mi}$ for the period 1997–2007 (no precipitation events).

Stand	Cumulative GW depletion (1997–2007)	Landsat $NDVI_{av}$	Landsat $NDVI_W$	Landsat $\Delta NDVI_{W-S}$	MODIS $\Delta NDVI_{W-S}$	MODIS $\Delta NDVI_{mo-mi}$
		n = 11	n = 10	n = 8	n = 4	n = 5
B1	0.45	<0.1	0.13	0.44*	0.55	0.29
B2	0.58	0.74***	0.76***	0.26	0.76	0.35
P1	0.90	0.58***	0.75***	0.60**	<0.1	<0.1
P2	1.06	0.27*	0.29	0.24	0.70	0.70
P3	0.61	0.82***	0.90***	0.70***	0.66	0.14
P4	0.89	0.77***	0.85***	0.74***	<0.1	0.52

Significative linear relationship with ***P<0,01; **P<0.05, *P<0.1.
av = average; W = winter; W-S = winter-summer; mo-mi = morning-midday.

We selected the Landsat $\Delta NDVI_{W-S}$ to map this effect in the study area because we believe it senses water stress earlier than $NDVI_W$ (see Discussion section for more details). We mapped the $\Delta NDVI_{W-S}$ for three different years: 1997, 2007, and 2011 as shown in Figure 7 averaged to a 1×1 km grid. The $\Delta NDVI_{W-S}$ difference between 1997 and 2007 can be explained by groundwater depletion since no precipitation events occurred in this period. For most of the stands, the $\Delta NDVI_{W-S}$ values in 2011 showed a recovery of the forest after the precipitation event of 2011 (7.9 mm), the most intense rain recorded in the last 25 years in Pampa del Tamarugal. The stands with more stable $\Delta NDVI_{W-S}$ through time were those located at the west border of the Bellavista plantation, close to the well W24-1, which reported very shallow GW depths in 1997 (2.6 m), 2007 (3.0 m), and 2011 (3.1 m). Furthermore, a $\Delta NDVI_{W-S}$ gradient can be observed in the Bellavista sector from east to west, showing a good spatial agreement with the increasing GW depletion towards the east (GW depth in well W26-8 was about 19.6 m in 1997, 20.3 m in 2007, and 21 m in 2011).

Discussion

Early stages of water stress in plants are associated with a lower leaf water potential, reduction in transpiration rate, and foliage water loss while late stages are associated with pigment degradation, biomass loss, and finally dying plants [32–34]. Although Tamarugo trees are naturally adapted to the predominant water scarcity of the Atacama desert, they can be affected by water stress due to GW depletion as shown in this paper. From previous papers [8,9], we know that Tamarugos show the typical water stress symptoms of most plants, but additionally water stress limits the normal functioning of the leaf pulvinar mechanism. These pulvinar movements are typical for heliotropic species. Furthermore, we have shown that leaf pulvinar movement can be remotely sensed by different metrics derived from the NDVI signal, allowing to understand both the temporal natural dynamic of the Tamarugo forest and the effects of GW depletion. In Table 3 we give an overview of the water stress symptoms of Tamarugo trees, the temporal scale at which they occur, and the NDVI derived metrics we can use to study these symptoms.

Diurnal leaf movements can be studied using $\Delta NDVI_{mo-mi}$ from MODIS Terra and Aqua satellites as shown in this paper. No significant differences have been found for the MODIS NDVI Terra and Aqua for other non-solar tracker vegetation [35,36]. Since these two satellites acquire data on a daily basis, it would be possible to map the $\Delta NDVI_{mo-mi}$ of the Pampa del Tamarugal basin every day at a spatial resolution of 250×250 m. This way, the effects of an abrupt GW depletion could be identified using MODIS data if the forest is dense enough to provide a sufficiently strong signal as well as large enough to cover one or more MODIS pixels [37,38]. In this paper we analysed averaged $\Delta NDVI_{mo-mi}$ values for the winter seasons and its relationship with annual records of GW depth. This time series was rather short, sometimes resulting in low R^2 values. Perhaps better results can be achieved when using the full temporal resolution (daily or 16 days) of the MODIS NDVI products and more detailed records of the water availability. This is an interesting topic for further research and not only for Tamarugo plants, but also for detecting short-term water stress in, e.g., bean crops, which also have documented paraheliotropic behaviour [5,6].

Seasonal differences of leaf pulvinar adjustments of Tamarugo vegetation can be studied at a large scale using the $\Delta NDVI_{W-S}$ as measured from Landsat (Figure 7) and MODIS Terra satellites. The advantage of using Landsat images is the possibility to map this variable at 30 meters pixel resolution and the disadvantage is that these satellites (Landsat 5, 7 and 8) have a revisit time of 16 days, increasing the chance of missing dates due to cloud cover. Although cloud cover is not such as problem in deserts, missing data can have an important impact on the calculation of the $\Delta NDVI_{W-S}$ if the NDVI values of winter or summer are not well represented by sufficient images. In this paper, we considered a minimum of three Landsat scenes for calculating a representative value of the summer or winter period. The NDVI signal of Tamarugo showed a strong seasonality (Figure 4c) and, for example, a calculation of the $NDVI_W$ using one or two images in May and a calculation of the $NDVI_S$ using one or two images in December may lead to a serious underestimation of the $\Delta NDVI_{W-S}$. This is not a problem for MODIS 16-day composites, which provide five or six images for the winter and summer period systematically distributed within the three months' timeframe. Therefore, there is a trade-off between temporal and spatial resolution when choosing Landsat or MODIS to detect the $\Delta NDVI_{W-S}$.

If the water stress persists, Tamarugo trees will react by selectively shutting down leaves, twigs and entire branches to reduce the transpiration surface while keeping the remaining foliage green with hydric parameters within normal ranges [8]. Foliage loss has been successfully assessed using NDVI for a wide range of vegetation types and it is especially accurate for LAI values <2 [39]. Such assessments are usually carried out at the

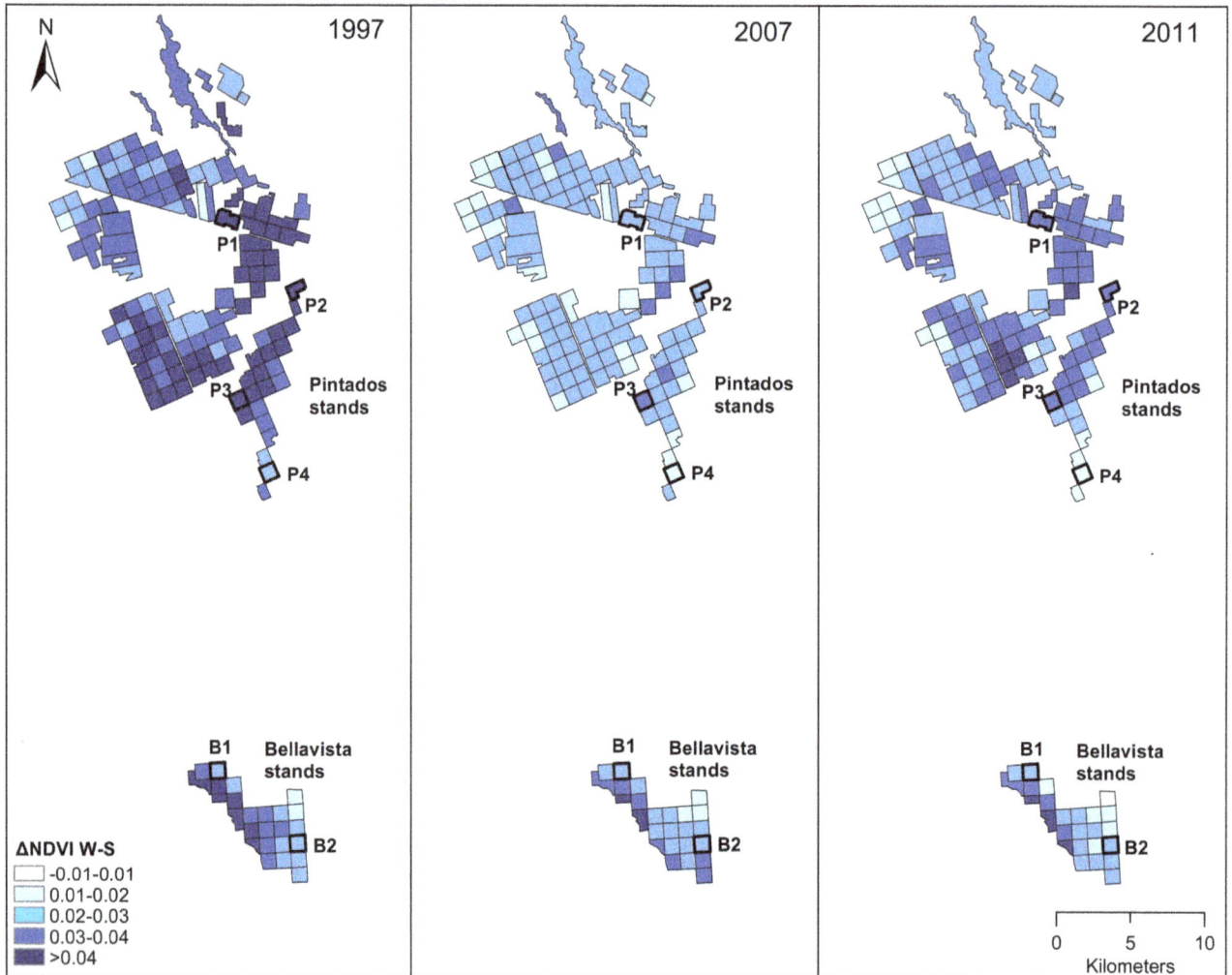

Figure 7. Landsat ΔNDVI$_{W-S}$ of all plantation stands in 1997, 2007, and 2011 (after a precipitation event).

peak of the vegetative period, usually in spring. In the case of Tamarugo the seasonal variation of the NDVI signal is mainly driven by the pulvinar movements, which are primarily driven by seasonal changes in solar irradiation. Thus, the 'pulvinar effect' on the NDVI signal is minimum in winter and therefore this is the best time to retrieve the NDVI for inter annual foliage loss estimations (Table 3).

The strong relationship between NDVI$_W$ and cumulative GW depletion observed for most of the Tamarugo stands is an indication that foliage is decreasing in the study area as a consequence of water extraction, in other words, the forest is reaching an advanced stage of water stress (Table 3). However, it was not possible to discriminate whether the decreasing NDVI$_W$ signal was because some trees were dying while others remained alive (intra species competition) or all trees were losing foliage gradually. The tree coverage played also an important role in the absolute value of the NDVI$_W$ signal and, therefore, it was not possible to directly compare different stands at a single point in time. In order to better interpret the Landat and MODIS NDVI$_W$ signal, we believe that high spatial resolution remote sensing data can provide complementary information about the actual tree coverage of the forest as well as the water status of single trees. This will be the topic for further research.

A recent publication entitled 'Remote sensing: A green illusion' [40] has drawn the attention of the scientific community and policy makers on the issue of the correct interpretation of remote sensing derived products for environmental applications. The authors reflected on this issue based on the results of Morton et al. [41] showing how the apparent canopy greenness of the Amazon forest, interpreted as a positive response to more sunlight in the dry season, was caused by a bidirectional reflectance effect. In other words, it was caused by an optical artefact due to seasonal changes of the sun-sensor geometry. In this paper, we also discussed the correct interpretation of remote sensing derived products, but this time for paraheliotropic vegetation. As shown for the case of Tamarugo in this study, the seasonal changes in NDVI were related to leaf pulvinar movements causing a change in the canopy structure. This change in canopy structure explained the observed seasonal changes (Figure 4). Three pieces of evidence support the hypothesis that pulvinar movements are responsible for NDVI diurnal and seasonal changes of Tamarugo vegetation and that this is not an optical artefact due to bidirectional reflectance effects:

i.   As shown in a previous paper [9], canopy spectral reflectance of Tamarugo plants simulated with the Soil-Leaf-Canopy

Table 3. Water stress stages of Tamarugo desert trees and NDVI based variables to assess their effects using satellite remote sensing.

Water stress			Remote sensing	
Stage	Effects on Tamarugo vegetation	Temporal scale to perceive the effects	Monitoring variable	Sensor useful to retrieve this variable
Instantaneous	Limitation of the diurnal photoinhibition control via pulvinar movements	Diurnal	$\Delta NDVI_{mo\text{-}mi}$	MODIS Terra and Aqua
Early	Limitation of the seasonal photoinhibition control via pulvinar movements	Seasonal	$\Delta NDVI_{W\text{-}S}$	MODIS Terra; Landsat
Advanced	Foliage loss	More than 1 year	$NDVI_W$	MODIS Terra; Landsat
Irreversible	Partial crown death and tree death	Several years	$NDVI_W$	MODIS Terra; Landsat, combined with very high spatial resolution sensors: Quickbird2, WorldView2 or GeoEye

(SLC) radiative transfer model showed that the SLC parameter LIDF (leaf inclination distribution function) could explain diurnal changes in canopy reflectance measured empirically with a spectroradiometer under laboratory conditions (lamp-sensor geometry was fixed). Thus, leaf movements, set in the SLC simulations as a 'random' LIDF in the morning and as an 'erectophile' LIDF after midday, explained diurnal changes in canopy reflectance in the absence of water stress.

ii. Another previous paper [8] showed a negative empirical relationship between diurnal values of NDVI, measured for single Tamarugo trees with a spectroradiometer, and solar irradiation under field conditions. In that study, the authors observed a predominantly erectophyle position of Tamarugo leaves around midday, corresponding to the diurnal peak of solar irradiation and the lowest values of NDVI. In the current paper, we showed a negative empirical relationship between seasonal NDVI values, measured by Landsat and MODIS satellites for Tamarugo stands, and solar irradiation (Figure 4). Furthermore, it is a known botanical fact that paraheliotropic movements are a response to increasing solar irradiation on the leaves [1]. Thus, pulvinar movements activated by changes in solar irradiation govern diurnal and seasonal changes in the NDVI signal of Tamarugo vegetation.

iii. This study provided evidence that the amplitude of the seasonal NDVI trend ($\Delta NDVI_{W\text{-}S}$) of Tamarugo stands declined with water stress (Figure 6). If the NDVI seasonal trend measured by satellite remote sensing was governed by a sun-sensor artefact, there is no reason why water stress would cause the amplitude of the NDVI signal to decline significantly.

In the southern hemisphere, more internal shadowing in satellite images (captured at nadir) is expected to occur in winter at lower solar elevation, and therefore, the bidirectional reflectance effect should cause an 'apparent greening' towards spring/summer [41]. However, as shown in Figure 4, the peak of the NDVI signal of Tamarugo stands does not occur in summer, but in winter. Although bidirectional reflectance effects may also occur in the case of Tamarugo vegetation, we believe that such effects are obscured by the stronger effect of seasonal pulvinar movement.

Conclusions

1. Monthly values of solar irradiation were negatively correlated to NDVI measured by the MODIS-Terra and Landsat satellites. Previous studies have shown that pulvinar movement causes the NDVI signal to drop from morning to midday as solar irradiation increases, and therefore, in the absence of water stress the seasonal variation of NDVI is also expected to be controlled by pulvinar movement.

2. The NDVI difference between midday and morning ($\Delta NDVI_{mo\text{-}mi}$), as measured by the difference of the NDVI signal from the MODIS Terra and Aqua satellites, can be used to detect the diurnal leaf pulvinar movement of Tamarugo plantation stands. This has not been reported in literature before, and therefore, this paper constitutes a proof of concept that MODIS images can be used to detect diurnal movements of paraheliotropic vegetation.

3. Similarly, the NDVI difference between winter and summer ($\Delta NDVI_{W\text{-}S}$), as measured by the Landsat or the MODIS Terra satellites, can be used to detect differences in seasonal pulvinar movements, associated to photoinhibition regulation.

4. Leaf pulvinar movements are triggered by changes in cell turgor and they can be limited by water stress. Thus, water stress in Tamarugo vegetation caused by groundwater overexploitation can be assessed and monitored using $\Delta NDVI_{mo\text{-}mi}$ and $\Delta NDVI_{W\text{-}S}$. For long time series (more than 10 years), Landsat $\Delta NDVI_{W\text{-}S}$ of Tamarugo stands showed a positive linear relationship with cumulative groundwater depletion.

5. Under water stress, a limitation of the pulvinar movement occurs in Tamarugo trees before they start losing foliage. For this reason, changes in $\Delta NDVI_{mo\text{-}mi}$ and $\Delta NDVI_{W\text{-}S}$ are expected to occur before NDVI decreases due to foliage loss, and therefore, $\Delta NDVI_{mo\text{-}mi}$ and $\Delta NDVI_{W\text{-}S}$ have potential for early water stress detection.

Acknowledgments

The authors would like to thank Digital Globe for providing the WorldView2 imagery and UNAP (Chile) for providing the solar irradiation records. Especial thanks to L. Dutrieux, B. de Vries, M. Schultz, and J. González de Tanago from Wageningen University for the valuable contribution of R code for the time series analysis. Finally, we thank V.

Urra and G. Valenzuela from Universidad de Chile for their collaboration on the anatomical description of Tamarugo leaves.

References

1. Ehleringer J, Forseth I (1980) Solar tracking by plants. Science 210: 1094–1098.
2. Koller D (1990) Light-driven leaf movements*. Plant Cell Environ 13: 615–632.
3. Moran MS, Pinter Jr PJ, Clothier BE, Allen SG (1989) Effect of water stress on the canopy architecture and spectral indices of irrigated alfalfa. Remote Sens Environ 29: 251–261.
4. Koller D (2001) Solar navigation by plants. In: Comprehensive Series in Photosciences. pp.833–895.
5. Pastenes C, Porter V, Baginsky C, Norton P, González J (2004) Paraheliotropism can protect water-stressed bean (*Phaseolus vulgaris* L.) plants against photoinhibition. J Plant Physiol 161: 1315–1323.
6. Pastenes C, Pimentel P, Lillo J (2005) Leaf movements and photoinhibition in relation to water stress in field-grown beans. J Exp Bot 56: 425–433.
7. Kimes DS, Kirchner JA (1983) Diurnal variations of vegetation canopy structure. Int J Remote Sens 4: 257–271.
8. Chávez RO, Clevers JGPW, Herold M, Acevedo E, Ortiz M (2013) Assessing water stress of desert tamarugo trees using in situ data and very high spatial resolution remote sensing. Remote Sens 5: 5064–5088.
9. Chávez RO, Clevers JGPW, Herold M, Ortiz M, Acevedo E (2013) Modelling the spectral response of the desert tree *Prosopis tamarugo* to water stress. Int J Appl Earth Obs Geoinf 21: 53–65.
10. Williams DL, Goward S, Arvidson T (2006) Landsat: Yesterday, today, and tomorrow. Photogramm Eng Remote Sensing 72: 1171–1178.
11. Birdsey R, Angeles-Perez G, Kurz WA, Lister A, Olguin M, et al. (2013) Approaches to monitoring changes in carbon stocks for REDD+. Carbon Manage 4: 519–537.
12. Cohen WB, Goward SN (2004) Landsat's role in ecological applications of remote sensing. Bioscience 54: 535–545.
13. Verbesselt J, Zeileis A, Herold M (2012) Near real-time disturbance detection using satellite image time series. Remote Sens Environ 123: 98–108.
14. McKay CP, Friedmann EI, Gómez-Silva B, Cáceres-Villanueva L, Andersen DT, et al. (2003) Temperature and moisture conditions for life in the extreme arid region of the atacama desert: Four years of observations including the El Niño of 1997–1998. Astrobiology 3: 393–406.
15. Navarro-González R, Rainey FA, Molina P, Bagaley DR, Hollen BJ, et al. (2003) Mars-Like Soils in the Atacama Desert, Chile, and the Dry Limit of Microbial Life. Science 302: 1018–1021.
16. Gajardo R (1994) La vegetación natural de Chile. Clasificación y distribución geográfica. Santiago (Chile): Editorial Universitaria. 165 p.
17. CONAMA (2008) Biodiversidad de Chile, patrimonio y desafíos. Santiago de Chile: Ocho Libros Editores. 640 p.
18. Estades CF (1996) Natural history and conservation status of the Tamarugo Conebill in northern Chile. Wilson Bull 108: 268–279.
19. Ramírez-Leyton G, Pincheira-Donoso D (2005) Fauna del Altiplano y Desierto de Atacama. Vertebrados de la Provncia de El Loa. Calama, Chile: Phrynosaura Ediciones. 396 p.
20. Rojas R, Dassargues A (2007) Groundwater flow modelling of the regional aquifer of the Pampa del Tamarugal, Northern Chile. Hydrogeol J 15: 537–551.
21. Zelada L (1986) The influence of the productivity of *Prosopis tamarugo* on livestock production in the Pampa del Tamarugal - a review. For Ecol Manage 16: 15–31.
22. Altamirano H (2006) *Prosopis tamarugo* Phil. Tamarugo. In: C. Donoso, editor. Las especies arbóreas de los bosques templados de Chile y Argentina Autoecología. Valdivia (Chile): Marisa Cuneo Ediciones. pp.534–540.
23. Riedemann P, Aldunate G, Teillier S (2006) Flora nativa de valor ornamental. Chile, Zona Norte. Identificación y propagación. Santiago, Chile: Productora Gráfica Andros Ltda. 404 p.
24. Trobok S (1985) Fruit and seed morphology of Chilean *Prosopis* (Fabaceae-Mimosoidae). In: M. Habit, editor. The current state of knowledge on *Prosopis tamarugo*. Rome: F.A.O. Available: http://www.fao.org/docrep/006/ad316e/AD316E13.htm#ch3.2. Accessed 2014 February 28.
25. Aravena R, Acevedo E (1985) The use of environmental isotopes Oxigen-18 and deuterium in the study of water relations of *Prosopis tamarugo* Phil. In: M. Habit, editor. The current state of knowledge on *Prosopis tamarugo*. Rome: F.A.O. Available: http://www.fao.org/docrep/006/ad316e/AD316E14.htm#ch3.4. Accessed 2014 February 28.
26. Mooney HA, Gulmon SL, Rundel PW, Ehleringer J (1980) Further observations on the water relations of *Prosopis tamarugo* of the northern Atacama Desert. Oecologia 44: 177–180.
27. Sudzuki F (1985) Environmental moisture utilization by *Prosopis tamarugo* Phil. In: M. Habit, editor. The current state of knowledge on *Prosopis tamarugo*. Rome: F.A.O. Available: http://www.fao.org/docrep/006/ad316e/AD316E04.htm#ch1.2. Accessed 2014 February 28.
28. Houston J (2006) Evaporation in the Atacama Desert: An empirical study of spatio-temporal variations and their causes. J Hydrol 3: 402–412.
29. Houston J, Hartley AJ (2003) The central andean west-slope rainshadow and its potential contribution to the origin of hyper-aridity in the Atacama desert. Int J Climatol 23: 1453–1464.
30. CONAF (1997) Plan de manejo reserva nacional Pampa del Tamarugal. Corporación Nacional Forestal (CONAF). Ministerio de Agricultura. Gobierno de Chile. 110 p.
31. Masek JG, Vermote EF, Saleous NE, Wolfe R, Hall FG, et al. (2006) A landsat surface reflectance dataset for North America, 1990–2000. IIEE Geosci Remote S 3: 68–72.
32. Baret F, Houlès V, Guérif M (2007) Quantification of plant stress using remote sensing observations and crop models: The case of nitrogen management. J Exp Bot 58: 869–880.
33. Alpert P, Oliver MJ (2002) Drying without dying. In: M. . Black and H. W. Pritchard, editors. Desiccation and survival in plants: drying without dying. Wallingford (UK): CABI. pp.4–31.
34. Taiz L, Zeiger E (2010) Plant physiology. Sunderland, MA: Sinauer Associates. 782 p.
35. Wang J, Guo N, Wang X, Yang J (2007) Comparisons of normalized difference vegetation index from MODIS Terra and Aqua data in northwestern China. Int Geosci Remote Se pp.3390–3393.
36. Wu A, Xiong X, Cao C (2008) Terra and Aqua MODIS inter-comparison of three reflective solar bands using AVHRR onboard the NOAA-KLM satellites. Int J Remote Sens 29: 1997–2010.
37. Verbesselt J, Robinson A, Stone C, Culvenor D (2009) Forecasting tree mortality using change metrics derived from MODIS satellite data. For Ecol Manage 258: 1166–1173.
38. Wolfe RE, Nishihama M, Fleig AJ, Kuyper JA, Roy DP, et al. (2002) Achieving sub-pixel geolocation accuracy in support of MODIS land science. Remote Sens Environ 83: 31–49.
39. Gamon JA, Field CB, Goulden ML, Griffin KL, Hartley AE, et al. (1995) Relationships between NDVI, canopy structure, and photosynthesis in three Californian vegetation types. Ecol Appl 5: 28–41.
40. Soudani K, Francois C (2014) Remote sensing: A green illusion. Nature 506: 165–166.
41. Morton DC, Nagol J, Carabajal CC, Rosette J, Palace M, et al. (2014) Amazon forests maintain consistent canopy structure and greenness during the dry season. Nature 506: 221–224.

Author Contributions

Conceived and designed the experiments: ROC JGPWC MH. Performed the experiments: ROC JV PIN. Analyzed the data: ROC JGPWC JV PIN MH. Contributed reagents/materials/analysis tools: ROC JV PIN. Wrote the paper: ROC JGPWC.

8

The Occurrence, Sources and Spatial Characteristics of Soil Salt and Assessment of Soil Salinization Risk in Yanqi Basin, Northwest China

Zhang Zhaoyong[1,2], Jilili Abuduwaili[1]*, Hamid Yimit[3]

1 State Key Laboratory of Desert and Oasis Ecology, Xinjiang Institute of Ecology and Geography, Chinese Academy of Sciences, Urumqi, China, **2** University of the Chinese Academy of Sciences, Beijing, China, **3** Key Laboratory of Xingjiang Arid Land Lake Environment and Resource, Xinjiang Normal University, Urumqi, China

Abstract

In order to evaluate the soil salinization risk of the oases in arid land of northwest China, we chose a typical oasis-the Yanqi basin as the research area. Then, we collected soil samples from the area and made comprehensive assessment for soil salinization risk in this area. The result showed that: (1) In all soil samples, high variation was found for the amount of Ca^{2+} and K^+, while the other soil salt properties had moderate levels of variation. (2) The land use types and the soil parent material had a significant influence on the amount of salt ions within the soil. (3) Principle component (PC) analysis determined that all the salt ion values, potential of hydrogen (pHs) and ECs fell into four PCs. Among them, PC1 (Cl^-, Na^+, SO_4^{2-}, EC, and pH) and PC2 (Ca^{2+}, K^+, Mg^{2+} and total amount of salts) are considered to be mainly influenced by artificial sources, while PC3 and PC4 (CO_3^- and HCO_3^{2-}) are mainly influenced by natural sources. (4) From a geo-statistical point of view, it was ascertained that the pH and soil salt ions, such as Ca^{2+}, Mg^{2+} and HCO_3^-, had a strong spatial dependency. Meanwhile, Na^+ and Cl^- had only a weak spatial dependency in the soil. (5) Soil salinization indicators suggested that the entire area had a low risk of soil salinization, where the risk was mainly due to anthropogenic activities and climate variation. This study can be considered an early warning of soil salinization and alkalization in the Yanqi basin. It can also provide a reference for environmental protection policies and rational utilization of land resources in the arid region of Xinjiang, northwest China, as well as for other oases of arid regions in the world.

Editor: Andrew C. Singer, NERC Centre for Ecology & Hydrology, United Kingdom

Funding: This study was supported by the Knowledge Innovation Program of the Chinese Academy of Sciences (KZCX2-EW-308; KZCX2-YW-GJ04). The funders had no role in study design, data collection and analysis, decision to publish, or preparation of the manuscript.

Competing Interests: The authors have declared that no competing interests exist.

* Email: jilil@ms.xjb.ac.cn

Introduction

Soil salinization is a global problem and it is a potential environmental problem in all continents with the exception of unassessed Antarctica. Soil saline levels are found within a wide range, and soil salinization occurs in much of the waterfront, arid and semi-arid zones of more than 100 countries and regions [1–3]. According to statistics done by the United Nations Educational, Scientific and Cultural Organization (UNESCO), and the Food and Agriculture Organization (FAO), salinized soil covers an area of about 9.543×10^6 km^2 on Earth [4–6]. In China alone, the area of salinized soil is about 3.693×10^5 km^2, which accounts for about a third of the total arable land [7–8]. The area of salinized soil in the oasis basin of Xinjiang in northwest China is about 1.05×10^4 km^2, which accounts for 33.4% of the total land in this area and research has found that the salinity of this area is trending upwards [9,10]. Soil salinization restricts agricultural development, especially when sustainable agricultural development and environmental quality improvement strategies are being considered. Studies have found that when the salt ions in the soil attain 8 g.kg^{-1}, they can greatly harm and even kill crops in farmland [11,12].

In oases of arid regions of northwest china, the environment is so weak, including a lack of precipitation and high envapotion, that economic activities such as fishing, agriculture, forestry and grassland farming of the oases have been strongly limited, especially for agriculture [13–15]. Therefore, it is necessary to identify the distribution characteristics, sources of the soil properties, such as salt ions, potential of hydrogen (pH) and electrical conductivity (EC), and also the status and causes of salinization of the land of the oasis, in order to provide a scientific basis for protection of the soil that sustains land plants.

Multivariate analyses and other statistical methods have been widely applied in studies to determine the sources of elements found in soil, such as total soil salt content and heavy metals [16–18]. The spatial variation model and spatial distribution are used to make a hazard risk map of soil salt properties in regions of interest. Correlation analysis, principal component (PC) analysis and cluster analysis are classic methods used to identify the natural and man-made sources of salt ions and to simplify data. Additionally, use of the comprehensive index results in a class of data with high correlations that better reflect the associations between the data.

Table 1. Indicators used for risk assessment of soil salinization.

Indicators	Class limits and their ratings score				
	None	**Slight**	**Moderate**	**Severe**	**Very severe**
*EC (dS.m^{-1}) [49]	<4	4–8	8–16	16–32	>32
**SAR [50]	<8	8–13	13–30	30–70	>70
Total salt content (%) (0–20 cm) [51]	<1	2~3	3–4	4–8	>8

*EC is electrical conductivity; **SAR is sodium adsorption ratio.

The Geostatistical Analyst is based on GIS technology [19,20]. Among these, the ordinary kriging is the most widely used one in the study of soil salt distributions [21,22,29]. In recent years, the Geostatistical Analyst method has been used in the field of hydrology and water resources, including studies of groundwater pollution risk, water potential research and spatial distribution of soil salinization in arid land [23–25,28,32]. Since the 1990s, geostatistical methods have been widely used to study spatial variability characteristics of soil salt properties (salt ions, EC and pH). Sylla et al. [26] studied the spatial variation characteristics of soil salt content of an agricultural ecosystem under different scales in West Africa. Ammari et al. [27] studied the soil salinity changes in the Jordan Valley and the potential threat against the sustainable irrigated agriculture. In China, Bai et al. [30] researched the spatial variation characteristics and composition of soil salt content in Huang Huai Hai plain, northeast China and found that the average influential range of the soil salt content was higher than 200 km, indicating the salt content of the soil is mainly gathered in a large area of the regions. In Xinjiang in northwest China, Lin et al. [33] researched the spatial variation characteristics of soil salt in the Wei Gan He irrigated area and found the agricultural irrigation has resulted in serious soil salinization in this area and these deserve serious attention.

In arid regions, oases in basins are the main places where humans live and life can survive [31]. Therefore, it is important to understand the spatial distribution characteristics of the soil salt properties, including total salt content, salt ions, pH and EC. A quantitative grasp of soil salinity levels could serve as a reference and a basis for maintaining soil quality, which would help to effectively control the human pollution and develop the regional economy in a reasonable and orderly fashion [34,35]. However, previous research has focused on rapidly developing areas, such as coastal plains and large irrigation areas in eastern china and elsewhere of the world with the purpose of assessing land usability, environmental effects and soil salinization risk [36–38]. Since the 1990s, implementation of the "western development policy of China" has led to prodigious economic development in many oases in Xinjiang, and the agriculture in these regions has undergone rapid progress. However, the rational irrigation of the agriculture, lack of precipitation and high envapotion of these regions have negatively influenced soil salt properties, resulting in increased soil salt contents, ECs and pHs, which can result in serious soil salinization [39]. Unfortunately, research on the soil salt property distribution characteristics and soil salinization risk assessment in the oases of arid regions of northwest of China is lacking.

The Yanqi basin is a typical oasis in a basin in the southern Tianshan Mountains, Xinjiang in northwest China. Since the 1990s, both the implementation of the "western development policy of China" and the development policy made by the Xinjiang Province, China, have led to prodigious economic development in the Yanqi basin, but regional economic development and associated human activity have left the current ecological environment fragile [39,40]. Together with economic development, the blind expansion of farmland and unrestrained surface water irrigation led to a rise in groundwater and an increase in soil salinization in the basin oasis. 64.12% of the area experienced mild soil salinization, 8.25% had moderate salinization, and 27.07% had severe salinization. Research has shown that excessive use of water resources by agriculture has made the soil salinization status severe and decreased agricultural production [41].

After a basic analysis of land use and soil parent materials types in the area, we created land use and soil type geological maps, and, using ArcGIS 10.0 software and combining the grid sampling method with 3S technology, we made sampling points to get soil samples across the whole area. We evaluated the soil salt properties in different land use types in the laboratory, and assessed the soil salinization status and the cause in the Yanqi basin. Then ordinary kriging of Geostatistical Analyst method was used to reveal the spatial distribution characteristics of the soil salt properties in this region. Then by combining these properties with the climate, precipitation, evaporation and temperature, we assessed the soil salinization risk of this area. From this we can provide helpful proposals to prevent the environmental risks that could lead to soil salinization in this area. This research can serve as a helpful reference for environmental protection in this region and for soil salinization prevention in arid regions of northwest China.

Materials and Methods

Study area

The area studied in this work is a desert basin oasis in the arid region of northwest China including four counties in the Yanqi basin: Yanqi County, Hejing County, Bohu County and Heshuo County. This region lies within the geographical coordinates of 85°50′–87°50′E and 41°40′–42°20′N with a length of about 85 km from north to south and width of 130 km from east to west, totaling an area of about 723100 km^2. The terrain slopes up from the northwest down to the southeast. The northwest is mountainous and the south is low-lying desert that is 1050–2000 m above sea level. The western area has extensive intrusive rock and metamorphic rock from the *Proterozoic era*, *Neoproterozoic* and *Cenozoic*. Weathering of this rock results in brown earth soil, acidic rocky soil and an acidic soil skeleton in the west. The east is primarily made up of quaternary sediments, which form *Takyic* (Calcisols), *Chemic* (Phaeozems), *Stagnic* (Gleysols), *Irragric* (Anthrosols), *Fragic* (Arenosols), *Eutric* (Gleysols) and *Yemic* (Solonchaks) [42]. The area researched is in a continental desert climate temperate zone with an annual mean temperature of 14.6°C, 186 frost-free days per year, and 50.7–79.9 mm of annual

Table 2. Descriptive statistics of the soil salt properties from Yanqi basin.

Elements	Ranges (g·kg⁻¹)	Contributions(%)	Median (g·kg⁻¹);EC (dS·m⁻¹)	Average (g·kg⁻¹)	Standard deviation (%)	Coefficient of variation (%)	Kurtosis (%)	Skewness (%)
HCO₃⁻	0.13-0.98	3.51	0.171	0.19	12.25	35.37	0.14	0.58
CO₃⁻	0.18-0.85	4.08	0.252	0.46	10.35	32.08	1.63	8.76
Ca²⁺	0.59-1.95	6.54	0.681	0.75	9.83	191.67	31.38	42.43
Na⁺	0.69-2.43	13.18	0.955	1.19	12.38	23.01	2.45	16.22
Mg²⁺	0.47-1.89	7.75	0.987	0.58	21.02	20.07	1.10	10.88
K⁺	0.49-2.13	12.36	0.855	0.69	23.04	226.25	23.47	51.81
SO₄²⁻	0.93-1.58	18.95	1.245	1.14	12.56	19.63	1.46	0.76
Cl⁻	0.75-2.36	25.63	1.167	0.98	25.7	12.94	2.30	14.79
SAR	3.41-33.241	-	22.417	10.51	121.45	148.56	11.54	15.78
EC	0.7-1.39	-	0.981	0.95	15.09	21.27	1.46	10.99
pH	7.85-8.55	-	8.141	8.15	16.39	30.32	1.17	0.73
Total salt	1.16-14.77	-	8.56	9.73	15.36	126.73	16.84	18.46

precipitation. By calculating the potential evaporation (ET_0) by the method of Hargreaves (1985) [43], we then got the annual average potential evaporation of this area as 2438.9 mm, the $\geq 10°C$ active accumulated temperature 3414.4–3694.1°C and an annual average relative humidity of 72%.

Soil sampling and analyses

In order to perform a basic analysis of the land use and soil type, geological maps were made of the study area using ArcGIS 10.0 software to lay out a grid of soil sampling points on a digital map of the Yanqi basin. All samples were acquired in July 2012 or July 2013 from a collection area. In order to best assess the ecological risk in the Yanqi basin, diverse land use types were encompassed in our study of salt ion distribution. Soil samples were collected at depths of 0–20 cm, where a hard plastic shovel was used to dig a vertical $20×20$ cm soil profile. 1 kg uniform samples were collected, and then they were put into a clean cloth, numbered and sealed. The collection position, date, sample vegetation types and surrounding vegetation conditions of each sampling area were recorded. After the soil samples were taken back to the laboratory, they were air dried and impurities, such as plant residues and rocks, were removed. The samples were then pushed through a 20 mesh nylon sieve (0.84 mm) to eliminate the plant residue and stones. We then used agate to grind the soil samples through 100 mesh nylon sieves (0.25 mm) to prevent contamination and then stored the samples in plastic bottles [46].

Total soil salt content, soil salt ions, pH and EC tested are as follows: 50 g of ground sample were removed from the plastic bottles, dissolved in 250 ml of deionized water (CO_2 has been removed) (1:5, soil:water) for 2 hours to fully dissolve salt ions contained in the soil. The samples were then put in a centrifuge tube, vibrated for 3 min with an oscillator and then centrifuged at a speed of 4,500–5,000 r·min⁻¹. To get the supernatant prepared for analysis of total soil salt content, salt ions content, pH and EC, the method described by Lu (2000) was followed [44].

The total salt content of the soil was determined by gravimetry of the evaporation residue. First, the supernatant was absorbed in a porcelain dish, and hydrogen peroxide (H_2O_2) was used to oxidize organic matter. Then, the samples were boiled in a water bath at 105–110°C until it dried, and weighed. The drying quality of the residue is expressed as total salt content of the soil. The pH of the extracted supernatant was tested using a Potentiometric Titrimeter (G20, METTLER, and TOLEDO). Burette drive resolution was 1/20000. Mv/pH electrode measurement range was ±2000 mv. The ECs were tested using a Conductivity Meter (DDSJ-308A, Shanghai, China) with a measurement range of 0–1.999×10⁵ μs/cm and a test error of ±0.5% (FS) ±1.

The supernatant was run through a 0.45 μm drainage cellulose acetate membrane. Then, the cation content (K^+, Na^+, Ca^{2+}, Mg^{2+}) of the solutions was determined using an inductively coupled plasma atomic emission spectrometer (Vista MPX, Varian, USA). The anion content (Cl^-, SO_4^{2-}, CO_3^{2-}, and HCO_3^-) of the solutions was determined using an Ion Chromatograph (ICS-90, Dionex, USA). All tests were conducted using the following protocol: a standard solution was prepared for Na^+, K^+, Ca^{2+}, Mg^{2+}, Cl^-, SO_4^{2-}, CO_3^{2-} and HCO_3^-. The salt ion content was determined by comparing each sample to the standard solution of known concentration. The standard solutions used for the salt ions in this study were national level standard material (Gss series, China). The coefficient of the best fitting curve was determined by the testing equipment based on the standard material and then the amount of the salt ions (Na^+, K^+, Ca^{2+}, Mg^{2+}, Cl^-, SO_4^{2-}, CO_3^{2-}, and HCO_3^-) in the liquid supernatant was tested. After all the samples had been tested for their salt ion

Table 3. Statistical parameters of the soil salt ions found at 0–20 cm depth within the investigated land use and land cover categories of the study area.

LUCC SPM	Parameters	HCO₃⁻ (g.kg⁻¹)	CO₃⁻ (g.kg⁻¹)	Na⁺ (g.kg⁻¹)	Mg²⁺ (g.kg⁻¹)	K⁺ (g.kg⁻¹)	Ca²⁺ (g.kg⁻¹)	SO₄²⁻ (g.kg⁻¹)	Cl⁻ (g.kg⁻¹)	Total salt (g.kg⁻¹)	EC (dS.m⁻¹)	pH
Farmland (n = 51)	Ranges	0.18–0.98	0.27–0.85	0.69–2.43	1.05–1.89	1.31–2.13	1.17–1.95	1.05–1.51	1.47–2.36	8.56–14.77	0.96–1.39	8.05–8.55
	Average	0.32a	0.62a	1.36a	1.15a	1.83a	1.71a	1.21a	1.68a	11.2a	1.02a	8.15a
	SD	22.45	15.23	31.35	23.3	27.57	37.73	24.57	19.19	23.97	18.64	14.77
Forest (n = 46)	Ranges	0.13–0.53	0.18–0.78	0.69–0.98	0.47–0.98	0.49–0.97	0.59–1.02	1.09–1.58	0.75–0.97	9.13–13.35	0.7–1.13	7.85–8.45
	Average	0.33a	0.41b	0.75b	0.87b	0.75a	0.77a	1.18b	0.82a	9.98b	0.94b	8.17b
	SD	14.75	17.73	23.53	25.55	23.74	23.73	17.86	28.33	16.79	18.71	14.37
Grassland (n = 63)	Ranges	0.21–0.73	0.25–0.82	1.03–1.63	0.54–1.02	0.61–1.31	0.89–1.14	0.93–1.19	0.96–1.61	9.08–11.24	0.76–1.25	7.92–8.42
	Average	0.42a	0.65b	1.32b	0.68b	1.18b	0.98b	1.04b	1.28b	9.45a	0.95b	8.27b
	SD	17.54	23.77	27.52	23.13	11.51	15.23	12.22	22.14	13.75	16.71	21.37
Desert (n = 71)	Ranges	0.26–0.83	0.19–0.76	0.98–1.57	0.47–1.44	0.95–1.56	0.99–1.19	1.05–1.24	1.08–2.29	9.31–13.89	0.78–1.34	8.06–8.44
	Average	0.58b	0.32b	1.31b	1.13b	1.21b	1.02b	1.11.6b	1.57b	10.56a	1.08b	8.28a
	SD	17.33	15.65	17.52	25.55	18.85	13.43	17.33	19.25	23.75	12.52	15.31
Urban construction areas (n = 40)	Ranges	0.49–0.93	0.33–0.83	0.91–1.54	0.65–1.03	0.51–1.51	0.88–1.21	0.94–1.18	0.79–1.31	6.16–14.13	0.75–1.33	7.94–8.51
	Average	0.53c	0.39c	1.21c	0.91c	0.97c	0.97c	1.06c	0.94c	10.52a	1.11c	8.31c
	SD	15.75	16.51	13.25	15.52	12.35	19.52	23.37	21.54	32.24	22.31	24.87
Sandy shale of weathered material (n = 66)	Ranges	0.13–0.86	0.18–0.38	1.19–2.39	0.47–1.75	0.58–2.11	0.59–1.36	0.93–1.53	0.75–2.36	8.69–14.77	0.79–1.26	7.86–8.55
	Average	0.45a	0.42a	1.51a	0.95a	0.91a	0.82a	1.01a	1.21a	10.88a	0.3a	7.93a
	SD	25.55	17.53	13.52	22.31	22.35	23.35	15.75	18.65	11.54	12.25	21.35
Coarse crystalline rock weathered material (n = 74)	Ranges	0.18–0.52	0.22–0.79	0.69–2.43	0.51–1.54	0.49–2.13	0.69–1.45	0.97–1.56	0.78–2.12	6.16–12.41	0.7–1.39	7.85–8.32
	Average	0.32b	0.38b	1.72b	0.81b	0.87b	0.95b	1.13b	1.46b	8.72a	1.14b	8.14b
	SD	13.35	11.41	21.25	23.74	11.29	12.52	21.57	22.53	15.54	12.57	24.58
Diluvial material (n = 68)	Ranges	0.21–0.93	0.24–0.85	0.81–2.23	0.65–1.89	0.64–1.72	0.71–1.56	0.95–1.58	0.85–2.26	7.89–13.25	0.84–1.28	7.89–8.19
	Average	0.67b	0.51b	1.42b	0.75b	0.98b	1.01b	1.05b	1.62b	9.82a	0.91b	8.01b
	SD	12.15	12.25	23.73	27.35	11.37	12.75	12.73	22.36	21.57	23.37	11.52
Lacustrine deposits (n = 63)	Ranges	0.33–0.98	0.23–0.79	0.85–2.35	0.52–1.49	0.71–1.98	0.62–1.95	0.96–1.51	0.96–2.19	8.98–11.51	0.85–1.32	7.97–8.37
	Average	0.74a	0.44b	1.16a	0.91c	1.02c	0.85c	1.06a	1.31b	9.24a	1.24a	8.23a
	SD	13.51	22.35	21.26	23.59	15.34	15.62	21.94	19.31	26.49	23.77	12.35
R²	LUCC (%)	31.21	9.74	13.49	8.4	71.45	37.85	11.3	10.8	56.71	12.64	12.43
	SPM (%)	58.79	34.59	16.47	28.93	32.7	11.4	17.9	15.9	23.51	17.53	15.2

Different small letters represent a significance of 0.05; LUCC represent land use types; SPM represent soil parent material types.

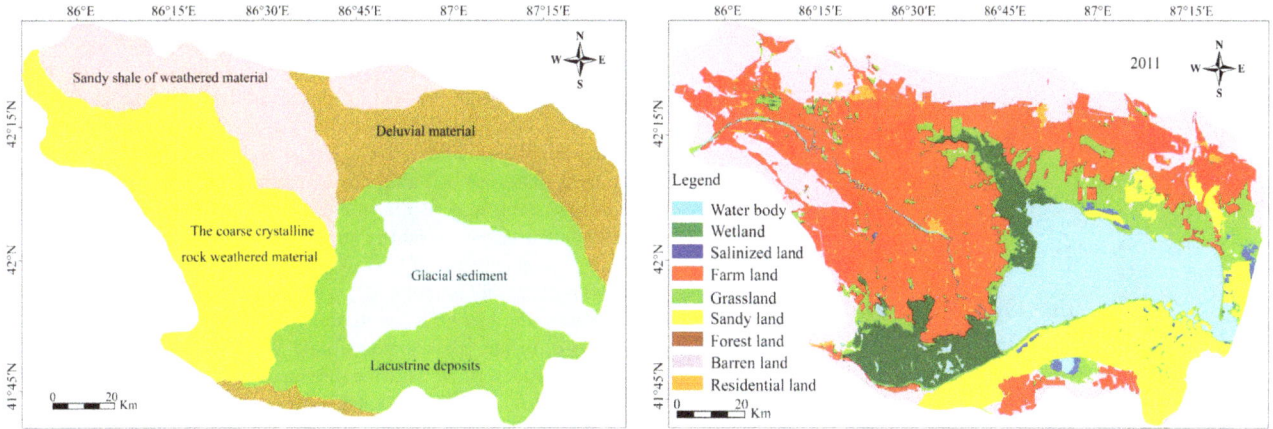

Figure 1. Land use types and parental material pattern in Yanqi basin.

content (Na^+, K^+, Ca^{2+}, Mg^{2+}, Cl^-, SO_4^{2-}, CO_3^{2-}, and HCO_3^-), we chose approximately 20% for retesting and found that 97.3% of the results were repeatable, inspiring confidence in the original data. After all the total salt ion contents (Na^+, K^+, Ca^{2+}, Mg^{2+}, Cl^-, SO_4^{2-}, CO_3^{2-}, and HCO_3^-) of the solution were determined, we recalculated them from unit of $\mu g/ml$ into mg/g (g/kg) using the method described by Bao (2005) [45]. To prevent contamination during the testing process, all glassware was soaked in 5% HNO_3 for 24 hours, rinsed and then dried.

Statistical analyses

Descriptive and multivariate statistical analysis. Descriptive statistical methods were used to analyze the range, mean, median, standard deviation, coefficient of variation, kurtosis and skewness of the total salt content, each salt ion, SAR, pH and EC of the soil samples. Correlation analysis, PC analysis and cluster analysis of the classic multivariate statistical method were used to process data and identify the soil salinity. Single factor analysis of variance (ANOVA) was used to analyze the differences in the amount of salt ions between different land use types. These analyses were all processed using the software SPSS 19.0.

Ordinary kriging method. Ordinary kriging (OK) is a commonly used linear spatial interpolation method that estimates variables at unsampled locations by using information from neighboring points and assigning weights to these points based on their distance from the point and the spatial variability structure. The OK method can be formulated as

$$Z^*_{OK}(x_0) = \sum_{i=1}^{n} w_i Z(x_i) \qquad (1)$$

where $Z^*_{OK}(x_0)$ is the OK estimation at an unsampled location (x_0), n is the number of samples in a search neighborhood, and w_i are the weights assigned to the ith observation $Z(x_i)$. Weights are assigned to each sample such that the estimation or kriging variance $E\left[\{Z^*(x_0) - Z(x_0)\}^2\right]$ is minimized and the estimates are unbiased [47]. Weights are determined after computing a semivariogram that models spatial correlation and covariance structure between data points for each variable using Eq. 1 [48].

$$\gamma(h) = \frac{1}{2N(h)} \sum_{i=1}^{N} [Z(x_i + h) - Z(x_i)]^2 \qquad (2)$$

where $\gamma(h)$ is the semivariance between two observation points $Z(x_i)$ and $Z(x_i + h)$ separated by a distance h, and N is number of observation pairs at the distance h.

Soil salinization evaluation criteria used in this research as under below (Table 1).

Results and Discussion

Descriptive statistical analysis of soil salinity

The descriptive statistics concerning the soil properties in Yanqi basin in Table 2 show that the maximum and average values of HCO_3^-, Ca^{2+}, CO_3^-, Na^+, Mg^{2+}, K^+, SO_4^{2-}, Cl^-, SAR, total amount of salts, EC, and pH were 0.98(0.19) $g.kg^{-1}$, 1.95(0.75) $g.kg^{-1}$, 0.85 (0.46) $g.kg^{-1}$, 2.43(1.19) $g.kg^{-1}$, 1.89(0.58) $g.kg^{-1}$, 2.13(0.69) $g.kg^{-1}$, 1.58(1.14) $g.kg^{-1}$, 2.36(0.98) $g.kg^{-1}$, 33.241(10.51), 14.77(9.73) $g.kg^{-1}$, 1.39(0.95) $dS.m^{-1}$, and 8.55(8.15), respectively. Within the analysis of the soil samples, a large amount of variation occurred, suggesting that the sources and influencing factors of the salt properties in Yanqi basin are complex. This work found that the main salt ions accounted for 84.41% of the total salt content and were K^+, Ca^{2+}, Na^+, Cl^-, Mg^{2+} and SO_4^2. Meanwhile the amount of HCO_3^- and CO_3^- was very low. The pH of the soil of the study areas ranged from 7.85 to 8.55, which had only a small variation. However, there was a dramatic change within the total salt content of the soil, ranging from 1.16 to 14.77 $g.kg^{-1}$.

The coefficient of variation is the ratio between the standard deviation and average, and it can be used to compare different dimensions of indicators. The coefficients of variation of HCO_3^-, CO_3^-, Na^+, Mg^{2+}, SO_4^{2-}, Cl^-, EC, and pH were 35.37%, 32.08%, 23.01%, 20.07%, 19.63%, 12.94%, 21.27%, and 30.32%, respectively, and were of medium variation (10%<CV<100%). However, the coefficients of variation of SAR, the total amount of salts, Ca^{2+} and K^+ were 148.56%, 126.73%, 191.67% and 226.25%, and, therefore, had high levels of variation (CV> 100%)[52]. In particular, Ca^{2+} and K^+ had higher coefficients of variation as compared to the other elements. From the perspective of skewness, the values of these ten soil properties are ordered as $K^+>Ca^{2+}>$total amount of salts$>Na^+>$SAR$>Cl^->$EC$>Mg^{2+}>$ $CO_3^->SO_4^{2-}>$pH$>HCO_3^-$.

The differences in soil salt properties from different land use types and soil parent materials

Land utilization types and soil parent materials are the main examples of human activity and geological background that

Table 4. The correction matrix of soil salt properties in Yanqi basin.

	EC	TSA	TSC	TS	Mg²⁺	Na⁺	K⁺	SO₄²⁻	Cl⁻	CO₃⁻	Ca²⁺	HCO₃⁻	pH
EC	1												
TSA	0.58	1											
TSC	0.40	0.30	1										
TS	0.41	0.52**	0.38**	1									
Mg²⁺	0.10	0.04	0.11	0.06**	1								
Na⁺	0.98**	0.34	0.96**	0.95**	0.01	1							
K⁺	0.65	0.48	0.78*	0.71**	0.51	0.64	1						
SO₄²⁻	0.87**	0.94**	0.42	0.93**	0.001	0.82**	−0.24	1					
Cl⁻	0.66**	0.55*	0.12	0.57**	0.13	0.69**	−0.15	0.24	1				
CO₃⁻	0.48*	0.49	0.27	0.48**	−0.24	0.57	0.12	0.27	0.17	1			
Ca²⁺	0.42	0.55	0.51*	0.54**	0.22	0.25	0.57*	0.23	0.03	−0.14	1		
HCO₃⁻	−0.25	−0.22	−0.20	0.21*	0.03	−0.16	−0.19	−0.42	0.28	0.16	−0.21	1	
pH	0.72**	0.62	0.14	0.82	0.42	0.32*	0.27	0.67**	0.58**	0.10	0.41	0.01	1

EC is electrical conductivity, TSA is the total anionic salt, TSC is the total cationic salt and TS is total salt content.

influence the salt ion content of soil. We analyzed the relationship between soil salt properties, human activity and geological background to further explore the distribution characteristics and sources of the soil salt properties of the Yanqi basin. The salt properties of the soil from each land use type and soil parent material of Yanqi basin are in Table 3. The land use types and soil parent materials of Yanqi basin are shown in Fig. 1. The analysis of these data suggests that the manner in which land is used has a significant influence on the amount of Ca^{2+} and K^+. In farmland, the average content of Ca^{2+} was 1.83 g.kg^{-1}, K^+ was 1.71 g.kg^{-1}, and the total amount of salts was 10.88 g.kg^{-1}. This was significantly higher than in the other land use types including areas of urban construction, forest, grassland, urban construction areas and desert. Meanwhile the maximum average values of Na^+, Mg^{2+}, SO_4^{2-}, Ca^{2+}, the total amount of salts, and K^+ found in farmland were higher than in grassland, desert, and areas of urban construction. The variance test attained a significant level of 0.05, indicating these elements and their distribution are mainly controlled by human activity. The activities that had a significant influence on these soil salt properties include agricultural activities, such as irrigating, fertilizing and farming. The CV calculated indicates that the pH, EC, total amount of salts and all the soil salt ions measured belong to medium variability categories (10–100%). Among the five land use types, the differences between these groups were small and, therefore, the classes of element enrichment were not obvious.

This research also found differences in the amount of organic soil salt ions. For example, HCO_3^- reached a maximum in lacustrine deposits of 0.74 mg.kg^{-1}. The maximum average values of Na^+, Mg^{2+}, K^+, SO_4^{2-}, total amount of salts, and Cl^- were found in coarse crystalline rock weathered material, sandy shale of weathered material and lacustrine deposits, while the maximum average values of CO_3^-, Ca^{2+}, and EC were found to be significantly higher in diluvial material than in sandy shale or coarse crystalline rock from weathered material or lacustrine deposits. For the five land use types, the class of element enrichment was not obvious as the differences between these groups were small.

R^2 represents the ratio of the sum of squares in groups and the total error of the sum of squares. It reflects the contribution of different factors on the soil salt properties [53]. For this study, the land use types explained the variances in Ca^{2+} (37.85%), K^+ (71.45%) and total amount of salts (56.71%), which were higher than that of the soil parent material. This demonstrates that the way land was used played a major role in the accumulation of Ca^{2+}, K^+ and total amount of salts in Yanqi basin. This analysis also found that the variances of HCO_3^-, CO_3^-, Mg^{2+}, Na^+, SO_4^{2-}, and Cl^- in the soil parent material are higher than those of the land use factors (Table 4), indicating that the soil parent material played a major role in the accumulation of these elements. However, there was little difference in the variances of EC and pH, indicating that these soil salt properties were mainly influenced by land use and soil parent material. Overall, the R^2 analysis fits well with the results of the multivariate statistical analysis.

Multivariable statistics

Correlation analysis. Table 4 shows the Pearson correlation coefficients between the soil salinity variables. There is a significant correlation of 0.96 (P<0.01) between the total amount of salt cations and Na^+, as well as the total amount of salt cations and Ca^{2+} at 0.51 (P<0.05) in the soil in the Yanqi basin, but there is no significant correlation with other salt cations. Furthermore, we found a significant correlation between the amount of salt

Table 5. Factors matrix of soil salt properties from Yanqi basin.

Soil salt properties	Principal components			
	PC 1	PC 2	PC 3	PC 4
K^+	−0.18	0.65	0.37	0.56
Mg^{2+}	0.37	0.80	−0.07	0.14
Ca^{2+}	0.63	0.74	0.50	0.45
CO_3^{2-}	−0.05	−0.02	0.59	−0.10
SO_4^{2-}	0.61	0.54	0.42	0.17
pH	0.49	0.15	0.06	−0.79
Cl^-	0.91	−0.04	0.29	0.13
Na^+	0.68	−0.08	0.56	0.31
EC	0.46	−0.01	−0.73	0.13
HCO_3^-	−0.09	−0.09	0.29	0.37
Total salt	0.21	0.78	0.14	0.62
Percentage of variance (%)	33.75	28.54	18.25	15.18
Percentage of cumulative variance (%)	33.75	62.29	80.54	95.72

anions and SO_4^{2-} of 0.94 (P<0.01), and the correlation coefficients between the total amount of salt anions and Cl^- or CO_3^{2-} are 0.55 and 0.49 (P<0.05), respectively. This indicates that SO_4^{2-} was the primary salt anion, Cl^- was the secondary and CO_3^{2-} was the tertiary. Together, the correlation coefficients between the total amount of salts and Cl^-, and the total amount of salts and SO_4^{2-} are 0.57 and 0.92, respectively (P<0.01), indicating that the main types of salt in the soil were sulfate and chloride. This study also found that there are close correlation coefficients between the EC and Na^+, the EC and Cl^-, and the EC and SO_4^{2-} content, where the coefficients are 0.98, 0.66, and 0.87, respectively (P<0.01). Additionally, the correlation coefficient between the soil EC and CO_3^{2-} is 0.48, which is also significant (P<0.05). Overall, a significant correlation was found between pH and EC, and total amount of salts (TS) and salt anions (Na^+, Ca^{2+}, Ca^{2+}, Ca^{2+}, Cl^-, SO_4^{2-}, HCO_3^-, and CO_3^{2-}) (P<0.01). Further analysis shows that the correlation coefficient between the pH and SO_4^{2-} is 0.67 and between pH and Cl^- is 0.58 (P<0.01), indicating the pH of the soil is primarily influenced by the SO_4^{2-} and Cl^- content.

Principal component analysis and clustering analysis. All of the elements studied were found to fall into four PCs (Table 5) with a cumulative variance of 95.72%. This is a reflection of their sources and main influences, particularly for the ten indicators of Cl^-, Na^+, EC, pH, K^+, Mg^{2+}, Ca^{2+}, HCO_3^-, CO_3^{2-}, total amount of salts, and SO_4^{2-}. Among these, the variance contribution rate of the first PC (Cl^-, Na^+, SO_4^{2-}, EC and pH) was 33.75% and the second PC (Ca^{2+}, K^+, Mg^{2+}, and total amount of salts) was 28.54%. The primary contribution was by agricultural development, in particular from herbicide application and acid salt fertilizer [54,55]. The variance contribution rates of the third (CO_3^-) and fourth PCs (HCO_3^{2-}) were at 18.25% and at 15.18%, respectively. Upon combining the sampling sites within their respective land use types, it was found that the samples that had a high content of salts were often taken from the desert, grassland and forest, and appear to have originated from natural sources. This analysis also showed there were larger loads of Ca^{2+} in the first PC, SO_4^{2-} in the second PC and K^+, total salt content in the fourth PC, indicating these elements were influenced by both artificial and natural sources. We used clustering analysis to see if the results are consistent with the results from the PC analysis

(Fig. 2). All the elements were classified into four categories: the first category consisted of Cl^-, Na^+, SO_4^{2-}, pH, and EC; the second of Mg^{2+}, Ca^{2+}, K^+, and the total amount of salts; the third of HCO_3; and the fourth of CO_3^{2-}.

Spatial distribution of soil salt ions in Yanqi basin. The spatial dependence of the soil salt properties was determined by semivariance analysis in order to quantitate the spatial variability. The parameters of the semivariogram included model type, nugget, sill and effective range. The nugget value (C_0) represents the random variation derived from measurement inaccuracy or variations in properties that cannot be detected in the sample range [54]. The sill value is the upper limit of the fitted semivariogram model [47]. The ratio of nugget to sill was a criterion to classify the spatial dependence of soil properties and it reflects the influence of regional factors (nature) and the role of the non-regional factors (human factors). The range of the semivariogram (A_0) represents the average distance through which the variable semivariance reaches its peak value. A small effective

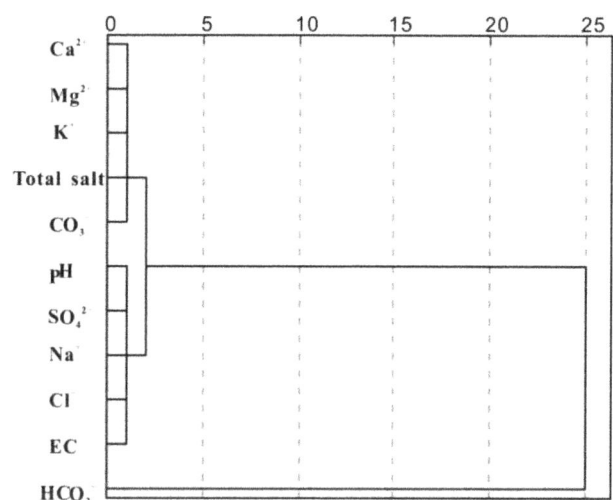

Figure 2. Clustering tree of soil salt properties of Yanqi basin.

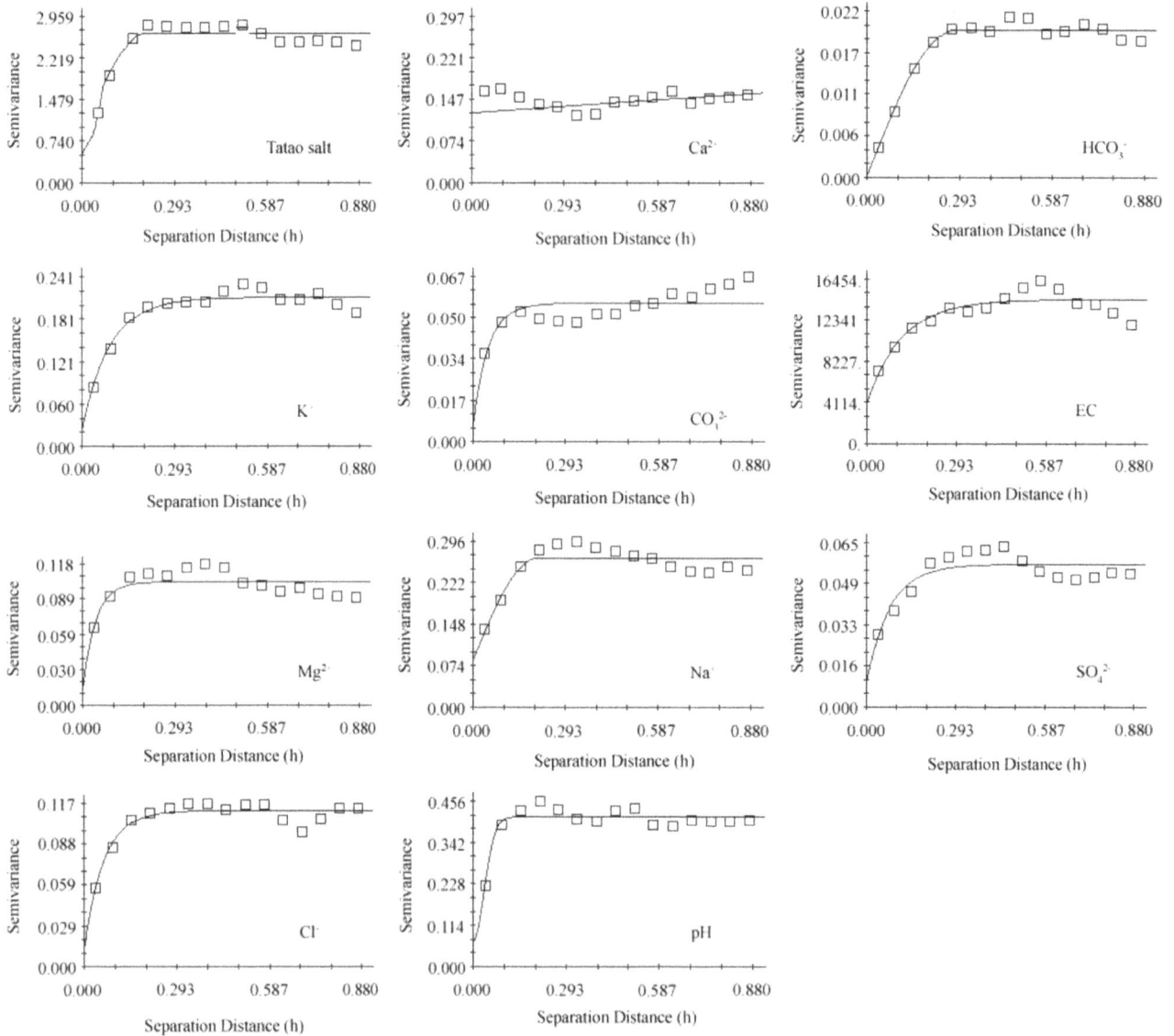

Figure 3. The semivariance function diagram of the soil salt properties of Yanqi basin.

range implies a distribution pattern composed of small patches. The cross-validation value is the coefficient of determination (R^2) of the correlation between the measured values and the cross-validation values, which were predicted based on the semivariogram and neighbor values [53]. Previous studies have shown that the semivariogram often differs considerably from its regional counterpart [33,53]. Fig. 3 and Table 6 show the tested variables over the study area that were modeled using spherical, gaussian and exponential semivariograms with the lower nugget effect based on the coefficients of determination (R^2) and residual sum of square (RSS). In this respect, a low ratio (less than 25% as was found for pH, Ca^{2+}, Mg^{2+}, and HCO_3^-) means that a large part of the variance is introduced spatially. This implies a strong spatial dependency of the variable, most likely due to intrinsic factors, including soil formation factors, such as soil parent materials, topography, and/or climate. A high ratio (more than 75% was found with Na^+ and Cl^-) often indicates a weak spatial dependency in the present sampling that was most likely due to extrinsic factors, including contamination, irrigation, and soil management

practices, where the fertilization and the soil chemical properties were continuously shifting. In this research the Nug/Sill ratios of CO_3^{2-}, and total salt content are 67.27 and 67.53% in the range between 25% and 75%, indicating the total salt content of the soil in Yanqi basin was influenced by the above factors. Additionally, all the other variables have a moderate spatial dependency for both intrinsic and extrinsic factors. The effective range calculated for variograms of different soil salt properties was 50–280 m, indicating that the sample distance was adequate for the characterization of the spatial variability of these properties.

The main application of geostatistics in soil science has been estimating and mapping the chemical properties in the soil of unsampled areas. Maps for each of the soil properties can be obtained using an ordinary kriging interpolation based on the best-fit semivariogram model. The skewness, defined as more than +1 or less than -1, indicates that some soil properties were not normally distributed. For these properties, it is difficult to estimate the semivariogram and doing so would result in a high value of kriging standard deviations [53]. Lognormal kriging with non-

Table 6. The spatial variation parameters of the soil salt properties of Yanqi basin.

Soil salt properties	Model	Nugget (Co)	Sill (Co+C)	Nug/Sill ratios $C_o/(C_o+C)$ (%)	Range A_o (m)	R^2	RSS
Na^+	Spherical	0.084	0.102	82.35	200	0.921	4.053E-03
K^+	Exponential	0.068	0.212	32.08	90	0.861	2.900E-03
Ca^{2+}	Exponential	0.034	0.217	15.67	101	0.954	1.131E-02
Mg^{2+}	Spherical	0.023	0.103	22.33	100	0.733	1.288E-03
Cl^-	Gaussian	0.087	0.111	78.38	100	0.891	4.104E-04
CO_3^{2-}	Exponential	0.037	0.055	67.27	50	0.914	4.014E-04
HCO_3^-	Spherical	0.004	0.021	19.05	280	0.927	2.446E-05
SO_4^{2-}	Exponential	0.012	0.041	29.27	80	0.823	1.826E-04
pH	Gaussian	0.071	0.415	17.11	50	0.872	5.215E-03
EC	Exponential	0.082	0.106	77.36	120	0.791	2.081E-05
Total salt	Spherical	0.497	0.836	59.45	180	0.833	1.472E-02

linear transformations is an alternative method for dealing with a data set with outliers or a non-normal distribution [31,33]. Since the concentration of the variables with the non-normal distribution had a lognormal distribution, their concentrations were log-transformed, resulting in more regular variograms. The kriging interpolations were performed on the log-concentrations and the estimated values were back-transformed by an exponential function. As seen in the results, total amounts of salt, pH, EC, and the concentrations of Na^+, K^+, Mg^{2+}, CO_3^{2-}, and Cl^- were relatively higher when from farmland and grassland. The spatial patterns of these variables had a significant geographical distribution with their primary occurrence in the western, north and central areas, which were higher than in other areas. This further proves the effect of the intrinsic factors of topography, soil forming factors and soil type, and extrinsic factors of soil management practices, such as fertilization, use of organic fertilizer and land management in farms, on the spatial distribution of soil chemical properties [33,53]. The soil salinity tended to increase from the margin to the center across the study area, where agricultural wells are denser. It is clear that substantial soil salinization has taken place in these areas due to the effects of land management, farming and climate conditions, such as rainfall and high evaporation. This shows that more attention should be paid to these areas to prevent future problems. Soil pH in the study area ranged from 7.95 to 8.55 in most parts of the Yanqi basin. This pH range falls into the middle level of values that meets the fundamental conditions for plant growth and fertility. The ECs of the soil in all sections were within the acceptable value of <4 dS/m as based on the soil quality standard given by Bao et al. [45]. Areas presently not affected by salinity, but near saline areas, are potential areas for the development of salinity in the future, especially if they are also low-lying. In this respect, it is important to take necessary precautions and implement proper land use plans and cultivation practices.

Assessment of soil salinization risk in Yanqi basin. Soil salinization occurs mainly in arid and semi-arid regions. It may arise due to climate, but is more likely to occur when irrigation practices alter the natural salt balance. Irrigation promotes soil salinization by raising the water table of the underlying aquifer, thus carrying salts upwards. Salts, unlike water, remain in the soil as evaporation and plant transpiration take place, thereby amplifying the salinity. The accumulation of salts in the surface and near-surface zones of soil is a major issue of environmental degeneration and is one of the main causes of low crop yields, loss of land and decreased production. The risk of soil salinization in the Yanqi basin is presented in Table 7. We observed that the whole area has a low salinization risk, as previously mentioned, mainly due to anthropogenic activities and climatic variation [39,40]. The overall salinization risks for farmland, grassland, forest, urban construction areas and desert were none, none, none, low, and moderate, respectively. The mean values of the salinization risk indicators, except the SAR in farmland and desert, are lower than the corresponding maximum limitations that are predicted for the risk (Table 7). This indicates that farmland and desert in the study area have a low salinization risk, but they have mean values of 15.2 and 32.7 for SAR, which showed a moderate grade of salinization (Table 1, Table 7). This indicates a potential risk for environment declination. This research showed the grassland and forest have no risk of salinization, while the areas of urban construction have a moderate risk of salinization. In terms of spatial distribution, the soil salinization risk and the soil salt ions present at higher concentrations in the study area were similar. Environment declination due to soil salinity within the study area almost all

Table 7. Classification of soil salinization risk taken at 0–20 cm depth within the investigated land use and land cover categories in Yanqi basin.

LUCC categories	EC	Total salt content	SAR	Levels of risk
Farmland	1.981	0.552	15.2	Low
Grassland	0.901	0.715	7.5	None
Forest	0.857	0.382	6.8	None
Urban construction areas	1.035	2.761	8.2	Low
Desert	1.024	3.657	32.7	Moderate

took place on the edge of a lake, pond or river. These areas are heavily affected by climate warming, which in turn results in an increase in the amount of evapotranspiration that exceeds 2438.9 mm/a and in the average annual rainfall <100 mm (the drought and waterlog) (Fig. 4). Apart from natural factors, the main driving factors that jointly determined how local dwellers changed the landscape pattern were land use policies, economic systems and population growth. Human activity increases salinization through excessive application of irrigation water without adequate drainage. In addition, the cultivation of grassland is another major cause of inland salinization [40,41]. On one side of the study area, plenty of grass landscape suffered damage, and, in some regions, the land salinization and desertification problem were serious enough to destroy the harmony between the material cycle and energy flow of the ecosystems [42]. On the other side of the study area, the habitat for wildlife was deteriorating, thus seriously threatening the biological diversity. Therefore, establishing and modifying policy, adjusting the irrigation system, improving drainage, dredging the surface water system, promoting circulation between surface water and underground water, and setting up wetland resource monitoring systems are necessary in order to restore damaged wetland and grassland [49,50,55].

Conclusions

(1) From all the soil samples taken in this study, it can be gathered that there are numerous salt properties that vary largely between different samples. This analysis shows that the main salt ions in the soil were K^+, Ca^{2+}, Na^+, Cl^-, Mg^{2+}, and SO_4^{2-}, which accounted for 84.41% of the total salt content of the soil samples. Conversely, the concentrations of HCO_3^- and CO_3^- were very low. Except for the high variation found for the amounts of Ca^{2+} and K^+ (191.67% and 226.25%, respectively), the other soil salt properties of Yanqi basin had moderate levels of variation (10%< CV<100%).

(2) From the analysis shown that the average values of Na^+, Mg^{2+}, SO_4^{2-}, Ca^{2+}, total amounts of salt and K^+ being higher in farmland than grassland, desert, and urban construction areas. This work also determined that the maximum average values of Na^+, Mg^{2+}, K^+, SO_4^{2-}, total amount of salts, and Cl^- were found in coarse crystalline rock weathered material, sandy shale of weathered material and lacustrine deposits. Within the five land use types examined, they were not obvious which classes of elements were enriched and the differences between these groups were small.

(3) PC analysis determined that PC1 (Cl^-, Na^+, SO_4^{2-}, EC, and pH) and PC2 (Ca^{2+}, K^+, Mg^{2+}, and total amount of salts) originated from artificial sources, while PC3 and PC4 (CO_3^- and HCO_3^{2-}) originated from natural sources. Together, this research

shows that Ca^{2+}, K^+, SO_4^2 and the total amount of salts were influenced by both artificial and natural sources. Clustering analysis is consistent with the results from the PC analysis.

(4) From the geo-statistical point of view, it can be speculated that pH and soil salt ions, such as Ca^{2+}, Mg^{2+} and HCO_3^-, had a strong spatial dependency. Meanwhile, Na^+ and Cl^- had only a weak spatial dependency, which was probably due to extrinsic factors, such as contamination, irrigation, and current soil management practices. We evaluated the EC, SAR and total salt content standard to reveal the risk of soil surface salinization. Soil salinization indicators suggest that the entire area had a low risk of salinization as mentioned previously, and this risk was mainly due to anthropogenic activities and climate variation. It is recommended that management of salinized land be preceded by an assessment of local factors and processes that may affect land composition.

Although the overall soil environment was healthy in the Yanqi basin, human activity, such as excessive groundwater pumping, have negatively impacted conditions by inducing soil salinization in the oasis. This matter deserves increased attention. This study can be considered an early warning of soil salinization and alkalization in the Yanqi basin. It can also provide a reference for environmental protection policies and for rational utilization of land resources in the arid region of Xinjiang, northwest China, as well as for other oases of arid regions in the world.

Acknowledgments

Many thanks to the members from the Institute of Geographical Science and Tourism of Xinjiang Normal University, Urumqi, China, and the

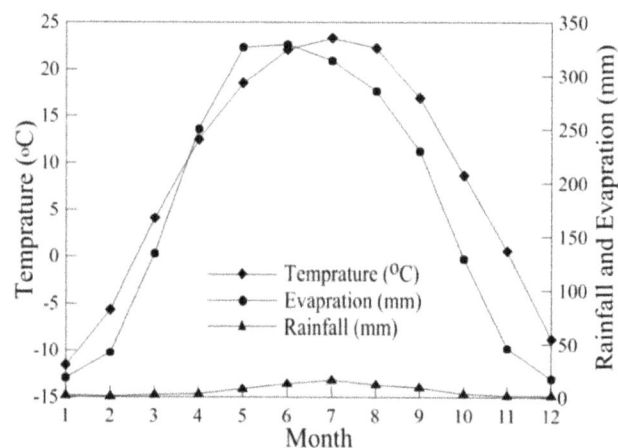

Figure 4. The monthly evaporation, precipitation and temperature of Yanqi basin in 2011.

members form College of Resources and Environment Sciences of Xinjiang University, Urumqi, China for the data collection and assistance in processing.

Author Contributions

Conceived and designed the experiments: ZZ JA. Performed the experiments: ZZ JA HY. Analyzed the data: ZZ JA. Contributed reagents/materials/analysis tools: ZZ JA HY. Contributed to the writing of the manuscript: ZZ.

References

1. Yang YJ, Yang SJ, Liu GM, Yang XY (2005) Space-Time Variability and Prognosis of Soil Salinization. Pedosphet 15(6):797–804.
2. Amezketa E (2006) An integrated methodology for assessing soil salinization, a pre-condition for land desertification. Journal of Arid Environments 67(4):594–606.
3. Masoud AA, Koike K (2006) Arid land salinization detected by remotely-sensed landcover changes: A case study in the Siwa region, NW Egypt. Journal of Arid Environments 66(1):151–167.
4. Yang JS (2008) Development and prospect of the research on salt-affected soils in China. Acta Pedologica Sinica 45(5):837-845. doi:10.3321/j.issn:0564-3929.2008.05.010
5. Stirzaker RJ, Cook FJ, Knight JH (1999) Where to plant trees on cropping land for control of dryland salinity: some approximate solutions. Agricultural Water Management 39(2):115–133.
6. Ghassemi F, Jakeman AJ, Nix HA (1995) Salinisation of land and water resources: human causes, extent, management and case studies. CAB international.
7. Chai S X, YangBZ, Wang XY, Wei L, Wang P, et al. (2008) Analysis of salinization of saline soil in west coast area of Bohai gulf. Rock and Soil Mechanics 29(5):1217–1221. (In Chinese).
8. Wang XL, Zhang FR, Wang YP, Feng T, Lian XJ, et al. (2013) Effect of irrigation and drainage engineering control on improvement of soil salinity in Tianjin. Transactions of the Chinese Society of Agricultural Engineering (20):82–88. (In Chinese).
9. Ren JG, Heng XL, Xi JM, Li JL (2005) Salinization characteristics of the soil in Yeerqiang river valley, Xinjiang. Soils 37 (6): 635–639. (In Chinese).
10. Chen XB, Yang JS, Liu CQ, Hu SJ (2007) Soil Salinization Under Integrated Agriculture and Its Countermeasures in Xinjiang. Soils 39(3):347–353. (In Chinese).
11. Wang YG, Xiao DN, Li Y (2008) Spatial and Temporal Dynamics of Oasis Soil Salinization in Upper and Middle Reaches of Sangonghe River, Northwest China. Journal of Desert Research 28(3):478–484. (In Chinese).
12. Jiang L, Li PC, Hu AY, Yi X (2009) Analysis and evaluation of soil salinization in oasis of arid region. Arid Land Geography 32(2):234–239. (In Chinese).
13. Sawut M, Eziz M, Tiyip T (2013) The effects of land-use change on ecosystem service value of desert oasis: a case study in Ugan-Kuqa River Delta Oasis, China. Canadian Journal of Soil Science 93(1): 99–108.
14. Ling H, Xu H, Fu J, Fan Z, Xu X (2013) Suitable oasis scale in a typical continental river basin in an arid region of China: A case study of the Manas River Basin. Quaternary International 286:116–125.
15. Eziz M, Yimit H, Mamat Z, Li JT (2013) Driving forces of farmland dynamics and its ecological effects in Keriya Oasis in recent 60 years. Agricultural Research in the Arid Areas 3:033. (In Chinese).
16. Triki I, Trabelsi N, Zairi M, Dhia HB (2014) Multivariate statistical and geostatistical techniques for assessing groundwater salinization in Sfax, a coastal region of eastern Tunisia. Desalination and Water Treatment 52(10-12):1980–1989.
17. Fu S, Wei CY (2013) Multivariate and spatial analysis of heavy metal sources and variations in a large old antimony mine, China. Journal of Soils and Sediments 13(1): 106–116.
18. Gil PM, Saavedra J, Schaffer B, Navarro R, Fuentealba C, et al. (2014) Quantifying effects of irrigation and soil water content on electrical potentials in grapevines (Vitis vinifera) using multivariate statistical methods.Scientia Horticulturae173:71–78.
19. Hu W, Shao MA, Wan L, Si BC (2014) Spatial variability of soil electrical conductivity in a small watershed on the Loess Plateau of China. Geoderma 230: 212–220.
20. Oliver MA, Webster R (2014) A tutorial guide to geostatistics: Computing and modelling variograms and kriging.Catena 113:56–69.
21. Emadi M, Baghernejad M (2014) Comparison of spatial interpolation techniques for mapping soil pH and salinity in agricultural coastal areas, northern Iran. Archives of Agronomy and Soil Science 60(9):1315–1327.
22. Elbasiouny H, Abowaly M, Abu-Alkheir A, Gad A (2014) Spatial variation of soil carbon and nitrogen pools by using ordinary Kriging method in an area of north Nile Delta, Egypt. Catena 113:70–78.
23. Wu Y, Wang Y, Xie X (2014) Spatial occurrence and geochemistry of soil salinity in Datong basin, northern China. Journal of Soils and Sediments 1–11.
24. Li SJ, Sun YN, Wang HB, Chen ZW (2013) Impact of Soil Nutrient Contents and Spatial Variability in Different Sampling Schemes under the Conservation Tillage.Advanced Materials Research 718:316–320.
25. Bilgili AV (2013) Spatial assessment of soil salinity in the Harran Plain using multiple kriging techniques. Environmental monitoring and assessment 185(1):777–795.
26. Sylla M, Stein A, Van Breemen N, Fresco LO (1995) Spatial variability of soil salinity at different scales in the mangrove rice agro-ecosystem in West Africa. Agriculture, ecosystems & environment 54(1), 1–15.
27. Ammari TG, Tahhan R, Abubaker S, Al-Zu'Bi Y, Tahboub A, et al. (2013) Soil salinity changes in the Jordan Valley potentially threaten sustainable irrigated agriculture.Pedosphere 23(3):376–384.
28. Walter C, McBratney AB, Douaoui A, Minasny B (2001) Spatial prediction of topsoil salinity in the Chelif Valley, Algeria, using local ordinary kriging with local variograms versus whole-area variogram. Soil Research 39(2), 259–272.
29. Jordán MM, Navarro-Pedreno J, García-Sánchez E, Mateu J, Juan P (2004) Spatial dynamics of soil salinity under arid and semi-arid conditions: geological and environmental implications. Environmental Geology 45(4): 448–456.
30. Bai YL, Li BG, Hu KL (1999) Spatial variability of soil salt and its composing ions in salt-affected soil in Huang-Huai-Hai plain. Soil and fertilizer (3): 22–26. (In Chinese).
31. Hu Kl, Li BG, Chen DL (2001) Spatial variability of soil water and salt in field and estimating soil salt using CoKriging. Advances in water science 12(4): 460–466. (In Chinese).
32. Xu Y, Chen YX, Shi HB (2004) Scale effect of spatial variability of soil water-salt. Transactions of the Chinese Society of Agricultural Engineering 20(2): 1–5. (In Chinese).
33. Lin J, Anwar M, Dilbar S (2007) Investigation of the spatial variability of soil salts in saline soil in Xinjiang. Research of Soil and Water Conservation 14(6): 189–192.
34. Aragüés R, Medina ET, Claverìa I, Martínez-Cob A, Faci J (2014) Regulated deficit irrigation, soil salinization and soil sodification in a table grape vineyard drip-irrigated with moderately saline waters. Agricultural Water Management 134, 84–93.
35. Bouksila F, Bahri A, Berndtsson R, Persson M, Rozema J, et al. (2013) Assessment of soil salinization risks under irrigation with brackish water in semiarid Tunisia. Environmental and Experimental Botany 92, 176–185.
36. Liang SY, Wu TN, Wu YS, Chou YC, Lee CH (2013) Assessment of Aquifer Salinization Beneath an Offshore Industrial Park Based on Solute Transport Calculation. Advanced Materials Research 779, 1285–1288.
37. Chen L, Feng Q (2013) Geostatistical analysis of temporal and spatial variations in groundwater levels and quality in the Minqin oasis, Northwest China.Environmental Earth Sciences 70(3):1367–1378.
38. Wang LC, Wu RW, Gao J (2014) Spatial coupling relationship between settlement and land and water resources–based on irrigation scale–A case study of Zhangye Oasis.Advanced Engineering and Technology 225.
39. Mamat Z, Yimit H, Eziz M, Ablimit A (2013) Analysis of the Ecology-Economy Coordination Degree in Yanqi Basin, Xinjiang, China. Asian Journal of Chemistry 25(16): 9034–9040.
40. Mamat Z, Yimit H, Eziz A (2014) Oasis land-use change and its effects on the eco-environment in Yanqi Basin, Xinjiang, China. Environment Monitoring and Assessment 186(1): 335–348.
41. Wu JS, Zhang YQ, Liu ZH, Peng J, He JF (2010) Land salnization monitorng with remote sensing on Yanqi County, Xin jiang. Arid Land Geography 32(7):251–257. (In Chinese).
42. IUSS Working Group, WRB (2006). World reference base for soil resources. World Soil Resources Report,103.
43. Hargreaves G H, Allen R G (2003) History and evaluation of hargreaves evapotranspiration equation. Journal of Irrigation and Drainage Engineering 129(1): 53–63.
44. Lu RK (2000) Soil and agricultural chemistry analysis. Beijing/China Agricultural.
45. Bao SD (2005) Soil Agricultural Chemistry Analysis. Beijing/China Agriculture Press. pp: 17–200.
46. Carter M R (1993) Soil sampling and methods of analysis. CRC Press.
47. Lark RM (2001) Geostatistics for environmental scientists.European Journal of Soil Science 52(3), 526–526.
48. Cressie N (1988) Spatial prediction and ordinary kriging. Mathematical Geology 20(4): 405–421.
49. Fan LQ, Yang JG, Xu X, Sun ZJ (2012) Salinity characteristics and correlation analysis of saline soil in irrigation area of Ningxia. Soil and Fertilizer Sciences in China 6: 003. (In Chinese).
50. Metternicht G, Zinck JA (1997) Spatial discrimination of salt-and sodium-affected soil surfaces. International Journal Remote Sensing 18(12): 2571–2586.
51. Agriculture Department of Xinjiang (1996) Soil Survey Office in Xinjiang. Xinjiang Soil. Beijing/Science press 458–464.
52. Wilding LP (1984) Spatial variability: Its documentation, accommodation and implication to soil surveys//MNielson D R, Bouma J. Soil Spatial Variability. Purdoc, Wageningen: 166–193.

53. Fan XM, Liu GH, Liu HG (2014) Evaluating the spatial distribution of soil salinity in the Yellow river delta based on Kriging and Cokriging Methods. Resources Science 36(2):0321–0327. (In Chinese).

54. Wang SX, Dong XG, Liu YF (2009) Spatio-Temporal variation of subsurface hydrology and groundwater and salt evolution of the oasis area of Yanqi Basin in 50 Years recently. Geological Science and Technology Information 28(5):101–108. (In Chinese).

55. Li XG, Lai N, Chen SJ, Mamattursun E (2014) Spatial variability of soil salt based on geostatistics and GIS in the oasis of the lower reaches of Kaidu River: a case study on Yanqi County. Gco-graphy and Gco-information Science 30(1):105–109. (In Chinese).

9

Linking *Populus euphratica* Hydraulic Redistribution to Diversity Assembly in the Arid Desert Zone of Xinjiang, China

Xiao-Dong Yang[1,2], Xue-Ni Zhang[3,4], Guang-Hui Lv[3,4]*, Arshad Ali[1,2]

1 College of Ecological and Environmental Sciences, East China Normal University, Shanghai, China, 2 Tiantong National Forest Ecosystem Observations and Research Station, Chinese National Ecosystem Observation and Research Network, Ningbo, China, 3 Institute of Resources and Environment Science, Xinjiang University, Urumqi, China, 4 Xinjiang Key Laboratory of Oasis Ecology, Ministry of Education, Urumqi, China

Abstract

The hydraulic redistribution (HR) of deep-rooted plants significantly improves the survival of shallow-rooted shrubs and herbs in arid deserts, which subsequently maintain species diversity. This study was conducted in the Ebinur desert located in the western margin of the Gurbantonggut Desert. Isotope tracing, community investigation and comparison analysis were employed to validate the HR of *Populus euphratica* and to explore its effects on species richness and abundance. The results showed that, *P. euphratica* has HR. Shrubs and herbs that grew under the *P. euphratica* canopy (under community: UC) showed better growth than the ones growing outside (Outside community: OC), exhibiting significantly higher species richness and abundance in UC than OC ($p < 0.05$) along the plant growing season. Species richness and abundance were significantly logarithmically correlated with the *P. euphratica* crown area in UC ($R^2 = 0.51$ and 0.84, $p < 0.001$). In conclusion, *P. euphratica* HR significantly ameliorates the water conditions of the shallow soil, which then influences the diversity assembly in arid desert communities.

Editor: Han Y.H. Chen, Lakehead University, Canada

Funding: This work was supported by the National Natural Science Foundation of China (Grant No. 31060061), and Doctoral Program Foundation of Institutions of Higher Education of China (Grant No. 20106501110001). The funders had no role in study design, data collection and analysis, decision to publish, or preparation of the manuscript.

Competing Interests: The authors have declared that no competing interests exist.

* Email: ler@xju.edu.cn

Introduction

The hydraulic redistribution (HR) is defined as the movement of water from the moist to dry soil portions through roots of deep-rooted plants [1–3]. Through this process, water moves upward or laterally among different soil layers, and subsequently changes the water spatial pattern of soil [3,4]. Thus, HR improves the fine roots survival rate of deep-rooted plants and protects the shallow-rooted plants growth in extremely dry environments [4–6]. In addition HR plays an important role in positive feedback between plants and soil moisture circulation [1,7–10], it also benefits the plant rhizosphere nutrient absorption, community stability, and biodiversity maintenance [3,11–20].

Populus euphratica Oliv., a deciduous plant with high tolerance to drought and saline conditions, and is a dominant species for the Tugai Forest (desert riparian forest) in the Ebinur desert. Pervious research findings have revealed that the water gradient of soil layers determines the occurrence of HR in deep-rooted plants [3,4,11], such as when the surface soils are extremely dry and the deeper soils are rich in available water [3,4,21]. *P. euphratica* developed a deep vertical root system for absorbing water in Chinese arid deserts due to effect of high wind [22–24]. Moreover, the annual precipitation of this site is less than 100 mm and the surface soils are extremely dry, whereas the groundwater level is relatively high because of glacial melt waters [22]. Hence, *P. euphratica* should possess HR.

In our previous study [22] and the one conducted by Hao et al. [23] employed the Ryel model [25], that was used to simulate HR of *P. euphratica*. According to the research findings of the above mentioned studies that *P. euphratica* has HR which reached a maximum point at 2:30 am for a given day, and the total amount of HR decreased with changes in the plant growing season [22,23]. But, those results were only based on water characteristics (water content and water potential) changes along the various soil layers, which are only obtained from model predictions. The experimental support of *P. euphratica* possessing HR is lacking. Here we investigate whether oxygen isotope tracking can be used to test whether the plants possess HR [1–4,11], and this could provide the experimental support to the studies based on model simulations. As oxygen isotope does not fractionate during water transport through xylem vessels, whereas the $^{18}O/^{16}O$ values varied across the different soil layers [1,26–28]. For example, the $\delta^{18}O$ content continues to increase from deep to shallow soils based on fractionation. Thus, we can predict that if *P. euphratica* possess HR, *P. euphratica* roots can lift water containing lower $\delta^{18}O$ content from groundwater and deep soils to shallow soils through xylem vessels, which can result in soil having lower $\delta^{18}O$ content under the *P. euphratica* canopy (under community: UC) rather than outside the *P. euphratica* canopy (outside community: OC).

Annual precipitation in Ebinur desert is very scarce, which is insufficient for the basic physiological activity of plants, while the deep soils and shallow groundwater contains abundant water for plant growth in this region [29]. However root density of most plants decreases with increasing soil depth [22,30,31], and only a small number of deep-rooted plants are able to penetrate to the deeper soils or the water table to absorb water which limits water utilization and subsequently, decreases the species richness and abundance in the arid desert [32,33]. So, if *P. euphratica* HR lifts water from the deeper soils or groundwater aquifer to shallow soils through its roots, this process can increase the water content of surface soils and afford more water for shallow-rooted shrubs and herbs. In this case, it can induce the plants assembly under *P. euphratica* canopy and subsequently maintain the biodiversity in arid deserts. Therefore, we hypothesize that, the shallow-rooted shrubs and herbs grow better in UC than in OC and there is a higher species richness and abundance in UC than in OC, because *P. euphratica* HR significantly improves the soil water content. But, the correlation between *P. euphratica* HR, plant growth condition and biodiversity in arid desert areas is poorly understood.

The objective of this study was to test for *P. euphratica* HR and to explain the effect of HR on plants growth condition and species biodiversity maintenance in the Ebinur desert. We predict the following: (1) $\delta^{18}O$ content of the shallow soils in UC are higher than in OC; (2) shrubs and herbs of UC have higher growth condition than those of OC; and (3) *P. euphratica* HR influence species abundance and richness in arid desert.

Materials and Methods

Study site and Ethics Statement

The experimental site is located in the Ebinur Lake Wetland Nature Reserve (ELWNR) in the western margin of the Gurbantonggut Desert in Xinjiang Uygur Autonomous Region of China (44° 30′–45° 09′ N, 82° 36′–83° 50′ E). This site belongs to a tuyere zone of Alashankou, the annual wind days (days with wind speed ≥ 17 m/s) are more than 164 d, and the annual fresh gale hours are approximately 241 hr. The annual sunshine hours reach approximately 2800 hr, and the annual precipitation is less than 100 mm, whereas the potential evaporation is more than 1600 mm. In addition, the temperature in this area ranges from 44 to −33°C, with an average temperature ranging from 6 to 8°C, and an average temperature of the growing season is roughly 25°C. Due to the extremely dry conditions and sparse rainfall, the climate is classified as typical temperate continental arid [22].

No specific permissions were required for the described field studies in ELWNR. The ELWNR is owned and managed by the local government and the location including the site used for our experiment are not privately owned or protected in any way and thus a specific permit for not for-profit research is not required. The field studies did not involve endangered or protected plant species in this area.

Study plot and samples collection

5×5 km (25 km²) typical plots of *P. euphratica* were chosen as our experimental plots, where 48 *P. euphratica* individuals are distributed and the vegetal coverage is 4%. This site has a moderate slope and sandy soil. The distribution of groundwater level in our experimental site ranges from 1.5 to 1.8 m. Moreover, sand dunes are present 3 km north of the site. Based on the terrain of the experimental site, we randomly selected three *P. euphratica* individuals with no differences in growth conditions (DBH, tree height and crown area) among those trees as the experimental

(UC) group. The selected plant individuals were at least 6 m apart from each other to prevent mutual water transfer process among them. For comparison, we randomly selected three plots located closely to the selected three *P. euphratica* individuals as the control (OC) group, where no *P. euphratica* were growing and no groundwater table difference with UC group.

To distinguish the difference in water transport from deeper soil layers to surface soil layers between the UC and OC groups, three points in each two types of plots were randomly selected and each point was divided into five soil layers (0 to 10 cm, 10 to 40 cm, 40 to 70 cm, 70 to 100 cm, and 100 to 150 cm, the deepest soil layer is 150 cm, as groundwater appeared approximately at 160 cm when the soil columns were dug out in the above selected OC and UC points). Each soil sample was dug with a soil auger between 4:30 AM and 5:30 AM (Xinjiang local time) in the middle of July, 2010. The nine soil samples for each soil layer from three points were mixed to makeup a composite soil samples. Meanwhile, on the periphery of selected plots, well water 3 m beneath ground level was collected from five sites, and subsequently mixed them to obtain a composite sample to replace groundwater. A river water sample was also mixed from three sites at least 1 km apart of the Aqikesu River. Each sample of river water was collected from 1 m beneath river surface. All the soil and water samples were immediately placed into glass bottles, sealed with Parafilm and stored in a mini refrigerator (4°C) and brought to the Xinjiang University Physiological Ecology Laboratory for further experimental analysis.

Oxygen isotope measurement

Methods for extraction of soil water are consistent with Allison et al. [34]. Water isotope content ($\delta^{18}O$) was measured using a DELTA V Advantage Isotope Ratio Mass Spectrometer (Thermo, Waltham, MA, USA) at the Chinese Academy of Forestry Stable Isotope Laboratory. Each sample was measured three times continuously with the third result as the experimental oxygen isotope value. Precision values of continuous measurements for standard sample were as follows: D, <3‰ and ^{18}O, <0.5‰. The isotopic abundance was expressed in delta notation (δ) in parts per thousand (‰) as

$$\delta = (R_{sample}/R_{standard} - 1) \times 1,000 \qquad (1)$$

Where Rsample and Rstandard are the molar ratios of heavy to light isotope of the sample and the international standard (Vienna standard mean ocean water for $^2H/^1H$ and $^{18}O/^{16}O$) [1].

Plots investigation

48 plots of UC were established having each plot size of 10×10 m (14 plots in June, 17 plots in August and October) in 5×5 km experimental plot from June to October, 2010. Consequently, 34 control plots with 10×10 m of OC (13 plots in June, 15 plots in August and 16 plots in October) were set at the same time. The species identification, richness, abundance, coverage, DBH, and crown area of each plant in each plot were investigated.

Data analysis

To determine whether *P. euphratica* undergoes HR, a comparison analysis was used to show the difference in $\delta^{18}O$ content for the five soil layers between the UC and OC groups. In addition, the paired-sample *t*-test was used to test the significant difference between above mentioned two groups.

Furthermore, based on the plots investigation data, the growth dominance index (Formulae 2 to 4) was used to show the growth

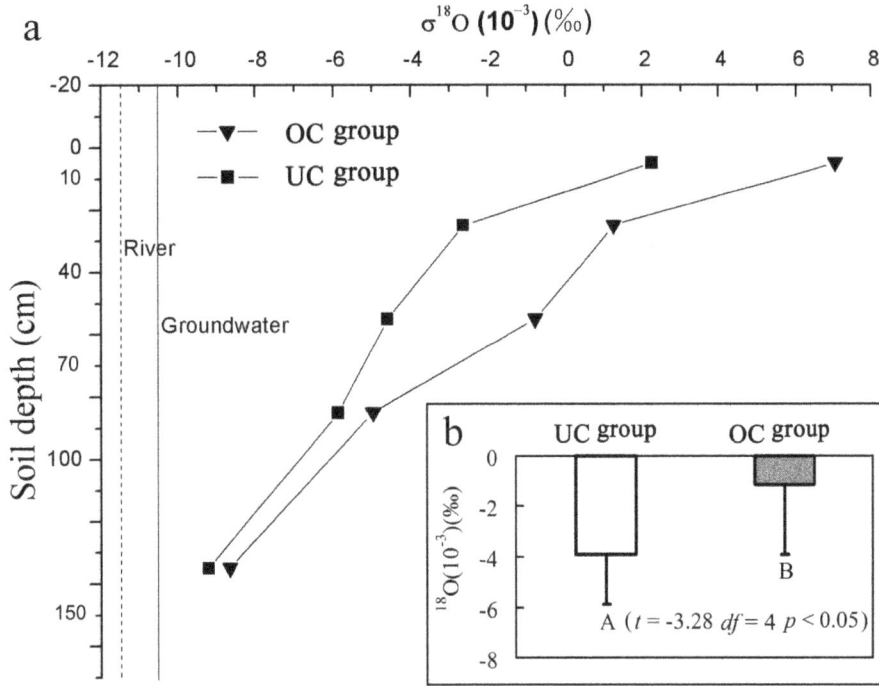

Figure 1. Difference in $\delta^{18}O$ value with soil depth between UC and OC groups. UC is the community under the *P. euphratica* canopy, whereas the OC is the community outside the *P. euphratica* canopy. The soils were divided into five layers (0 to 10 cm, 10 to 40 cm, 40 to 70 cm, 70 to 100 cm, and 100 to 150 cm) based on *P. euphratica* roots and underground water level distributions. Each point in the Fig. 1a showing $\delta^{18}O$ for each soil layer sample, which was measured for three times continuously with the third result as the experimental oxygen isotope value. Nine soil samples for each soil layer from three points were mixed to get one composite soil in each group. ^{18}O contents of river ($-11.55‰$) and underground water ($-10.59‰$) are show through vertical dashed and solid lines respectively in Fig. 1a. Fig. 1b shows the mean comparison between UC and OC groups and the data in parenthesis showing the Paired-Sample t-test result. (Mean ± SD) for significance difference between two groups.

condition, i.e. difference between UC and OC.

$$x_s = 1 - \frac{x_{max} - x}{x_{max} - x_{min}} \qquad (2)$$

$$DI = \frac{x_{sc} + x_{sa} + x_{sh} + x_{sf}}{400} \qquad (3)$$

$$GDI = \frac{\sum_{i=1}^{N} DI_i}{N} \qquad (4)$$

Where x_s, x_{sc}, x_{sa}, x_{sh} and x_{sf} are the standardized value, coverage, abundance, height and frequency for a given species in a given plot respectively. x_{max} and x_{min} are the maximum and minimum values for a given species in a given plot while x is the actual investigative value of each individual belonging to the same species in a given plot. DI is the dominance index in a given plot, and GDI is the growth dominance index of all plots. N is the count of plots, and DI_i is the dominance index of a given species in i th plot.

Individual size has positive correlations with the amount of HR which can generally represent the water lift capacity of HR. It is well understood that the crown area of arid desert vegetation is significantly linked with adult size (tree height) and root biomass [35,36]. Thus in this study, we used the individual crown area (CA, Formula 5) to represent water lifting capacity of *P. euphratica* HR. Maximum vegetation crown diameter (CD1)

and its perpendicular diameter (CD2) were measured on each plant individual, which were used to calculate CA as follows (Formula 5).

$$CA(m^2) = CD1(m) \times CD2(m) \times \frac{\pi}{4} \qquad (5)$$

In order to explain the effects of HR on species richness and abundance in the arid desert community, an independent sample t-test was used between UC and OC to determine the diversity difference for shrubs and herbs. Finally, the logarithmic regression analysis was applied to explore the relationship of HR capacity (CA) to species richness (species number in 10×10 m) and abundance (count of individual plants in 10×10 m) for arid desert community.

All statistical tests were conducted using SPSS 11.5 while related figures were drawn using Origin 8.0. All statistical tests were considered significant at the $p < 0.05$ level.

Results

Comparisons of $\delta^{18}O$ values between the UC and OC

$\delta^{18}O$ content of the OC were generally higher than that of the UC among five soil layers. In addition, the maximum difference between the two groups were observed in the surface soil layer (0 to 10 cm), and then decreased with soil depth (Fig. 1a). Therefore, a paired-sample t-test exhibited that the mean of five soil layer $\delta^{18}O$ contents of OC was significantly higher than that of UC ($t = -3.28$, $df = 4$, $p < 0.05$, Fig. 1b). The general comparison

Table 1. Differences of species growth dominance index between the UC and the OC.

Life form	Species	Growth dominance index					
		June		August		October	
		UC	OC	UC	OC	UC	OC
Shrub	Kalidium foliatum	0.29	0.27	0.31	0.19	0.19	0.12
	Reaumuria soogarica	0.54	0.35	0.53	0.33	0.41	0.41
	Alhagi sparsifolia	0.67	0.50	0.29	0.22	0.34	0.25
	Lycium ruthenicum	0.23	0.21	0.35	0.32	0.26	0.09
	Halostachys caspica	–	0.06	0.26	0.33	0.16	0.21
	Calligonum L	0.15	0.08	0.05	–	0.05	0.05
	Populus euphratica	0.41	0.18	0.06	–	0.05	–
	Nitraria schoberi	0.20	0.13	0.12	–	–	0.05
	Haloxylon persicum	0.22	–	0.31	0.05	0.32	0.21
	Salsola passerina	0.05	–	0.20	0.03	0.15	0.05
	Nitraria sibirica	–	–	0.15	0.10	0.12	0.12
	Tamarix ramosissina	–	–	0.14	–	0.13	–
Herbage	Karelinia caspicas	0.42	0.25	0.14	0.11	0.05	0.05
	Salsola soda	0.67	0.51	0.54	0.45	–	–
	Salsola nitraria	0.31	0.31	0.30	0.43	0.18	0.14
	Poacynum henclersoni	0.55	0.44	0.53	0.41	–	–
	Spriphidium.	0.34	0.11	0.09	–	0.05	–
	Suaeda glauca	–	–	0.22	0.10	0.17	0.05
	Aeluropus littoralis	0.07	0.06	0.05	–	0.05	–
	Phragm ites australis	–	–	0.17	0.30	0.15	0.05
	Scorzonera austriaca	0.11	0.18	0.10	–	–	–
	Carpesium abrotanoides	0.20	0.06	–	–	–	–
	Afriplex patens	0.08	0.06	–	–	–	–
	Petrosimonia sibirica	0.09	–	–	–	–	–
	Malcolmia africana	0.38	0.12	–	–	–	–

UC is the community under the *P. euphratica* canopy, whereas the OC is the community outside the *P. euphratica* canopy. For UC, *P. euphratica* is the only tree species (woody plant. For UC, *P. euphratica* is the only tree species (woody plant, height >6 m) while the others are shrub (woody plant, height <6 m) and herbage (herbaceous plants, height <1 m). For OC, all plants are shrubs and herbages. "–" in table indicating no values because of the ephemeral plants life history turnoff and randomly setting samples along June (early growth period of *P. euphratica*), August (middle growth period of *P. euphratica*) and October (defoliating period of *P. euphratica*). The figures in table showing Growth dominance indices of species belonged to UC and OC. All plots are established in 5×5 km area of Ebinur Lake Wetland Nature Reserve in Xinjiang Uygur Autonomous Region of China.

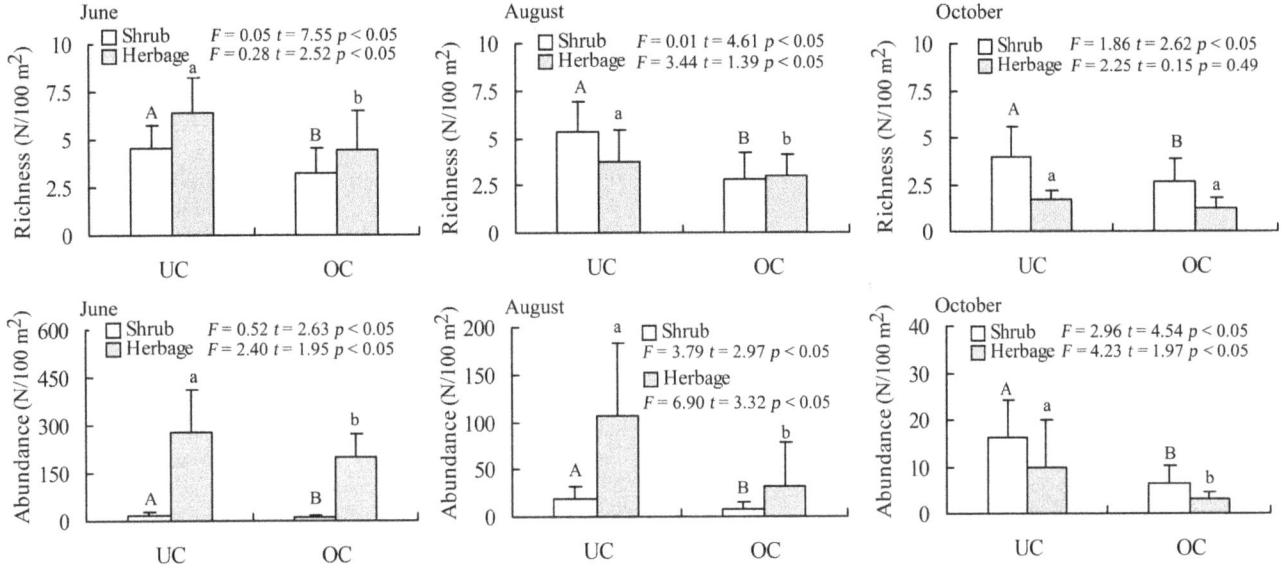

Figure 2. Difference in richness and abundance along plant growing season between UC and OC. UC is the community under the *P. euphratica* canopy, whereas the OC is the community outside the *P. euphratica* canopy. Blank and grid boxes indicate shrub and herbage, respectively. The Independent sample *t*-test is used to analyze the differences of richness and abundance between UC and OC. Different capital letters on each blank box indicate significant differences of shrub richness or abundance between UC and OC. Different lowercase letters on each grid box indicate significant differences of herbage richness or abundance between UC and OC. $p<0.05$. Numbers in figure are the results of Independent sample *t*-test. (Mean ± *SD*) for significance difference between UC and OC.

showed that $\delta^{18}O$ content was the highest in soil, medium in groundwater and the lowest in river (Fig. 1a).

Difference in species dominance index and biodiversity between the UC and OC

Both the shrubs and herbs growth dominance index (*GDI*) of UC were higher than that of OC in June, August and October (Table 1). This finding indicates that the shrubs and herbs species grew better in UC than that in OC. Furthermore, compared with OC, the shrubs richness and abundance of the UC were significantly higher across three months ($p<0.05$) (Fig. 2). But

for the herbs, this pattern of richness and abundance varies along the plant growing season. Between UC and OC, the species richness and abundance of herbs and shrubs were significantly similar in June and August, while non-significant differences were found in October (Fig. 2). Further, based on all UC plots investigated, the logarithmic regression analysis was used to analyze the relationship of *P. euphratica* individual HR capacities (individual crown area) with species richness and abundance. The results show that, species richness and abundance exhibited a significant exponent correlation with *P. euphratica* CA ($p<0.05$) (Fig. 3).

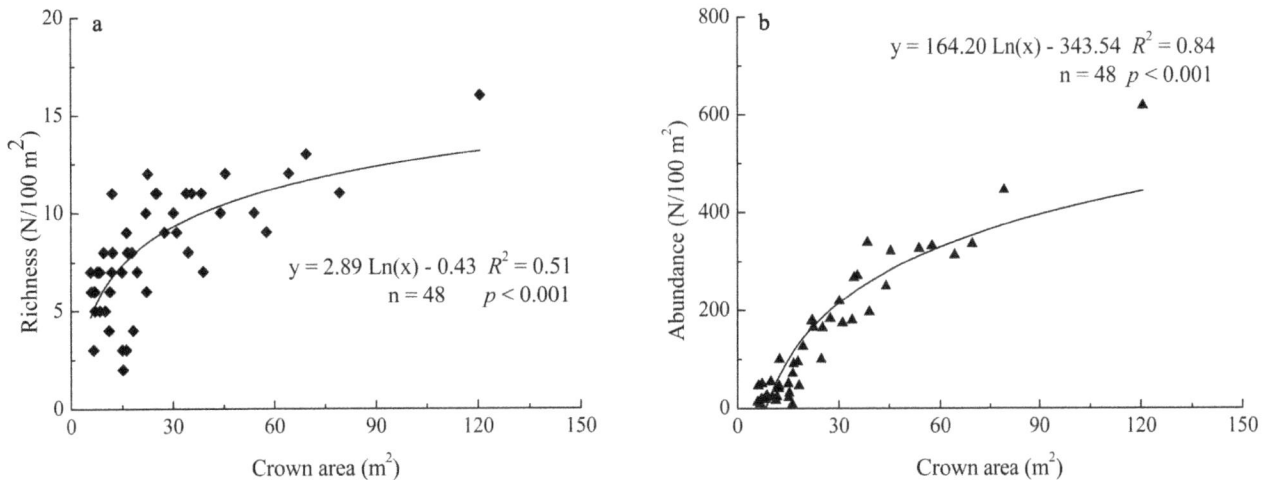

Figure 3. Logarithmic regression of *P. euphratica* crown areas against species abundance and richness. In regression line (Fig. 3a), each point indicating the richness of all species (no including *P. euphratica*) under the *P. euphratica* canopy. In regression line (Fig. 3b), each point indicating the abundance of all species (no including *P. euphratica*) under the *P. euphratica* canopy. The crown area of each sample is the sum of *P. euphratica* individuals.

Table 2. Variation in soil volumetric water content across soil depth and $\delta^{18}O$ content in UC and OC.

Soil layers (cm)	$\delta^{18}O$ (‰)		Soil volumetric water contents (Mean ± SD) (cm³/cm³)		F	t	P
	UC	OC	UC	OC			
0-10	2.27	7.10	8.41±0.38A	3.86±1.67B	45.60	21.79	<0.001
10-40	-2.62	1.25	29.72±0.41A	18.86±0.26B	1.95	181.58	<0.001
40-70	-4.60	-0.77	12.63±0.45A	20.26±0.35B	16.03	-109.46	<0.001
70-100	-5.87	-4.98	23.07±0.40A	21.40±0.07B	2.08	34.07	<0.001
100-150	-9.21	-8.67	27.98±0.03B	27.55±0.02B	66.78	7.49	<0.05

UC is the community under the *P. euphratica* canopy, whereas the OC is the community outside the *P. euphratica* canopy. UC and OC have three measuring plots, respectively. The independent sample *t*-test is used to analyze the differences of soil volumetric water content between UC and OC. Different capital letters in each row indicate significant differences of soil volumetric water content between UC and OC. The measurement period of soil volumetric water content lasts from 5th to 8th August, 2010. During the time between 2:30 am to 6:30 am, 11:30 am to 5:00 pm at each experimental day, the soil volumetric water content recorded manually at intervals of 2 hours, while the time between 6:30 am to 9:30 am, 9:30 pm to 12:30 am, the soil volumetric water content recorded manually once an hour.

Discussion

Experimental validation of *P. euphratica* HR

Oxygen isotope fractionates during water transport between the different soil layers because of physical and chemical adsorption and conduction processes, which changes the $^{18}O/^{16}O$ values in different soil layers. However, the fractionation of oxygen isotopes does not occur during water transport through the xylem vessels [1,26-28]. So the water originating from the deeper soils or groundwater in the shallow soils is an indicator for HR occurrence [1-3,11,37]. In the present study, the soil samples of UC and OC groups were collected at between 4:30 AM and 5:30 AM (local time), during which the fractionation not occur due to soil evaporation. Therefore, if *P. euphratica* HR does not occur, as the environmental conditions were same between UC and OC groups, the $\delta^{18}O$ content of the corresponding soil layers did not differ significantly between UC and OC groups. However, in this study, all $\delta^{18}O$ content of UC were lower than those of OC group among five soil layers (Fig. 1a), and the mean of UC was also significantly lower than OC group (Fig. 1b). This overall pattern exhibited that the water of all UC soil layers partially originated from deeper soil or groundwater through plants vessel transport. In other words, the deep-rooted *P. euphratica* exhibited HR, which transported and lifted water from deeper to shallow soil layers, proof for the experimental support to Ryel model based studies [22].

It is well understood that changes in soil evaporation depends on vegetation coverage, for example, the higher vegetation coverage will has more shade on the soil surface and hence evaporation will decrease, and vice versa [38,39]. Furthermore, Allison [40] and Kim [41] confirmed that $\delta^{18}O$ content of soil samples in UC is lower as compared to OC, because high vegetation coverage decreased the evaporation due to shading. Specifically, higher evaporative demand in OC could easily drive a greater upward movement of water from the groundwater or the deeper soils to surface soils, which then enriched the $\delta^{18}O$ content due to evaporative fractionation. However, pan evaporation was not significantly different between UC and OC ($UC_{soil\ evaporation} = 0.91 \pm 0.52$ cm·m^{-2}·d^{-1}, $OC_{soil\ evaporation} = 0.93 \pm 0.55$ cm·m^{-2}·d^{-1}, $t = -1.02$, $df = 6$, $p = 0.34$, which was tested through paired-sample *t*-test, and its supportive data was measured by 255 Series Evaporation Stations (EP255, Novalynx Inc, OR, USA) for three paired groups in UC and OC), because the sparse vegetation coverage of *P. euphratica* community (approximate 4%) has little shading influence on the evaporative demand of UC and OC in Ebinur desert [22]. In addition, this desert is a part of tuyere zone having high wind, and thus other evaporative environmental factors (*e.g.*, air temperature and moisture) homogenously influence soil evaporation, subsequently caused relatively no difference of soil evaporation between UC and OC plots. Hence, soil evaporation not likely differed $\delta^{18}O$ content between OC and UC in the Ebinur desert and further proved that *P. euphratica* HR occurs.

P. euphratica HR and water transport direction along soil profile can also be judged by moisture differences among soil layers. In this study, Thetaprobe ML2 soil moisture sensors (Delta-T Devices, Cambridge, UK) were installed respectively in five soil layers of UC and OC to measure the variation in soil volumetric water content (SVWC) along soil depth. The results showed that, (1) SVWCs of 0 to 10 cm, 10 to 40 cm, 70 to 100 cm and 100 to 150 cm layers in UC were significantly higher than those in OC, except in 40 to 70 cm soil layer; (2) SVWC increases with soil depth but decreases with $\delta^{18}O$ content in OC, while no significant trend was found in UC; and (3) SVWC of 100 to 150 cm soil layer was a little higher in OC than in OC (Table 2). These results suggested that groundwater supplies soil water, and the water's table in both UC and OC were similar in Ebinur desert. SVWC increased with soil depth due to the extraction of soil evaporation in OC. But HR can lift water from groundwater and deep soils to shallow soils and then decrease the effect of evaporation on water extracting in UC. Previous studies showed that the location and the amount of HR releasing water were significantly depended on plant fine roots distribution [11,22,23]. Also *P. euphratica*'s fine roots was mainly growing within 0 to 70 cm soil depth in arid area [22,29]. Thus, HR may cause no change in SVWC across soil depth in UC and SVWC of 40 to 70 cm layer in UC less than that in OC.

The greatest $\delta^{18}O$ content existed in soil while the intermediate in groundwater and the lowest in river (Fig. 1a). This pattern indicates that river is the initial source of water for arid desert area. Our results were supported by Zhao et al. [42] reported that river largely originated from glacial melting and has the lowest $\delta^{18}O$ content in arid desert. Based on oxygen isotope fractionation theory [26–28,40], initial water resource has lowest $\delta^{18}O$ content and then increased with transmittal distance and pathways. In this study, the river is the main source for groundwater and then supply to soil water through water transmittal ways. i.e. underground or surface runoff, evaporation and HR. All of those can led to oxygen isotope fractionation and then resulted in soil having the greatest value of $\delta^{18}O$ content.

Effects of *P. euphratica* HR on species growth condition

The importance value index [importance value index = (relative abundance + relative frequency + relative coverage + relative height)/400] and its deduced dominance index (dominance index $= \sum_{i=1}^{n} \left(\frac{n_i}{N_i} \right)^2$, N is the count of plots and N_i is the importance value of a certain species in i th plot) were traditional ecological methods in generally using to evaluate species growth conditions within a specific community type [43,44]. Specifically, the importance value is commonly assumed as 1, and then based on individual relative statistical values, such as individual numbers, relative abundance, relative coverage, relative frequency, and relative height. 1 was divided into different components values that show the different species growth conditions within a given community. Nevertheless, many studies also needed to compare the difference of same species growth conditions among environmental sites, such as the same species between UC and OC. According to this situation, considering *P. euphratica* can account for the largest partition of the importance value in UC, if we used the traditional methods above to evaluate the difference of species growth conditions between UC and OC, a lower importance value could be found in UC than in OC among other species, even these species have the same abundance, coverage, frequency and height between two environmental types.

In this study, the growth dominance index was structured by the actual values of coverage, abundance, height, and frequency of a given species in a given plot to avoid the components influence of importance value, and to compare the difference of species growth conditions between the UC and OC (Formulae 2 to 4). The results showed that UC species grow better than OC along plant growing seasons (Table 1). Additionally SVWC increased with soil depth in OC while no change in UC (Table 2). These suggest that *P. euphratica* HR increased significantly water content of shallow

soils and benefited the growth condition and survival of shallow-rooted plants [16,17,24,37,45–50].

Relationship between *P. euphratica* HR and arid desert diversity assembly

Water is the main limitation for arid desert plant communities [1,51]. Thus, species coexistence pattern in this community directly depends on spatial availability and distribution of water [32,33]. In this study, the species richness and abundance in the UC were significantly higher than those in the OC ($p<0.05$), i.e., UC has higher species diversity (Fig. 2).

Furthermore, based on the hypothesis of this study that a tree with a larger crown area has higher root biomass and transports more water from the deep to shallow soils, the logarithmic regression was used to analyze the relationship between the *P. euphratica* HR capacity (CA) and the species richness and abundance in the UC. The results showed that the *P. euphratica* CA was significantly logarithmically correlated with richness ($R^2 = 0.51$) and abundance ($R^2 = 0.84$) ($p<0.001$) (Fig. 3). These results indicated that species diversity was influenced significantly by HR. Similarly, this conclusion was also reflected in other studies. For example, herbaceous plants under tree canopies have higher abundance and richness [46,52], and the presence of trees promoted the survival of shrubs and seedlings grown under tree canopies [37,45,48–50,53].

Shallow-rooted herbs and shrubs that absorb water from *P. euphratica* HR provided more organic matter and mineral elements to the soil of the UC than to those of the OC because of the "fertile island" effect. It implies that more individuals of the UC are intercepted and sequester larger amounts of organic matter and minerals from surface winds because of increased total crown area [54–59]. Meanwhile, the more numerous individuals also produced more litter into the UC soils than OC soils. Thus, these processes contributed to the nutrient absorption and growth of *P. euphratica*, which further increased its HR capacity. In turn, these processes also increased the shading in understory plants, which can decrease the transpiration of herbs and shrubs, and further improve its survival. Therefore, there appears to be a species coexistence pattern of water sharing and resource complementation between the deep-rooted *P. euphratica* and other shallow-rooted species, as well as positive feedback between *P. euphratica* HR and biodiversity maintenance in arid deserts community.

Acknowledgments

The authors thank Yang jun, Sun Jingxing, Tian Youhua, Sun Lijun, Zhang Xuemei, Li Changjun, He Xueming, Qin Lu, Luo Chong, Ren Manli, Zhu Ya for their help in the field and laboratory. The authors also thank Qao Qiang and Dearlyn Fernandes for English grammar checking of the manuscript.

Author Contributions

No other contributions. Conceived and designed the experiments: XDY GHL. Performed the experiments: XDY XNZ. Analyzed the data: XDY. Contributed reagents/materials/analysis tools: XDY GHL. Wrote the paper: XDY AA.

References

1. Armas C, Padilla F, Pugnaire F, Jackson R (2010) Hydraulic lift and tolerance to salinity of semiarid species: consequences for species interactions. Oecologia 162: 11–21.

2. Haase P, Pugnaire FI, Fernández EM, Puigdefábregas J, Clark S, et al. (1996) An investigation of rooting depth of the semiarid shrub *Retama sphaerocarpa* (L.) Boiss. by labelling ground water with a chemical tracer. Journal of Hydrology 177: 23–31.

3. Richards JH, Caldwell MM (1987) Hydraulic lift: substantial nocturnal water transport between soil layers by Artemisia tridentata roots. Oecologia 73: 486–489.

4. Caldwell MM, Dawson TE, Richards JH (1998) Hydraulic lift: consequences of water efflux from the roots of plants. Oecologia 113: 151–161.

5. Hacker SD, Bertness MD (1995) Morphological and physiological consequences of a positive plant interaction. Ecology 76: 2165–2175.

6. Vetterlein D, Marschner H (1993) Use of a microtensiometer technique to study hydraulic lift in a sandy soil planted with pearl millet (*Pennisetum americanum* [L.] Leeke). Plant and Soil 149: 275–282.

7. Filella I, Penuelas J (2003) Indications of hydraulic lift by Pinus halepensis and its effects on the water relations of neighbour shrubs. Biologia Plantarum 47: 209–214.

8. Franco A, Nobel P (1990) Influences of root distribution and growth on predicted water uptake and interspecific competition. Oecologia 82: 151–157.

9. Kurz-Besson C, Otieno D, do Vale RL, Siegwolf R, Schmidt M, et al. (2006) Hydraulic lift in cork oak trees in a savannah-type Mediterranean ecosystem and its contribution to the local water balance. Plant and Soil 282: 361–378.

10. Williams DG, Ehleringer JR (2000) Intra-and interspecific variation for summer precipitation use in pinyon-juniper woodlands. Ecological Monographs 70: 517–537.

11. Burgess SS, Adams MA, Turner NC, Ong CK (1998) The redistribution of soil water by tree root systems. Oecologia 115: 306–311.

12. Dawson TE (1993) Hydraulic lift and water use by plants: implications for water balance, performance and plant-plant interactions. Oecologia 95: 565–574.

13. Lee J-E, Oliveira RS, Dawson TE, Fung I (2005) Root functioning modifies seasonal climate. Proceedings of the National Academy of Sciences of the United States of America 102: 17576–17581.

14. Lehto T, Zwiazek JJ (2011) Ectomycorrhizas and water relations of trees: a review. Mycorrhiza 21: 71–90.

15. Neumann RB, Cardon ZG (2012) The magnitude of hydraulic redistribution by plant roots: a review and synthesis of empirical and modeling studies. New Phytologist 194: 337–352.

16. Prieto I, Martínez-Tillería K, Martínez-Manchego L, Montecinos S, Pugnaire FI, et al. (2010) Hydraulic lift through transpiration suppression in shrubs from two arid ecosystems: patterns and control mechanisms. Oecologia 163: 855–865.

17. Warren JM, Brooks JR, Dragila MI, Meinzer FC (2011) In situ separation of root hydraulic redistribution of soil water from liquid and vapor transport. Oecologia 166: 899–911.

18. Zou C, Barnes P, Archer S, McMurtry C (2005) Soil moisture redistribution as a mechanism of facilitation in savanna tree–shrub clusters. Oecologia 145: 32–40.

19. David TS, Pinto CA, Nadezhdina N, Kurz-Besson C, Henriques MO, et al. (2013) Root functioning, tree water use and hydraulic redistribution in *Quercus suber* trees: A modeling approach based on root sap flow. Forest Ecology and Management 307: 136–146.

20. Sardans J, Peñuelas J (2014) Hydraulic redistribution by plants and nutrient stoichiometry: Shifts under global change. Ecohydrology 7: 1–20.

21. Horton JL, Hart SC (1998) Hydraulic lift: a potentially important ecosystem process. Trends in Ecology & Evolution 13: 232–235.

22. Yang XD, Lv GH (2011) Establishment and analysis of Populus euphratica'root Hydraulic redistribution model. Chinese journal of Plant Ecology 35: 816–824 (in Chinese with English abstract).

23. Hao XM, Li WH, Guo B, Ma JX (2013) Simulation of the effect of root distribution on hydraulic redistribution in a desert riparian forest. Ecological research 28: 653–662.

24. Yu T, Feng Q, Si J, Xi H, Li Z, et al. (2013) Hydraulic redistribution of soil water by roots of two desert riparian phreatophytes in northwest China's extremely arid region. Plant and soil 372: 297–308.

25. Ryel R, Caldwell M, Yoder C, Or D, Leffler A (2002) Hydraulic redistribution in a stand of Artemisia tridentata: evaluation of benefits to transpiration assessed with a simulation model. Oecologia 130: 173–184.

26. Dawson TE, Mambelli S, Plamboeck AH, Templer PH, Tu KP (2002) Stable isotopes in plant ecology. Annual review of ecology and systematics 33: 507–559.

27. Ehleringer J, Dawson T (1992) Water uptake by plants: perspectives from stable isotope composition. Plant, Cell & Environment 15: 1073–1082.

28. Ellsworth PZ, Williams DG (2007) Hydrogen isotope fractionation during water uptake by woody xerophytes. Plant and Soil 291: 93–107.

29. Zhao F, Jin HL (2011) Study on characteristics of groundwater and its impact on Populus euphratica along the banks of Aqikesu River. Journal of Arid Land Resources and Environment 8: 156–161 (in Chinese with English abstract).

30. Hamblin AP (1985) The influence of soil structure on water movement, crop root growth and water uptake. Advances in Agronomy 38: 95–158.

31. Jackson R, Canadell J, Ehleringer J, Mooney H, Sala O, et al. (1996) A global analysis of root distributions for terrestrial biomes. Oecologia 108: 389–411.

32. Li XR, Tan HJ, He MZ, Wang XP, Li XJ (2009) Patterns of shrub species richness and abundance in relation to environmental factors on the Alxa Plateau: Prerequisites for conserving shrub diversity in extreme arid desert regions. Science in China Series D: Earth Sciences 52: 669–680.

33. Xia Y, Moore DI, Collins SL, Muldavin EH (2010) Aboveground production and species richness of annuals in Chihuahuan Desert grassland and shrubland plant communities. Journal of arid environments 74: 378–385.

34. Allison GB, Hughes MW (1983) The use of natural tracers as indicators of soil-water movement in a temperate semi-arid region. Journal of Hydrology 60: 157–173.

35. King DA (1996) Allometry and life history of tropical trees. Journal of tropical ecology 12: 25–44.

36. Poorter L, Bongers F, Sterck FJ, Wöll H (2003) Architecture of 53 rain forest tree species differing in adult stature and shade tolerance. Ecology 84: 602–608.

37. Callaway RM (1992) Effect of shrubs on recruitment of Quercus douglasii and Quercus lobata in California. Ecology 73: 2118–2128.

38. Monteith JL (1965) Evaporation and environment. Symposia of the Society for Experimental Biology 19: 205–234.

39. Raz-Yaseef N, Rotenberg E, Yakir D (2010) Effects of spatial variations in soil evaporation caused by tree shading on water flux partitioning in a semi-arid pine forest. Agricultural and Forest Meteorology 150: 454–462.

40. Allison GB, Barnes CJ, Hughes MW (1983) The distribution of deuterium and 18O in dry soils 2. Experimental. Journal of Hydrology 64: 377–397.

41. Kim K, Lee X (2011) Isotopic enrichment of liquid water during evaporation from water surfaces. Journal of hydrology 399: 364–375.

42. Zhao L, Xiao H, Cheng G, Song Y, Zhao L, et al. (2008) A preliminary study of water sources of riparian plants in the lower reaches of the Heihe basin. Acta Geoscientica Sinica 29: 709–718.

43. Zhang JT (2004) Quantitive Ecology. Beijing: Science Press (in Chinese).

44. Numata M (1979) Methods for Ecological Study. Kokin Shou-in, Tokyo (in Japanese).

45. Barchuk A, Valiente-Banuet A, Díaz M (2005) Effect of shrubs and seasonal variability of rainfall on the establishment of Aspidosperma quebracho-blanco in two edaphically contrasting environments. Austral ecology 30: 695–705.

46. Belsky A, Amundson R, Duxbury J, Riha S, Ali A, et al. (1989) The effects of trees on their physical, chemical and biological environments in a semi-arid savanna in Kenya. Journal of Applied Ecology 26: 1005–1024.

47. Belsky AJ (1994) Influences of trees on savanna productivity: tests of shade, nutrients, and tree-grass competition. Ecology 75: 922–932.

48. Franco A, Nobel P (1989) Effect of nurse plants on the microhabitat and growth of cacti. The Journal of Ecology 77: 870–886.

49. Maestre FT, Bautista S, Cortina J (2003) Positive, negative, and net effects in grass-shrub interactions in Mediterranean semiarid grasslands. Ecology 84: 3186–3197.

50. Shumway SW (2000) Facilitative effects of a sand dune shrub on species growing beneath the shrub canopy. Oecologia 124: 138–148.

51. Huxman TE, Snyder KA, Tissue D, Leffler AJ, Ogle K, et al. (2004) Precipitation pulses and carbon fluxes in semiarid and arid ecosystems. Oecologia 141: 254–268.

52. Weltzin JF, Coughenour MB (1990) Savanna tree influence on understory vegetation and soil nutrients in northwestern Kenya. Journal of Vegetation Science 1: 325–334.

53. Rousset O, Lepart J (1999) Shrub facilitation of Quercus humilis regeneration in succession on calcareous grasslands. Journal of Vegetation Science 10: 493–502.

54. Armas C, Pugnaire FI (2005) Plant interactions govern population dynamics in a semi-arid plant community. Journal of Ecology 93: 978–989.

55. Jackson L, Strauss R, Firestone M, Bartolome J (1990) Influence of tree canopies on grassland productivity and nitrogen dynamics in deciduous oak savanna. Agriculture, Ecosystems & Environment 32: 89–105.

56. Reynolds JF, Virginia RA, Kemp PR, de Soyza AG, Tremmel DC (1999) Impact of drought on desert shrubs: effects of seasonality and degree of resource island development. Ecological Monographs 69: 69–106.

57. Schade JD, Hobbie SE (2005) Spatial and temporal variation in islands of fertility in the Sonoran Desert. Biogeochemistry 73: 541–553.

58. Segoli M, Ungar ED, Shachak M (2012) Fine-Scale Spatial Heterogeneity of Resource Modulation in Semi-Arid "Islands of Fertility". Arid Land Research and Management 26: 344–354.

59. Walker LR, Thompson DB, Landau FH (2001) Experimental manipulations of fertile islands and nurse plant effects in the Mojave Desert, USA. Western North American Naturalist 61: 25–35.

10

Transcriptome Analysis of the Desert Locust Central Nervous System: Production and Annotation of a *Schistocerca gregaria* EST Database

Liesbeth Badisco, Jurgen Huybrechts, Gert Simonet, Heleen Verlinden, Elisabeth Marchal, Roger Huybrechts, Liliane Schoofs, Arnold De Loof, Jozef Vanden Broeck*

Department of Animal Physiology and Neurobiology, Katholieke Universiteit Leuven, Leuven, Belgium

Abstract

Background: The desert locust (*Schistocerca gregaria*) displays a fascinating type of phenotypic plasticity, designated as 'phase polyphenism'. Depending on environmental conditions, one genome can be translated into two highly divergent phenotypes, termed the solitarious and gregarious (swarming) phase. Although many of the underlying molecular events remain elusive, the central nervous system (CNS) is expected to play a crucial role in the phase transition process. Locusts have also proven to be interesting model organisms in a physiological and neurobiological research context. However, molecular studies in locusts are hampered by the fact that genome/transcriptome sequence information available for this branch of insects is still limited.

Methodology: We have generated 34,672 raw expressed sequence tags (EST) from the CNS of desert locusts in both phases. These ESTs were assembled in 12,709 unique transcript sequences and nearly 4,000 sequences were functionally annotated. Moreover, the obtained *S. gregaria* EST information is highly complementary to the existing orthopteran transcriptomic data. Since many novel transcripts encode neuronal signaling and signal transduction components, this paper includes an overview of these sequences. Furthermore, several transcripts being differentially represented in solitarious and gregarious locusts were retrieved from this EST database. The findings highlight the involvement of the CNS in the phase transition process and indicate that this novel annotated database may also add to the emerging knowledge of concomitant neuronal signaling and neuroplasticity events.

Conclusions: In summary, we met the need for novel sequence data from desert locust CNS. To our knowledge, we hereby also present the first insect EST database that is derived from the complete CNS. The obtained *S. gregaria* EST data constitute an important new source of information that will be instrumental in further unraveling the molecular principles of phase polyphenism, in further establishing locusts as valuable research model organisms and in molecular evolutionary and comparative entomology.

Editor: Jialin Zheng, University of Nebraska Medical Center, United States of America

Funding: This work was supported by grants from the K.U. Leuven Research Foundation (GOA 2005/06, www.kuleuven.be/bof), the Research Foundation of Flanders (FWO-Flanders, www.fwo.be) and the Interuniversity Attraction Poles program (Belgian Science Policy Grant P6/14, www.belspo.be). L.B., H.V. and E.M. were supported by a scholarship from the 'Agentschap voor Innovatie door Wetenschap en Technologie' (IWT) and J.H. obtained a postdoctoral research fellowship from the FWO. The funders had no role in study design, data collection and analysis, decision to publish, or preparation of the manuscript.

Competing Interests: The authors have declared that no competing interests exist.

* E-mail: jozef.vandenbroeck@bio.kuleuven.be

Introduction

For many decades, locusts have proven to be important model organisms for insect physiological research, in particular for the study of endocrinological and neurobiological processes. Their relatively large size has enabled the purification and identification of an extensive repertoire of nearly a hundred biologically active regulatory peptides [1–5]. Unlike many other insect model species, such as the fruit fly, the honey bee and the silk worm, locusts belong to the hemimetabolous branch of insects. This subgroup comprises insects that undergo an incomplete metamorphosis, lacking the formation of a pupal stage. The rapidly expanding genome and transcriptome data have given research in several holometabolous model insects an unprecedented impetus. Al-

though genome data have recently become available for the hemimetabolous species *Acyrthosiphon pisum* (pea aphid) [6] and *Pediculus humanus* (human body louse) [7], sequence information from this branch of insects is still lagging behind. In addition, locusts appear to have a very big genome (estimated 2–3 times larger than the human genome [8]), which constitutes a major hurdle to completely sequence it.

In addition, the desert locust, *Schistocerca gregaria*, is particularly well known as a notorious swarm-forming insect species, which can inflict devastating damage to the agricultural production in large areas of the world (*cf*. FAO website: http://www.fao.org/ag/locusts/). Intriguingly, the same species can also occur in a more harmless solitarious form, which tends to avoid the company of other locusts. Besides this very prominent behavioral distinction, solitarious and gregarious locusts also differ in many other traits such as coloration, morphology, developmental and reproductive physiology. The existence of these two extremely different forms or phases, also designated as phase polyphenism, is a fascinating example of phenotypic plasticity, whereby two obviously different phenotypes are encoded by the same genome [9–11]. [In laboratory conditions the two locust phases are referred to as isolated-reared (solitarious) and crowded-reared (gregarious).] Conversion between the two phases is termed phase transition, which is a reversible, continuous process that is accompanied by the occurrence of several intermediate forms [9,10]. Development towards the gregarious phase is triggered by an increase in population density. Remarkably, a behavioral shift can be observed from mutual aversion to aggregation within a few hours of crowding [12–14].

The central nervous system (CNS) plays a crucial role in these early gregarization effects. Sensory stimuli generated by the presence of other locusts can induce changes in the titers of several neurotransmitters [14–16]. The involvement of the CNS is not surprising since it constitutes the primary systemic control center that is integrating sensory input, generates behavioral responses and regulates many physiological processes. In addition, although crowded-reared locusts are on average smaller, their brain was found to be 30% larger than that of isolated-reared animals and to be differently proportionated [17]. Elevated population density leads to increased competition for food and forces the locusts to alter their foraging strategy. Since foraging behavior and social life style have already been associated with differences in the brain volume of insects [18–23], these may also be involved in distinguishing gregarious from solitarious brain size [17]. Furthermore, serotonin has been demonstrated to be a crucial central mediator of the behavioral phase transformation [13]. During the first hours of forced crowding a temporary increase in serotonin has been observed in the thoracic ganglia [15]. However, development towards the gregarious phase is not only character-ized by a behavioral shift. In later stages of gregarization (which can comprise several generations) multiple physiological processes are affected. These include reproduction, development and determination of life span [11,24]. However, to a great extent, the molecular basis underlying all these phenotypic changes still remains elusive. Therefore, additional sequence information (both nucleotide and protein sequence information) is currently needed to allow further investigations of the mechanisms underlying phase-dependent physiological processes in locusts.

In order to compensate for the absence of locust genome data (and for the difficulty to obtain such data, given the huge estimated size of their genome), we have currently produced an EST ('Expressed Sequence Tags') database representing transcripts expressed in the CNS of the desert locust, S. gregaria. Furthermore, to our knowledge, we hereby present the first EST database derived exclusively from a complete insect CNS (head ganglia and ventral nerve cord).

An EST database (LocustDB) derived from various body parts was already publicly available for a related locust species, Locusta migratoria [25,26]. Within the family of Acrididae S. gregaria and L. migratoria belong to the subfamilies of Cyrtacanthacridinae and Oedipodinae, respectively. Not all genera or species classified under both subfamilies display phase polyphenism, indicating that the desert locust and the migratory locust must have developed this phenotypic plasticity independently and/or that other species may have lost this ability. Most probably, the truth lies somewhere in between. Song suggested that phase polyphenism is a very complex character that results from interactions between several density-dependent plastic traits, which may have evolved differ-

ently. Behavioral phenotypic traits seem to be less well conserved than physiological traits, which may explain why not all grasshopper species within the subfamilies of Cyrtacanthacridinae or Oedipodinae display phase polyphenism [27–29]. In addition, the neural mechanisms inducing the behavioral switch may differ among different locust species. Visual and olfactory incentives and tactile stimulation of the hind legs act synergistically to trigger gregarization behavior in S. gregaria [30,31]. However, it was recently demonstrated that these stimuli do probably not result in a behavioral change in the Australian plague locust (Chortoicetes terminifera), which, like L. migratoria, belongs to the subfamily of the Oedipodinae. In these animals, tactile stimulation of the antennae is suggested to be the primary gregarizing input [32]. This example emphasizes that phase polyphenism research in different locust species may have to be approached differently.

Therefore, by specifically focusing on the desert locust CNS, information in the obtained database is not only expected to complement previously obtained orthopteran transcript sequence data. In addition, the S. gregaria database will also be instrumental for a more detailed molecular dissection of neuronal and neuro-endocrine control mechanisms and signaling pathways responsible for many physiological processes, including the very fascinating process of locust phase transition. Information in this S. gregaria EST database is compared to that in other organisms' databases (both genome and EST). Annotation of the EST sequences and assignment of 'Gene Ontology' (GO) terms demonstrates that a broad range of GO categories is represented in this novel database. Since many transcripts encode products that are predicted to play a role in neuronal signaling or in signal transduction, this paper will focus on these particular sequences and provide an overview of their possible role(s). Furthermore, a selection of genes showing differential expression in isolated- and crowded-reared desert locusts is retrieved from the database.

Results and Discussion

1. An EST database for *Schistocerca gregaria* CNS

Production of the EST database. Normalized cDNA libraries were derived from microdissected CNS of isolated- and crowded-reared, larval and adult desert locust males and females. In total, 34,672 raw EST data have been generated from these libraries. These ESTs were further assembled in 4,785 contigs and 7,924 singletons, resulting in a total of 12,709 unique sequences. A sequence length distribution analysis (Figure 1) shows that for most of these uniques between 600–900 nucleotides of high quality information (≥99% confidence level) is available, although several longer sequences were also observed. Sequence data have been integrated in a desert locust 'Expressed Sequence Tag Information Management and Annotation' (ESTIMA) database (http://titan. biotec.uiuc.edu/locust/), which has several functionalities for retrieval of EST-related information. An overview of what these functionalities accept as query/input and what they produce as output is given in Table 1.

Containing a total of 12,709 unique sequences, this EST database probably represents a large part of the desert locust CNS transcriptome. However, the order of magnitude of the full transcriptome is currently unknown, since the desert locust genome has not been sequenced yet. The average length of the EST sequences indicates that there is a relatively high degree of high quality sequence information available per transcript. This information will prove to be useful in bioinformatic analyses and their downstream applications, such as homology searches, prediction of the biological function of the encoded products, design of primers for 'Rapid Amplification of cDNA Ends' (RAcE)

Number of sequences with length(x)

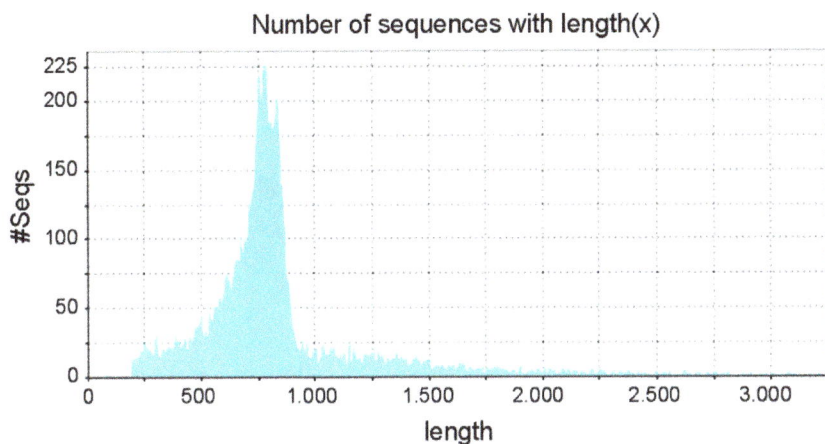

Figure 1. Sequence length distribution of the 12,709 unique *S. gregaria* EST sequences (4,785 contigs/7,924 singletons).

or for 'quantitative real-time reverse transcriptase polymerase chain reaction' (qRT-PCR) analysis and production of microarrays for transcriptome-wide expression profiling studies.

Blastx searches in protein databases. The *S. gregaria* sequences were used as a query for blastx searches in the *Anopheles gambiae*, *Tribolium castaneum*, *Drosophila melanogaster*, *Apis mellifera*, *Caenorhabditis elegans*, *Homo sapiens* and the NCBI nr.aa protein databases. These searches produced 4,891 hits in the *A. gambiae* database, 4,780 hits in the *T. castaneum* database, 4,693 hits in the *D. melanogaster* database, 4,735 hits in the *A. mellifera* database, 3,612 hits in the *C. elegans* database, 4,492 hits in the *H. sapiens* database and 5,779 hits in the nr.aa database. An important fraction of these hits consists of (hypothetical) proteins without known function. Furthermore, in the protein databases for various insect species with a completely sequenced genome (*A. gambiae*, *T. castaneum*, *D. melanogaster*, *A. mellifera*), no obvious orthologs were found for *ca.* 60% of the *S. gregaria* EST sequences. This observation may be explained by the fact that EST data do not necessarily cover the entire protein encoding region of a gene. Also, the applied search criteria need to be rather stringent to avoid mistakes in the annotation process. Moreover, locusts are phylogenetically distant from these holometabolous insect models. It can also not be excluded that there may be many species-specific genes that are absent in these model species. These considerations underline the need for additional, functionally analyzed hemimetabolous/orthopteran insect genome/transcriptome data, which may add to the current knowledge of homologous or species-specific genes in desert locusts and other Hemimetabola.

Gene Ontology annotations. In the functional annotation process, 3,887 *S. gregaria* sequences were classified according to the GO vocabulary [33]. Figure 2A shows the distribution of EST sequences among the different subcategories (level 2) of the main ontology *Biological Process*. The best represented GO-groups were *Cellular Process* and *Metabolic Process*. To obtain a more detailed view on the functional differentiation in these GO groups, an overview of the subgroups is presented (Figure 2B), showing that nearly 60% of the ESTs within the GO group *Metabolic process* are classified in *Primary metabolic process* and *Cellular metabolic process*. With regard to cellular processes, more than 50% of these ESTs are classified in three subgroups: *Cellular metabolic process*, *Regulation of cellular process* and *Cellular component organization*.

Next, the distribution of GO-annotated transcripts, classified under the main ontology *Biological Process*, was analyzed in more detail. For all ESTs the GO-term corresponding to the lowest node in the GO hierarchy with an annotation score above the threshold (>45) [34,35] was used to calculate 100 GO-terms representing the highest number of ESTs. As shown in Figure S1, numerous ESTs were classified in groups describing processes occurring in virtually all cell types, such as cell cycle, transcription/translation events, metabolic processes and transport. Nevertheless, many other ESTs are predicted to be involved in processes such as development and functioning of the nervous system, reproduction, determination of life span and growth.

Table 1. Overview of the six main ESTIMA functionalities for retrieval of *S. gregaria* EST-related information.

Functionality	Input	Output
GO browser	Key word	GO tree
	GO term ID	EST count per term
	Root category	
Sequence ID	Sequence ID	Chromatogram
		Download raw/trimmed sequence
		Alignment length (only contigs)
		GenBank accession ID
		Contig structure (only contigs)
		GO browser
		Annotation
Gene Association	Gene Symbol	EST ID
	Sequence ID	GO annotation
	Unigene Number	Blast hits
Blast	Query (protein/DNA)	Standard blast output [216]
	Database	
Annotations	Keyword	Blastx hits in the searched protein databases
	Sequence ID	Assigned GO terms
Contig Viewer	Contig ID	Schematic overview of contig assembly
		Composing ESTs' IDs
		Contig sequence

A

B

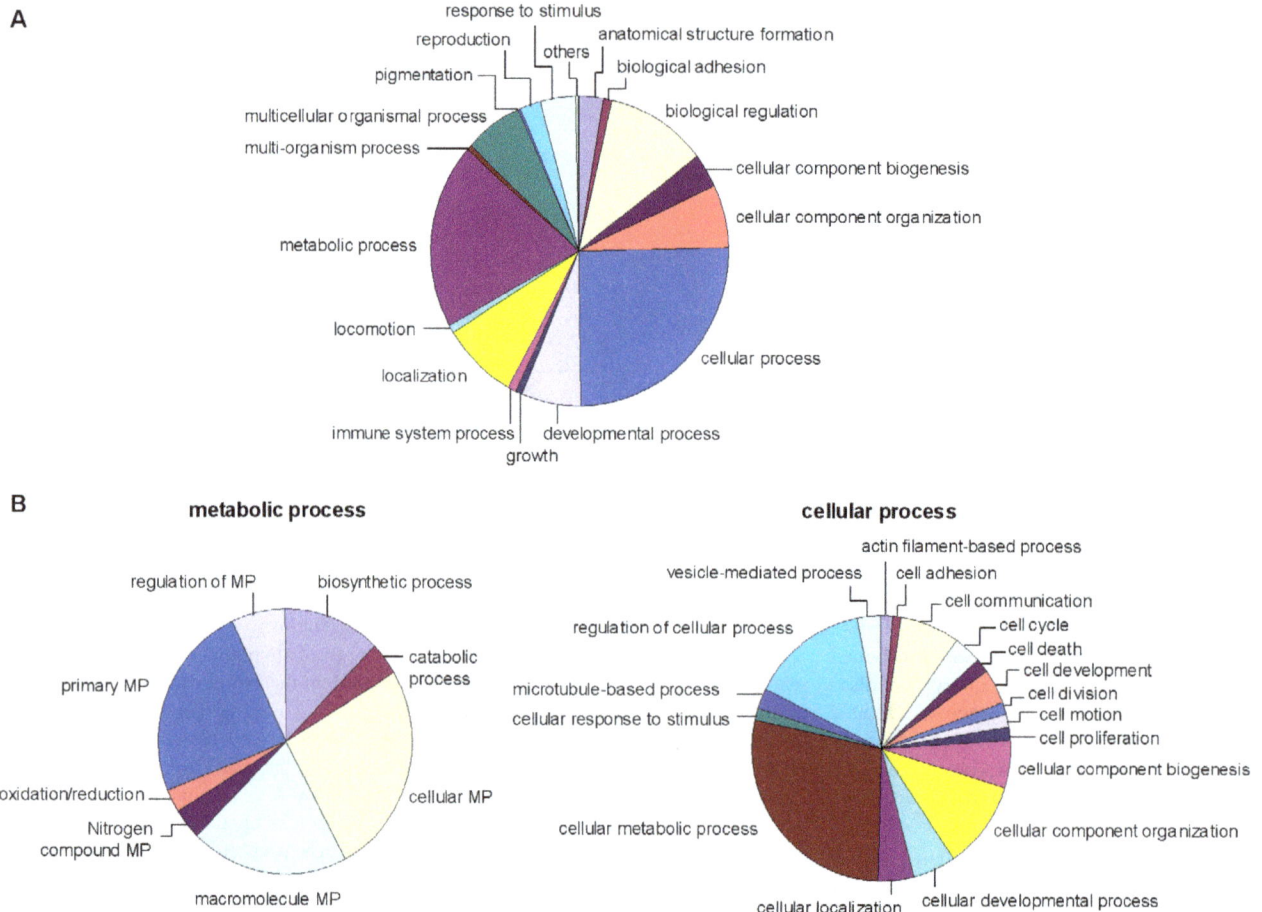

Figure 2. Second level GO distribution of the *S. gregaria* EST sequences (*Biological Process*). Distribution of the *S. gregaria* EST sequences in the major subclasses of the main ontology *Biological Process* (A) and a more detailed overview of sequence distribution in the two subclasses *Metabolic process* and *Cellular process* (B).

It is not very surprising that many sequences are classified under metabolic and cellular processes. These are more general GO terms which comprise basic processes, needed to maintain a living organism. Of particular interest are those ESTs that are predicted to be involved in processes such as development and functioning of the nervous system, reproduction, determination of life span and growth. Study of these transcripts may provide information about how the CNS controls reproduction, growth and adult life span and how these processes are regulated in a phase-dependent way.

A classification of EST sequences according to the GO vocabulary [33] represents a convenient starting point for analysis of large amounts of data, such as microarray data. In addition, a GO-based classification also proves to be useful when studying a certain biological process, molecular function or cellular component represented by a specific GO term. For example, *S. gregaria* EST sequences classified under a specific term are easily retrieved from the database and may add to the current knowledge of the studied process. However, the GO terms are only 'labels' and should not be taken for granted. Therefore, the user should remain critical when performing GO-based analysis of data. Assignment of GO terms is not only based on experimental evidence and is very often derived from homology searches. In case of the latter, the GO terms have been transferred from homologous sequences to the newly annotated sequence (these assignments are mostly based on a certain score that considers degree of similarity to the

homologous sequence, evidence code of how the GO term has been assigned to the homologous sequence, *etc*...). However, the biological function of the homologous sequences may for example not be evolutionary conserved. Also when focusing on a specific tissue or condition, it should be considered that the GO terms assigned to a specific factor may apply to biological roles in other tissues or conditions. It can be summarized that GO terms offer a good starting point for analyzing large sets of sequence data, but that the further analysis process should be performed critically in order to eliminate potential noise disturbing the biological relevance of the data.

Comparison to LocustDB. At present, another locust EST database is publicly available. It encompasses 12,161 unique sequences from *L. migratoria* (LocustDB) [25,26]. Although it is similar in its total transcript sequence number output, this other database was generated from different tissue sources. Whereas LocustDB contains EST data derived from primary (*i.e.* non-normalized) cDNA libraries for head, hind leg, midgut and whole organisms (5th larval stage), the novel *S. gregaria* database entirely focused on transcript information from the CNS (3–5th larval stage and adults). Therefore, it is not surprising that this different approach has been responsible for a very different representation of locust transcript sequences. This is nicely illustrated by the result of comparative blastn searches between both databases showing that only 4,189 sequences (*ca.* 1/3) can be considered as orthologs (cut-off at E-value <1E-10) occurring in both databases. A 'Blast2GO' analysis of both databases showed that in total only 5,701 of the 20,681 unique locust (either *L. migratoria* or *S. gregaria*) sequences can be annotated, indicating that the (possible) function of the majority of locust sequences remains unknown. That there is little overlap between two different EST databases may have various reasons. First, as mentioned before, *S. gregaria* and *L. migratoria* belong to different subfamilies within the family of Acrididae. Therefore, the phylogenetic distance between these two species may explain part of the observed lack of homology. Second, the two databases were derived from different tissue sources. By focusing on the CNS, the *S. gregaria* EST database is more specialized, whereas the *L. migratoria* database may be more general. Third, EST data represent only partial sequences of most transcripts. Moreover, the average EST sequence length is larger for the *S. gregaria* sequences. Therefore, it is not unlikely that identification of orthologous sequences may be hampered by a lack of overlapping sequence information. Fourth, EST databases generally do not cover the complete transcriptome of the respective tissues and species. And finally, it should also be emphasized that, in contrast to LocustDB, the *S. gregaria* EST database is derived from a normalized cDNA library, thereby compensating for very abundant transcripts and increasing the odds of sequencing less abundant ones, for which the orthologs may not have been included in the *L. migratoria* database.

To further compare the functional variety of the transcript data represented in both locust EST databases, the distribution of the annotated sequences in the different *Biological Process* GO classes was calculated (where each sequence is classified under the most detailed GO term). The number of sequences linked to a particular class was used to rank the '100 best represented' classes (Figures S1 and S2). Gene products involved in processes occurring in virtually all cell types, such as *Transport*, general metabolic processes (including the GO-classes *Oxidation reduction*, *Proteolysis* and *Metabolic process*) and transcription/translation events (including the GO-classes *Ribosome biogenesis* and *Translation*) are well represented in both databases. On the contrary, the distribution of transcripts in biological processes occurring in specific tissue types tends to be more dependent on the cDNA source of the respective database. The GO-groups *Carbohydrate/starch metabolic process* and *Sucrose metabolic process*, as well as *Digestion* and *Catabolic protein processes* are highly enriched in the *L. migratoria* database (Figure S2). Transcripts associated with the CNS (e.g. *Axon guidance*, *Nervous system development* and *Synaptic transmission*) are predominantly present in the *S. gregaria* EST database (Figure S1). In addition, sequences classified under *Spermatogenesis* and *Growth* are also more represented in the desert locust EST database. This observation can probably be explained by the fact that these processes are either directly or indirectly regulated by factors produced by the CNS and/or by the fact that many factors having a role in development also have a function in the CNS.

It can be summarized that the current generation of *S. gregaria* ESTs adds a very significant amount of novel information, when compared to previously available transcript sequence data from other insect species. It should also be emphasized that, in addition to the phylogenetic distance between *S. gregaria* and *L. migratoria*, there are several technical differences between the two locust databases that may account for the observed lack of homology between the ESTs of both species. Generally, GO-annotated sequences that are linked to the function of a specific tissue/organ are differentially distributed among both databases, providing a complementary set of locust transcriptome information.

2. Neuronal signaling and signal transduction components

Virtually all ongoing physiological processes in a living organism are initiated or regulated by a signal from the environment or from other cells. Depending on the incoming signal, a specific receptor (cell-surface or intracellular receptor) is activated and subsequently induces a downstream signal transduction pathway, leading to the cellular response. There are numerous signal transduction pathways, leading to an abundance of cellular responses and cross-talk between different pathways is possible. Signaling pathways make use of a range of molecular components, which may require activation in order to transfer the signal to downstream effectors. The CNS is expected to contain a large number of transcripts coding for components that are involved in neuronal signaling and signal transduction. For a first analysis of the contents of the *S. gregaria* EST database, we therefore have searched for sequences coding for compounds predicted to play a role in these processes. Here, we give an overview of the result of this analysis.

Of the sequences classified in the database under the GO term *Cellular process*, 108 are associated with *Signal transduction* (*Biological process* > *Cellular process* > *Cell communication* > *Signal transduction*). When all existing GO terms, at all levels, are considered, the biological process of signal transduction is the second best represented GO category in the *S. gregaria* EST database. The majority of sequences classified under the GO term *Signal transduction* can be divided under three GO terms, namely *Cell surface receptor-linked signal transduction*, *Intracellular signaling cascade* and *Regulation of signal transduction* (Figure 3). Since signal transduction in the nervous system entails more aspects than the ones classified under this specific GO term, this overview is extended towards neuronal signaling processes. Therefore, several transcripts that are predicted to be involved in *Peptide hormone processing* or in *Biogenic amine synthesis*, as well as more than twenty neuropeptide precursors, which have been annotated manually, will also be discussed. A list with the sequence IDs of components discussed in the context of (neuronal) signaling is made available in Table 2 and Table S1 (neuropeptides).

Cell surface receptor-linked signal transduction (GO:0007166). More than 80% of the uniques involved in signal transduction through cell surface receptors are predicted to play a role in enzyme-linked receptor protein signaling, Notch and Wnt signaling or G protein-coupled receptor (GPCR) signaling (Figure 3). Given its importance in neuronal signaling (while the former may be mainly involved in developmental processes), the latter category is now discussed in more detail.

A very important group of cell surface receptors are the seven transmembrane segments containing GPCRs [36–38]. Many GPCRs play a role in sensory perception and typical ligands are odor molecules, pheromones or light-sensitive components [39,40]. Other GPCRs function as receptors for a wide range of hormones and neurotransmitters [41,42], indicating that these receptors have versatile functions in a large variety of physiological processes. The fact that human GPCRs are the target for a vast number of drugs is yet another indication for their importance [43].

In the *S. gregaria* database several ESTs were annotated to be involved in *GPCR signaling pathways* (GO: 0007186). An opsin-like GPCR was found in the EST database that displays similarity to the previously identified desert locust opsin-1 protein, which is involved in perception of long wavelength light [44].

In addition to short neuropeptide F (sNPF) receptors from *D. melanogaster* [45,46], the red fire ant *Solenopsis invicta* [47] and the mosquito *A. gambiae* [48,49], a novel putative sNPF receptor has

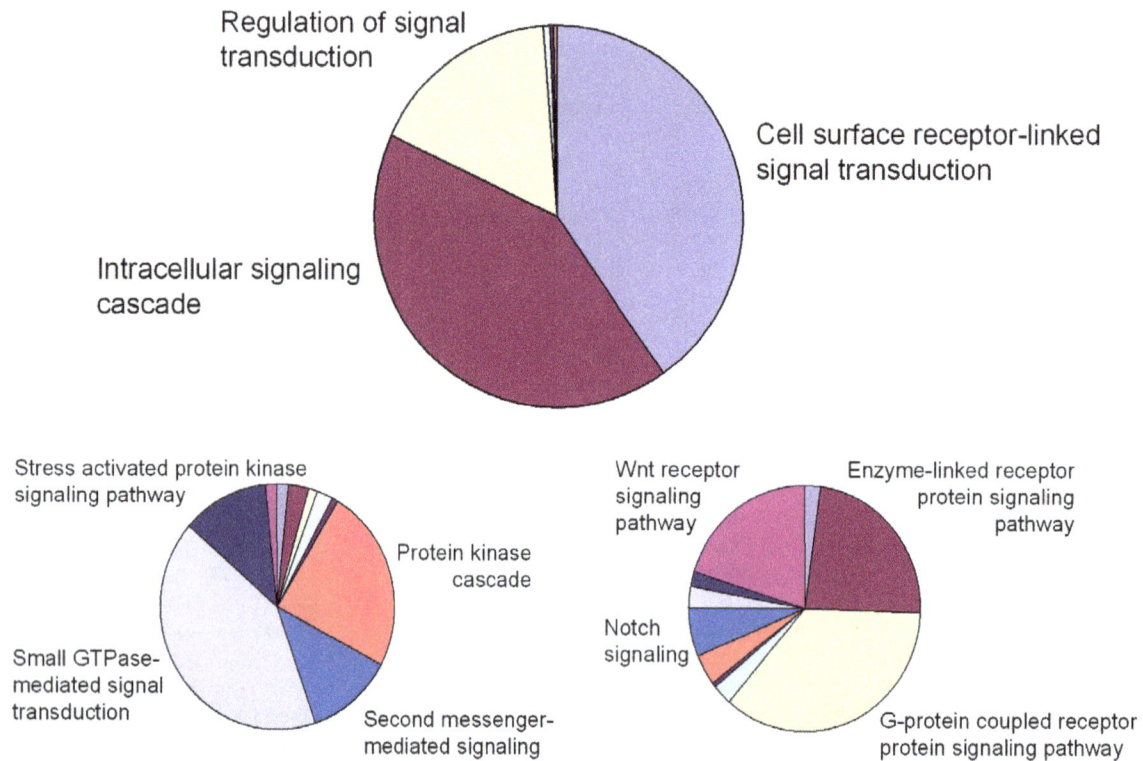

Figure 3. GO distribution of the *S. gregaria* EST sequences classified under *Signal Transduction*. Distribution of the *S. gregaria* EST sequences classified under the GO term *Signal transduction* and a more detailed overview of the sequence distribution in the two subcategories *Intracellular signaling cascade* and *Cell surface receptor-linked signal transduction*.

now been identified from the *S. gregaria* database. Recently, two sNPF peptides were identified from both *S. gregaria* and *L. migratoria*, and these peptides were shown to be widespread in the locust neuroendocrine system [50]. In *D. melanogaster* sNPF regulates food intake and body size [51] and it was shown to affect growth by a functional interaction with extracellular regulated mitogen-acitvated kinase (ERK)-mediated insulin signaling [52]. Furthermore, in the fruit fly it also has an effect on foraging, social and aggressive behavior [53,54].

Another gene product from the *S. gregaria* database shows significant homology with predicted orexin receptors from *Pediculus humanus corporis* (Kirkness *et al.*, unpublished), *Nasonia vitripennis*, *T. castaneum* and *A. mellifera*. Orexin is a mammalian neuropeptide involved in the control of several processes, including food intake. Interestingly, a *Bombyx mori* homolog of mammalian orexin receptors was recently demonstrated to be capable of responding to *Manduca sexta* allatotropin (AT), for which the receptor had long remained unknown [55]. AT has recently been associated with food intake control in insects [56]. Since solitarious and gregarious locusts show different foraging strategies, it is possible that AT and its receptor may be regulators of this phase-dependently regulated process. Furthermore, AT was initially described, at least in some insect species, as a potent stimulator of juvenile hormone (JH) synthesis [57]. JH has previously been demonstrated to induce certain phase-specific characteristics [58], again suggesting that AT and its receptor may possibly be involved in the control or establishment of phase-dependently regulated physiological processes. This hypothesis deserves further attention.

The *S. gregaria* EST database also contains Methuselah (Mth)-like GPCRs. Mth was first identified from *D. melanogaster* as a secretin-like receptor [59], belonging to the B-family of GPCRs

[38]. Interestingly, downregulation of Mth in the fruit fly resulted in an increased life span [60]. Although Stunted was originally identified as the ligand for Mth and mutants also resulted in increased life span [61], it was recently demonstrated that Mth displays a promiscuous response to several non-homologous peptides, including the *Drosophila* sex peptide and the newly identified 'Serendipitous Peptide Activator of Mth' (SPAM). However, mutations in Mth did not result in changes in sex peptide-induced behavior, questioning the biological significance of its Mth activation [62].

Finally, based on sequence similarity to (putative) serotonin receptors from *Culex quinquefasciatus*, *Aedes aegypti* and *T. castaneum*, one of the *S. gregaria* ESTs has been annotated as a putative serotonin receptor, while it also has some similarity to adenosine receptors. As serotonin has been demonstrated to be a crucial mediator for gregarization behavior [13], study of this putative receptor and its activated signaling pathway(s) might advance our knowledge of locust phase transition and could constitute a useful basis for the development of a new generation of products for a 'locust-specific' pest control. Although some other neurotransmitter receptors are also GPCRs, they are discussed in the following paragraph.

Neurotransmitter receptors are, unlike the previous GO terms, classified as a separate GO term (GO:0030594: *Neurotransmitter receptor activity*) under the main GO ontology *Molecular Function*. In general, two types of neurotransmitter receptors can be discriminated, namely ionotropic and metabotropic receptors. Whereas ionotropic receptors form an ion channel pore upon activation, metabotropic receptor activity results in indirect activation of a plasma membrane ion channel through a series of intracellular events [63,64]. Transcripts encoding both types of neurotransmit-

Table 2. Sequence ID and annotation of EST sequences associated with neuronal signaling.

Cell surface receptor-linked signal transduction	
G protein-coupled receptor protein signaling pathway	
LC.1391.C1.Contig1536	opsin
LC01037B2B08.f1	short neuropeptide F receptor
LC01003X1D07.r1	orexin receptor
LC02006X1H05.f1	Methuselah-like
LC01008B2B10.f1	Methuselah-like
LC01013A1E06.f1	putative serotonin receptor
Neurotransmitter receptor activity	
LC.1031.C1.Contig1170	nicotinic acetylcholine receptor alpha subunit
LC.907.C1.1039	N-methyl-D-aspartate glutamate receptor
LC01009A2C01.f1	GABA-A receptor
LC01019B2E05.f1	GABA-B receptor
LC01049B1E11.f1	metabotropic glutamate receptor
Neurotransmitters and (neuro)peptides	
Biogenic amine synthesis	
LC01054B2B09.f2	dopa decarboxylase
LC.243.C1.Contig314	tyramine beta hydroxylase
LC01006B2E02.f1	tryptophan hydroxylase
Peptide hormone processing	
LC.2118.C1.Contig2277	furin 2
LC.381.C1.Contig436	neuroprotein 7B2 precursor
LC.1129.C1.Contig1270	angiotensin converting enzyme
LC02007B2B06.f1	angiotensin converting enzyme
LC.1302.C1.Contig1447	endothelin converting enzyme
LC.3441.C1.Contig3579	endothelin converting enzyme
LC01071B1B03.f1	endothelin converting enzyme
LC.410.C1.Contig495	STE24 homolog

Sequence ID and annotation for the *S. gregaria* EST sequences encoding GPCRs and the EST sequences classified under the GO-terms *Neurotransmitter receptor activity*, *Biogenic amine synthesis* and *Peptide hormone processing*.

ter receptors are found in the desert locust EST database. These include an NMDA (N-methyl-D-aspartate) type of ionotropic glutamate receptor, which is co-activated by glutamate and glycine [65]. Second, the database also contains a metabotropic glutamate receptor 1, a GPCR which stimulates NMDA receptor activity through activation of phospholipase C, a rise in intracellular Ca^{2+} and activation of protein kinase C [66]. In addition, sequences resembling an ionotropic 'γ-aminobutyric acid' (GABA)-A and a metabotropic GABA-B receptor have been annotated. Finally, a transcript is present that corresponds to the previously reported α-subunit of the nicotinic acetylcholine ionotropic receptor [67,68].

Although receptor transcripts tend to be rare and are generally poorly represented in EST databases, this overview demonstrates that the *S. gregaria* EST database nevertheless contains several receptor encoding sequences, which may be involved in a range of important processes in the CNS. Study of the proteins encoded by these transcripts will provide us with new functional information which may constitute a scientific basis for the development of novel pest control agents or strategies. Insect pest control is nowadays often based on the use of chemicals that interact with a very limited number of neuronal targets, such as acetylcholinesterase, the voltage-gated sodium channel, the acetylcholine receptor, or the GABA receptor [69]. However, due to resistance phenomena and an increasing demand for ecologically considered pest control strategies, there is a constant need for new and more species-specific compounds (generating fewer side effects). Because GPCRs play such an important role in novel drug discovery programmes of pharmaceutical companies, they may constitute interesting targets in the context of pest control strategies as well. Therefore, the study of GPCRs may possibly broaden the range of species-specific pest control targets. The wider the range of targets, the more easily resistance phenomena can be avoided. However, not all GPCRs will prove to be excellent targets for pest control. An important step forward in locust pest management would be to identify specific receptor ligands that are capable of disturbing the process of phase change.

Biogenic amine synthesis (GO:0042401). Neuronal signaling processes involve neurotransmitter (and/or neuromodulator) molecules responsible for communication between neurons or communication between neurons and other target cells. Neurotransmitters are synthesized in neurons and vary in structure, ranging from single amino acids and mono-amines to peptides (which may also function as hormones). In what follows, an overview is given of transcripts related to the synthesis of biogenic amines and the processing of neuropeptide precursors. In addition, sequences corresponding to neuropeptide precursors are also presented.

Various neurotransmitters are biogenic amines, which are derived from amino acids. Several enzymes involved in the synthesis of these biogenic amines were predicted from desert locust EST sequences. Both dopamine and octopamine are synthesized from L-tyrosine. The dopamine synthesis pathway starts with hydroxylation of L-tyrosine to L-dihydroxyphenylalanine (L-DOPA) by the enzyme L-tyrosine hydroxylase [70,71]. L-DOPA is subsequently converted to dopamine by removal of the carboxyl group by DOPA decarboxylase [71,72]. Synthesis of octopamine is initiated by decarboxylation of L-tyrosine (L-tyrosine decarboxylase) [73]. Next, the resulting L-tyramine (which may also function as a neuronal signal [74]) is hydrolyzed by tyramine β-hydroxylase and the end product is termed octopamine (the role of octopamine in locusts and other arthropods was recently reviewed by Verlinden *et al.* [75]). The precursor for serotonin, L-tryptophan, is converted by subsequent actions of tryptophan hydroxylase and 5-hydroxytryptophan decarboxylase into 5-hydroxy-L-tryptophan and serotonin, respectively [76–78]. Sequences coding for tyramine β-hydroxylase, DOPA decarboxylase and tryptophan hydroxylase orthologs are present in the *S. gregaria* database.

Peptide hormone processing (GO:0016486). Neuropeptides generally require processing from a larger precursor polypeptide to become biologically active. The factors that aid in this processing process are proprotein or prohormone convertases and many of these are members of the subtilisin family. Subtilisin-like convertases generally cleave the peptide precursor at dibasic (Lys-Arg or Arg-Arg) or monobasic (Arg) amino acid sites [79]. Furin, which is a subtilisin-type of enzyme, cleaves specifically at Arg-Xaa-Yaa-Arg sites (Xaa: any amino acid, Yaa: Arg or Lys) [80,81]. Substrates for furin include polypeptide hormones, growth factors, growth factor receptors, neuropeptides and viral envelope glycoproteins. Furin has been characterized in mammalian species, but also in several invertebrate species. The desert locust EST database includes a homolog for *D. melanogaster* furin 2.

In addition, the desert locust EST database includes a transcript encoding a factor which is homologous to the mammalian neuro-endocrine protein 7B2. Although this factor does not possess protein processing activity itself, it is a regulator of the maturation of prohormone convertase 2 (PC2) [82–85]. PC2 transcript levels in mammals are most abundant in neuro-endocrine cells [86–89], which explains why it is involved in the processing of important peptide hormone precursors. *D. melanogaster* homologs for PC2 (dPC2) and 7B2 (d7B2) have also been identified, and it was demonstrated that dPC2 requires d7B2 for both maturation and secretion [90]. Interestingly, the desert locust transcript for the 7B2 homolog appears to be upregulated in isolated-reared animals, as demonstrated in the present study (*cf. infra*).

Insect homologs for the zinc metalloproteases 'angiotensin converting enzyme' (ACE) and 'endothelin converting enzyme' (ECE) [91,92] have also been suggested to play a role in peptide processing and/or degradation. Mammalian ACE is needed to process angiotensin I into angiotensin II, which is a peptide displaying vasoconstrictor activity and hence increases blood pressure. This conversion involves release of the C-terminal dipeptide from angiotensin I, which explains why ACE is also termed dipeptidyl carboxypeptidase [93]. A homolog of mammalian ACE was previously characterized in *L. migratoria*, and an *S. gregaria* EST encodes an almost identical enzyme. Immunolocalization studies in *L. migratoria* already demonstrated the presence of ACE in neurosecretory cells from brain and suboesophageal ganglion, in the storage part of the *corpora cardiaca*, in both the *nervi corporis cardiaci* and the *nervi corporis allati* and in the testes [94,95]. In three groups of neurosecretory cells, ACE was found to co-localize with locustamyotropins, suggesting that these factors may be ACE substrates. Locustamyotropins extended with Gly-Arg-Arg or Gly-Lys-Arg can indeed be hydrolyzed by recombinant *D. melanogaster* ACE [94]. Furthermore, *L. migratoria* ACE was also demonstrated to be involved in degradation of locustatachykinin-1 [96]. Finally, qRT-PCR studies revealed that *L. migratoria* ACE transcripts are found in virtually all tissues. Interestingly, ACE transcript levels in haemocytes increase in response to bacterial lipopolysaccharide administration, suggesting that ACE might play a role in immunity as well [97]. An insect ECE was first identified from *L. migratoria*. In vertebrates, ECE catalyzes the conversion from big endothelin into the vaso-active endothelin-1 by cleaving the -Trp_{21}-Val- bond [98]. In *L. migratoria*, ECE activity was demonstrated in neuronal membranes [99] and in the reproductive system, although specific substrates for ECE remain to be identified. Three ESTs were predicted to encode an ECE-like protein, one of which encodes a protein almost identical to *L. migratoria* ECE.

Another putative processing enzyme represented in the desert locust EST database is a STE24 endopeptidase. This type of enzyme was first characterized in the yeast *Saccharomyces cerevisae*. It was shown to remove aaX from the C-terminal CaaX-motif found in the a-factor, which is a mating pheromone (C: cysteine, a: aliphatic amino acid, X: any amino acid) [100,101]. Generally, CaaX-containing proteins are processed as follows: 1) a prenyl group is attached to the cysteine residue, 2) the aaX group is proteolytically removed and 3) the cysteine α-carboxyl group is methyl esterified. The prenylated and methyl esterified cysteine residue is now the new C-terminus and may form a hydrophobic membrane anchor. At present, STE24 endopeptidases have also been identified in bacteria, archaea, and virtually all eukaryotic kingdoms. Although STE24 endopeptidases have been predicted from insect genome data, no studies of their substrates have been performed so far.

Neuropeptide precursors. Neuropeptides are a versatile class of signaling molecules produced in the nervous system. Mostly, the bioactive peptide is processed from a larger precursor polypeptide (*cf.* reviews on neuropeptides and their precursors in locusts and other insects [4,5,102,103]). An *in silico* analysis of the novel *S. gregaria* EST database resulted in the (manual) annotation of more than twenty neuropeptide precursor encoding transcript

sequences (Table S1). Thirteen sequences appear to cover the entire reading frame (from start to stop codon) coding for the precursor polypeptide, while nine others represent partial sequences coding for a large, but still incomplete portion of the peptide precursor. Six of these precursor transcript sequences (present as four complete and two partial sequences in the EST database) were previously identified in the desert locust, *S. gregaria*, by conventional cDNA cloning strategies, *i.e.* the ones encoding the following peptides: adipokinetic hormone I (AKH-I) [104], short and long ion transport peptides (ITP-S and ITP-L) [105], neuroparsins 1 and 2 (NP-1 and NP-2) [106], and allatostatins (AST-A) [107]. The current study thus reveals sixteen novel sequences (eight complete and eight partial ones) for peptide precursor transcripts in the desert locust *S. gregaria*. Moreover, the characterization of these precursor sequences also results in the prediction of several (putative) neuropeptides which had as yet not been identified before (neither at the protein nor at the nucleic acid level) in *S. gregaria*. An overview of these novel desert locust peptides is given in Table 3.

The complete precursor sequences for the adipokinetic hormones, AKH I and AKH II, were retrieved from the *S. gregaria* EST database. The AKHs were among the first insect peptide hormones to be described, and the first fully characterized AKH was AKH-I from *L. migratoria* [1]. AKHs are important in releasing energy from reserves by increasing lipid and trehalose levels in the haemolymph [108–110]. Hypertrehalosemic hormone, HrTH, is a related neuropeptide which stimulates trehalose synthesis and release from the fat body. The migratory locust HrTH peptide has previously been identified by Siegert [111]. A transcript sequence encoding the entire HrTH precursor occurs in the *S. gregaria* EST database and reveals that both locust HrTH peptides are identical.

The allatostatins (AST-A family) were originally isolated from brain tissue of the cockroach *Diploptera punctata* as peptides inhibiting JH biosynthesis by the *corpora allata* [112,113]. Besides their allatostatic activity, these peptides appear to be pleiotropic in function [114,115]. Members of the AST-A peptide family have also been characterized in the desert locust [116] and their precursor cDNA, which codes for ten distinct AST-A related peptides (also designated as 'schistostatins') has been cloned [107]. After the discovery of additional neuropeptides with JH biosynthesis inhibiting activities in a variety of insect orders, 'allatostatins' were further classified in three distinct families (AST-A, B and C) based on sequence characteristics. Recently, 'AST-CC' neuropeptides were identified as a novel group of peptides displaying some sequence similarities to AST-C-type allatostatins. In the current study, a transcript sequence coding for a complete AST-CC precursor is found in the desert locust EST database. It is highly similar to a partial *L. migratoria* sequence that has been detected in LocustDB [117,118].

Allatotropin (AT) was initially purified from the moth *M. sexta* as a 13-residue amidated peptide that strongly activates JH synthesis [57]. A locust member of the AT peptide family was identified by Paemen *et al.* [119] as a myotropin from *L. migratoria* male accessory gland extracts (the initial name of this peptide was *Lom*-AG-MT I). Recently, this AT was also found in several tissues of *S. gregaria* and sequenced partially by means of mass spectrometry [5]. By analyzing the transcript sequence in the EST database, the full AT precursor has now been unveiled in the desert locust.

Insectatachykinins (TKs), also designated as tachykinin-related peptides, form an evolutionary conserved family of brain-gut peptides characterized by a common C-terminal sequence -FXGXRamide (X = variable amino acid residue) [102,103]. The first members of this family were isolated as myotropic peptides from the CNS of *L. migratoria* [120,121]. The desert locust database

Table 3. Overview of the newly predicted *S. gregaria* neuropeptides.

Neuropeptide	ID	Amino acid sequence *S. gregaria*
AST CC	LC.407.C1.Contig492	SYWKQCAFNAVSCFamide
Burs-β	LC01070B1E02.f1	VVRAPLEVDGIDKLDIEFRCCRCQWACNSQVQPSVTTPTGFLKECYCC RESFLRERTVTLSHCYDPDGARLTAEGTATMDIRLREPAECKCFKCGDFSR
CCHa	LC01024B1A06.f1	GCMAFGHSCFGGHamide
DH	LC.1479.C1.Contig1625	MGMGPSXXIVNPMDVLRQRLLLEIARRRLRDAEEQIKANKDFLQQIamide
GPA	LC.3116.C1.Contig3268	MVPPSSRSALHFFALAVALCLSAVSAGMDGERDAWEKPGCHRVGHTRKISIPDCIEFPITTNACRGF CESWSVPSALNTLRVNPHQAITSIGQCCNIMETEDVEVRVMCLDGPRDLVFKSAKSCQCYHC
HrTH	LC.3853.C1.Contig3980	QVTFSRDWSPamide
ITG	LC.817.C1.Contig945	ITGKVASFNHI
NPF	LC.1768.C1.Contig1921	QQAAADGNKLEGLADALKYLQELDRY**YSQVARPRFamide***
TK-2	LC.108.C1.Contig162	APLSGFYGVRamide
TK-3	LC.108.C1.Contig162	APQAGFYGVRamide
TK-4	LC.108.C1.Contig162	APSLGFHGVRamide
TK-5	LC.108.C1.Contig162	APLLGFHGVRamide
TK-6	LC.108.C1.Contig162	APLRGFQGVRamide
TK-7	LC.108.C1.Contig162	ALKGFFGTRamide
TK-8	LC.108.C1.Contig162	GNT*KK*APVGFYGTRamide **
Predicted neuropeptides that do not show homology to any known peptide		
PVK-4	LC.2414.C1.Contig2580	KGLVANARVamide
PVK-5	LC.2414.C1.Contig2580	DSLWFGPRVamide
MT-2	LC.2414.C1.Contig2580	TSSLFPHPRIamide
MT-3	LC.2414.C1.Contig2580	SLRL*RL*PAAAWLAAGDVGNGKGDFTPRLamide **
PVKDP	LC.2414.C1.Contig2580	AGLGQDETRAGTK
NLP-1	LC.1768.C1.Contig1921	YLASLVRSHGLPYPLT
NLP-2	LC.1768.C1.Contig1921	EDDGPGEI
NLP-3	LC.1768.C1.Contig1921	NVGALARNWMLPSamide
NLP-4	LC.1768.C1.Contig1921	ASDDDQEVD
NLP-5	LC.1768.C1.Contig1921	YLASVLRQamide
NLP-6	LC.1768.C1.Contig1921	HLGSLAKSGMAIH
NLP-7	LC.1768.C1.Contig1921	FLGVPPAAADYamide
NLP-8	LC.1768.C1.Contig1921	HIGALARLGWLPSFRAASA*RS*G*RS*AGSRSamide **

This table shows an overview of the newly predicted *S. gregaria* neuropeptides (*i.e.* which had not been identified previously, neither at the protein nor at the nucleic acid level) and the sequence ID of the precursor encoding transcript. It needs to be emphasized that some of these peptides are predicted on the basis of possible cleavage sites within the precursor but do not show homology to any known peptide. Their presence and function(s) *in vivo* have not been demonstrated yet. In addition, the precursor sequences are derived from EST sequence information and a limited degree of sequencing errors cannot be fully excluded. Abbreviations: AST CC: allatostatin double C; Burs-β: bursicon β-subunit; DH: diuretic hormone; GPA: glycoprotein hormone α; HrTH: hypertrehalosemic hormone; NPF: neuropeptide F; TK: tachykinin (TK 2–4 had previously been demonstrated by means of mass spectrometry, but their complete amino acid sequence had so far not been completely determined [5]); PVK: periviscerokinin; PPDP: PVK precursor-derived peptide; NLP: neuropeptide-like peptide. The nomenclature of CCHa is based on the fact that this neuropeptide has two conserved cysteine residues and an amidated histidine residue. The nomenclature of Apis ITG is based on a pattern of three amino acid residues in the sequence.

*: The amino acid residues printed in bold represent the truncated form of NPF, which had previously been demonstrated by means of mass-spectrometry [50].

**: A possible cleavage site within the predicted peptide is printed in bold italics. Different cleavage forms of a certain peptide may occur in a tissue-dependent manner. For instance, TK-8 may occur as GNTKKAPVGFYGTRa or APVGFYGTRa; MT-3 may occur as SLRLRLPAAAWLAAGDVGNGKGDFTPRLa or LPAAAWLAAGDVGNGKGDFTPRLa or GDFTPRLa. For NPLP-8, different cleavage patterns might perhaps result in HIGALARLGWLPSFRAASARSGRSAGSRSa, LGWLPSFRAASARSGRSAGSRSa, AASARSGRSAGSRSa, HIGALARLGWLPSFRAASA, SGRSAGSRSa, HIGALARLGWLPSFRAASARSa or SAGSRSa.

contains a transcript sequence encoding a precursor that contains a secretory signal peptide and nine distinct TK related peptides. The peptides are flanked by dibasic cleavage sites (mostly Lys-Arg or KR) and each peptide has a C-terminal amidation signal (Gly or G). Five of these peptides have previously been demonstrated in desert locust nervous tissue extracts by means of mass spectrometry and were termed *Scg*-TK-1-4 (two peptides on the precursor are identical and correspond to *Scg*-TK-4) [5]. The amino acid sequences of *Scg*-TK-2-4 could in that study only partially be determined, but the current study reveals that these peptides are

identical to the corresponding TKs from *L. migratoria*. A partial *L. migratoria* TK precursor sequence was previously retrieved from Locust DB [122].

Neuropeptide Y (NPY), the most abundant neuropeptide in the mammalian nervous system, is a highly conserved 36 amino acid neuromodulator [123]. NPFs are considered as invertebrate homologs of vertebrate NPY. Whereas the vertebrate NPYs have a C-terminal amidated tyrosine (Y) residue, NPFs end with an amidated phenylalanine (F). NPFs have been found in several invertebrate phyla, such as flatworms [124,125], mollusks

[126,127] and insects [128]. The *L. migratoria* EST database, LocustDB [25,26], provided the first evidence for the existence of NPF in locusts. In addition, a complete NPF precursor encoding transcript has now also been identified from the desert locust EST database. The obtained precursor sequence encodes (long) NPF, which is nearly identical to *L. migratoria* NPF. A truncated form of NPF (YSQVARPRFamide) has previously been demonstrated in both locust species [50].

Ion transport peptide (ITP) was first characterized from the desert locust, *S. gregaria*, as a peptide that stimulated transport of Cl^- ions across the hindgut epithelium [129,130]. Two different locust cDNAs were cloned, encoding a short (ITP-S) and a long variant of ITP (ITP-L) [105,131]. Both transcript sequences (coding for the complete precursors) were also encountered in the desert locust EST database. ITP-S and ITP-L peptides are highly similar in their N-terminal region, but differ in their C-terminal part. In addition, ITP-L has four more amino acid residues. Both locust ITP precursors also contain an ITP co-peptide between the signal peptide and ITP. Although this co-peptide has been confirmed by mass spectrometry, its function is as yet unknown [5,132].

Neuroparsin (NP-1) was initially identified from *L. migratoria* as a peptide having antigonadotropic, as well as several other pleiotropic functions [133–139]. Later, a similar peptide was also characterized in *S. gregaria* [140]. Cloning studies in the desert locust further revealed the existence of four distinct NP precursors (NPPs), two of which (NPP-1 and NPP-2) were more prominently expressed in the CNS [106,141]. At present, NPs and NP-like peptides have meanwhile been identified from several other invertebrate species [142]. Interestingly, NPs display sequence similarity to the hormone binding domain of vertebrate 'insulin-like growth factor binding proteins' (IGFBPs) and a recombinant NP was demonstrated to be capable of interacting with the *S. gregaria* insulin-related peptide *in vitro* [143]. Moreover, desert locust NPP transcripts have been shown to be expressed in a (isolated/crowded-reared) phase and reproduction cycle dependent manner [144,145]. The *S. gregaria* EST database contains a complete and a partial sequence coding for the neuroparsin precursors NPP1 and NPP2, respectively.

Another transcript in the *S. gregaria* database was found to encode a complete CCHamide-like neuropeptide precursor. CCHamides were first identified in the silkworm *B. mori* [146] and their name refers to two conserved cysteine residues (CC) and a C-terminal amidated histidine residue (Ha).

The last neuropeptide for which a complete precursor sequence was retrieved shows sequence similarity to a peptide previously identified from *A. mellifera* by mass spectrometry [147]. It was termed *Apis* ITG peptide because of its first three amino acids, but so far nothing is known about its physiological function.

In this study, a large part of the 'crustacean cardioactive peptide' (CCAP) precursor transcript sequence was identified from the *S. gregaria* EST database. The CCAP was first identified from the crab *Carcinus maenas* as a heart contraction accelerating peptide [148]. CCAP has also been isolated from the locusts *L. migratoria* and *S. gregaria* [116,149]. In locusts, CCAP has been shown to act as a pleiotropic factor, stimulating hindgut and oviduct contractions [149,150] and triggering the release of AKH from *corpora cardiaca* [116].

Another cardioactive peptide identified from *M. sexta* was designated as 'Cap-2b' [151–155]. This peptide was later shown to be a member of the periviscerokinin (PVK) family (so-called because of their presence in abdominal perivisceral organs) [156,157]. PVKs have also been identified from locust abdominal ganglia and perivisceral organs and display similarity to the

peptides derived from the *D. melanogaster capability* gene encoded precursor [158]. EST sequence information reveals that the precursor encoded by an *S. gregaria* transcript sequence contains the periviscerokinins *Scg*-PVK-2 (GLLAFPRVa) and *Scg*-PVK-3 (DGAETPGAAASLWFGPRVa). The periviscerokinin/myotropin-like peptide *Scg*-MT-2 (TSSLFPHPRLa) was suggested to be encoded by the same precursor [158]. However, we only found the nearly identical sequence TSSLFPHPRIa in the current precursor. Nevertheless, it should be emphasized that *Scg*-MT-2 was initially identified by means of mass spectrometry which does not make the distinction between leucine and isoleucine residues. In addition, the PVK precursor encodes four (extra) newly predicted peptides. Two of these show a leucine at position n−7, an arginine at position n−2 and a C-terminal amidation which are typical characteristics of insect PVKs [159]. These two peptides are therefore termed *Scg*-PVK-4 and *Scg*-PVK-5. Another peptide contains the -FXPRLamide C-terminus that is typical for mytropins and is designated as *Scg*-MT-3. The last peptide shows no homology to any known insect peptide and is designated as 'PVK precursor-derived peptide' (PVKDP). In addition, a blast search revealed that this *S. gregaria* transcript displays highest sequence similarity to the (computationally predicted) capability-like precursor from the aphid, *A. pisum*.

Locust diuretic hormone (DH) was isolated from brain and *corpora cardiaca* as a peptide that stimulated urine production in *L. migratoria* [160,161]. An *S. gregaria* transcript sequence encoding a precursor for a closely related DH is found in the EST database. Both locust peptides seem to be identical, although two amino acids of the *S. gregaria* DH could so far not be determined. Insect DH displays a motif that is typical for members of the vertebrate corticotropin releasing factor (CRF) family [162]. Because of the dissimilar gene structures, it is however still under discussion whether insect DHs and vertebrate CRFs are true orthologs. Similarly to vertebrate IGFs, the bioavailability and bioactivity of CRFs is regulated by binding proteins. CRF binding proteins (CRF-BPs) and related factors appear to be remarkably well conserved in metazoan evolution and occur in both the deuterostomian and protostomian lineages. Because of the similar gene structure and conserved cysteine pattern, the CRF-BP-like proteins of insects have been described as true orthologs of vertebrate CRF-BPs [163]. Although the biological role of insect DHs and vertebrate CRFs seems to differ, the above example may represent an extra argument suggesting that DH and CRF are both part of an evolutionary conserved system. In this context, it is interesting to observe that a CRF-BP ortholog is also predicted from the *S. gregaria* EST database (LC.2107.C1.Contig2266).

The 'ovary maturing parsin' (OMP) is a locust gonadotropic peptide that is produced in the *pars intercerebralis*, hence the name 'parsin' [140,164]. The identification of its precursor cDNA has been unsuccessful until this study. Two *S. gregaria* EST sequences encode two (slightly) different OMP precursors, hereby representing the first (partial) OMP precursor transcript sequences identified in insects. Each of the two identified precursors encodes a different isoform of OMP, the difference between both mainly resides in an insertion of three amino acids (Pro-Ala-Ala or PAA) [140].

Bursicon is a neurohormone initiating cuticular tanning and wing spreading immediately after eclosion of an adult fly [165–168]. The active hormone is a heterodimer consisting of two cysteine knot containing subunits [169,170]. Although bursicon subunits have been identified in several insect orders and even in other invertebrate phyla [171], this study is the first to reveal a partial peptide sequence of the β-subunit of bursicon in locusts.

Members of the glycoprotein (GP) hormone family, such as vertebrate thyroid-stimulating hormone (TSH) and gonadotropins [follicle-stimulating hormone (FSH), luteinizing hormone (LH) and chorionic gonadotropin (CG)], are also heterodimeric factors that consist of two cysteine knot containing subunits, an α (GPA) and a β subunit (GPB) [172–176]. Later, based on vertebrate genome data, an extra α (GPA2) and β subunit (GPB5) have been predicted [177,178]. Interestingly, orthologs of GPA2 and GPB5 have also been identified in invertebrates [177]. In *D. melanogaster*, a heterodimer of these subunits was shown to be capable of activating the 'leucine-rich repeat-containing GPCR 1' (dLGR1) [179]. In the *S. gregaria* EST database, a transcript sequence is found that encodes a GPA-like peptide, which displays highest similarity to putative GPAs from *P. humanis corporis* and *T. castaneum*.

Furthermore, a partial *S. gregaria* transcript sequence appears to encode a 'neuropeptide-like precursor 1' (NPLP1)-like sequence. Previously, three neuropeptides showing no homology to other known neuropeptides were identified from a *D. melanogaster* CNS extract. It appeared that these peptides were all encoded by the same precursor, which was consequently termed 'neuropeptide-like precursor' (NPLP) [180]. Later, similar precursors were identified from *A. mellifera* [181], *T. castaneum* [182], *A. gambiae* [183] and *Neobellieria bullata* [184]. Although the predicted *S. gregaria* precursor displays putative hallmarks of an NPLP precursor, comparative sequence analyses of NPLP precursors have revealed substantial sequence heterogeneity among different species, impeding unambiguous sequence-based identification of NPLP orthologs. The eight predicted *S. gregaria* NPLP-encoded peptides are referred to as 'neuropeptide-like (precursor derived) peptides' (NLP) 1–8.

The last decades, a wide range of neuropeptides have been isolated from locust nervous tissue extracts, illustrating the advantage of the big size of locusts for neurobiological and endocrinological studies [1–5]. However, the lack of gene sequence information often proved to be a drawback for the application of novel techniques. The above findings show that this EST database constitutes a very important step to bridge this gap allowing for the implementation of new post-genomic research strategies (e.g. peptidomics, proteomics and transcriptomics) techniques. This is also nicely illustrated by a recent neuropeptide search performed in the shrimp *Litopenaeus vannamei*. Based on the availability of an EST database, peptide precursors were predicted and by combining the obtained information with mass spectrometry, the *in vivo* occurrence of many *L. vannamei* peptides could be confirmed [185].

3. Genes differentially regulated in the two phases

The novel EST database will constitute a valuable tool for future studies analyzing and comparing transcript profiles in desert locusts under different developmental or physiological conditions. In our lab, microarray studies are being prepared for a transcriptome-wide analysis of phase-dependent gene expression in the desert locust. Meanwhile, since specific sequence tags had been incorporated in the cDNA inserts derived from isolated- and crowded-reared animals during library construction, an analysis of contigs consisting of at least five tagged ESTs resulted in the selection of an initial set of transcripts for a qRT-PCR analysis of their relative abundance in isolated- and crowded-reared animals' nervous systems (Table 4). Three additional transcripts, for which corresponding EST sequences were identified in the *S. gregaria* database, were also evaluated as (putative) positive controls in this study (Table 5). Relative levels for all transcripts which were statistically confirmed by qRT-PCR as differentially expressed in isolated- and crowded-reared locusts are shown in Figures 4, 5 and 6.

Confirmation by qRT-PCR of previous studies. 'Phase-related peptide' (PRP) and 'secreted protein acidic and rich in cysteine' (SPARC) transcript (which was initially selected by differential display PCR) more abundantly occur in crowded-reared desert locust haemolymph and CNS, respectively [186–188], while the 'solitarious phase specific gene' (SSG) transcript was previously identified by differential display PCR and found to be more abundant in isolated-reared locust CNS [187]. In the present study, qRT-PCR indeed confirmed that the transcripts encoding SSG and SPARC are more abundant in the nervous system of isolated- and crowded-reared locusts, respectively (Figure 4). While PRP has been detected as a highly abundant peptide in crowded-reared locust haemolymph extracts, only relatively low quantities were found in extracts of isolated-reared animals [186,188]. In line with these initial observations, which were all performed at the peptide/protein level, this study now shows that the corresponding transcript levels are also significantly higher in nervous tissue from crowded-reared desert locusts. Although their biological role(s) remain(s) to be elucidated, PRP and SSG transcripts may serve as molecular markers for phase transition.

SPARC is an extracellular matrix-associated Ca^{2+}-binding glycoprotein and has pleiotropic functions in vertebrates, mainly in embryonic tissues and adult tissues undergoing remodeling [189–191]. These functions include regulation of cell shape [192,193], cell adhesion [194,195], cell cycle [196], extracellular matrix [197,198], cell proliferation [196,199] and cell migration [200,201]. Furthermore, SPARC was demonstrated to interact with certain growth factors, thereby influencing the cellular response to these factors [194,199,202]. Interestingly, a SPARC-like factor was shown to have a very particular effect on neuronal cell migration, regulating radial glia-guided neuron migration in the rat cerebral cortex [203]. Although SPARC has been identified in many insect species, its involvement in neuronal migration has hitherto not been studied.

Differentially expressed genes: crowded > isolated. By means of qRT-PCR, the levels of nine other transcripts were shown to significantly differ between crowded- (Figure 5) and isolated-reared (Figure 6) animals, either in head ganglia (HG) or ventral nerve cord (VNC) or in both CNS parts. That not all tested transcripts show significant differences in relative abundance in the isolated- and crowded-reared locust CNS may have two possible reasons. First, it needs to be emphasized that the approach of tag-based evaluation of contigs should rather not be seen as a truly quantitative method. It is also only of use for the most abundant transcripts (for which contigs could be created). In this context it needs to be emphasized that the *S. gregaria* EST database has been derived from normalized cDNA libraries, thereby compensating for very abundant transcripts. Since only a limited number of tagged ESTs is available per contig the factor coincidence should not be completely excluded. Therefore, based on this initial analysis, a selection of 16 transcripts was made, for which the expression level in both phases was further investigated by qRT-PCR. In the end, the qRT-PCR results allowed us to formulate more solid statements about differences in expression levels of the tested genes between locusts reared in the isolated or crowded condition. Second, the qRT-PCR assays were performed starting from nervous tissue derived from 4- to 10-day old adults. Since sequence information in the EST-database has been derived from locusts in several developmental stages, a stage-dependency of certain phase-related differences should not be excluded. Nevertheless, we feel that this independently obtained evidence strengthens our conclusions.

In the context of the previous paragraph, it may also appear of interest to notice that two transcripts which show significantly higher levels in crowded-reared desert locusts, code for factors

Table 4. Overview of the contigs that have been selected by tag-based evaluation for qRT-PCR analysis.

EST ID	C	I	Annotation	Forward primer (5'→3')	Reverse primer (5'→3')
Crowded>Isolated					
LC.228.C1.Contig298	9	0	**Fasciclin(-like) precursor**	GAATCACTTGGTGGGCCTCTT	CCATATGAATGCGACCCTCAT
LC.308.C2.Contig390	8	0	*No annotation*	CAAAAATCTGTGCCAAGGAACTG	GCGCTTCAACAACAGCAATC
LC.393.C1.Contig477	8	0	Similar to CG12163 (Cys-protease inhibitor)	CAAGCGTGAGATTACGGAAATACA	AGGTAGTGCTGCTCGTCTCGTT
LC.1955.C1.Contig2112	7	0	Similar to T-complex protein 1 subunit gamma	TGCGTGTGGTGCTACTATTGTG	TGTTCCCACGTCATCTTCCTT
LC.4273.C1.Contig4391	7	0	*No annotation*	GCATAGGAGAGTGAAGCATTCACA	ACAAGAATGCAGACAAAAACTACACA
LC.446.C2.Contig534	5	0	Similar to 14-3-3 protein (leonardo protein)	CGTGTCAGTTGGCGAAACAG	CTTCGTTTAGCGTATCCAGTTCAG
LC.129.C1.Contig185	6	1	**Slit homologue**	GGCGCACCTCAAATTGGA	TCCACCGTGAAGCTGTCTTG
LC.1849.C1.Contig2006	6	1	Signal peptidase complex subunit 2	AAACATTGTGTTTGCATGGGTAAG	GGCTGCGCTTCCTTGCT
LC.392.C1.Contig474	5	1	**Probable cytochrome P450**	GAGGTGAACCGTGGAGAAAGTT	CGGCTCGCTGTGAAGGA
Isolated>Crowded					
LC.1602.C1.Contig1749	0	15	**RNA helicase Ddx1**	TGTCCTAGTCGAGGACGGAATT	TGCAACAGCCTCCTTCATCA
LC.587.C1.Contig691	0	11	G-protein gamma subunit	CGTACGGCTGCATTAAATTCTG	AGGAACGGGCGACTGAATC
LC.1473.C1.Contig1619	0	7	*No annotation*	GCTGTGACATTTCTGGCCTCTT	TTCATTAGAGGGATACTCTTTCAAGCTA
LC.312.C1.Contig394	0	7	Glutamine synthetase	CGCGCGAATCTGCAAGA	TGTCGGGCTGCGGAAGT
LC.4308.C1.Contig4427	0	7	*No annotation*	AGTCATTCTGAGAGAGACAAAGTTTCTTAT	AACACTGCAATTCGCTTCGA
LC.1603.C1.Contig1750	0	5	*No annotation*	CCCCCTGGTGGACAGTCAT	TGCCAATACGTGCACAGAATC
LC.733.C2.Contig853	1	5	**Similar to 7B2 precursor**	ACCTCATTCAGCGCCAAAAT	CCCAGCCATGCTGGAGTCT

The different columns show the sequence ID of the contig, the number of tags referring to the origin (C: crowded-reared; I: isolated-reared) of the composing ESTs per contig, the annotation of the sequence and the primers that have been used for the qRT-PCR assays. The contigs printed in bold black are those for which the corresponding transcript levels were significantly different in nervous tissue of crowded- and isolated-reared locusts. The transcript levels corresponding to the contigs printed in normal black were not found to be significantly different in nervous tissue from locusts in the two phases.

predicted to be involved in development and/or modeling of the nervous system. One of these transcripts encodes a locust homolog of Slit, a factor known to be involved in axon guidance in both vertebrates and invertebrates. To assure that growing neurons connect the two brain halves in *D. melanogaster*, Slit will prevent these neurons from crossing back over the midline [204]. Dimitrova *et al.* [205] also demonstrated that fruit flies mutant in Slit or in its receptor Robo (roundabout), displayed space-filling neurons with longer, but less branched dendrites. Recently, midline signaling molecules, including Slit and Robo, were shown to play a crucial role in the formation of a neural map by dendritic targeting of motor neurons in the *D. melanogaster* nervous system [206]. The second transcript encodes a homolog of fasciclin-I, which was shown to play a role in neuron path finding. This cell adhesion molecule is involved in certain cell-cell interactions necessary for development of the nervous system. In addition, it

was shown previously that fasciclin plays an important role in axonogenesis during embryonic brain development in *S. gregaria* [207]. These findings suggest that the gregarious CNS is possibly in a more plastic state than the solitarious CNS. It will thus be of interest to evaluate whether SPARC also has a role in neural plasticity in *S. gregaria*.

Crowding-induced aggregation behavior of locusts implicates learning (habituation) and memory related processes [208]. Both non-associative learning and memory require the acquisition and retention of neuronal representations of new informational input and are typically associated with neuroplasticity. While a small brain size and a relatively simple neuronal organization in combination with a short life span used to be common arguments for rejecting an important role for brain plasticity in insects, several neurogenetic studies have overthrown this assumption (as reviewed by Dukas. 2008 [209]). Accordingly, we have characterized at least

Table 5. Overview of the EST/contig sequences corresponding to PRP, SPARC and SSG.

EST ID	Annotation	Reference	Forward primer (5'→3')	Reverse primer (5'→3')
LC.3176.C1.Contig3329	Phase-related peptide	[186,188]	CCGTCTGAAATTCAAAGATGGAA	CGCGACGCATATGGTATCC
LC.463.C1.Contig551	SPARC (secreted protein, acidic, rich in cystein)	[187]	GGCGAGTCCGCACTACAACT	CACCATCCTCTGCTCCTTGAA
LC01031A1A08.f1	Solitary phase specific gene	[187]	CGACTGACGCCTGATTTTCC	CACAGTGACTGGGCGTGAGA

Overview of the EST/contig sequences corresponding to PRP, SPARC and SSG and the primers that have been used for their qRT-PCR assays.

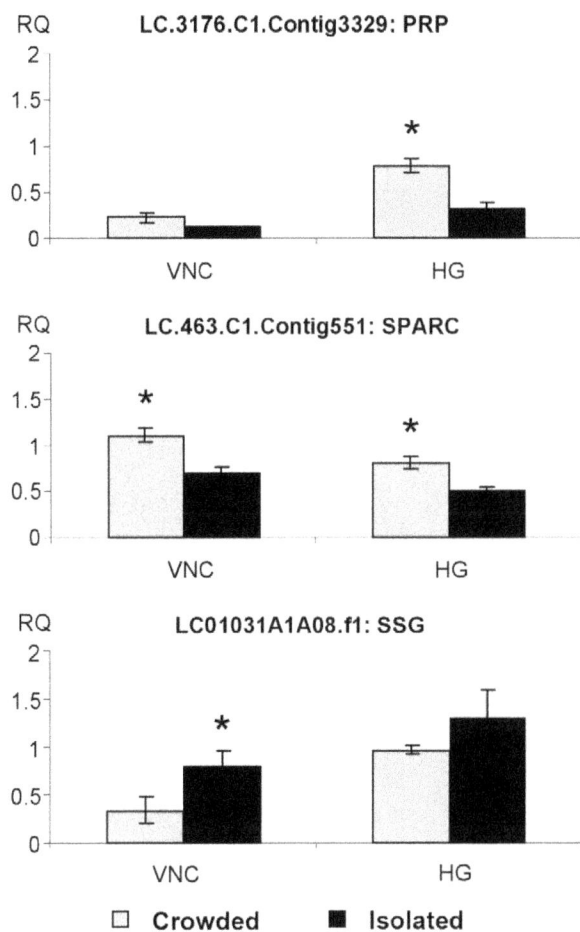

Figure 4. Relative transcript levels for PRP, SPARC and SSG in desert locusts in the two phases. Relative transcript quantity (RQ) for PRP, SPARC and SSG in isolated- and crowded-reared desert locust ventral nerve cords (VNC) and head ganglia (HG). Results were obtained by analyzing four independent pooled samples of ten individuals per condition and are represented as means ± standard error. Statistical analysis consisted of a Student's t-test for comparing two independent groups. Significantly higher transcript levels (p<0.05) are indicated by an asterisk (*).

three transcripts (*slit, fasciclin-1* and *SPARC*) that may be involved in the modeling and/or organization of the CNS and are upregulated in adult crowded-reared locusts, as compared to their isolated-reared conspecifics. It should also be noted that the *S. gregaria* EST database contains 210 transcripts classified under the GO term *Nervous system development* and that 179 (85%) of these transcripts are involved in neurogenesis and CNS development. Therefore, the necessary information is now available to perform a wide-scale comparative transcriptomic analysis of the molecular mechanisms underpinning the neurobiological dynamics of the locust CNS during phase transition.

Although the outcome of this analysis is much awaited, accumulating evidence points to a crucial role for serotonin. Not only was this biogenic amine demonstrated to be a necessary and sufficient trigger for gregarization behavior [13], it had earlier been shown to be a regulator of neuroplasticity in a number of different animals. In the *M. sexta* olfactory system, serotonin has an effect on excitability of neurons and it was suggested to play a role in structural plasticity of these neurons (more specifically in neuron growth and establishment of new neuron connections) [210]. The

latter is also supported by previous observations, describing a positive effect of serotonin on the growth of olfactory neurons *in vitro* [211]. In the marine mollusk *Aplysia* it was shown that the learning process of gill and siphon withdrawal reflexes is accompanied by formation of new connections between siphon sensory neurons and their target cells. Intriguingly, this process of new synapse formation could also be mimicked by serotonin, which was moreover demonstrated to influence the number of fasciclin II-like cell adhesion proteins [212]. Although functional serotonin-fasciclin interactions have hitherto not been studied in insects, it is plausible that both factors may also act together in regulation of neuroplasticity in insects.

We also identified a putative cytochrome P450 encoding transcript as being more abundantly present in the nervous tissue of crowded-reared locusts. The superfamily of cytochrome P450 proteins comprises heme-containing enzymes that oxidize diverse organic substrates [213]. However, no specific functional information is as yet available about this transcript or the protein it encodes. Therefore, future studies will be required to determine its role in phase-dependently regulated processes.

Differentially expressed genes: crowded < isolated. A transcript that is less abundant in crowded- than in isolated-reared locusts, encodes a DEAD-box protein Ddx1 homolog. DEAD-box proteins are involved in RNA splicing and early translation events [214,215]. This may possibly indicate that some phase-related differences arise from the very basic level of RNA processing and translation. Accordingly, a similar study in *L. migratoria* predicted that transcripts classified under the GO-terms *Ribosome* (*Cellular Component*), *Nucleic acid metabolism* and *Protein biosynthesis* are more abundantly present in isolated-reared locusts [25].

Another transcript that was significantly more common in isolated-reared locusts encodes a homolog of the PC2-supporting protein 7B2 (*cf. supra*) [82–85]. Neuropeptides and peptide hormones are typically derived from larger precursors and act as regulators in a wide range of physiological processes, such as growth, development, reproduction, ecdysis, diuresis, feeding behavior and metabolism. Since the activity of prohormone processing enzymes is essential for the production of (most) bioactive peptides, the neuropeptidergic regulation of (many) physiological processes may also be indirectly linked to the level of precursor processing. Differences in 7B2 transcript levels in the two locust phases may therefore be one of the underlying mechanisms leading to a differential regulation of a wide range of physiological processes during phase transition.

For several other transcripts that are significantly more abundant in isolated- or crowded-reared locusts, it is not yet clear what their role in phase transition might be or why they are more abundant in one of the two phases. Many sequences in the database could not (yet) be annotated, since clearly orthologous sequences could not (yet) be identified. For example, LC.4273.C1.Contig4391, strongly upregulated in crowded-reared locusts, and LC.4308.C1.Contig4427, increased in isolated-reared locusts, are certainly worth further investigation, but at present there is no information about their possible biological activities. Nevertheless, several of the newly identified phase-dependently regulated transcripts may constitute promising (candidate) markers for monitoring the process of locust phase transition.

It needs to be mentioned that the solitarious (isolated) locust colony was reared in isolation for many generations. By separating the isolated- and crowded-reared locusts for such a long time we aimed at establishing genuine long-term phases (when only studying early behavioral changes a lower number of isolated generations can be applied). Since the colonies for both phases were quite large (hundreds of individuals per generation for the isolated-reared

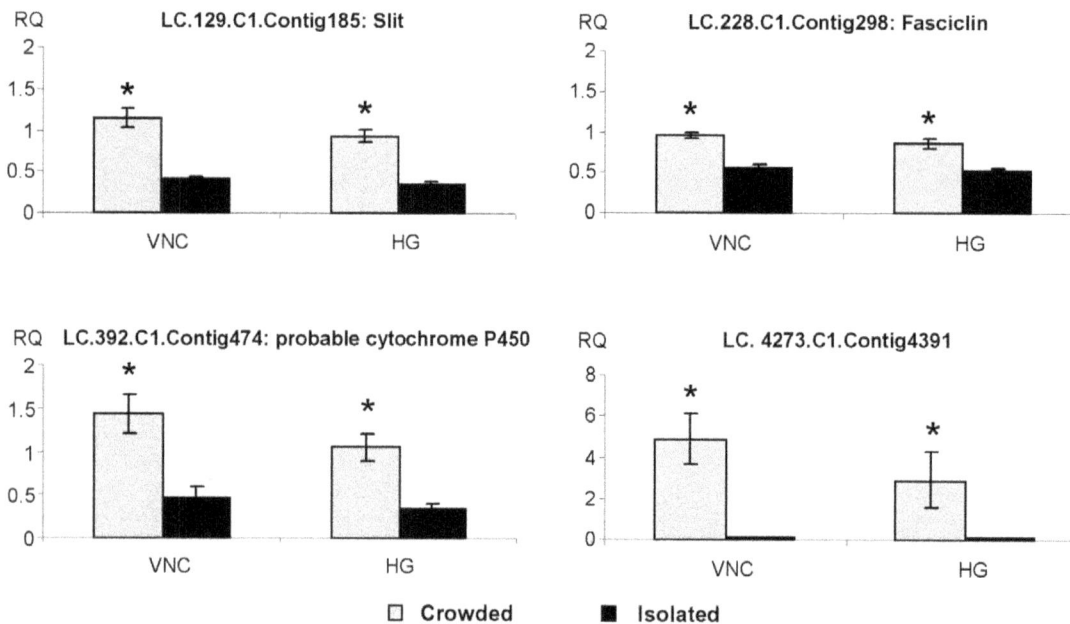

Figure 5. Relative transcript levels for genes showing higher expression levels in nervous tissue of crowded-reared locusts. Relative transcript quantity (RQ) for genes showing higher expression levels in crowded-reared desert locusts (VNC: ventral nerve cord, HG: head ganglia). Results were obtained by analyzing four independent pooled samples of ten individuals per condition and are represented as means ± standard error. Statistical analysis consisted of a Student's t-test for comparing two independent groups. Significantly higher transcript levels ($p < 0.05$) are indicated by an asterisk (*).

colony), genetic drift and inbreeding phenomena were expected to remain limited. In addition, by washing egg pods from crowded-reared females and introducing the newly emerged hoppers into the isolated-reared colony, 'fresh blood' has from time to time been brought in. Nevertheless, genetic drift and inbreeding cannot be fully ruled out and, at this time, the role of genetic divergence *versus* environmental effects cannot be conclusively determined [216]. We intend to study these phenomena in more detail via microarray and/or deep sequencing analyses. Nonetheless, we are convinced that further exploration and thorough analysis of the EST database will most certainly contribute to future investigations on the process of phase polyphenism and its accompanying neuroplasticity, as well as to many other research themes in the biology of the desert locust.

In summary, by developing an *S. gregaria* EST database we met the need for extra hemimetabolous insect transcriptomic data. The current report describes the construction of an *S. gregaria* EST database containing 12,709 unique transcript sequences. In addition, we demonstrated that construction of this database did not result in a high degree of redundancy of locust transcriptomic data. For now, analysis of the database already allowed us to functionally annotate 3,887 sequences, many of which are annotated as involved in neuronal signaling and signal transduction. Finally, several genes displaying significantly differential transcript levels in isolated- and crowded-reared desert locusts were identified. Interestingly, some of these are predicted to be involved in development and modeling of the nervous system. These observations contribute to the view that density-dependent behavioral plasticity in locusts is not only defined by innate signaling pathways, but represents a more sophisticated adaptation for coping with complex differences in environmental situations, including neural plasticity.

By specifically focusing on the CNS, this *S. gregaria* EST database will most certainly contribute to future studies unraveling the complex regulation of phase transition and allow to study

neuro-endocrine control mechanisms of certain physiological processes. Furthermore, parallel studies focusing on phase polyphenism and factors involved in nervous system development will most probably lead to novel insights in phenomena of neuroplasticity in general.

Materials and Methods

1. Preparation of the central nervous system samples

Animals. The *S. gregaria* colony in our laboratory originated from the Aquazoo in Düsseldorf (Germany, 1985). To start their breeding programme, the Aquazoo had collected animals from the field in Nigeria, Africa. Crowded-reared *S. gregaria* were kept under a 13 h light, 11 h dark photoperiod at a temperature which was maintained at 32°C. Each cage also contained at least 1 light bulb (25 Watt) to create a light and temperature gradient within the cage. To sustain the gregarious characteristics the animals were kept in Plexiglas cages (40×32×48 cm) at a density of 500–1000 newly emerged hoppers per cage. They were fed daily with fresh cabbage and oat flakes *ad libitum*.

Isolated-reared locusts were kept individually for 25 successive generations in plastic rearing containers (14×8.5×7.5 cm). They received the same food as the crowded-reared animals. The breeding rooms for isolated-reared locusts have 20 exchanges of air volume per hour, a constant temperature of 32°C and an identical light-dark cycle as in crowd-reared rooms. (For more detailed information on solitarious (isolated) locust breeding, see Hoste *et al.*, 2002 [217].)

The main objective of this project was to determine an as large as possible part of the CNS transcriptome covering different stages. Third to fifth larval instars were selected randomly for dissection to get an equal distribution of age. Adults were staged at eclosion and dissected at regular time intervals (1, 5, 7, 9, 13, 15, 17, 19 and 21 days after eclosion) to cover the sexual maturation

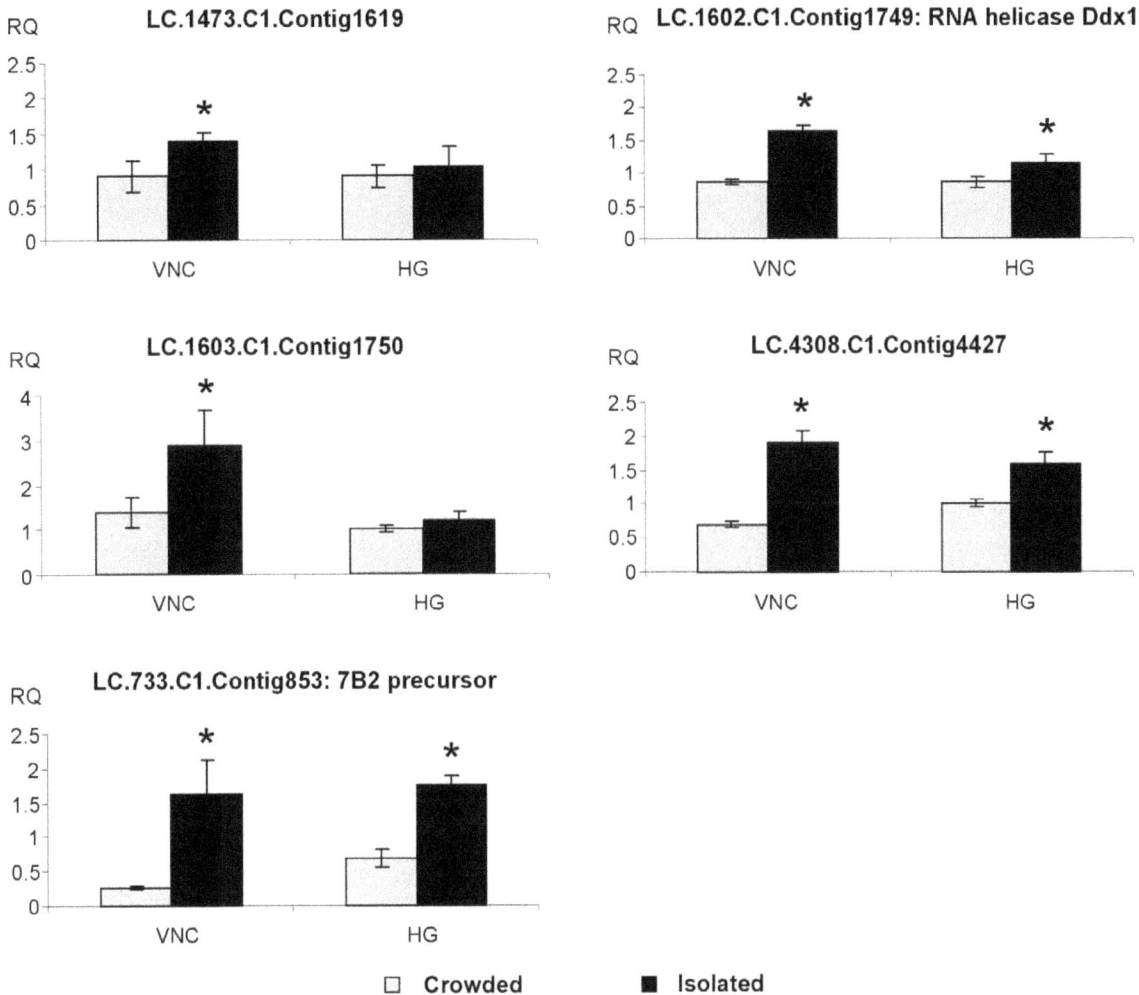

Figure 6. Relative transcript levels for genes showing higher expression levels in nervous tissue of isolated-reared locusts. Relative transcript quantity (RQ) for genes showing higher expression levels in isolated-reared desert locusts (VNC: ventral nerve cord, HG: head ganglia). Results were obtained by analyzing four independent pooled samples of ten individuals per condition and are represented as means ± standard error. Statistical analysis consisted of a Student's t-test for comparing two independent groups. Significantly higher transcript levels (p<0.05) are indicated by an asterisk (*).

and initial reproduction period. In total, head ganglia (HG) and ventral nerve cords (VNC) were dissected from 1,431 animals, distributed as evenly as possible over different developmental stages, sexes and phases.

RNA extractions. Desert locust nervous tissues were microdissected under a binocular microscope, cleaned and rinsed in Ringer solution (1 L: 8.766 g NaCl; 0.188 g $CaCl_2$; 0.746 g KCl; 0.407 g $MgCl_2$; 0.336 g $NaHCO_3$; 30.807 g sucrose; 1.892 g trehalose; pH 7.2) and immediately collected in RNA*later* solution (Ambion) to prevent degradation. HG and VNC were collected separately. Until further processing, pooled tissue samples (each sample consisted of a certain stage, phase and origin) were stored at −20°C. Samples were added to MagNa Lyser Green Beads (2 ml screw tubes filled with 1.4 mm ceramic beads) and homogenized in the MagNA Lyser instrument (Roche) (6,500 rpm during 30 s). Subsequently, total RNA was extracted from the resulting homogenates utilizing the RNeasy Lipid Tissue Mini Kit (Qiagen), following the manufacturer's instructions. In combination with this extraction procedure, a DNase treatment (RNase-free DNase set, Qiagen) was performed to eliminate potential genomic DNA contamination. Each extraction was

followed by a spectrophotometric quantification and quality control with the Agilent 2100 Bioanalyzer (Agilent Technologies). Hence, only those samples were used for which RNA degradation proved to be minimal, thereby excluding contamination with small degraded pieces of RNA during the further workflow. For each stage, sufficient RNA of optimal quality was prepared. All samples were divided into four categories: 1) isolated-reared HG, 2) isolated-reared VNC (thoracic and abdominal ganglia), 3) crowded-reared HG and 4) crowded-reared VNC (thoracic and abdominal ganglia).

2. Construction of the EST database

Construction and normalization of cDNA libraries, sequencing and subsequent development of the EST database were performed at the W.M. Keck Center for Comparative and Functional Genomics (University of Illinois, Urbana-Champaign). Normalization of the cDNA library was performed to compensate for very abundant transcripts, thereby increasing the odds of sequencing rare transcripts. Two normalized libraries were constructed using different methods, hence both protocols are described below. The aim was to generate an EST database containing at least 10,000 unique transcript sequences. When the sequencing redundancy for ESTs derived from the first library became higher than 50% before reaching this number, we decided to construct a second normalized library [redundancy was calculated as: 100 * (total number of sequences − number of unique sequences)/total number of sequences]. Both methods involved denaturation of double stranded (ds) cDNA or vector, followed by a hybridization step which was limited in time. The resulting, newly formed ds DNAs were then eliminated, while the sequences that remained single stranded (ss) were further used in the normalization protocol. The common concept of both alternative methods is that abundantly present transcripts will more easily find a complementary sequence, whereas rare transcripts have a much higher probability of remaining ss.

Construction of the first normalized library. Poly(A)$^+$mRNA was isolated from total RNA using the Oligotex Direct mRNA kit (Qiagen). The poly(A)$^+$mRNA was converted to double-stranded (ds) cDNA by using tagged primers (tag underlined) which contain a NotI restriction site. The primers were used as follows (V = A,C,G):

CNS head (HG) isolated-reared:

5'- AACTGGAAGAATTCGCGGCCGCACG-CATTTTTTTTTTTTTTTTTTTV -3'

CNS body (VNC) isolated-reared:

5'- AACTGGAAGAATTCGCGGCCGCACC-GATTTTTTTTTTTTTTTTTTTV -3'

CNS body (VNC) crowded-reared:

5'- AACTGGAAGAATTCGCGGCCGCTCG-CATTTTTTTTTTTTTTTTTTTV -3'

CNS head (HG) crowded-reared:5'- AACTGGAA-GAATTCGCGGCCGCTCCGATTTTTTTTTTTT-TTTTTV -3'

Ds cDNAs were size selected (>600 bp). An equal quantity of the cDNA from each of the different samples was pooled, ligated to *EcoRI* adaptors (5'-AATTCCGTTGCTGTCG-3', Promega #C1291) and digested with *NotI*. This pooled cDNA was then directionally cloned into *EcoRI-NotI* digested pBluescript II SK+ phagemid vector (Stratagene). The total number of white colony forming units (cfu) before amplification was 3×10^6. Blue colonies (empty vectors) were less than 2%.

Normalization of the primary library was performed as previously described by Bonaldo and co-workers [218]. Purified plasmid DNA from the primary library was converted to single-stranded (ss) plasmids and used as a template for a polymerase chain reaction (PCR) by using the T7 and T3 priming sites flanking the cloned cDNA insert. The purified PCR products, representing the entire cloned cDNA population, were used as a driver for normalization. Hybridization between the ss library (50 ng) and the PCR products (500 ng) was carried out for 44 hr at 30°C. By means of hydroxyapatite chromatography [218] unhybridized ssDNA plasmids were separated from hybridized DNA rendered partially ds and electroporated into DH10B cells (Invitrogen) to generate the normalized library. The total number of clones with insert was 2×10^6. Background of empty clones was less than 2%.

Construction of a second normalized library. Poly-(A)$^+$mRNA from each of the tissues was reverse transcribed into ds cDNA by using the Creator Smart cDNA Library Construction kit (Clontech) following the manufacturer's instructions. The following 5' tagged oligos (tag underlined) were used for cDNA synthesis:

CNS Head (HG) isolated-reared: Creator SMART IV original

5' AAG CAG TGG TAT CAA CGC AGA GTG GCC ATT ACG GCC GGG 3'

CNS Body (VNC) isolated-reared: Creator AA

5' AAG CAG TGG TAT CAA CGC AGA GTG GCC ATT ACG GCC AAG GG 3'

CNS Head (HG) crowded-reared: Creator TT

5' AAG CAG TGG TAT CAA CGC AGA GTG GCC ATT ACG GCC TTG GG 3'

CNS Body (VNC) crowded-reared: Creator GA

5' AAG CAG TGG TAT CAA CGC AGA GTG GCC ATT ACG GCC GAG GG 3'

Equal amounts of ds cDNA from each of the tissues were mixed and a total of 300 ng of cDNA was denatured at 98°C for 2 min and allowed to renature at 68°C for 5 hr in 50 mM Hepes (pH 7.5) and 0.5 mM NaCl. Ds cDNAs (i.e. abundant transcripts) were degraded by the addition of Duplex-Specific Nuclease (Evrogen). The ss fraction (which constituted the normalized library) was converted to ds cDNA by specific-suppression PCR as described in the Trimmer-direct kit (Evrogen). The normalized cDNA was directionally cloned into a pDNR-LIB vector (Clontech) and transformed using DH10B electrocompetent cells (Invitrogen). The total number of white colony forming units (cfu) was 2×10^6.

Plasmid isolation, sequencing and sequence processing. Libraries were plated on agar and random clones were picked with the Genetix Q-Pix robot and racked as glycerol stocks in 384-well plates. After overnight growth, bacteria were inoculated into 96-well deep cultures with Luria Bertani Medium and 100 mg/ml of Carbenicillin (libraries cloned in pBluescriptII SK+) or 30 mg/ml of Chloramphenicol (libraries cloned in pDNR-LIB). Plasmid DNA was purified from the bacterial cultures after 24 hr of growth at 37°C with Qiagen 8000 and Qiagen 9600 robots.

Sequencing reactions of the 5' ends of the inserts (standard T7 primer for libraries cloned in pBluescriptII SK+, primer 5'-CGAGCGCAGCGAGTCAGT-3' for libraries cloned in pDNR-LIB) were performed using BigDye terminator (Applied Biosystems) on two ABI 3730XL capillary system sequencers (Applied Biosystems). Phred quality scores were calculated for each base call [219,220]. Bases with a score ≥20 (equivalent to 99% confidence) were considered of high quality. High quality sequence regions were determined by Qualtrim (W.M. Keck Center for Comparative and Functional Genomics, University of Illinois, Urbana-Champaign). In order to maximize the sequence length, some bases with a Phred score <20 were tolerated. Sequences were considered successful if the high quality region was ≥200 bases, otherwise they were termed 'low quality'. Vector sequences were detected and masked by using Cross_Match (http://www.phrap.org/phredphrapconsed.html). Both vector and low quality sequences were trimmed off the original sequences. Sequences with a length of ≥200 bases after trimming were considered 'clean' sequences, otherwise they were called 'short insert'. Repeat and low complexity sequences were identified and masked by using RepeatMasker Open-3.0 (http://www.repeatmasker.org/). Finally a screen was performed for unwanted sequences such as *Escherichia coli* genome, vector, mitochondrial and viral DNA and ribosomal RNA by using blastn [221]. Sequences having significant similarities with these DNA sequences were considered as contaminants and excluded from the final 'clean' sequence set.

Contig assembly. If two or more sequences represented the same transcript they were assembled into a 'contig'. This means that the overlapping EST sequences were aligned and joined in an as complete as possible representation of the corresponding transcript sequence (which is here termed 'contig'). Assembly of sequences into contigs was performed by using the Phrap software package (http://www.phrap.org/phredphrapconsed.html). When transcripts were represented by only one sequence, the EST was referred to as a 'singleton'.

3. Annotations

Blastx searches in protein databases. All obtained EST sequences (contigs and singletons) were used as query for a blastx search [221] in protein databases from *A. gambiae*, *T. castaneum*, *D. melanogaster*, *A. mellifera*, *B. mori* (all of which are insects with a completely sequenced genome), *C. elegans* and *H. sapiens*, and in the 'National Center for Biotechnology Information' (NCBI) nr.aa protein database. Sequences producing an e-value$<$1E-7 were considered a hit.

Gene Ontology annotations. All EST information is integrated in an 'Expressed Sequence Tag Information Management and Annotation' (ESTIMA) database (http://titan.biotec.uiuc.edu/locust/) [222]. Functional GO annotations [33] of *S. gregaria* sequences in ESTIMA were based on the blastx hits from *A. gambiae*, *T. castaneum*, *D. melanogaster*, *A. mellifera*, *B. mori*, *C. elegans* and *H. sapiens*, which are reference organisms that had previously been GO annotated. These GO annotations can be easily retrieved from ESTIMA.

Apart from ESTIMA, functional annotation of the *S. gregaria* EST sequences was also performed by means of the software package 'Blast2GO' [34,35]. This facilitates high-throughput functional annotation by assigning GO terms [33] to the input sequences. All further GO-based analyses described in this study were performed by means of the 'Blast2GO' software. The rationale for doing so was (i) to easily analyze the GO annotations (as 'Blast2GO' includes several functionalities for graphical representations and statistical analysis) and (ii) to compare data from *S. gregaria* CNS with publicly available *L. migratoria* EST sequences by performing the same analysis for both, based on the same criteria (*cf. infra*). Functional annotation in 'Blast2GO' was performed in several steps. First, a blastx search [221] in the NCBI nr.aa database was performed, and all sequences producing an E-value$<$1E-3 and a minimal alignment length of 33 residues were considered a hit. Based on the GO annotations for the hit sequences, an annotation score for the candidate GO terms was calculated [34,35]. This score takes into account the degree of similarity to the blast hits, the evidence code for how the GO term had been assigned to the blast hit sequences, the number of blast hits displaying the GO term and a factor that determines the weight of parent GO terms. If this score was $>$45, the GO term was assigned to the query sequence. If both a parent and a child term had a score $>$45, the lowest node in the hierarchy (i.e. the child term, most detailed node) was chosen. Next, all sequences were used as a query to search in the InterPro database [223,224]. GO terms were derived from the obtained hits and for each sequence they were merged with the already assigned terms. Finally, an 'Annex' step was performed, which resulted in an annotation augmentation for the already annotated sequences [34,35].

Study of signal transduction components. A more detailed study of components involved in signal transduction was performed by analyzing ESTs classified under the biological process term *Signal transduction*. The GO term *Signal transduction* is a level 4 term under *Cellular process* (1: *Biological process* $>$2: *Cellular process* $>$3: *Cell communication* $>$4: *Signal transduction*). Given the importance of neuronal signaling in the CNS, a similar analysis

was done for sequences classified under *Biogenic amine biosynthesis* and *Peptide hormone processing*.

In order to identify neuropeptide precursors from the EST database, sequences available from different insect species, such as *S. gregaria*, *L. migratoria*, *D. melanogaster*, *A. mellifera* and *A. pisum*, were used in a homology search. Previously identified locust neuropeptides, without known precursor information, were used as well. For this purpose, tblastn [221], adapted for short sequences, was employed. The retrieved results were analyzed using SignalP 3.0 to predict potential signal peptide cleavage sites. Alignments of putative neuropeptide precursors were performed by means of CLUSTALW.

4. Comparison to LocustDB

All unique sequences available from the *L. migratoria* EST-database (LocustDB) [25,26] were used as a query to do a blastn search in the *S. gregaria* EST-database, and *vice versa*. Sequences producing an E-value$<$1E-10 were considered a hit.

Next, all unique *L. migratoria* sequences were entered in a new 'Blast2GO' functional annotation project (*cfr. supra*). The annotation steps were performed by using the same parameters as for the *S. gregaria* sequences. In order to compare functional annotations in both EST databases, the 100 best represented biological processes (i.e. the best represented GO terms classified under the main ontology 'Biological Process') were retrieved for both databases.

5. Genes differentially regulated in the two phases

Selection of candidate differentially regulated genes. Since different inserts were provided with a specific tag during the cDNA synthesis procedure, ESTs could be identified as derived from head ganglia (HG) or ventral nerve cord (VNC), and from isolated- or crowded-reared locusts. In practical terms, this tag identification was only feasible for the 5′ ESTs derived from the second library (where the tag was incorporated at the 5′ end of the insert), as well as for a limited number of sequences (i.e. 5′ ESTs including the tag-containing 3′ end) derived from the first library (where the tag was incorporated at the 3′ end of the insert). Therefore, contig sequences composed by at least five separate, tag containing ESTs were evaluated for the presence of an isolated- or crowded-reared specific tag, allowing us to classify them as carrying substantially more (i.e. each with a ratio ≥five to one) (i) crowded- or (ii) isolated-reared tags. Based on this initial analysis, a selection of 16 transcripts was made, for which the expression level in both phases was further investigated by qRT-PCR.

In addition to this selection, the transcripts representing the factors 'secreted protein acidic and rich in cysteine' (SPARC) (LC.3176.C1.Contig3329) and 'phase-related peptide' (PRP) (LC.463.C1.Contig551), which were both previously shown to be upregulated in crowded-reared locusts, and the 'solitarious phase specific gene' (SSG) (LC01031A1A08.f1), which was previously found to be upregulated in isolated-reared locusts [186–188] were included as (putative) positive controls in the subsequent qRT-PCR analysis.

Quantitative real-time reverse transcription PCR. To investigate whether the above selected sequences indeed correspond to phase-dependently regulated genes, their transcript levels were evaluated in locust nervous tissue. Sampling was done from adult males and females staged between 4 and 10 days after ecdysis. HG and VNC from crowded- and isolated-reared animals were microdissected and pooled as separate samples. Each condition was represented by four biological repeats (each containing pooled tissues from 10 animals). RNA extraction was performed as described in the above section. Subsequent cDNA synthesis was done with SuperScript

III Reverse Transcriptase (Invitrogen Life Sciences) by using 1 μg of the total RNA and random hexamers as described in the manufacturer's protocol. All primers for these qRT-PCR analyses were designed by means of the Primer Express software (Applied Biosystems) and their sequences are represented in Table 4 and Table 5. QRT-PCR was performed in a 20 μl reaction volume, as described in the Fast SYBR Green Master Mix protocol (Applied Biosystems). The final concentration of the primers was 300 nM. All reactions were run in duplicate on a StepOnePlus Real-Time PCR system (Applied Biosystems), using the following thermal profile: holding stage at 95°C (10 min), followed by 40 cycles of 95°C (3 s) and 60°C (30 s). Then a dissociation protocol was performed allowing for a melt curve analysis. In order to compensate for small differences in reverse transcription efficiency, we also evaluated the transcript levels for two endogenous controls in the samples under study. A random selection of samples covering all conditions and tissues was used to determine the most stable endogenous control genes. The tested genes were the desert locust orthologs of β-actin, elongation factor 1α (EF1α), ribosomal protein Rp49, glyceraldehyde-3-phosphate dehydrogenase, tubulin A1, ubiquitin and CG13220 [225]. EF1α and the S. gregaria ortholog for CG13220 were selected by the GeNorm software [226] for having the most stable transcript levels in the studied samples. Relative quantities of transcript levels were calculated as described by Vandesompele et al., 2002 [226]. One of the pooled HG samples derived from crowded-reared locusts was used in every qRT-PCR run as a calibrator cDNA sample. Transcript levels for specific genes in the samples under study were calculated relatively to levels for the same transcript in the calibrator sample. Statistical analysis was performed by means of Microsoft Excel analysis software and consisted of a Student's t-test for comparing two independent groups. A level of p<0.05 was considered as significant.

Supporting Information

Figure S1   The 100 best represented GO terms (*Biological Process*) in the *S. gregaria* EST database. Overview of the 100 best represented GO terms classified under the main ontology *Biological Process* in the *S. gregaria* EST database.

Figure S2   The 100 best represented GO terms (*Biological Process*) in the *L. migratoria* EST database. Overview of the 100 best represented GO terms classified under the main ontology *Biological Process* in the publicly available *L. migratoria* EST database LocustDB [25,26].

Acknowledgments

The authors gratefully thank S. Van Soest, L. Vanden Bosch and J. Van Duppen for technical support; R. Jonckers for taking care of the insect cultures; T. Vandersmissen, F. Sas, G. Perée and I. Claeys for assisting at the microdissection of locust tissues; the W.M. Keck Center for Comparative and Functional Genomics (University of Illinois, Urbana-Champaign) for their delivery of the ESTIMA database based on high-throughput sequencing of normalized cDNA libraries. In addition, all collaborators who were regularly involved in taking care of the gregarious and solitarious locust cultures are acknowledged.

Author Contributions

Conceived and designed the experiments: LB GS RH LS ADL JVB. Performed the experiments: LB JH HV EM. Analyzed the data: LB JH GS HV. Contributed reagents/materials/analysis tools: RH LS ADL JVB. Wrote the paper: LB JH GS JVB. Senior academic authors who designed the study and, as principal investigators, were responsible for the management of the projects that provided financial support: RH LS ADL JVB.

References

1. Stone JV, Mordue W, Batley KE, Morris HR (1976) Structure of locust adipokinetic hormone, a neurohormone that regulates lipid utilisation during flight. Nature 263: 207–211.
2. Gade G, Goldsworthy GJ, Kegel G, Keller R (1984) Single step purification of locust adipokinetic hormones I and II by reversed-phase high-performance liquid chromatography, and amino-acid composition of the hormone II. Hoppe-Seyler's Z Physiol Chem 365: 393–398.
3. Schoofs L, Vanden Broeck J, De Loof A (1993) The myotropic peptides of *Locusta migratoria*: structures, distribution, functions and receptors. Insect Biochem 23: 859–881.
4. Schoofs L, Veelaert D, Vanden Broeck J, De Loof A (1997) Peptides in the locusts, *Locusta migratoria* and *Schistocerca gregaria*. Peptides 18: 145–156.
5. Clynen E, Schoofs L (2009) Peptidomic survey of the locust neuroendocrine system. Insect Biochem 39: 491–507.
6. The International Aphid Genomics Consortium (2010) Genome sequence of the pea aphid *Acyrthosiphon pisum*. PLoS Biol 8: e1000313.
7. Kirkness EF, Haas BJ, Sun W, Braig HR, Perotti MA, et al. (2010) Genome sequences of the human body louse and its primary endosymbiont provide insights into the permanent parasitic lifestyle. Proc Natl Acad Sci USA 107: 12168–12173.
8. Wilmore PJ, Brown AK (1975) Molecular properties of orthopteran DNA. Chromosoma 51: 337–345.
9. Uvarov B (1966) Grasshoppers and locusts. Cambridge: Cambridge University Press.
10. Uvarov B (1977) Grasshoppers and locusts. London: Centre for overseas pest research.
11. Pener MP, Yerushalmi Y (1998) The physiology of locust phase polymorphism: an update. J Insect Physiol 44: 365–377.
12. Simpson SJ, Despland E, Hagele BF, Dodgson T (2001) Gregarious behavior in desert locusts is evoked by touching their back legs. Proc Natl Acad Sci USA 98: 3895–3897.
13. Anstey ML, Rogers SM, Ott SR, Burrows M, Simpson SJ (2009) Serotonin mediates behavioral gregarization underlying swarm formation in desert locusts. Science 323: 627–630.
14. Rogers SM, Matheson T, Despland E, Dodgson T, Burrows M, et al. (2003) Mechanosensory-induced behavioural gregarization in the desert locust *Schistocerca gregaria*. J Exp Biol 206: 3991–4002.
15. Rogers SM, Matheson T, Sasaki K, Kendrick K, Simpson SJ, et al. (2004) Substantial changes in central nervous system neurotransmitters and neuro-modulators accompany phase change in the locust. J Exp Biol 207: 3603–3617.
16. Lester LR, Grach C, Pener MP, Simpson SJ (2005) Stimuli inducing gregarious colouration and behaviour in nymphs of *Schistocerca gregaria*. J Insect Physiol 51: 737–747.
17. Ott SR, Rogers SM (2010) Gregarious desert locusts have substantially larger brains with altered proportions compared with the solitarious phase. Proc Biol Sci 277: 3087–3096.
18. Withers GS, Fahrbach SE, Robinson GE (1993) Selective neuroanatomical plasticity and division of labour in the honeybee. Nature 364: 238–240.
19. Durst C, Eichmuller S, Menzel R (1994) Development and experience lead to increased volume of subcompartments of the honeybee mushroom body. Behav Neural Biol 62: 259–263.
20. Gronenberg W, Heeren S, Holldobler B (1996) Age-dependent and task-related morphological changes in the brain and the mushroom bodies of the ant *Camponotus floridanus*. J Exp Biol 199: 2011–2019.
21. Riveros AJ, Gronenberg W (2010) Brain allometry and neural plasticity in the bumblebee *Bombus occidentalis*. Brain Behav Evol 75: 138–148.
22. Molina Y, O'Donnell S (2007) Mushroom body volume is related to social aggression and ovary development in the paperwasp *Polistes instabilis*. Brain Behav Evol 70: 137–144.
23. Molina Y, O'Donnell S (2008) Age, sex, and dominance-related mushroom body plasticity in the paperwasp *Mischocyttarus mastigophorus*. Dev Neurobiol 68: 950–959.

24. Verlinden H, Badisco L, Marchal E, Van Wielendaele P, Vanden Broeck J (2009) Endocrinology of reproduction and phase transition in locusts. Gen Comp Endocrinol 162: 79–92.

25. Kang L, Chen X, Zhou Y, Liu B, Zheng W, et al. (2004) The analysis of large-scale gene expression correlated to the phase changes of the migratory locust. Proc Natl Acad Sci USA 101: 17611–17615.

26. Ma ZY, Yu J, Kang L (2006) LocustDB: a relational database for the transcriptome and biology of the migratory locust (Locusta migratoria). BMC Genomics 7: 11.

27. Song H (2005) Phylogenetic perspectives on the evolution of locust phase polyphenism. J Orthopt Res 14: 235–245.

28. Song H, Wenzel JW (2008) Phylogeny of bird-grasshopper subfamily Cyrtacanthacridinae (Orthoptera: Acrididae) and the evolution of locust phase polyphenism. Cladistics 24: 515–542.

29. Song H (2011) Density-dependent phase polyphenism in nonmodel Locusts: a minireview. Psyche (In press).

30. Simpson SJ, Bouai chiA, Roessingh P (1998) Effects of sensory stimuli on the behavioural phase state of the desert locust, Schistocerca gregaria. J Insect Physiol 44: 883–893.

31. Hagele BF, Simpson SJ (2000) The influence of mechanical, visual and contact chemical stimulation on the behavioural phase state of solitarious desert locusts (Schistocerca gregaria). J Insect Physiol 46: 1295–1301.

32. Cullen DA, Sword GA, Dodgson T, Simpson SJ (2010) Behavioural phase change in the Australian plague locust, Chortoicetes terminifera, is triggered by tactile stimulation of the antennae. J Insect Physiol 56: 937–942.

33. Ashburner M, Ball CA, Blake JA, Botstein D, Butler H, et al. (2000) Gene ontology: tool for the unification of biology. The Gene Ontology Consortium. Nat Genet 25: 25–29.

34. Conesa A, Gotz S, Garcia-Gomez JM, Terol J, Talon M, et al. (2005) Blast2GO: a universal tool for annotation, visualization and analysis in functional genomics research. Bioinformatics 21: 3674–3676.

35. Gotz S, Garcia-Gomez JM, Terol J, Williams TD, Nagaraj SH, et al. (2008) High-throughput functional annotation and data mining with the Blast2GO suite. Nucleic Acids Res 36: 3420–3435.

36. Claeys I, Poels J, Simonet G, Franssens V, Van Loy T, et al. (2005) Insect neuropeptide and peptide hormone receptors: current knowledge and future directions. Vitam Horm (NY) 73: 217–282.

37. Vanden Broeck J (1996) G-protein-coupled receptors in insect cells. Int Rev Cytol 164: 189–268.

38. Vanden Broeck J (2001) Insect G protein-coupled receptors and signal transduction. Arch Insect Biochem Physiol 48: 1–12.

39. Spehr M, Munger SD (2009) Olfactory receptors: G protein-coupled receptors and beyond. J Neurochem 109: 1570–1583.

40. Morris MB, Dastmalchi S, Church WB (2009) Rhodopsin: structure, signal transduction and oligomerisation. Int J Biochem Cell Biol 41: 721–724.

41. Hauser F, Cazzamali G, Williamson M, Park Y, Li B, et al. (2008) A genome-wide inventory of neurohormone GPCRs in the red flour beetle Tribolium castaneum. Front Neuroendocrinol 29: 142–165.

42. Hauser F, Cazzamali G, Williamson M, Blenau W, Grimmelikhuijzen CJ (2006) A review of neurohormone GPCRs present in the fruitfly Drosophila melanogaster and the honey bee Apis mellifera. Prog Neurobiol 80: 1–19.

43. Filmore D (2004) It's a GPCR world. Modern Drug Discovery 7: 24–28.

44. Towner P, Harris P, Wolstenholme AJ, Hill C, Worm K, et al. (1997) Primary structure of locust opsins: a speculative model which may account for ultraviolet wavelength light detection. Vision Res 37: 495–503.

45. Mertens I, Meeusen T, Huybrechts R, De Loof A, Schoofs L (2002) Characterization of the short neuropeptide F receptor from Drosophila melanogaster. Biochem Biophys Res Commun 297: 1140–1148.

46. Garczynski SF, Brown MR, Shen P, Murray TF, Crim JW (2002) Characterization of a functional neuropeptide F receptor from Drosophila melanogaster. Peptides 23: 773–780.

47. Chen ME, Pietrantonio PV (2006) The short neuropeptide F-like receptor from the red imported fire ant, Solenopsis invicta Buren (Hymenoptera: Formicidae). Arch Insect Biochem Physiol 61: 195–208.

48. Garczynski SF, Crim JW, Brown MR (2007) Characterization and expression of the short neuropeptide F receptor in the African malaria mosquito, Anopheles gambiae. Peptides 28: 109–118.

49. Garczynski SF, Crim JW, Brown MR (2005) Characterization of neuropeptide F and its receptor from the African malaria mosquito, Anopheles gambiae. Peptides 26: 99–107.

50. Clynen E, Husson SJ, Schoofs L (2009) Identification of new members of the (short) neuropeptide F family in locusts and Caenorhabditis elegans. Ann NY Acad Sci 1163: 60–74.

51. Lee KS, You KH, Choo JK, Han YM, Yu K (2004) Drosophila short neuropeptide F regulates food intake and body size. J Biol Chem 279: 50781–50789.

52. Lee KS, Kwon OY, Lee JH, Kwon K, Min KJ, et al. (2008) Drosophila short neuropeptide F signalling regulates growth by ERK-mediated insulin signalling. Nat Cell Biol 10: 468–475.

53. Wu Q, Wen T, Lee G, Park JH, Cai HN, et al. (2003) Developmental control of foraging and social behavior by the Drosophila neuropeptide Y-like system. Neuron 39: 147–161.

54. Dierick HA, Greenspan RJ (2007) Serotonin and neuropeptide F have opposite modulatory effects on fly aggression. Nat Genet 39: 678–682.

55. Yamanaka N, Yamamoto S, Zitnan D, Watanabe K, Kawada T, et al. (2008) Neuropeptide receptor transcriptome reveals unidentified neuroendocrine pathways. PLoS One 3: e3048.

56. Audsley N, Weaver RJ (2009) Neuropeptides associated with the regulation of feeding in insects. Gen Comp Endocrinol 162: 93–104.

57. Kataoka H, Toschi A, Li JP, Carney RL, Schooley DA, et al. (1989) Identification of an allatotropin from adult Manduca sexta. Science 243: 1481–1483.

58. Pener MP (1991) Locust phase polymorphism and its endocrine relations. In: Advances in Insect Physiology. London: Academic Press Limited.

59. West AP, Jr., Llamas LL, Snow PM, Benzer S, Bjorkman PJ (2001) Crystal structure of the ectodomain of Methuselah, a Drosophila G protein-coupled receptor associated with extended lifespan. Proc Natl Acad Sci USA 98: 3744–3749.

60. Ja WW, West AP, Jr., Delker SL, Bjorkman PJ, Benzer S, et al. (2007) Extension of Drosophila melanogaster life span with a GPCR peptide inhibitor. Nat Chem Biol 3: 415–419.

61. Cvejic S, Zhu Z, Felice SJ, Berman Y, Huang XY (2004) The endogenous ligand Stunted of the GPCR Methuselah extends lifespan in Drosophila. Nat Cell Biol 6: 540–546.

62. Ja WW, Carvalho GB, Madrigal M, Roberts RW, Benzer S (2009) The Drosophila G protein-coupled receptor, Methuselah, exhibits a promiscuous response to peptides. Protein Sci 18: 2203–2208.

63. Erreger K, Chen PE, Wyllie DJ, Traynelis SF (2004) Glutamate receptor gating. Crit Rev Neurobiol 16: 187–224.

64. Knopfel T, Uusisaari M (2008) Modulation of excitation by metabotropic glutamate receptors. Results Probl Cell Differ 44: 163–175.

65. Johnson JW, Ascher P (1987) Glycine potentiates the NMDA response in cultured mouse brain neurons. Nature 325: 529–531.

66. Skeberdis VA, Lan J, Opitz T, Zheng X, Bennett MV, et al. (2001) mGluR1-mediated potentiation of NMDA receptors involves a rise in intracellular calcium and activation of protein kinase C. Neuropharmacology 40: 856–865.

67. Marshall J, Buckingham SD, Shingai R, Lunt GG, Goosey MW, et al. (1990) Sequence and functional expression of a single alpha subunit of an insect nicotinic acetylcholine receptor. EMBO J 9: 4391–4398.

68. Marshall J, Darlison MG, Lunt GG, Barnard EA (1988) Cloning of putative nicotinic acetylcholine receptor genes from the locust. Biochem Soc Trans 16: 463–465.

69. Casida JE (2009) Pest toxicology: the primary mechanisms of pesticide action. Chem Res Toxicol 22: 609–619.

70. Nagatsu T, Levitt M, Udenfriend S (1964) Tyrosine hydroxylase. The initial step in norepinephrine biosynthesis. J Biol Chem 239: 2910–2917.

71. Mena MA, Casarejos MJ, Solano RM, de Yebenes JG (2009) Half a century of L-DOPA. Curr Top Med Chem 9: 880–893.

72. Lovenberg W, Weissbach H, Udenfriend S (1962) Aromatic L-amino acid decarboxylase. J Biol Chem 237: 89–93.

73. Axelrod J, Saavedra JM (1977) Octopamine. Nature 265: 501–504.

74. Vanden Broeck J, Vulsteke V, Huybrechts R, De Loof A (1995) Character-ization of a cloned locust tyramine receptor cDNA by functional expression in permanently transformed Drosophila S2 cells. J Neurochem 64: 2387–2395.

75. Verlinden H, Vleugels R, Marchal E, Badisco L, Pfluger HJ, et al. (2010) The role of octopamine in locusts and other arthropods. J Insect Physiol 56: 854–867.

76. Ichiyama A, Nakamura S, Nishizuka Y, Hayaishi O (1970) Enzymic studies on the biosynthesis of serotonin in mammalian brain. J Biol Chem 245: 1699–1709.

77. Christenson JG, Dairman W, Udenfriend S (1972) On the identity of DOPA decarboxylase and 5-hydroxytryptophan decarboxylase (immunological titra-tion-aromatic L-amino acid decarboxylase-serotonin-dopamine-norepineph-rine). Proc Natl Acad Sci USA 69: 343–347.

78. Wang L, Erlandsen H, Haavik J, Knappskog PM, Stevens RC (2002) Three-dimensional structure of human tryptophan hydroxylase and its implications for the biosynthesis of the neurotransmitters serotonin and melatonin. Biochemistry 41: 12569–12574.

79. Seidah NG, Chretien M (1994) Pro-protein convertases of subtilisin/kexin family. Methods Enzymol 244: 175–188.

80. Van de Ven WJ, Voorberg J, Fontijn R, Pannekoek H, van den Ouweland AM, et al. (1990) Furin is a subtilisin-like proprotein processing enzyme in higher eukaryotes. Mol Brain Res 14: 265–275.

81. Hatsuzawa K, Murakami K, Nakayama K (1992) Molecular and enzymatic properties of furin, a Kex2-like endoprotease involved in precursor cleavage at Arg-X-Lys/Arg-Arg sites. J Biochem 111: 296–301.

82. Zhu X, Lindberg I (1995) 7B2 facilitates the maturation of proPC2 in neuroendocrine cells and is required for the expression of enzymatic activity. J Cell Biol 129: 1641–1650.

83. Benjannet S, Savaria D, Chretien M, Seidah NG (1995) 7B2 is a specific intracellular binding protein of the prohormone convertase PC2. J Neurochem 64: 2303–2311.

84. Benjannet S, Lusson J, Hamelin J, Savaria D, Chretien M, et al. (1995) Structure-function studies on the biosynthesis and bioactivity of the precursor convertase PC2 and the formation of the PC2/7B2 complex. FEBS Lett 362: 151–155.

85. Lamango NS, Zhu X, Lindberg I (1996) Purification and enzymatic characterization of recombinant prohormone convertase 2: stabilization of activity by 21 kDa 7B2. Arch Biochem Biophys 330: 238–250.

86. Seidah NG, Day R, Marcinkiewicz M, Benjannet S, Chretien M (1991) Mammalian neural and endocrine pro-protein and pro-hormone convertases belonging to the subtilisin family of serine proteinases. Enzyme 45: 271–284.

87. Seidah NG, Day R, Benjannet S, Rondeau N, Boudreault A, et al. (1992) The prohormone and proprotein processing enzymes PC1 and PC2: structure, selective cleavage of mouse POMC and human renin at pairs of basic residues, cellular expression, tissue distribution, and mRNA regulation. NIDA Res Monogr 126: 132–150.

88. Smeekens SP, Avruch AS, LaMendola J, Chan SJ, Steiner DF (1991) Identification of a cDNA encoding a second putative prohormone convertase related to PC2 in AtT20 cells and islets of Langerhans. Proc Natl Acad Sci USA 88: 340–344.

89. Day R, Schafer MK, Watson SJ, Chretien M, Seidah NG (1992) Distribution and regulation of the prohormone convertases PC1 and PC2 in the rat pituitary. Mol Endocrinol 6: 485–497.

90. Hwang JR, Siekhaus DE, Fuller RS, Taghert PH, Lindberg I (2000) Interaction of Drosophila melanogaster prohormone convertase 2 and 7B2. Insect cell-specific processing and secretion. J Biol Chem 275: 17886–17893.

91. Macours N, Poels J, Hens K, Francis C, Huybrechts R (2004) Structure, evolutionary conservation, and functions of angiotensin- and endothelin-converting enzymes. Int Rev Cytol 239: 47–97.

92. Macours N, Hens K (2004) Zinc-metalloproteases in insects: ACE and ECE. Insect Biochem 34: 501–510.

93. Corvol P, Michaud A, Soubrier F, Williams TA (1995) Recent advances in knowledge of the structure and function of the angiotensin I converting enzyme. J Hypertens Supplement 3 13: S3–S10.

94. Isaac R, Schoofs L, Williams TA, Veelaert D, Sajid M, et al. (1998) A novel peptide-processing activity of insect peptidyl-dipeptidase A (angiotensin I-converting enzyme): the hydrolysis of lysyl-arginine and arginyl-arginine from the C-terminus of an insect prohormone peptide. Biochem J 330(Pt 1): 61–65.

95. Schoofs L, Veelaert D, De Loof A, Huybrechts R, Isaac E (1998) Immunocytochemical distribution of angiotensin I-converting enzyme-like immunoreactivity in the brain and testis of insects. Brain Res 785: 215–227.

96. Isaac RE, Parkin ET, Keen JN, Nassel DR, Siviter RJ, et al. (2002) Inactivation of a tachykinin-related peptide: identification of four neuropeptide-degrading enzymes in neuronal membranes of insects from four different orders. Peptides 23: 725–733.

97. Macours N, Hens K, Francis C, De Loof A, Huybrechts R (2003) Molecular evidence for the expression of angiotensin converting enzyme in hemocytes of Locusta migratoria: stimulation by bacterial lipopolysaccharide challenge. J Insect Physiol 49: 739–746.

98. Xu D, Emoto N, Giaid A, Slaughter C, Kaw S, et al. (1994) ECE-1: a membrane-bound metalloprotease that catalyzes the proteolytic activation of big endothelin-1. Cell 78: 473–485.

99. Macours N, Poels J, Hens K, Luciani N, De Loof A, et al. (2003) An endothelin-converting enzyme homologue in the locust, Locusta migratoria: functional activity, molecular cloning and tissue distribution. Insect Mol Biol 12: 233–240.

100. Fujimura-Kamada K, Nouvet FJ, Michaelis S (1997) A novel membrane-associated metalloprotease, Ste24p, is required for the first step of NH2-terminal processing of the yeast a-factor precursor. J Cell Biol 136: 271–285.

101. Tam A, Nouvet FJ, Fujimura-Kamada K, Slunt H, Sisodia SS, et al. (1998) Dual roles for Ste24p in yeast a-factor maturation: NH2-terminal proteolysis and COOH-terminal CAAX processing. J Cell Biol 142: 635–649.

102. Vanden Broeck J (2001) Neuropeptides and their precursors in the fruitfly, Drosophila melanogaster. Peptides 22: 241–254.

103. Nassel DR (2002) Neuropeptides in the nervous system of Drosophila and other insects: multiple roles as neuromodulators and neurohormones. Prog Neurobiol 68: 1–84.

104. Schulz-Aellen MF, Roulet E, Fischer-Lougheed J, O'Shea M (1989) Synthesis of a homodimer neurohormone precursor of locust adipokinetic hormone studied by in vitro translation and cDNA cloning. Neuron 2: 1369–1373.

105. Meredith J, Ring M, Macins A, Marschall J, Cheng NN, et al. (1996) Locust ion transport peptide (ITP): primary structure, cDNA and expression in a baculovirus system. J Exp Biol 199: 1053–1061.

106. Janssen I, Claeys I, Simonet G, De Loof A, Girardie J, et al. (2001) CDNA cloning and transcript distribution of two different neuroparsin precursors in the desert locust, Schistocerca gregaria. Insect Mol Biol 10: 183–189.

107. Vanden Broeck J, Veelaert D, Bendena WG, Tobe SS, De Loof A (1996) Molecular cloning of the precursor cDNA for schistostatins, locust allatostatin-like peptides with myoinhibiting properties. Mol Cell Endocrinol 122: 191–198.

108. Goldsworthy GJ, Lee MJ, Luswata R, Drake AF, Hyde D (1997) Structures, assays and receptors for locust adipokinetic hormones. Comp Biochem Physiol B, Biochem Mol Biol 117: 483–496.

109. Van der Horst DJ, Van Marrewijk WJ, Diederen JH (2001) Adipokinetic hormones of insect: release, signal transduction, and responses. Int Rev Cytol 211: 179–240.

110. Gade G (2009) Peptides of the adipokinetic hormone/red pigment-concentrating hormone family: a new take on biodiversity. Ann NY Acad Sci 1163: 125–136.

111. Siegert KJ (1999) Locust corpora cardiaca contain an inactive adipokinetic hormone. FEBS Lett 447: 237–240.

112. Woodhead AP, Stay B, Seidel SL, Khan MA, Tobe SS (1989) Primary structure of four allatostatins: neuropeptide inhibitors of juvenile hormone synthesis. Proc Natl Acad Sci USA 86: 5997–6001.

113. Pratt GE, Farnsworth DE, Siegel NR, Fok KF, Feyereisen R (1989) Identification of an allatostatin from adult Diploptera punctata. Biochem Biophys Res Commun 163: 1243–1247.

114. Bendena WG, Donly BC, Tobe SS (1999) Allatostatins: a growing family of neuropeptides with structural and functional diversity. Ann NY Acad Sci 897: 311–329.

115. Gade G, Hoffmann KH (2005) Neuropeptides regulating development and reproduction in insects. Physiol Entomol 30: 103–121.

116. Veelaert D, Passier P, Devreese B, Vanden Broeck J, Van Beeumen J, et al. (1997) Isolation and characterization of an adipokinetic hormone release-inducing factor in locusts: the crustacean cardioactive peptide. Endocrinology 138: 138–142.

117. Veenstra JA (2009) Allatostatin C and its paralog allatostatin double C: the arthropod somatostatins. Insect Biochem 39: 161–170.

118. Weaver RJ, Audsley N (2009) Neuropeptide regulators of juvenile hormone synthesis: structures, functions, distribution, and unanswered questions. Ann NY Acad Sci 1163: 316–329.

119. Paemen L, Tips A, Schoofs L, Proost P, Van Damme J, et al. (1991) Lom-AG-myotropin: a novel myotropic peptide from the male accessory glands of Locusta migratoria. Peptides 12: 7–10.

120. Schoofs L, Holman GM, Hayes TK, Kochansky JP, Nachman RJ, et al. (1990) Locustatachykinin III and IV: two additional insect neuropeptides with homology to peptides of the vertebrate tachykinin family. Regul Pept 31: 199–212.

121. Schoofs L, Holman GM, Hayes TK, Nachman RJ, De Loof A (1990) Locustatachykinin I and II, two novel insect neuropeptides with homology to peptides of the vertebrate tachykinin family. FEBS Lett 261: 397–401.

122. Clynen E, Huybrechts J, Verleyen P, De Loof A, Schoofs L (2006) Annotation of novel neuropeptide precursors in the migratory locust based on transcript screening of a public EST database and mass spectrometry. BMC Genomics 7: 201.

123. Minth CD, Bloom SR, Polak JM, Dixon JE (1984) Cloning, characterization, and DNA sequence of a human cDNA encoding neuropeptide tyrosine. Proc Natl Acad Sci USA 81: 4577–4581.

124. Curry WJ, Shaw C, Johnston CF, Thim L, Buchanan KD (1992) Neuropeptide F: primary structure from the tubellarian, Artioposthia triangulata. Comp Biochem Physiol C, Comp Pharmacol Toxicol 101: 269–274.

125. Maule AG, Shaw C, Halton DW, Brennan GP, Johnston CF, et al. (1992) Neuropeptide F (Moniezia expansa): localization and characterization using specific antisera. Parasitology 105(Pt 3): 505–512.

126. Leung PS, Shaw C, Maule AG, Thim L, Johnston CF, et al. (1992) The primary structure of neuropeptide F (NPF) from the garden snail, Helix aspersa. Regul Pept 41: 71–81.

127. Rajpara SM, Garcia PD, Roberts R, Eliassen JC, Owens DF, et al. (1992) Identification and molecular cloning of a neuropeptide Y homolog that produces prolonged inhibition in Aplysia neurons. Neuron 9: 505–513.

128. Brown MR, Crim JW, Arata RC, Cai HN, Chun C, et al. (1999) Identification of a Drosophila brain-gut peptide related to the neuropeptide Y family. Peptides 20: 1035–1042.

129. Audsley N, McIntosh C, Phillips JE (1992) Isolation of a neuropeptide from locust corpus cardiacum which influences ileal transport. J Exp Biol 173: 261–274.

130. Phillips JE, Wiens C, Audsley N, Jeffs L, Bilgen T, et al. (1996) Nature and control of chloride transport in insect absorptive epithelia. J Exp Zool 275: 292–299.

131. Macins A, Meredith J, Zhao Y, Brock HW, Phillips JE (1999) Occurrence of ion transport peptide (ITP) and ion transport-like peptide (ITP-L) in orthopteroids. Arch Insect Biochem Physiol 40: 107–118.

132. Huybrechts J, Clynen E, De Loof A, Schoofs L (2003) Osmotic lysis of corpora cardiaca using distilled water reveals the presence of partially processed peptides and peptide fragments. Physiol Entomol 28: 46–53.

133. Boureme D, Fournier B, Matz G, Girardie J (1989) Immunological and functional cross-reactivities between locust neuroparsins and proteins from cockroach corpora cardiaca. J Insect Physiol 35: 265–271.

134. Girardie J, Girardie A, Huet JC, Pernollet JC (1989) Amino acid sequence of locust neuroparsins. FEBS Lett 245: 4–8.

135. Hietter H, Vandorsselaer A, Luu B (1991) Characterization of 3 structurally related 8–9 Kda monomeric peptides present in the corpora cardiaca of Locusta - A revised structure for the neuroparsins. Insect Biochem 21: 259–264.

136. Lagueux M, Kromer E, Girardie J (1992) Cloning of a Locusta cDNA encoding neuroparsin-A. Insect Biochem 22: 511–516.

137. Fournier B, Herault JP, Proux J (1987) Antidiuretic factor from the nervous corpora cardiaca of the migratory locust - Improvement of an existing in vitro bioassay. Gen Comp Endocrinol 68: 49–56.

138. Vanhems E, Delbos M, Girardie J (1990) Insulin and neuroparsin promote neurite outgrowth in cultured locust CNS. Eur J Neurosci 2: 776–782.

139. Moreau R, Gourdoux L, Girardie J (1988) Neuroparsin - a new energetic neurohormone in the African locust. Arch Insect Biochem Physiol 8: 135–145.

140. Girardie J, Huet JC, Atay-Kadiri Z, Ettaouil S, Delbecque JP, et al. (1998) Isolation, sequence determination, physical and physiological characterization

of the neuroparsins and ovary maturing parsins of *Schistocerca gregaria*. Insect Biochem 28: 641–650.

141. Claeys I, Simonet G, Van Loy T, De Loof A, Vanden Broeck J (2003) cDNA cloning and transcript distribution of two novel members of the neuroparsin family in the desert locust, *Schistocerca gregaria*. Insect Mol Biol 12: 473–481.

142. Badisco L, Claeys I, Van Loy T, Van Hiel MB, Franssens V, et al. (2007) Neuroparsins, a family of conserved arthropod neuropeptides. Gen Comp Endocrinol 153: 64–71.

143. Badisco L, Claeys I, Van Hiel MB, Clynen E, Huybrechts J, et al. (2008) Purification and characterization of an insulin-related peptide in the desert locust, *Schistocerca gregaria*: immunolocalization, cDNA cloning, transcript profiling and interaction with neuroparsin. J Mol Endocrinol 40: 137–150.

144. Claeys I, Simonet G, Breugelmans B, Van Soest S, Franssens V, et al. (2005) Quantitative real-time RT-PCR analysis in desert locusts reveals phase dependent differences in neuroparsin transcript levels. Insect Mol Biol 14: 415–422.

145. Claeys I, Breugelmans B, Simonet G, Van Soest S, Sas F, et al. (2006) Neuroparsin transcripts as molecular markers in the process of desert locust (*Schistocerca gregaria*) phase transition. Biochem Biophys Res Commun 341: 599–606.

146. Roller L, Yamanaka N, Watanabe K, Daubnerova I, Zitnan D, et al. (2008) The unique evolution of neuropeptide genes in the silkworm *Bombyx mori*. Insect Biochem 38: 1147–1157.

147. Boerjan B, Cardoen D, Bogaerts A, Landuyt B, Schoofs L, et al. (2010) Mass spectrometric profiling of (neuro)-peptides in the worker honeybee, *Apis mellifera*. Neuropharmacology 58: 248–258.

148. Stangier J, Hilbich C, Beyreuther K, Keller R (1987) Unusual cardioactive peptide (CCAP) from pericardial organs of the shore crab *Carcinus maenas*. Proc Natl Acad Sci USA 84: 575–579.

149. Stangier J, Hilbich C, Keller R (1989) Occurrence of crustacean cardioactive peptide (CCAP) in the nervous system of an insect, *Locusta migratoria*. J Comp Physiol B, Biochem Syst Environ Physiol 159: 5–11.

150. Donini A, Agricola H, Lange AB (2001) Crustacean cardioactive peptide is a modulator of oviduct contractions in *Locusta migratoria*. J Insect Physiol 47: 277–285.

151. Tublitz NJ, Truman JW (1985) Intracellular stimulation of an identified neuron evokes cardioacceleratory peptide release. Science 228: 1013–1015.

152. Tublitz NJ, Truman JW (1985) Identification of neurones containing cardioacceleratory peptides (CAPs) in the ventral nerve cord of the tobacco hawkmoth, *Manduca sexta*. J Exp Biol 116: 395–410.

153. Tublitz NJ, Truman JW (1985) Insect cardioactive peptides. I. Distribution and molecular characteristics of two cardioacceleratory peptides in the tobacco hawkmoth, *Manduca sexta*. J Exp Biol 114: 365–379.

154. Tublitz NJ, Truman JW (1985) Insect cardioactive peptides. II. Neurohormonal control of heart activity by two cardioacceleratory peptides in the tobacco hawkmoth, *Manduca sexta*. J Exp Biol 114: 381–395.

155. Huesmann GR, Cheung CC, Loi PK, Lee TD, Swiderek KM, et al. (1995) Amino acid sequence of CAP2b, an insect cardioacceleratory peptide from the tobacco hawkmoth *Manduca sexta*. FEBS Lett 371: 311–314.

156. Wegener C, Herbert Z, Eckert M, Predel R (2002) The perivisccerokinin (PVK) peptide family in insects: evidence for the inclusion of CAP(2b) as a PVK family member. Peptides 23: 605–611.

157. Wegener C, Predel R, Eckert M (1999) Quantification of perivisccerokinin-1 in the nervous system of the American cockroach, *Periplaneta americana*. An insect neuropeptide with unusual distribution. Arch Insect Biochem Physiol 40: 203–211.

158. Clynen E, Huybrechts J, De Loof A, Schoofs L (2003) Mass spectrometric analysis of the perisympathetic organs in locusts: identification of novel perivisccerokinins. Biochem Biophys Res Commun 300: 422–428.

159. Predel R, Wegener C (2006) Biology of the CAPA peptides in insects. Cell Mol Life Sci 63: 2477–2490.

160. Lehmberg E, Ota RB, Furuya K, King DS, Applebaum SW, et al. (1991) Identification of a diuretic hormone of *Locusta migratoria*. Biochem Biophys Res Commun 179: 1036–1041.

161. Kay I, Wheeler CH, Coast GM, Totty NF, Cusinato O, et al. (1991) Characterization of a diuretic peptide from *Locusta migratoria*. Bio Chem Hoppe-Seyler 372: 929–934.

162. Kataoka H, Troetschler RG, Li JP, Kramer SJ, Carney RL, et al. (1989) Isolation and identification of a diuretic hormone from the tobacco hornworm, *Manduca sexta*. Proc Natl Acad Sci USA 86: 2976–2980.

163. Huising MO, Flik G (2005) The remarkable conservation of corticotropin-releasing hormone (CRH)-binding protein in the honeybee (*Apis mellifera*) dates the CRH system to a common ancestor of insects and vertebrates. Endocrinology 146: 2165–2170.

164. Girardie J, Richard O, Huet JC, Nespoulous C, Van Dorsselaer A, et al. (1991) Physical characterization and sequence identification of the ovary maturing parsin. A new neurohormone purified from the nervous *corpora cardiaca* of the African locust (*Locusta migratoria migratorioides*). Eur J Biochem 202: 1121–1126.

165. Fraenkel G, Hsiao C, Seligman M (1966) Properties of bursicon: an insect protein hormone that controls cuticular tanning. Science 151: 91–93.

166. Dewey EM, McNabb SL, Ewer J, Kuo GR, Takanishi CL, et al. (2004) Identification of the gene encoding bursicon, an insect neuropeptide responsible for cuticle sclerotization and wing spreading. Curr Biol 14: 1208–1213.

167. Honegger HW, Dewey EM, Ewer J (2008) Bursicon, the tanning hormone of insects: recent advances following the discovery of its molecular identity. J Comp Physiol A, Neuroethol Sens Neural Behav Physiol 194: 989–1005.

168. Luan H, Lemon WC, Peabody NC, Pohl JB, Zelensky PK, et al. (2006) Functional dissection of a neuronal network required for cuticle tanning and wing expansion in *Drosophila*. J Neurosci 26: 573–584.

169. Mendive FM, Van Loy T, Claeysen S, Poels J, Williamson M, et al. (2005) *Drosophila* molting neurohormone bursicon is a heterodimer and the natural agonist of the orphan receptor DLGR2. FEBS Lett 579: 2171–2176.

170. Luo CW, Dewey EM, Sudo S, Ewer J, Hsu SY, et al. (2005) Bursicon, the insect cuticle-hardening hormone, is a heterodimeric cystine knot protein that activates G protein-coupled receptor LGR2. Proc Natl Acad Sci USA 102: 2820–2825.

171. Van Loy T, Van Hiel MB, Vandersmissen HP, Poels J, Mendive F, et al. (2007) Evolutionary conservation of bursicon in the animal kingdom. Gen Comp Endocrinol 153: 59–63.

172. Corless CL, Matzuk MM, Ramabhadran TV, Krichevsky A, Boime I (1987) Gonadotropin beta subunits determine the rate of assembly and the oligosaccharide processing of hormone dimer in transfected cells. J Cell Biol 104: 1173–1181.

173. Boime I, Ben-Menahem D (1999) Glycoprotein hormone structure-function and analog design. Recent Prog Horm Res 54: 271–288.

174. Querat B, Sellouk A, Salmon C (2000) Phylogenetic analysis of the vertebrate glycoprotein hormone family including new sequences of sturgeon (*Acipenser baeri*) beta subunits of the two gonadotropins and the thyroid-stimulating hormone. Biol Reprod 63: 222–228.

175. Kwok HF, So WK, Wang Y, Ge W (2005) Zebrafish gonadotropins and their receptors: I. Cloning and characterization of zebrafish follicle-stimulating hormone and luteinizing hormone receptors–evidence for their distinct functions in follicle development. Biol Reprod 72: 1370–1381.

176. So WK, Kwok HF, Ge W (2005) Zebrafish gonadotropins and their receptors: II. Cloning and characterization of zebrafish follicle-stimulating hormone and luteinizing hormone subunits–their spatial-temporal expression patterns and receptor specificity. Biol Reprod 72: 1382–1396.

177. Hsu SY, Nakabayashi K, Bhalla A (2002) Evolution of glycoprotein hormone subunit genes in bilateral metazoa: identification of two novel human glycoprotein hormone subunit family genes, GPA2 and GPB5. Mol Endocrinol 16: 1538–1551.

178. Macdonald LE, Wortley KE, Gowen LC, Anderson KD, Murray JD, et al. (2005) Resistance to diet-induced obesity in mice globally overexpressing OGH/GPB5. Proc Natl Acad Sci USA 102: 2496–2501.

179. Sudo S, Kuwabara Y, Park JI, Hsu SY, Hsueh AJ (2005) Heterodimeric fly glycoprotein hormone-alpha2 (GPA2) and glycoprotein hormone-beta5 (GPB5) activate fly leucine-rich repeat-containing G protein-coupled receptor-1 (DLGR1) and stimulation of human thyrotropin receptors by chimeric fly GPA2 and human GPB5. Endocrinology 146: 3596–3604.

180. Baggerman G, Cerstiaens A, De Loof A, Schoofs L (2002) Peptidomics of the larval *Drosophila melanogaster* central nervous system. J Biol Chem 277: 40368–40374.

181. Hummon AB, Richmond TA, Verleyen P, Baggerman G, Huybrechts J, et al. (2006) From the genome to the proteome: uncovering peptides in the *Apis* brain. Science 314: 647–649.

182. Li B, Predel R, Neupert S, Hauser F, Tanaka Y, et al. (2008) Genomics, transcriptomics, and peptidomics of neuropeptides and protein hormones in the red flour beetle *Tribolium castaneum*. Genome Res 18: 113–122.

183. Riehle MA, Garczynski SF, Crim JW, Hill CA, Brown MR (2002) Neuropeptides and peptide hormones in *Anopheles gambiae*. Science 298: 172–175.

184. Verleyen P, Chen X, Baron S, Preumont A, Hua YJ, et al. (2009) Cloning of neuropeptide-like precursor 1 in the gray flesh fly and peptide identification and expression. Peptides 30: 522–530.

185. Ma M, Gard AL, Xiang F, Wang J, Davoodian N, et al. (2009) Combining *in silico* transcriptome mining and biological mass spectrometry for neuropeptide discovery in the Pacific white shrimp *Litopenaeus vannamei*. Peptides 31: 27–43.

186. Rahman MM, Breuer M, Begum M, Baggerman G, Huybrechts J, et al. (2008) Localization of the phase-related 6-kDa peptide (PRP) in different tissues of the desert locust *Schistocerca gregaria*-immunocytochemical and mass spectrometric approach. J Insect Physiol 54: 543–554.

187. Rahman MM, Vandingenen A, Begum M, Breuer M, De Loof A, et al. (2003) Search for phase specific genes in the brain of desert locust, *Schistocerca gregaria* (Orthoptera: Acrididae) by differential display polymerase chain reaction. Comp Biochem Physiol A, Comp Physiol 135: 221–228.

188. Rahman MM, Vanden Bosch L, Baggerman G, Clynen E, Hens K, et al. (2002) Search for peptidic molecular markers in hemolymph of crowd-(gregarious) and isolated-reared (solitary) desert locusts, *Schistocerca gregaria*. Peptides 23: 1907–1914.

189. Sage H, Vernon RB, Funk SE, Everitt EA, Angello J (1989) SPARC, a secreted protein associated with cellular proliferation, inhibits cell spreading in vitro and exhibits Ca+2-dependent binding to the extracellular matrix. J Cell Biol 109: 341–356.

190. Sage H, Vernon RB, Decker J, Funk S, Iruela-Arispe ML (1989) Distribution of the calcium-binding protein SPARC in tissues of embryonic and adult mice. J Histochem Cytochem 37: 819–829.

191. Sage H, Decker J, Funk S, Chow M (1989) SPARC: a Ca2+-binding extracellular protein associated with endothelial cell injury and proliferation. J Mol Cell Cardiol 21 Suppl 1: 13–22.

192. Golembieski WA, Ge S, Nelson K, Mikkelsen T, Rempel SA (1999) Increased SPARC expression promotes U87 glioblastoma invasion in vitro. Int J Dev Neurosci 17: 463–472.

193. Brekken RA, Sage EH (2001) SPARC, a matricellular protein: at the crossroads of cell-matrix communication. Matrix Biol 19: 816–827.

194. Murphy-Ullrich JE, Lightner VA, Aukhil I, Yan YZ, Erickson HP, et al. (1991) Focal adhesion integrity is downregulated by the alternatively spliced domain of human tenascin. J Cell Biol 115: 1127–1136.

195. Rosenblatt S, Bassuk JA, Alpers CE, Sage EH, Timpl R, et al. (1997) Differential modulation of cell adhesion by interaction between adhesive and counter-adhesive proteins: characterization of the binding of vitronectin to osteonectin (BM40, SPARC). Biochem J 324(Pt 1): 311–319.

196. Funk SE, Sage EH (1991) The Ca2(+)-binding glycoprotein SPARC modulates cell cycle progression in bovine aortic endothelial cells. Proc Natl Acad Sci USA 88: 2648–2652.

197. Tremble PM, Lane TF, Sage EH, Werb Z (1993) SPARC, a secreted protein associated with morphogenesis and tissue remodeling, induces expression of metalloproteinases in fibroblasts through a novel extracellular matrix-dependent pathway. J Cell Biol 121: 1433–1444.

198. Barker TH, Baneyx G, Cardo-Vila M, Workman GA, Weaver M, et al. (2005) SPARC regulates extracellular matrix organization through its modulation of integrin-linked kinase activity. J Biol Chem 280: 36483–36493.

199. Kupprion C, Motamed K, Sage EH (1998) SPARC (BM-40, osteonectin) inhibits the mitogenic effect of vascular endothelial growth factor on microvascular endothelial cells. J Biol Chem 273: 29635–29640.

200. Hasselaar P, Sage EH (1992) SPARC antagonizes the effect of basic fibroblast growth factor on the migration of bovine aortic endothelial cells. J Cell Biochem 49: 272–283.

201. Wu RX, Laser M, Han H, Varadarajulu J, Schuh K, et al. (2006) Fibroblast migration after myocardial infarction is regulated by transient SPARC expression. J Mol Med 84: 241–252.

202. Raines EW, Lane TF, Iruela-Arispe ML, Ross R, Sage EH (1992) The extracellular glycoprotein SPARC interacts with platelet-derived growth factor (PDGF)-AB and -BB and inhibits the binding of PDGF to its receptors. Proc Natl Acad Sci USA 89: 1281–1285.

203. Gongidi V, Ring C, Moody M, Brekken R, Sage EH, et al. (2004) SPARC-like 1 regulates the terminal phase of radial glia-guided migration in the cerebral cortex. Neuron 41: 57–69.

204. Kidd T, Bland KS, Goodman CS (1999) Slit is the midline repellent for the robo receptor in Drosophila. Cell 96: 785–794.

205. Dimitrova S, Reissaus A, Tavosanis G (2008) Slit and Robo regulate dendrite branching and elongation of space-filling neurons in Drosophila. Dev Biol 324: 18–30.

206. Brierley DJ, Blanc E, Reddy OV, Vijayraghavan K, Williams DW (2009) Dendritic targeting in the leg neuropil of Drosophila: the role of midline signalling molecules in generating a myotopic map. PLoS Biol 7: e1000199.

207. Boyan G, Therianos S, Williams JL, Reichert H (1995) Axogenesis in the embryonic brain of the grasshopper Schistocerca gregaria: an identified cell analysis of early brain development. Development 121: 75–86.

208. Geva N, Guershon M, Orlova M, Ayali A (2009) Memoirs of a locust: density-dependent behavioral change as a model for learning and memory. Neurobiol Learn Mem 93: 175–182.

209. Dukas R (2008) Evolutionary biology of insect learning. Annu Rev Entomol 53: 145–160.

210. Kloppenburg P, Mercer AR (2008) Serotonin modulation of moth central olfactory neurons. Annu Rev Entomol 53: 179–190.

211. Mercer AR, Kirchhof BS, Hildebrand JG (1996) Enhancement by serotonin of the growth in vitro of antennal lobe neurons of the sphinx moth Manduca sexta. J Neurobiol 29: 49–64.

212. Mayford M, Barzilai A, Keller F, Schacher S, Kandel ER (1992) Modulation of an NCAM-related adhesion molecule with long-term synaptic plasticity in Aplysia. Science 256: 638–644.

213. Ortiz de Montellano PR, De Voss JJ (2005) Substrate oxidation by Cytochrome P450 enzymes. In: Cytochrome P450: Structure, Mechanism, and Biochemistry Ortiz de Montellano PR, ed. New York: Kluwer Academic/Plenum Publishers. pp 183–246.

214. Wassarman DA, Steitz JA (1991) RNA splicing. Alive with DEAD proteins. Nature 349: 463–464.

215. Schmid SR, Linder P (1992) D-E-A-D protein family of putative RNA helicases. Mol Microbiol 6: 283–291.

216. Berthier K, Chapuis MP, Simpson SJ, Ferenz HJ, Habib Kane CM, et al. (2010) Laboratory populations as a resource for understanding the relationship between genotypes and phenotypes: A global case study in locusts. Advances in Insect Physiology 39: 1–37.

217. Hoste B, Luyten L, Claeys I, Clynen E, Rahman MM, et al. (2002) An improved breeding method for solitarious locusts. Entomol Exp Appl 104: 281–288.

218. Bonaldo MF, Lennon G, Soares MB (1996) Normalization and subtraction: two approaches to facilitate gene discovery. Genome Res 6: 791–806.

219. Ewing B, Green P (1998) Base-calling of automated sequencer traces using phred. II. Error probabilities. Genome Res 8: 186–194.

220. Ewing B, Hillier L, Wendl MC, Green P (1998) Base-calling of automated sequencer traces using phred. I. Accuracy assessment. Genome Res 8: 175–185.

221. Altschul SF, Madden TL, Schaffer AA, Zhang J, Zhang Z, et al. (1997) Gapped BLAST and PSI-BLAST: a new generation of protein database search programs. Nucleic Acids Res 25: 3389–3402.

222. Kumar CG, LeDuc R, Gong G, Roinishivili L, Lewin HA, et al. (2004) ESTIMA, a tool for EST management in a multi-project environment. BMC Bioinformatics 5: 176.

223. Quevillon E, Silventoinen V, Pillai S, Harte N, Mulder N, et al. (2005) InterProScan: protein domains identifier. Nucleic Acids Res 33: W116–W120.

224. Hunter S, Apweiler R, Attwood TK, Bairoch A, Bateman A, et al. (2009) InterPro: the integrative protein signature database. Nucleic Acids Res 37: D211–D215.

225. Van Hiel MB, Van Wielendaele P, Temmerman L, Van Soest S, Vuerinckx K, et al. (2009) Identification and validation of housekeeping genes in brains of the desert locust Schistocerca gregaria under different developmental conditions. BMC Mol Biol 10: 56.

226. Vandesompele J, De Preter K, Pattyn F, Poppe B, Van Roy N, et al. (2002) Accurate normalization of real-time quantitative RT-PCR data by geometric averaging of multiple internal control genes. Genome Biol 3: RESEARCH0034.

11

Finding a Fox: An Evaluation of Survey Methods to Estimate Abundance of a Small Desert Carnivore

Steven J. Dempsey[1], Eric M. Gese[2]*, Bryan M. Kluever[1]

1 Department of Wildland Resources, Utah State University, Logan, Utah, United States of America, 2 United States Department of Agriculture, Animal and Plant Health Inspection Service, Wildlife Services, National Wildlife Research Center, Department of Wildland Resources, Utah State University, Logan, Utah, United States of America

Abstract

The status of many carnivore species is a growing concern for wildlife agencies, conservation organizations, and the general public. Historically, kit foxes (*Vulpes macrotis*) were classified as abundant and distributed in the desert and semi-arid regions of southwestern North America, but is now considered rare throughout its range. Survey methods have been evaluated for kit foxes, but often in populations where abundance is high and there is little consensus on which technique is best to monitor abundance. We conducted a 2-year study to evaluate four survey methods (scat deposition surveys, scent station surveys, spotlight survey, and trapping) for detecting kit foxes and measuring fox abundance. We determined the probability of detection for each method, and examined the correlation between the relative abundance as estimated by each survey method and the known minimum kit fox abundance as determined by radio-collared animals. All surveys were conducted on 15 5-km transects during the 3 biological seasons of the kit fox. Scat deposition surveys had both the highest detection probabilities (p = 0.88) and were most closely related to minimum known fox abundance ($r^2 = 0.50$, $P = 0.001$). The next best method for kit fox detection was the scent station survey (p = 0.73), which had the second highest correlation to fox abundance ($r^2 = 0.46$, $P < 0.001$). For detecting kit foxes in a low density population we suggest using scat deposition transects during the breeding season. Scat deposition surveys have low costs, resilience to weather, low labor requirements, and pose no risk to the study animals. The breeding season was ideal for monitoring kit fox population size, as detections consisted of the resident population and had the highest detection probabilities. Using appropriate monitoring techniques will be critical for future conservation actions for this rare desert carnivore.

Editor: Jesus E. Maldonado, Smithsonian Conservation Biology Institute, United States of America

Funding: Funding and logistical support were provided by the Department of Defense, United States Army Dugway Proving Ground, Environmental Programs, Dugway, Utah, and the United States Department of Agriculture, Wildlife Services, National Wildlife Research Center, Utah State University, Logan, Utah. Additional funding was provided by the Quinney College of Natural Resources, Utah State University, Logan, Utah, and the Endangered Species Mitigation Fund of the Utah Department of Natural Resources, Division of Wildlife Resources, Salt Lake City, Utah. The funders had no role in study design, data collection and analysis, decision to publish, or preparation of the manuscript.

Competing Interests: The authors have declared that no competing interests exist.

* Email: eric.gese@usu.edu

Introduction

Populations of large and small carnivores are threatened or imperiled throughout the world [1]. With increasing human populations and subsequent habitat loss and fragmentation, declines in natural prey, increased human persecution and illegal poaching, many carnivore species have declined in number and now occupy a fragment of their former range. Paramount to species management and conservation is knowledge about the status and distribution of many carnivore species. A question often facing wildlife agencies and conservation groups is how many animals are there and what is the population trend? However, many carnivore species are difficult to survey due to their low densities, are generally nocturnal and elusive, and wary of humans [2–6].

Kit foxes (*Vulpes macrotis*) are a slim, small canid (1–3 kg body mass) with ears that are relatively larger than those of other North American canids, is considered to be monestrus and socially monogamous, and is a dietary generalist feeding on rodents, insects, lagomorphs, ground-nesting birds, and reptiles [7]. Historically, kit foxes were once abundant and distributed throughout the desert and semi-arid regions of southwestern North America, ranging from Idaho to central Mexico [7]. Their range-wide decline has warranted the kit fox to be state-listed as endangered in Colorado, threatened in California and Oregon, and designated as a state sensitive species in Idaho and Utah [8]. However, a comprehensive study of kit fox abundance across its range is lacking, with the majority of studies focused on the endangered subspecies, the San Joaquin kit fox (*V. macrotis mutica*). In Utah, where kit foxes were once considered the most abundant carnivore in the west desert [9,10], the kit fox has been in steep decline over the past decade [2,11,12].

Current methods used for surveying kit foxes and their close relative the swift fox (*V. velox*), include capture-recapture [2,13–17], spotlight surveys [2,13–15,18], scent station surveys [2,13–15], scat deposition transects with and without scat detection dogs [2,14,15,17,19], track counts [14], activity index [15], and howling response [14]. Generally these methods have been evaluated in study areas with a relatively high fox density. How well these methods will perform for monitoring fox abundance in a low-density, widely dispersed kit fox population is unknown. We tested 4 survey methods (scat deposition, scent station, spotlight, trapping) on the U.S. Army Dugway Proving Ground (DPG), Utah, with the primary objectives to (1) determine detection probabilities for each method, and (2) evaluate how well the indices of relative abundance for each survey method correlate with known kit fox abundance as determined from available radio-collared animals. The kit fox population in the west desert of Utah is considered low density, declining in abundance, and widely dispersed [11,20,21].

Methods

Ethics Statement

Fieldwork was approved and sanctioned by the United States Department of Agriculture's National Wildlife Research Center and the United States Army's Dugway Proving Ground. Permission to access land on the Dugway Proving Ground was obtained from the United States Army; permission to access Bureau of Land Management property was obtained from the Bureau of Land Management.

Capture and handling protocols were reviewed and approved by the Institutional Animal Care and Use Committees (IACUC) at the United States Department of Agriculture's National Wildlife Research Center (QA-1734) and Utah State University (#1438). Permits to capture, handle, and radio-collar kit foxes were obtained from the Utah Division of Wildlife Resources (COR #4COLL8322). Data is archived and available from the United States Department of Agriculture's National Wildlife Research Center (QA-1734).

Study Area

We conducted this study on 879 km^2 of the eastern portion of the DPG and the adjoining land managed by the Bureau of Land Management, located approximately 128 km southwest of Salt Lake City, in Tooele County, Utah. Elevations ranged from 1302 m to 2137 m. The study site was in the Great Basin Desert and was characterized as a cold desert. Winters were cold, and summers were hot and dry with the majority of precipitation occurring in the spring [11]. The study area consisted of predominately flat playa punctuated with steep mountain ranges. We classified the landscape into 7 vegetation communities: chenopod, greasewood, pickle weed, grassland, stable dune, shrub-steppe, and urban; see [20,21] for a detailed description of vegetation communities.

Animal Capture and Handling

Beginning in December 2009, we captured kit foxes via transect trapping [15] and at known den sites [20,22], using box traps (25×25×80 cm; Model 107; Tomahawk Live Trap LLC, Hazelhurst, Wisconsin) baited with hot dogs. Trapping transects were distributed to provide maximum coverage of the area and allow for increased likelihood of capturing most of the kit foxes occupying the study area [11,15,20]. We deployed traps in the evening and checked them early morning each day. We coaxed captured foxes into a canvas bag placed at the edge of the trap, then restrained by personnel wearing thick leather gloves [22]. We weighed, sexed, ear tagged, and fitted each fox with a 30–50 g radio-collar (Model M1930; Advanced Telemetry Systems, Isanti, Minnesota). Collars included a mortality sensor that activated after 6 hours of non-motion and weighed <5% of body mass

[15,23,24]. We handled all foxes without the use of immobilizing drugs and released them at the capture site.

Radio-telemetry and Home Range Determination

We collected animal locations >3 times per week using a portable receiver (Model R1000; Communications Specialists, Inc., Orange, California) and a handheld 3-element Yagi antenna. We triangulated an animal's location using ≥2 compass bearings, each >20° but <160° apart, for each animal within 20 minutes [11,20]. We then calculated their location using program Locate III (Pacer Computing, Tatamagouche, Nova Scotia). For each week, we temporally distributed telemetry sampling by collecting two crepuscular (hunting) locations and one den (resting) location. To reduce auto-correlation and retain temporal independence between locations, we separated each crepuscular sample by > 12 hours and a difference of >2 hours in the time of day of each location [25–27]. We collected one weekly den location for each animal by homing in on the signal during daylight hours. We attempted to locate each fox ≥3 times weekly in order to obtain 40 locations for each fox for each biological season as the minimum number of locations needed to adequately describe the home range of a fox [27].

To determine space use of kit foxes, we created seasonal home ranges for all kit foxes with ≥30 locations within the season [27,28]. We defined the biological seasons based on the behavior and energetic needs of kit foxes: breeding 15 December – 14 April, pup-rearing 15 April – 14 August and dispersal 15 August – 14 December [10,24,29]. We created home range polygons using the Home-Range Analysis and Estimation (HoRAE) toolbox for the OpenJump geographic information system [30]. We created 95% point kernel density estimates (KDE) using a fixed kernel (standard sextante biweight) and the ad hoc method [31,32] for determination of the smoothing parameter h (e.g., h_{ref}, 90%h_{ref}, 80%h_{ref}, 70%h_{ref}, etc.). This method was designed to prevent over/under-smoothing and selection of the tightest fitting contiguous home range polygon before developing discrete patches [32–34]. We then loaded these polygons into ArcMap 10.0 (Environmental Systems Research Institute Inc., Redlands, CA) to calculate kit fox home range size.

Surveys

From March 2010 to April 2012, we attempted to conduct four different surveys (scat deposition, scent station, spotlight, and trapping) during each of the three biological kit fox seasons (breeding, pup-rearing, and dispersal) for two years. The surveys were initiated after the trapping and collaring effort due to our needing to know the number of foxes available along the transects prior to a survey period; thus the surveys did not begin until the pup rearing season of 2010. We conducted each survey along 15 5-km established transects (Figure 1). We distributed transects randomly along available roads with the constraints of being as linear as possible and having year-round access (limitations included military closures and low lying seasonally inundated greasewood areas). We attempted to conduct 4 consecutive nights of scent-station, spotlighting, and trapping surveys during each of the biological seasons along all 15 transects. Each survey was conducted separately along the transects; i.e., surveys were not conducted simultaneously on the same transect but one type of survey was conducted simultaneously over multiple transects. Due to concerns of overheating and the demands of natal care of female foxes, we did not conduct the trapping survey during the pup-rearing season. High winds, snowfall, and melting and freezing cycles limited our ability to complete some surveys during the winter months; scent stations were the most affected by

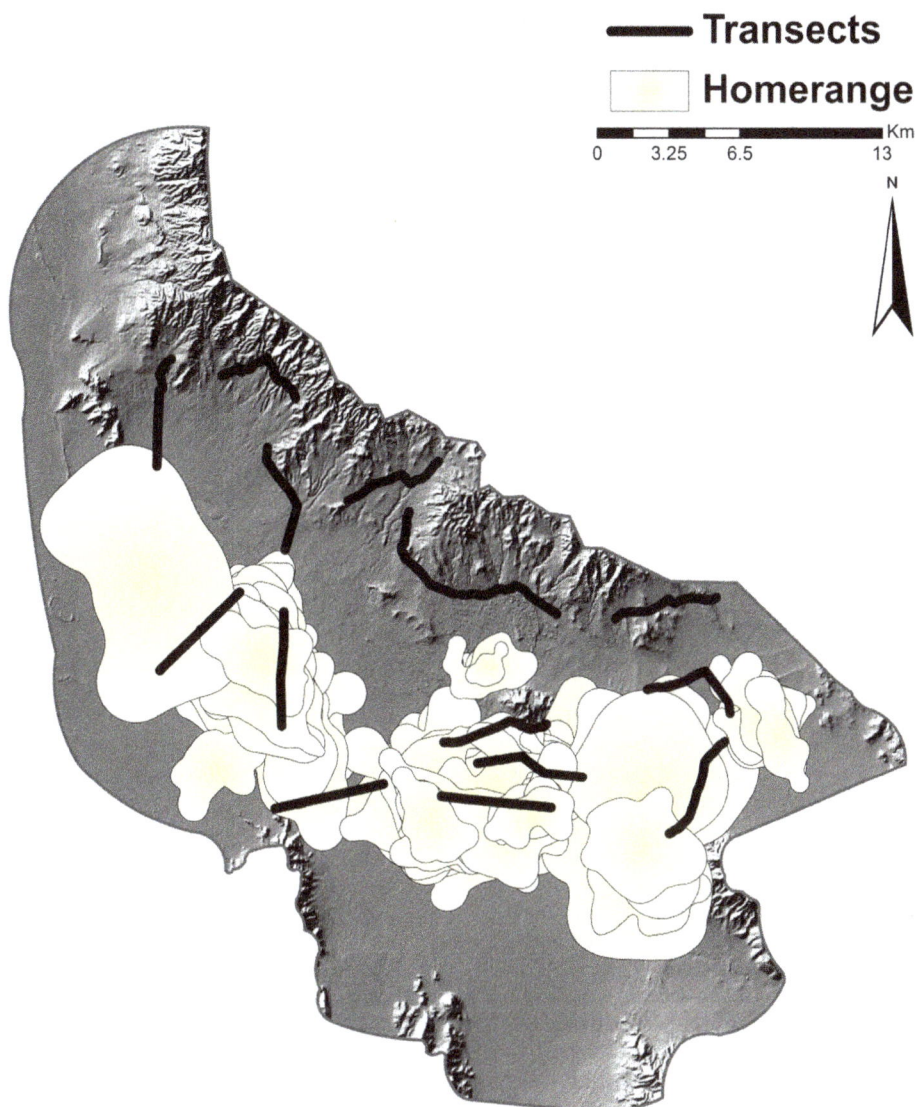

Figure 1. Transects and all kit fox home ranges created from telemetry locations, Dugway Proving Ground, Utah, 2010–2012; two transects were conducted along the same road and appear to be connected, but the end points were independent.

weather. Most notably, of the attempted 660 survey nights possible during the breeding season of 2011, only 462 (70%) scent stations were operable.

Initially designed for lagomorph counts, we conducted spotlight surveys for 3 consecutive nights during the pup rearing season of 2010 and the dispersal season of 2010. We modified our methodology and performed the spotlight surveys over 4 nights the remainder of the study. Two methods (scent stations and trapping surveys) were point sampling techniques with 11 discrete locations for detection. The remaining 2 techniques (scat deposition and spotlight surveys) allowed for detection along the entire length of a transect. Additionally, 2 techniques (scat deposition and scent stations) allowed for an individual animal to be detected multiple times along a transect, while during trapping or spotlight surveys an individual may only be detected once.

Scat deposition survey. We conducted scat deposition surveys by initially walking the transect to clear any scat from the road surface, then returning approximately 14 days later to walk and count the number of scats deposited [13,15]. Following recommendations [15,35], we walked each transect in both directions to reduce missed detections of scats. We recorded the scat location and type (species) on a handheld GPS unit, and collected the scat. This provided a count of the total number of scats per transect (surveys were a constant 5-km length and 14-day duration) as a measure of relative abundance [3].

Scent station survey. We placed scent stations at 0.5 km intervals on alternating sides along each 5 km transect [13,15]. A scent station consisted of a cleared 1-m circle of lightly sifted sand [36] with a Scented Predator Survey Disk (SPSD; United States Department of Agriculture's Pocatello Supply Depot, Pocatello, Idaho) with Fatty Acid Scent (FAS) placed in the center. The SPSD with FAS was recommended for "ease of use, attractiveness to kit fox, and their low cost" [2]. FAS saturated SPSD's are preferred over the use of liquid lures because they allow for control of a consistent attractiveness between batches [37]. We checked stations each morning for tracks of kit foxes, coyotes (*Canis latrans*), bobcats (*Lynx rufus*), leporids, small mammals, and other potential prey species. We then resifted each station and replaced the SPSD. To help maintain consistent attractiveness, we removed SPSD's from use once they were noticeably deteriorated, broken, or after a full season of use. We resampled inoperable station nights (due to inclement weather) for an additional 1–2 days in an attempt to complete the 4 nights of surveying. If transects remained inoperable after the additional days, we ceased the survey along that transect and results for that transect were not used in subsequent analysis. This survey provided a proportion of visited scent stations (i.e., total number of visits or detections divided by the number of operable stations) as a measure of relative abundance [3].

Spotlight survey. While driving a vehicle along the transect route at approximately 10–15 km/hr, 2 observers scanned their respective side of the road with a 3 million candlepower spotlight [13,15,18]. Once an animal was sighted, we stopped the vehicle and the species was identified. We recorded the species, location, distance, and bearing to the animal for kit foxes, coyotes, bobcats, and leporids. The survey provided a count of the total number of foxes detected divided by the number of nights surveyed as a measure of relative abundance [3].

Trapping survey. We conducted a trapping survey with box traps placed at 0.5 km intervals along each 5 km transect [15]. We baited traps with half of a hot dog, wired down towards the rear of the trap. We partially covered each trap with vegetation to deter a kit fox from digging under the trap for the bait. We checked traps daily, and re-baited them after two days or when a significant portion of the bait had deteriorated or had been eaten by small mammals. We deployed traps in the evening and closed them during the day to limit the amount of exposure to the animals [2]. We processed animals captured in this survey following the handling protocol previously described. We restricted trapping until late in the pup-rearing season to allow the foxes to mature enough to permit radio-collaring (i.e., they were old enough to be within our <5% body mass requirement for radio-collaring). This survey provided an index of foxes captured divided by the number of operable trap nights as a measure of relative abundance [3].

Detection probability. For each biological season we computed detection probabilities of each survey method with the occupancy estimator in Program MARK [38] that accommodates covariate information and missed observations [39]. To account for a measure of space use, we buffered each transect by one-third of the average radius of kit fox home range during each season [15]. A fox was considered available for detection if it was alive during the survey dates and it had locations within the transect buffer during that biological season. We fit models using 4 encounter occasions of 4 groups of survey methods (scat deposition, scent station, spotlight, and trapping), along with 3 covariates (survey year, number of radio-marked foxes available for detection, fox presence or absence). Fox presence or absence was binary and determined as presence if ≥ 1 fox was available for detection based on the criteria above. All possible models were examined and we selected the best model by AIC ranking [40].

Correlation between Survey Indices and Fox Abundance

In addition to determining detection probabilities for each survey method, we also examined the correlation between the index of relative abundance for each survey method and the minimum number of known kit foxes along each transect (i.e., minimum abundance), similar to the evaluation conducted for swift foxes [15]. As described above, we determined the minimum number of known foxes along a transect by buffering each transect by one-third of the average radius of kit fox home range during each season [15]. A fox was considered available for sampling along that transect if it was alive during the survey dates and it had locations within the transect buffer during that biological season.

Results

Capture and Telemetry

From December 2009 to April 2012, we accumulated 6,221 trap nights and captured 45 (26 females, 19 males) foxes across the study area 106 times. During the study we obtained 4,498 fox locations (1,487 in breeding, 1,464 in dispersal, 1,547 in pup-rearing) allowing for the calculation of 66 seasonal home ranges (21 in breeding, 24 in dispersal, and 21 in pup rearing) (Figure 1). However, due to mid-season dispersal events, 2 foxes with >30 locations were not included in home range determinations.

Home Range Estimation

We found seasonal 95% KDE home range sizes for kit foxes averaged 20.5 km^2 ($n = 64$, $SD = 15.1$). For both years combined, average home range size of kit foxes during the dispersal season was 23.3 km^2 ($n = 23$, $SD = 16.1$), followed by the breeding season ($\bar{x} = 20.8$ km^2, $n = 20$, $SD = 17.8$) and pup-rearing ($\bar{x} = 17.2$ km^2, $n = 21$, $SD = 9.4$). These home range sizes (Table 1) were then used to buffer the transects to determine the known number of kit foxes available for each survey. The number of foxes available for detection along transects varied by survey type, season, and year, from a maximum of 9 foxes available along one transect during the dispersal season of 2010 to 5 transects on which there never was a known fox present during any season or year. Although individual transects may have not had known foxes present along them, there were always known foxes available for detection along some proportion of transects.

Surveys

Detection probability. Detection probabilities were calculated for each transect to determine which survey method was best at detecting fox presence while controlling for differences in occupancy rates. For all biological seasons, the best model for detection probability (p) included differences across survey type (i.e., group) and the fox presence covariate. The corresponding best model for occupancy (Ψ) was constant across groups and the number of foxes available. This model fitted the expectation that each survey method would have a different p given the presence or absence of a fox available to be detected. Additionally, by holding Ψ constant and including a covariate for the minimum number of foxes available, we were able to include a known minimum number of foxes as determined through the space use information.

Scat deposition. We conducted 75 scat deposition surveys with 136 scat detections. Scat deposition produced the most detections (29) along an individual transect. Scat deposition surveys consistently had the highest detection probabilities (p = 0.88; Figure 2). Scat deposition surveys had the highest correlation (based on r^2 value) with kit fox abundance (r$^2 = 0.50$, P = 0.001). The correlation between scat detection rate and known fox abundance was linear and positive (Figure 3A).

Scent stations. Even with logistical difficulties due to weather, scent stations had the second highest detection probabilities (p = 0.73; Figure 2). Over the 3,718 operable station nights, we collected 159 fox detections. Scent stations had the second highest correlation with fox abundance which was linear and positive (r$^2 = 0.46$, P<0.001; Figure 3B).

Spotlight surveys. The spotlight survey was the only method that did not detect fox presence during a complete biological season. During the dispersal season of 2011, the spotlight survey produced 0 detections although 18 radio-collared foxes were

Table 1. Mean home range size (km^2) for kit foxes, range of home range size, and number of foxes monitored for each biological season and year, Dugway Proving Ground, Utah, 2010–2012.

Season	Mean home range size (km^2)	Range (km^2)	n foxes
Pup-rearing 2010	18.8	1.7–28.0	12
Dispersal 2010	26.1	3.1–80.1	14
Breeding 2011	20.6	2.2–66.1	14
Pup-rearing 2011	15.0	2.4–38.9	9
Dispersal 2011	18.8	7.9–47.8	9
Breeding 2012	21.2	6.6–71.6	6

known to be available along the 15 transects. We completed 327 spotlight survey nights with 15 detections. Spotlight surveys had the lowest number of detections and the lowest overall detection probabilities (p = 0.52; Figure 2). The relationship of spotlight surveys to fox abundance was positive, but not significant (r^2 = 0.21, P = 0.195; Figure 3C).

Trapping surveys. During the trapping surveys, we accumulated 2,640 capture nights and had 16 captures. Trapping had the second lowest detection probabilities (p = 0.59; Figure 2). The correlation between the indices from trapping surveys was significantly positive with kit fox abundance (r^2 = 0.45, P = 0.017; Figure 3D).

Costs of surveys. The costs to perform our four surveys varied. Considering the costs of the initial supplies required (e.g., spotlights, scent-station tabs, traps), labor, and gas for the field trucks, the total cost to conduct one full survey along all 15 5-km transects during one biological season was $898 for the scat

deposition transects, $940 for the spotlight survey, $2,406 for the trapping survey, and $2,760 for the scent-station survey. These costs would vary among study areas depending on differences in gasoline prices, labor costs, and the distance between transects.

Discussion

Although once abundant on the DPG, kit fox abundance was low during this study. Capture success during trapping on the DPG was the second lowest reported in the literature at 0.017 (106 captures/6221 total trap nights) which is within the range of reported capture rates of 0.013 to 0.173 [41,42]. This low capture rate may partially be due to our attempt to apply equal trapping effort across the entire study area, including areas known to be poor habitat for kit foxes and low numbers of foxes. During trapping at den sites, foxes were readily captured in one trap night and on most occasions we were able to capture the entire family

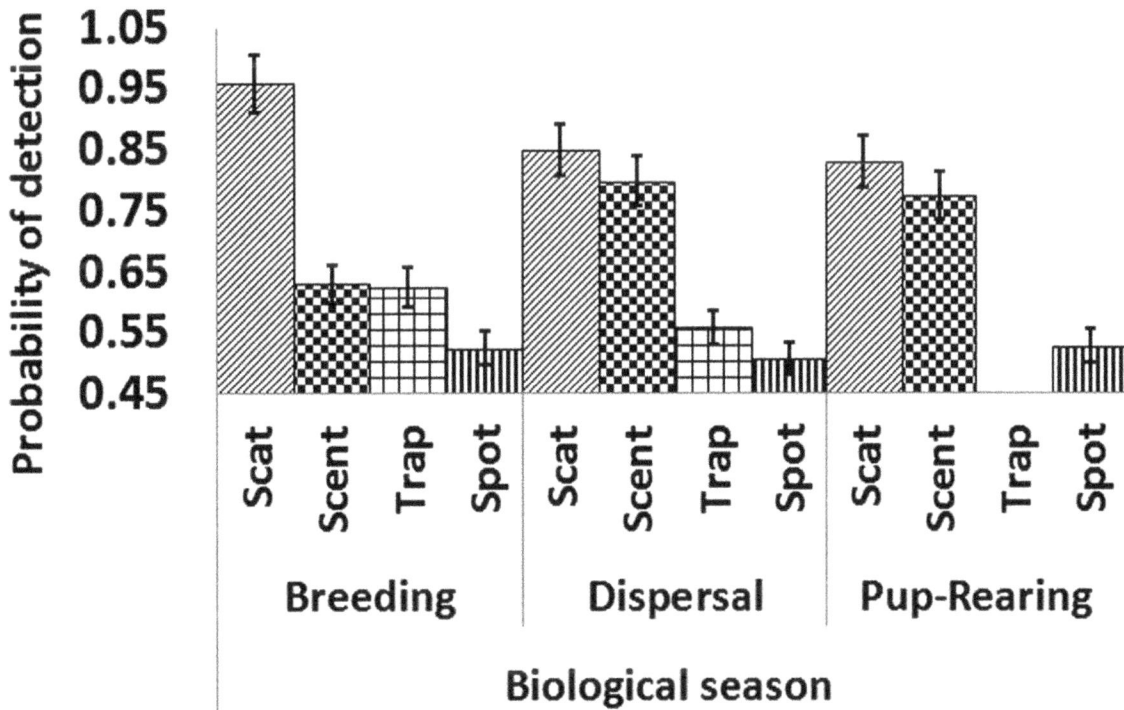

Figure 2. Probability of detection for scat deposition (Scat), scent station (Scent), trapping (Trap), and spotlight (Spot) surveys during 3 biological seasons for kit foxes on the Dugway Proving Ground, Utah, 2010–2012. Standard error bars included for each method.

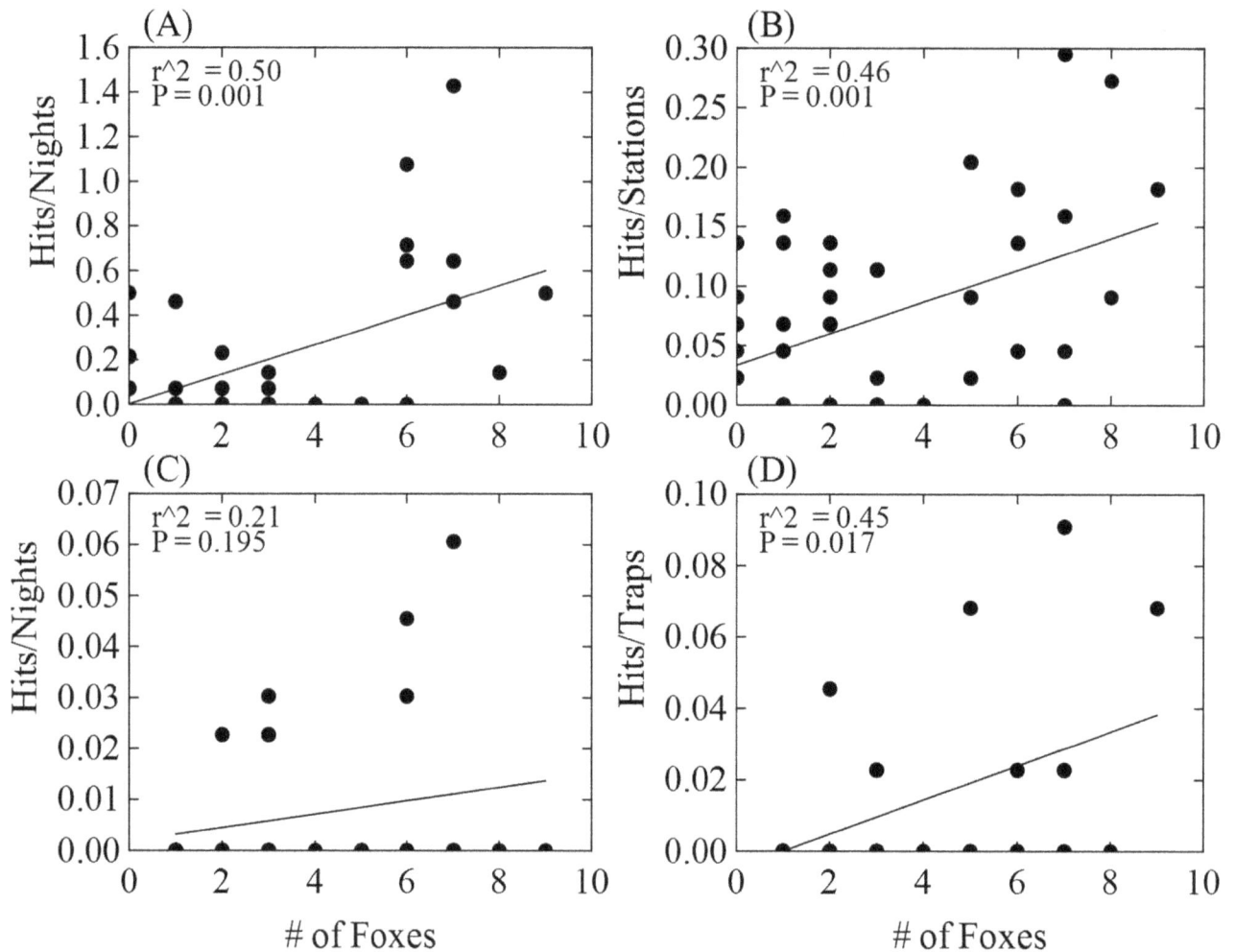

Figure 3. Relationship between the minimum number of known available foxes along the transects and indices of relative abundance for (A) scat deposition transects, (B) scent station surveys, (C) spotlight counts, and (D) trapping index, Dugway Proving Ground, Utah, 2010–2012.

group in a single night. One fox became 'trap shy' and the use of the tunnel trap [29] was successfully deployed to capture that individual for changing its radio-collar.

Low prey abundance and high intraguild predation by coyotes may be limiting kit fox density on the DPG [11,20]. Since the 1950's, much of the DPG has been converted from native Great Basin Desert shrub communities to grasslands [11] which support reduced small mammal diversity and abundance [22]. Fox home range size was largely dependent on prey availability [41,43–46]. In addition to this habitat change, the DPG has seen an increase in coyote abundance [11]. Predator-caused mortality was the highest cause of death for kit foxes during this study and coyotes have been shown to limit kit fox density [20,22,41,45]. Mean home range size for kit foxes on the DPG was large (20.1 km^2); similar to earlier study [22] with a mean home range size of 22.6 km^2 for kit foxes on the DPG. Studies of kit foxes in other regions have reported much smaller home ranges between 4.6 km^2 [46] and 13.7 km^2 [47] with an average of 11.4 km^2 [22,43,44,46–50].

Scat deposition surveys consistently had the highest detection probability, were most closely related to fox abundance, and were relatively inexpensive to perform. Scat deposition transects required the greatest period of surveying (i.e., 14 days) which

likely increased the chance of a sample being deposited and subsequently detected for this rare and widely dispersed species. Additionally, as a passive technique, scat deposition surveys do not require the target species to behave unnaturally (e.g., enter a trap or investigate a scent tab [15]). Our results corresponded with another study [6] finding scat deposition surveys to be the best method for detection of carnivores in the northeastern United States, and was similar to results for swift foxes in New Mexico [14].

Where misidentification and overlap with non-target species are a concern, training observers for accurate scat identification was critical [4,14,15]. But if multiple species are of concern, it would be possible to use this technique to efficiently identify multiple target species [19] with proper training. DNA analysis could also be used for verification of species and/or determining species abundance [4,14,17,19,51]. The use of scat detection dogs may increase detections rates [19,51] and if the dog is trained to detect a particular species, it could assist in the proper identification of the target species [19]. During this study, misidentification of scat may be the cause of detections along transects without known foxes. The risk of misidentification was highest during the pup-rearing season when juvenile coyotes and red foxes have the highest overlap in scat diameter with kit foxes. Before conducting scat deposition surveys, one should be aware of seasonal defecation patterns and related concerns when estimating site occurrence or abundance [17].

Scent-station surveys had the second highest detection probability of the 4 techniques compared and the second highest correlation with the number of available foxes, but were the most expensive on all four techniques evaluated. Snow and freeze/thaw cycles during the winter months on the DPG could freeze the sifted sand, thereby diminishing any tracks left by a visiting fox. Also, periods of high winds were more common during the winter, thereby erasing any tracks. We found making a small imprint in the sand helpful in determining if a scent station was still operable. During the breeding season we completed 85% of survey nights; surveys were more reliably completed during the dispersal and pup-rearing season with operable station nights of 100% and 97%, respectively.

Similar to scat deposition transects, more than one target species may be detected at a scent station [36], although wariness of a species to sifted sand on the station should be considered [3]. This technique has the highest potential for observer bias and possible misidentification of tracks. Training of the observer at track identification is crucial to avoid misidentification, especially when there are multiple canids on the landscape. We found that 1–2 cm of sifted sand left the most discernible tracks. Our results were consistent with other studies showing a positive correlation of scent station detections to fox abundance [13–15], although one study reported fairly erratic results and suggested scent station surveys were only able to detect large changes in the population [13].

In this widely dispersed kit fox population, spotlight surveys were found ineffective at detecting fox presence and failed to detect a single fox during the dispersal season of 2011, although 18 known foxes were available. During 327 survey nights, we only detected fox presence 15 times. Spotlighting had the lowest detection probability and was not significantly correlated to fox abundance. Obstruction of view from vegetation and topography [4,13,15] are concerns when using this technique. In addition, highly mobile, wary species may actively evade detection [52]. This technique failed to detect kit foxes twice when foxes were known to be available for detection and was the weakest performing technique, similar to a study on swift foxes [15]. Spotlight surveys were found to be inefficient at detecting swift foxes in New Mexico [14].

The trapping survey was only slightly less correlated to fox abundance than the scent station survey, but had a much lower detection probability than both the scat deposition transects and scent station surveys. One of the main benefits from this technique was the ability to add ear tags to captured foxes to conduct capture-mark-recapture estimation of abundance [15]. Due to a low capture rate, low numbers of foxes, and high mortality from predation, we had very few recaptures and therefore could not perform mark-recapture abundance analyses. Because of concerns for the safety of trapped individuals (high summer temperatures) and possible effects on natal young, trapping surveys were not conducted during the pup-rearing season. Trapping posed the highest risk to the animal of all methods used as we did have 3 minor foot injuries and 4 mouth injuries. We recommend modifying the mesh size to a mesh size of 1–2 cm [15]. The effect of repeated trapping of foxes should also be considered [15]. We had a few animals become trap happy and were repeatedly captured, while one fox became trap shy and could only be recaptured using a tunnel trap [29].

For detecting kit foxes in a low density population we suggest using scat deposition transects during the breeding season. This method had both the highest detection probability and highest correlation to kit fox abundance. This method also resulted in lower costs and labor requirements, was resilient to weather, and entailed no risk to the study animals [15]. The breeding season was ideal for monitoring kit fox population size, as detections consisted of primarily the resident population and we had the highest detection probabilities during this season. In areas where overlap with other sympatric canids occurs, careful training of technicians may be required, but the risk of overlapping scat dimensions should be lowest during the breeding season as most sympatric canids are also fully grown by the subsequent breeding season.

Acknowledgments

We thank R. Knight for providing continual support for the project, G. Smith, A. Hodge, C. Crawford, D. Page, J. Linnell, N. Mesce, L. Card, A. Reyer, M. Cent, K. Crowson, C. Hansen, and M. Cannan for field assistance, and P. Gipson, R. Knight, D. Koons, L. Šver, and J. Shivik for review of the manuscript.

Author Contributions

Conceived and designed the experiments: SJD EMG BMK. Performed the experiments: SJD EMG BMK. Analyzed the data: SJD EMG BMK. Contributed reagents/materials/analysis tools: SJD EMG BMK. Contributed to the writing of the manuscript: SJD EMG BMK.

References

1. Schaller GB (1996) Introduction: carnivores and conservation biology. In: Gittleman JL, editor. Carnivore behavior, ecology, and evolution, vol. 2. Ithaca: Cornell University Press.
2. Thacker RK, Flinders JT, Blackwell BH, Smith HD (1995) Comparison and use of four techniques for censusing three sub-species of kit fox. Utah Division of Wildlife Resources, Salt Lake City, Utah.
3. Gese EM (2001) Monitoring of terrestrial carnivore populations. In: Gittleman JL, Funk SM, Macdonald DW, Wayne RK, editors. Carnivore conservation. Cambridge: Cambridge University Press. pp. 372–396.
4. Gese EM (2004) Survey and census techniques for canids. In: Sillero-Zulbiri Hoffman CM, Macdonald DW, editors. Canids: foxes, wolves, jackals, and dogs. Glan, Switzerland: IUCN World Conservation Union. pp. 273–279.
5. Gompper ME, Kays RW, Ray JC, Lapoint SD, Bogan DA, et al. (2006) A comparison of noninvasive in techniques to survey carnivore communities in Northeastern North America. Wildl Soc Bull 34: 1142–1151.
6. Long RA, Donovan TM, Mackay P, Zielinski WJ, Buzas JS (2007) Comparing scat detection dogs, cameras, and hair snares for surveying carnivores. J Wildl Manage 71: 2018–2025.
7. McGrew JC (1976) Vulpes macrotis. Mammalian Species 123: 1–6.
8. Meaney CA, Reed-Eckert M, Beauvais GP (2006) Kit fox (Vulpes macrotis): a technical conservation assessment. USDA Forest Service, Golden, Colorado.
9. Egoscue HJ (1956) Preliminary studies of the kit fox in Utah. J Mammal 37: 351–357.
10. Egoscue HJ (1962) Ecology and life history of the kit fox in Tooele County, Utah. Ecology 43: 481–497.
11. Arjo WM, Gese EM, Bennett TJ, Kozlowski AJ (2007) Changes in kit fox – coyote – prey relationships in the Great Basin Desert, Utah. West N Amer Nat 67: 389–401.
12. Utah Department of Natural Resources (2011) Utah Sensitive Species List. Utah Division of Wildlife Resources, Salt Lake City, Utah.
13. Warrick GD, Harris CE (2001) Evaluation of spotlight and scent-station surveys to monitor kit fox abundance. Wildl Soc Bull 29: 827–832.
14. Harrison RL, Barr DJ, Dragoo JW (2002) A comparison of population survey techniques for swift foxes (Vulpes velox) in New Mexico. Amer Midl Nat 148: 320–337.
15. Schauster ER, Gese EM, Kitchen AM (2002) An evaluation of survey methods for monitoring swift fox abundance. Wildl Soc Bull 30: 464–477.
16. Finley DJ, White GC, Fitzgerald JP (2005) Estimation of swift fox population size and occupancy rates in eastern Colorado. J Wildl Manage 69: 861–873.
17. Ralls K, Sharma S, Smith DA, Bremner-Harrison S, Cypher BL, et al. (2010) Changes in kit fox defecation patterns during the reproductive season: implications for noninvasive surveys. J Wildl Manage 74: 1457–1462.

18. Ralls K, Eberhardt LL (1997) Assessment of abundance of San Joaquin kit foxes by spotlight surveys. J Mammal 78: 65–73.

19. Smith DA, Ralls K, Cypher BL, Maldonado JE (2005) Assessment of scat-detection dog surveys to determine kit fox distribution. Wildl Soc Bull 33: 897–904.

20. Kozlowski AJ, Gese EM, Arjo WM (2008) Niche overlap and resource partitioning between sympatric kit foxes and coyotes in the Great Basin Desert of Western Utah. Amer Midl Nat 160: 191–208.

21. Kozlowski AJ, Gese EM, Arjo WM (2012) Effects of intraguild predation: evaluating resource competition between two canid species with apparent niche separation. Int J Ecol 2012:1–12. doi:10.1155/2012/629246.

22. Arjo WM, Bennett TJ, Kozlowski AJ (2003) Characteristics of current and historical kit fox (*Vulpes macrotis*) dens in the Great Basin Desert. Can J Zool 81: 96–102.

23. Eberhardt LE, Hanson WC, Bengtson JL, Garrott RA, Hanson EE (1982) Arctic fox home range characteristics in an oil-development area. J Wildl Manage 46: 183–190.

24. Schauster ER, Gese EM, Kitchen AM (2002b) Population ecology of swift foxes (*Vulpes velox*) in southeastern Colorado. Can J Zool 80: 307–319.

25. Swihart RK, Slade NA (1985) Influence of sampling interval on estimates of home-range size. J Wildl Manage 49: 1019–1025.

26. Swihart RK, Slade NA (1985) Testing for independence of observations in animal movements. Ecology 66: 1176–1184.

27. Gese EM, Andersen DE, Rongstad OJ (1990) Determinging home-range size of residents coyotes from point and sequential locations. J Wildl Manage 54: 501–506.

28. Aebischer NJ, Robertson PA, Kenward RE (1993) Compositional analysis of habitat use from animal radio-tracking data. Ecology 74: 1313–1325.

29. Kozlowski AJ, Bennett TJ, Gese EM, Arjo WM (2003) Live capture of denning mammals using an improved box-trap enclosure: kit foxes as a test case. Wildl Soc Bull 31: 630–633.

30. Steiniger S, Hunter AJS (2012) OpenJUMP HoRAE-A free GIS and toolbox for home-range analysis. Wildl Soc Bull doi:10.1002/wsb.168.

31. Worton BJ (1989) Kernel methods for estimating the utilization distribution in home-range studies. Ecology 70: 164–168.

32. Berger KM, Gese EM (2007) Does interference competition with wolves limit the distribution and abundance of coyotes? J Anim Ecol 76: 1075–85.

33. Jacques CN, Jenks JA, Klaver RW (2009) Seasonal movement and home-range use by female pronghorns in sagebrush-steppe communities of western South Dakota. J Mammal 90: 433–441.

34. Kie JG, Matthiopoulos J, Fieberg J, Powell RA, Cagnacci F, et al. (2010) The home-range concept: are traditional estimators still relevant with modern telemetry technology? Royal Soc London (Series B, Biol Sci) 365: 2221–2231.

35. Knowlton FF (1984) Feasibility of assesssing coyote abundance on small areas. Final report, Denver Wildlife Research Center, Denver, Colorado.

36. Linhart SB, Knowlton FF (1975) Determining the relative abundance of coyotes by scent station lines. Wildl Soc Bull 3: 119–124.

37. Roughton RD, Sweeny MW (1982) Refinements in scent-station methology for assessing trends in carnivore populations. J Wildl Manage 46: 217–229.

38. White GC, Burnham KP (1999) Program MARK: survival estimation from populations of marked animals. Bird Study Supplement 46: 120–139.

39. Mackenzie DI, Nichols JD, Lachman GB, Droege S, Royle JA, et al. (2002) Estimating site occupancy rates when detection probabilities are less than one. Ecology 83: 2248–2255.

40. Burnham KP, Anderson DR (2002) Model selection and multimodel inference: a practical information-theoretic approach, second edition. New York: Springer-Verlag.

41. Cypher BL, Warrick GD, Otten MRM, O'Farrell TP, Berry WH, et al. (2000) Population dynamics of San Joaquin kit foxes at the Naval Petroleum Reserves in California. Wildl Monogr 145: 1–43.

42. Fitzgerald JP (1996) Status and distribution of the kit fox (*Vulpes macrotis*) in western Colorado. Final report, University of Northern Colorado, Greeley, Colorado.

43. Zoellick BW, Smith N (1992) Size and spatial organization of home ranges of kit foxes in Arizona. J Mammal 73: 83–88.

44. White PJ, Ralls K (1993) Reproduction and spacing patterns of kit fox relative to changing prey availability. J Wildl Manage 57: 861–867.

45. White PJ, Garrott RA (1997) Factors regulating kit fox populations. Can J Zool 75: 1982–1988.

46. Zoellick BW, Harris CE, Kelly BT, O'Farrell TP, Kato TT, et al. (2002) Movements and home ranges of San Joaquin Kit foxes (*Vulpes macrotis mutica*) relative to oil-field development. West N Amer Nat 62: 151–159.

47. White PJ, Ralls K, Garrott RA (1994) Coyote - kit fox interactions as revealed by telemetry. Can J Zool 72: 1831–1836.

48. Koopman ME, Cypher BL, Scrivner JH (2000) Dispersal patterns of San Joaquin kit foxes (*Vulpes macrotis mutica*). J Mammal 81: 213–222.

49. List R, Macdonald DW (2003) Home range and habitat use of the kit fox (*Vulpes macrotis*) in a prairie dog (*Cynomys ludovicianus*) complex. J Zool 259: 1–5.

50. Moehrenschlager A, List R, Macdonald DW (2007) Escaping intraguild predation: Mexican kit foxes survive while coyotes and golden eagles kill Canadian swift foxes. J Mammal 88: 1029–1039.

51. Long RA, Donovan TM, Mackay P, Zielinski WJ, Buzas JS (2007b) Effectiveness of scat detection dogs for detecting forest carnivores. J Wildl Manage 71: 2007–2017.

52. Ruette S, Stahl P, Albaret M (2003) Applying distance-sampling methods to spotlight counts of red foxes. J Appl Ecol 40: 32–43.

12

Prevalence and Characterization of *Escherichia coli* and *Salmonella* Strains Isolated from Stray Dog and Coyote Feces in a Major Leafy Greens Production Region at the United States-Mexico Border

Michele T. Jay-Russell*, Alexis F. Hake, Yingjia Bengson, Anyarat Thiptara, Tran Nguyen

Western Center for Food Safety, University of California Davis, Davis, California, United States of America

Abstract

In 2010, Romaine lettuce grown in southern Arizona was implicated in a multi-state outbreak of *Escherichia coli* O145:H28 infections. This was the first known Shiga toxin-producing *E. coli* (STEC) outbreak traced to the southwest desert leafy green vegetable production region along the United States-Mexico border. Limited information exists on sources of STEC and other enteric zoonotic pathogens in domestic and wild animals in this region. According to local vegetable growers, unleashed or stray domestic dogs and free-roaming coyotes are a significant problem due to intrusions into their crop fields. During the 2010–2011 leafy greens growing season, we conducted a prevalence survey of STEC and *Salmonella* presence in stray dog and coyote feces. Fresh fecal samples from impounded dogs and coyotes from lands near produce fields were collected and cultured using extended enrichment and serogroup-specific immunomagnetic separation (IMS) followed by serotyping, pulsed-field gel electrophoresis (PFGE), and antimicrobial susceptibility testing. A total of 461 fecal samples were analyzed including 358 domestic dog and 103 coyote fecals. STEC was not detected, but atypical enteropathogenic *E. coli* (aEPEC) strains comprising 14 different serotypes were isolated from 13 (3.6%) dog and 5 (4.9%) coyote samples. *Salmonella* was cultured from 33 (9.2%) dog and 33 (32%) coyote samples comprising 29 serovars with 58% from dogs belonging to Senftenberg or Typhimurium. PFGE analysis revealed 17 aEPEC and 27 *Salmonella* distinct pulsotypes. Four (22.2%) of 18 aEPEC and 4 (6.1%) of 66 *Salmonella* isolates were resistant to two or more antibiotic classes. Our findings suggest that stray dogs and coyotes in the desert southwest may not be significant sources of STEC, but are potential reservoirs of other pathogenic *E. coli* and *Salmonella*. These results underscore the importance of good agriculture practices relating to mitigation of microbial risks from animal fecal deposits in the produce production area.

Editor: Dongsheng Zhou, Beijing Institute of Microbiology and Epidemiology, China

Funding: Funding for this project was provided by the Center for Produce Safety (http://cps.ucdavis.edu) and University of California Agriculture and Natural Resources (UCANR) United States Department of Agriculture/National Institute of Food and Agriculture (http://www.csrees.usda.gov/) grant #2010-34608-20768 (SA7670). Additionally, JV Farms is a Center for Produce Safety contributor (https://cps.ucdavis.edu/campaign_contributors.php). Except JV Farms, the funders had no role in study design or data collection. JV Farms food safety staff assisted in enrolling local animal shelters and collecting fecal samples for shipment to the authors. All funders had no role in the analysis, decision to publish, or preparation of the manuscript.

Competing Interests: This study was funded in part by JV Farms (Yuma, Arizona) through a contribution to the Center for Produce Safety Campaign for Research.

* Email: mjay@ucdavis.edu

Introduction

Foodborne disease illnesses caused by pathogen contamination of fresh produce are being recognized in greater numbers in the United States (U.S.) and abroad [1], [2]. An analysis of Centers for Disease Control and Prevention (CDC) data on reported foodborne illnesses from 1973 to 1997 indicated that outbreaks associated with fresh produce accounted for 6% of all reported foodborne disease outbreaks in the 1990s compared with just 0.7% in the 1970s [3], [4]. A more recent survey of CDC outbreak data from 1998–2008 showed these numbers are still rising, with 46% of foodborne illnesses being attributed to produce and 22% specifically attributed to leafy greens [5]. While norovirus infections transmitted downstream during post-harvest handling are likely the major driver of these statistics, reports of fresh produce-associated outbreaks from zoonotic agents potentially spread by domestic and wild animal reservoirs in the pre-harvest environment are clearly contributing to this disease burden [6], [7].

Approximately 90% of commercial lettuce produced for the U.S. market is grown in two major produce production regions that rotate seasonally [8]: the Salinas Valley in the central California coast (April through October) and the desert southwest at the U.S.-Mexico border (November through March). The desert southwest growing region includes Yuma, Arizona, California's Imperial Valley, and northern Mexico. The role of domestic animals and wildlife as potential sources and transmitters of zoonotic bacterial pathogens to lettuce and other leafy greens and agriculture water has been studied at length in the central California coast [9], [10], [11], [12], [13]. There is limited information, however, on the importance of animal reservoirs in the pre-harvest bacterial contamination of fresh produce in other parts of the country [14]. The desert presents unique pre-harvest food safety challenges including urban encroachment where produce fields and irrigation canals may be adjacent to housing developments and recreation vehicle (RV) parks. In addition to concerns about human sources of foodborne pathogens near leafy green production areas of the desert, growers report problems with unleashed, free-roaming domestic dogs (*Canis familiaris*) entering their fields. Off-leash or stray dog intrusions into produce fields and the surrounding production area may result in damage to crops and destruction of potentially contaminated plants (Figure 1). Growers also report frequent coyote (*Canis latrans*) sightings and signs (tracks, scat, feces) on roads adjacent to produce fields where tractors and other equipment are used.

In spring 2010, an outbreak of *Escherichia coli* O145:H28 infections involving 27 confirmed and 4 probable case-patients from 5 states was linked to Romaine lettuce grown in southern Arizona [15], [16]. This was the first known leafy green-related Shiga toxin-producing *E. coli* (STEC) outbreak traced to the desert growing region. Based on investigations by the U.S. Food and Drug Administration (FDA), pre-harvest contamination of irrigation canals possibly due to sewage runoff from a nearby RV park could have caused the contamination, although no laboratory-confirmed environmental source of the outbreak was determined [17]. Following this outbreak, the present study was funded as a Center for Produce Safety "Rapid Response Project" with the purpose to determine prevalence and characterize pathogenic *Escherichia coli* and *Salmonella* strains isolated from dog and coyote fecal samples collected in the southwest desert during the 2010 to 2011 leafy green vegetable growing season.

Methods

Ethics Statement

Permission to access privately owned lands was obtained from the produce companies enrolled in the study. Animal shelter administrative directors in the U.S. and Mexico approved participation in the study. Dog fecal samples from the shelter in

Figure 1. Examples of animal intrusions into produce production areas of the desert southwest: a stray dog traveling next to an irrigation canal in northern Mexico (A); coyote feces adjacent to a lettuce field in southern California (B); dog feces on a lettuce plant in southern Arizona (C); areas of intentionally destroyed lettuce crop (arrow) following evidence of animal intrusion.

Mexico were transported by vehicle across the Mexico-US border by one of our industry collaborators. A permit for importation of dog feces was not required per the United States Department of Agriculture (USDA) Animal and Plant Health Inspection Service (APHIS) "Animal and Animal Product" import guidelines (#1102 Feline and Canine Material). Wildlife scientific collection permits and university animal care and use approval were not necessary in this study because fecal samples were collected from the ground and no animals were handled.

Sampling

Three animal shelters were enrolled in the study, one each in Yuma, Arizona, Imperial Valley, California, and northern Mexico. These facilities were chosen because animal control officers had worked historically with leafy green growers in the region to remove stray dogs from agriculture fields. We aimed to sample once monthly during the desert southwest leafy greens growing season (November to March) with a goal to collect ~300 samples based on sample size calculations. Due to limits in the number of impounded dogs available each month and logistics with the shelter personnel, each facility was sampled six times spread variably from November 3, 2010 to May 5, 2011 (Table 1).

A standardized questionnaire was used to collect demographic data (location and date found, breed, sex, age, and reason the dog was impounded) from records at the facilities. Dog fecal samples (n = 358) were collected by industry cooperators after training by University of California, Davis (UC Davis) veterinarians in aseptic fecal sample collection, storage and shipping. Freshly deposited feces from animals caged individually at the U.S. shelters were taken from the kennel floor using a sterile tongue depressor and placed in a sterile 227 gram fecal cup with a snap-cap lid (National Scientific Supply Co., Claremont, CA). Dogs at the shelter in Mexico were caged in groups, thus there was potential for cross-contamination. To minimize the risk of cross-contamination, care was taken to collect only freshly deposited feces from individual dogs impounded in the past 24 hours.

Fresh coyote fecal material (n = 103) found on the roads in and up to 1 mile from leafy green vegetable fields located in Yuma, Arizona and Imperial Valley, California were collected by industry cooperators using a sterile tongue depressor and fecal cup as described above. Industry personnel routinely survey fields for animal intrusions and have training in the identification of sign (sightings, tracks, feces) of wildlife species including coyotes common in agriculture areas. In order to ensure the fecal material was fresh, the sites were walked by grower personnel the evening prior to sampling, and any existing coyote feces was removed. The fields were then surveyed again at dawn the following morning to collect fresh feces. Wildlife trail cameras (Cuddeback Digital, Green Bay, WI), physical sightings, and other sign were used to confirm the presence of coyotes at the study areas.

Samples were shipped to UC Davis overnight on blue-ice on the day of collection and processed on the day of arrival at the laboratory (approximately 24 hours).

Laboratory

Our overall goal was to culture Shiga toxin-producing *E. coli* belonging to serogroups STEC O103, O145, O157, O26, and non-typhoidal *Salmonella enterica* using a combined pre-enrichment step followed by immunomagnetic separation (IMS), selective plating, latex agglutination, and PCR or biochemical confirmation as described below. Isolates were then characterized by presence of virulence factors, genetic relatedness, and antibiotic resistance.

Table 1. Monthly prevalence of atypical enteropathogenic *Escherichia coli* (aEPEC) and *Salmonella enterica* isolated from coyote and dog fecal samples, southwestern desert, November 3, 2010 through May 5, 2011.

Source	Animal	Location	No. positive/No. tested (%)[a]						
			Nov	Dec	Jan	Feb	Mar	May	Total
aEPEC									
	COYOTES								
		Arizona	0/10	0/11	1/11 (9.1)	0/7	0/0	0/0	1/39 (2.6)
		California	0/10	3/16 (18.8)	0/13	0/11	1/14 (7.1)	0/0	4/64 (6.3)
		Subtotal	0/20	3/27 (11.1)	1/24 (4.2)	0/18	1/14 (7.1)	0/0	5/103 (4.9)
	DOGS								
		Arizona	0/16	1/21 (4.8)	0/0	0/17	3/45 (6.7)	0/25	4/124 (3.2)
		California	1/24 (4.2)	3/18 (16.7)	0/23	1/18 (5.6)	1/17 (5.9)	0/0	6/100 (6.0)
		Mexico	0/30	0/28	1/27 (3.7)	2/49 (4.1)	0/0	0/0	3/134 (2.2)
		Subtotal	1/70 (1.4)	4/67 (6.0)	1/50 (2.0)	3/84 (3.6)	1/62 (1.6)	0/25	13/358 (3.6)
		Total	**1/90 (1.1)**	**7/94 (7.4)**	**2/74 (2.7)**	**3/102 (2.9)**	**2/76 (2.6)**	**0/25**	**18/461 (3.9)**
Salmonella enteric									
	COYOTES								
		Arizona	6/10 (60.0)	2/11 (18.2)	5/11 (45.5)	0/7	0/0	0/0	13/39 (33.3)
		California	5/10 (50.0)	7/16 (43.6)	3/13 (23.1)	3/11 (27.3)	2/14 (14.3)	0/0	20/64 (31.3)
		Subtotal	11/20 (55.0)	9/27 (33.3)	8/24 (33.3)	3/18 (16.7)	2/14 (14.3)	0/0	33/103 (32.0)
	DOGS								
		Arizona	0/16	2/21 (9.5)	0/0	0/17	2/45 (4.4)	0/25	4/124 (3.2)
		California	4/24 (16.7)	1/18 (5.6)	1/23 (4.3)	0/18	3/17 (17.6)	0/0	9/100 (9.0)
		Mexico	13/30 (43.3)	1/28 (3.6)	3/27 (11.1)	3/49 (6.1)	0/0	0/0	20/134 (14.9)
		Subtotal	17/70 (24.3)	4/67 (6.0)	4/50 (8.0)	3/84 (3.6)	5/62 (8.1)	0/25	33/358 (9.2)
		Total	28/90 (31.1)	13/94 (13.8)	12/74 (16.2)	6/102 (5.9)	7/76 (9.2)	0/25	66/461 (14.3)

[a]The Arizona shelter was sampled twice in March and in May. The California shelter was sampled twice in December and twice in February. The shelter in Mexico was sampled twice in November. The California shelter was sampled twice in November.

Pre-enrichment. Initially, non-selective pre-enrichment for the simultaneous culture of STEC and *Salmonella* was performed by adding 10 grams of feces to 100 mL of universal pre-enrichment broth (UPB; Difco, Becton Dickinson, Sparks, MD) and incubating for 20 hours at 35°C using a protocol modified in our laboratory for dog fecal material [18]. One milliliter of enriched UPB was then transferred to 9 mL tryptic soy broth (TSB; Becton Dickinson, Sparks, MD) for STEC detection and incubated for 2 hours at 25°C with shaking at 100 rpm (Innova 44, Eppendorf North America, Hauppauge, NY), followed by 8 hours at 42°C with shaking at 100 rpm, then at 6°C without shaking until processing the following day [10], [19]. One milliliter of enriched UPB was also transferred to 9 mL of buffered peptone water (BPW; Hardy Diagnostics, Santa Maria, CA) for *Salmonella* detection and incubated for 24 hours at 37°C with shaking at 50 rpm [11]. In 2011, we discontinued the use of the UPB pre-enrichment step to streamline the protocol. Instead, pre-enrichment was performed by adding 10 grams of feces directly into a WhirlPak bag containing 100 mL TSB followed by the same incubation parameters as just described [10], [19]. Spiking experiments comparing the UPB and TSB pre-enrichment methods revealed no statistical difference (p = 0.32) in recovery of STEC or *Salmonella* (data not shown). As such, we completed this study using the streamlined protocol without the UPB step.

Aliquots of the primary enrichment broths were mixed with sterile glycerol to a final concentration of 14.3%, and stored at −20°C [11].

Escherichia coli. IMS using Dynal anti-*E. coli* O157, O26, O103, and O145 beads (Invitrogen, Grand Island, NY) was performed on TSB enrichment broths with the automated Dynal BeadRetriever (Invitrogen) per the manufacturer's instructions. Following incubation and washing, 50 µL of the resuspended beads were plated onto Rainbow agar (Biolog, Hayward, CA) with novobiocin (20 mg/L) and tellurite (0.8 mg/L) (MP Biomedicals, Solon, OH) and streaked for isolation [10], [19]. Another 50 µL were plated onto Sorbitol MacConkey Agar (BD Becton, Sparks, MD) with cefixime (0.05 mg/L) (USP, Rockville, MD) and tellurite (2.5 mg/L), streaked for isolation, and incubated at 37°C overnight. *E. coli* O157:H7 RM1484 and three non-O157 STEC strains (O103, O145, O26) were used as positive controls to observe the expected phenotype on the selective agars. Up to 10 colonies exhibiting characteristic morphology were subcultured to both agar types and incubated again at 37°C overnight for purification. Up to four pure colonies were then streaked onto Luria-Bertani (LB; Fisher Scientific, Waltham, MA) agar and incubated at 37°C for 24 hours. Suspect colonies from LB agar were screened using ImmuLex commercial latex slide agglutination assays (Statens Serum Institut, Denmark) with pooled STEC antisera (*E. coli* OK O antiserum to detect EPEC and STEC) followed by O-group specific antisera (O103, O145, O157, O26) according to the manufacturer's instructions. All assays included a negative control (saline) to check for non-specific agglutination. Two bacterial colonies from each positive plate were banked onto Cryobeads (ProLab Diagnostics, Round Rock, TX) and stored at −80°C until characterized further.

All isolates (n = 278) positive for STEC O-groups by latex agglutination were submitted to the Pennsylvania State University *E. coli* Reference Center to confirm O-type using a multiplex PCR that detects eight major STEC O-groups (O26, O45, O103, O111, O113, O121, O145, O157) as described previously [20]. H-antigens were identified by the same lab using PCR-RFLP [21]. The isolates were also tested by PCR at the reference laboratory for the presence of virulence factors including *stx*1, *stx*2, *eaeA*, and *hlyA* genes [22]. A subset of aEPEC isolates (n = 12) positive for

the *eaeA* gene, but not belonging to the 8 STEC O-groups identified by the reference laboratory's multiplex PCR, were serotyped conventionally by agglutination reactions against antisera developed for each of the O serogroups [20].

Because all isolates were Shiga toxin-negative despite many belonging to STEC-associated O-groups, a retrospective analysis of banked TSB enrichment broths (n = 461) frozen at -20°C with glycerol was performed at the United States Department of Agriculture's Agricultural Research Service (ARS) Western Regional Research Center laboratory. Briefly, template DNA was prepared by boiling followed by detection of *stx*1 and *stx*2 virulence genes by multiplex qPCR; the method was validated previously for screening TSB pre-enrichment broths from environmental samples including coyote feces [19].

Salmonella. IMS using anti-*Salmonella* Dynabeads (Invitrogen, Grand Island, NY) was performed on BPW broths as described previously using the Dynal BeadRetriever (Invitrogen) [11]. Following incubation and washing, 100 µL of separated broth was further enriched in 3 mL Rappaport-Vassiliadis Soya Peptone (RVS; Difco, Becton Dickinson, Franklin Lakes, NJ) broth for 48 hours at 42°C [23]. Ten microliters of RVS broth were then plated on xylose lysine deoxycholate (XLD; Difco, Becton Dickinson, Franklin Lakes, NJ) agar and incubated at 37°C overnight. Samples with growth of hydrogen sulfide positive colonies were confirmed for *Salmonella* by performing biochemical profiles (triple sugar iron, urea, citrate, and lysine decarboxylase) on up to six individual colonies from each XLD agar. *Salmonella* Enteritidis ATCC BAA1045 was used as a positive control to observe the expected phenotype. The same colonies used for biochemical profiling were also streaked onto LB agar and incubated at 37°C overnight. Two bacterial colonies from each positive plate were banked onto Cryobeads (ProLab Diagnostics, Round Rock, TX) and stored at −80°C. Serotyping using the Kauffmann-White scheme was conducted by the United States Department of Agriculture (USDA) National Veterinary Services Laboratories in Ames, Iowa [24].

Pulse field Gel Electrophoresis. *E. coli* and *Salmonella* isolates were retrieved from frozen storage at −80°C and clonal relationships were assessed by pulsed-field gel electrophoresis (PFGE) according to the CDC's PulseNet standard procedure using *Salmonella* Braenderup ATCC BAA664 as the molecular size standard [25]. Briefly, bacterial isolates were suspended in buffer containing 10 mM Tris pH 8 and 10 mM EDTA for DNA isolation. DNA was digested in enzyme buffer with restriction enzyme *Xba*I. Images were analyzed, and the similarity among different strains was characterized using Bionumerics version 7.1 software (Applied Maths, Austin, TX). Pattern comparisons were made using the software cluster analysis tool and confirmed by visual examination to assign pulsotypes [26], [27].

Antimicrobial Susceptibility testing. Frozen *E. coli* and *Salmonella* isolates were thawed and streaked onto trypticase soy agar (TSA; Difco, Becton Dickinson, Franklin Lakes, NJ) with 5% sheep blood agar, then incubated at 37°C for 24 hours. The broth microdilution method for antimicrobial susceptibility testing was performed in accordance with the Clinical and Laboratory Standards Institute (CLSI) guidelines [28], [29]. Isolates were evaluated for susceptibility to 12 antimicrobial drugs (ampicillin, amoxicillin/clavulanic acid, ceftriaxonem azithromycin, chloramphenicol, sulfisoxazole, cefoxitin, kanamycin, streptomycin, trimethoprim/sulfamethoxazole, tetracycline, ceftiofur) using the National Antimicrobial Resistance Monitoring System (NARMS) Gram negative tray (Trek Diagnostic Systems, Westlake, OH). *E. coli* ATCC 25922, *E. coli* ATCC 35218, and *Pseudomonas aeruginosa* ATCC 27853 were used as quality control organisms for MIC determination in accordance with CLSI guidelines. Breakpoints guidelines were adopted from the NARMS report, with the exception of azithromycin, for which no breakpoints have been published [30]. Based on a recent publication proposing an epidemiologic cut-off for wild-type *Salmonella* of ≤16 μg/ml, isolates with MIC values >16 μg/ml were considered resistant to azithromycin for the purpose of this study [31].

Statistical Analysis

WinEpi online software (http://www.winepi.net/uk/index.htm) was used to calculate sample size with a confidence level set at 95% and a population size of 1,000–10,000. Prevalence estimates were based on data from a longitudinal study of coyote populations in the central California coast in 2008–2010 [11], [19], with 1% for *E. coli* O157, 5% for non-O157 STEC, and 10% for *Salmonella enterica*; using these estimates, the required number of samples would be 258–294 for *E. coli* O157, 57–59 for non-O157 STEC, and 29 for *Salmonella* detection. Based on the expected low prevalence of *E. coli* O157, our goal was to collect at least 300 fecal samples.

Data were entered in Microsoft Excel 2007 spreadsheets and exported for analysis in STATA (Stata 11.0, College Station, TX). McNemar's chi-square test was used to compare the sensitivities of pre-enrichment methods (UPB and TSB) and O-group testing methods (latex agglutination, multiplex PCR). Univariate logistic regression was conducted to identify statistical associations between enteric pathogen status and covariates (age, sex, etc.). Covariates with p-values equal or less than 0.20 were considered for inclusion in an exact logistic regression model. A p-value ≤ 0.05 was used to detect significantly associated factors for pathogen presence in the regression model.

Results

Prevalence and risk factors

A total of 461 fecal samples were collected from November 3, 2010 through May 5, 2011, including 358 domestic dog and 103 coyote fecals (Table 1). Descriptive characteristics of the shelter dog population are shown in Table 2. Shiga toxin-producing *E. coli* was not detected in any fecal samples, but aEPEC (*eaeA*+) strains were isolated from 13 (3.6%) dog and 5 (4.9%) coyote fecal samples. *Salmonella* was detected in 33 (9.2%) dog and 33 (32.0%) coyote fecal samples.

Univariate analysis revealed no significant association between the relative number of aEPEC or *Salmonella* positive and negative dogs by age, gender, breed, or reason that the dog was impounded. There was a significant difference in the number of dog fecal samples positive for *Salmonella* by shelter location. Exact logistic regression revealed that, after adjusting for age, the odds of a dog from the shelter in Mexico being *Salmonella* positive was 4.88 (95% CI: 1.60, 20.31, P <0.01) times higher than for dogs at the Arizona shelter. In November 2010, a higher seasonal prevalence of *Salmonella* was observed at the California and Mexico shelters, primarily due to a specific serovar (Senftenberg) as described below.

For coyote fecal samples, univariate logistic regression showed no significant difference in the odds of aEPEC or *Salmonella* isolation in fecal samples collected in Arizona compared with California (OR = 1.10, 95% CI: 0.43, 2.78, P = 0.99). There was a significantly lower odds of *Salmonella* isolation from coyote scat sampled in February (OR = 0.16, 95% CI: 0.04, 0.70, P = 0.02) and March (OR = 0.1, 95% CI: 0.02, 0.80, P = 0.03) compared with November; but, unlike dogs, this seasonal observation was not due to any specific serovar.

Phenotypic and molecular characterization of *E. coli* isolates

We screened initially for suspect STEC isolates by using O-specific (O103, O145, O157, O26) IMS and commercial latex agglutination. A total of 278 isolates presumptively belonging to these four serogroups were submitted to the Pennsylvania State University *E. coli* reference laboratory for confirmation using their multiplex PCR that detects 8 major STEC O-groups [20]. We found discordance between the IMS-latex agglutination classification of O-groups compared with the multiplex PCR (Table 3). Specifically, if the multiplex PCR at the reference laboratory is used as the standard, the sensitivities of O-specific latex agglutination (31.2% for O103, 8.5% for O145, and 26.3% for O26) were significantly different from the multiplex PCR method (p <0.01).

All 278 isolates were negative by PCR for genes encoding *stx*1 and *stx*2. Additionally, 461 frozen fecal TSB-enrichment broths were negative using a multiplex qPCR assay to detect *stx*1 and *stx*2 genes [19]. Isolates (n = 187) lacking any virulence factors and not belonging to one of the eight serogroups identified by the reference laboratory's multiplex PCR were not further characterized. Among the remaining 91 isolates, a total of 29 different *E. coli* serotypes were identified (Table 4). Excluding clones from the same samples, there were 18 isolates comprising 14 serotypes with genes encoding *eaeA*, including one (O26:H11) with both *eaeA* and *hlyA* genes; these isolates were classified as aEPEC (Table 4) [32], [33]. There was more diversity among serotypes from coyotes compared with dogs. Specifically, 11 of 15 coyote fecal isolates were different serotypes with none being dominant. In contrast, almost half of the dog samples contained two dominant serotypes that were negative for virulence markers, O103:H16 and O103:H49 (Table 4).

A total of 17 pulsotypes (PT) were found among 18 aEPEC strains from dogs and coyote feces (Figure 2). Two non-pathogenic *E. coli* O145:H11 isolates were included in the dendogram for comparison with the *E. coli* O145:H28 human clinical strain (PT-8) associated with the 2010 outbreak linked to Romaine lettuce. Of note, *E. coli* O145 strains isolated from dog feces during the study had different H types (O145:H11 and O145:H34) and were genetically unrelated to the 2010 outbreak strain based on PFGE analysis (Table 4, Figure 2).

All aEPEC isolates were tested for antibiotic resistance, and 6 (33.3%) were found to be pansusceptible to the antimicrobial drugs we used in the NARMS panel (Table 5). One isolate from coyote and 3 isolates from dog feces were resistant to two or more antibiotics. Two aEPEC isolates, serotypes O167:H9 and O114:H8, from dog feces collected in Arizona and Mexico, respectively, were resistant to four antibiotics including ampicillin, ceftriaxone, chloramphenicol, tetracycline (O167:H9), and chloramphenicol, sulfisoxazole, streptomycin, and trimethoprim/sulfamethoxazole (O114:H8).

Phenotypic and molecular characterization of *Salmonella* isolates

Overall, 29 different *Salmonella enterica* serovars were identified with 46 (70%) of 66 isolates belonging to subspecies Group I (Table 6). Two dominant serovars, Senftenberg and Typhimurium comprised 58% of the isolates from dog samples. In contrast, no predominant *Salmonella* serovars were identified among strains isolated from coyotes. A significant association was observed between serovar Senftenberg and the date of sample collection (P = 0.03), with 10 of the 14 S. Senftenberg isolations occurring in the month of November, primarily from the shelter in Mexico.

Table 2. Summary of population characteristics from domestic dogs sampled in a southwest United States and northern Mexico produce production region, November 3, 2010 through May 5, 2011 (N = 358).

Demographic	Number Sampled (%)
Shelter	
Arizona	124 (34.6)
California	100 (27.9)
Mexico	134 (37.4)
Reason Impounded	
Stray	297 (83.0)
Other[a]	44 (12.3)
Unknown	17 (4.7)
Age	
Puppy	58 (16.2)
Adult	279 (77.9)
Unknown	21 (5.9)
Sex	
Male	165 (46.1)
Female	186 (52.0)
Unknown	7 (2.0)
Breed	
Chihuahua/Mix	41 (11.5)
Labrador/Shepherd Mix	56 (15.6)
Pit Bull Terrier/Mix	60 (16.8)
Other	37 (10.3)
Unknown	140 (39.1)

[a]Includes all dogs born in shelter, relinquished by owner, confiscated from owner, or dogs being kept for quarantine or treatment purposes.

As shown in Figure 3, PFGE analysis of *Salmonella* isolates revealed 27 distinct pulsotypes. *S.* Senftenberg isolates (n = 8) from dog feces collected on two dates at the shelter in Mexico belonged to four different but closely related pulsotypes (PT-9, 10, 11, 12). In contrast, *S.* Senftenberg isolates (n = 6) from California shelter dogs collected on three sampling dates belonged to a single pulsotype (PT-13). *S.* Typhimurium PT-18 and PT-21 were the only shared pulsotypes among samples from different locations and species including 2 California coyote and an Arizona dog isolate (PT-18), and two dogs from Mexico and a California coyote (PT-21).

Antibiotic resistance testing revealed that 58 (87.9%) of the 66 *Salmonella* isolates, evenly distributed between dogs and coyotes, were susceptible to the antibiotics tested (Table 5). Of the 8 *Salmonella* isolates with resistance to at least one antibiotic, 4 were resistant to 2 or more drugs. An *S.* Newport isolate from coyote feces collected in California showed the most antibiotic resistant phenotype in this study including resistance to ampicillin, amoxicillin/clavulanic acid, cefriaxone, chloramphenicol, and trimethoprim/sulfamethoxazole.

Discussion

In this study, we show that stray dogs and free-roaming coyotes in the southwest desert leafy greens production region at the U.S.-Mexico border do not appear to be significant reservoirs of *E. coli* O157 and other STEC, but aEPEC and *Salmonella* were prevalent in fecal samples using the methods described herein.

E. coli detection and characterization

Previous studies of STEC occurrence in domestic dogs in the U.S. have focused on detection of Shiga toxin genes among animals with and without gastroenteritis. In one survey, a higher prevalence of *stx1* (3% and 15%) and *stx2* (36% and 23%) was found in diarrheic and non-diarrheic greyhounds, respectively [34]. In another study, there was no occurrence of *stx1* or *stx2* in 52 healthy Midwestern research colony dogs [35]. For logistical reasons, we were not able to sample unhealthy dogs with diarrhea in isolation wards at the U.S. shelters, and health status information was not available at the shelter in Mexico; thus, we collected only fresh, normally formed fecal samples from dogs. Outside the U.S., a survey in Japan revealed an extremely low prevalence of *E. coli* O157:H7 in dogs and cats, with only 1 of 614 (0.2%) fecal samples testing positive [36]. If we over-estimated *E. coli* O157:H7 prevalence at 1% in our sample size calculations, it is possible we would have needed to collect more than 300 fecal samples to detect *E. coli* O157:H7 in the shelter dog population. In another Japanese survey, a positive association between the presence of dogs or cats on beef cattle farms and prevalence of O157 in cattle was found [37]. This association, however, could be attributed to the hygiene practices of the farm (e.g., farms that allow dogs to run loose may have poorer hygiene and biosecurity practices than farms that do not), rather than evidence of colonization. In Argentina, *stx1* and *stx2* were detected in 3.7% and 4.2% of dog samples, respectively, and STEC was culture confirmed in 4% of the samples [38]. While our results suggest that STEC is rare in southwest dog populations, there are caveats to consider when comparing our results with prevalence surveys in

Table 3. Comparison of results from Shiga toxin-producing *E. coli* (STEC) O-group-specific (O103, O145, O157, O26) commercial latex agglutination screening tests and a multiplex PCR confirmatory test to detect 8 major STEC O-groups (O103, O111, O113, O121, O145, O157, O26, O45) used to serotype isolates cultured from fecal samples by selective enrichment and serogroup-specific (O103, O145, O157, O26) immunomagnetic separation (IMS).

Latex agglutination	No. isolates[a]	Multiplex STEC PCR								
		O103	O111	O113	O121	O145	O157	O26	O45	Other
O103	199	62	0	0	0	0	2	1	0	134[b]
O145	59	0	0	3	0	5	0	0	0	51[c]
O157	1	0	0	0	0	0	0	0	0	1
O26	19	0	0	0	0	0	0	5	0	14
Total	278	62	0	3	0	5	2	6	0	200[d]

[a]All isolates were stx1 and stx2 negative.
[b]Ten *eae+* isolates classified as O103 by latex agglutination screening and negative by multiplex STEC PCR belonged to serotypes O114 (n = 1), O123 (n = 2), O126 (n = 1), O128 (n = 1), O167 (n = 1), O64 (n = 1) and O- (n = 3) using O-antisera.
[c]Two *eae+* isolates classified as O145 by latex agglutination screening and negative by multiplex STEC PCR belonged to serotypes O153 (n = 1) and O- (n = 1) using O-antisera.
[d]Serotyping using O-antisera was not performed on isolates (n = 187) negative for virulence factors and not belonging to STEC O-groups identified by multiplex PCR.

other locations. For example, in Argentina, dog rectal swabs were taken and screened for Shiga toxin genes, which may have lead to improved detection compared with our pen floor samples.

Data on STEC occurrence in coyotes is even sparser in the literature. *E. coli* O157:H7 prevalence surveys in the Midwest and Washington State found the pathogen in 0/100 and 0/7 coyote fecal samples, respectively [39], [40]. We tested a comparable number of coyote samples (n = 103), which may have been too few to detect *E. coli* O157:H7. In contrast, during a longitudinal study of various domestic and wild animals in the central California coast, *E. coli* O157:H7 was isolated from 2 of 145 (1.4%) colonic-fecal samples from hunted coyotes [19]. Interestingly, 2 (1.9%) of 103 coyote fecal samples in our study contained Shiga toxin-negative *E. coli* O157:H+ isolates positive for *eaeA*. Bentancor and colleagues (2010) have characterized non-Shiga toxin producing *E. coli* O157 strains from dogs in Argentina and speculated that strains with the *eae* gene may represent a potential human health threat [41].

The finding of aEPEC shedding in dog feces was not unexpected as others have isolated this pathotype in domestic dogs with and without diarrhea [42], [43], [44]. Typical EPEC is the leading cause of infantile diarrhea in developing countries, while atypical EPEC is considered an emerging zoonoses [32]. Indeed, it has been speculated that aEPEC is genetically related to STEC and isolates may share O:H serotypes [32], [33], [40]. In our study, we found several previously described aEPEC serotypes including O-:H2 (dog), O128:H2 (coyote), O145:H34 (dog), and O26:H11 (dog) (Table 4). Serotype O26:H11 is well-described in calves, while serotype O128:H2 is known to occur in dog and rabbit populations [32], [33], [45]. Interestingly, Trabulsi et al. (2002) suggest that O26:H11 and O128:H2 are not aEPEC, but rather a heterogeneous serotype of STEC and aEPEC [32].

We also endeavored to determine if the *E. coli* O145:H28 human clinical outbreak strain linked to Romaine lettuce grown in southern Arizona in spring 2010 was present in our fecal samples, but found that all strains in the *E. coli* O145 serogroup isolated during this study were phenotypically and genetically unrelated to the 2010 outbreak strain (Figure 2). The public health significance of Shiga toxin-negative *E.coli* strains belonging to serogroup O145, as well as the other Shiga toxin-negative isolates belonging to "top 6" STEC-associated serogroups found in our study, is still somewhat unclear. Ultimately, additional molecular studies are needed to better understand the human virulence potential of these strains. The most clinically significant strain we isolated may be Shiga toxin-negative *E. coli* O26:H11 *eaeA+/hlyA+* cultured from dog feces at the Mexican shelter in February (Table 4). Loss and gain of *stx* genes by STEC O26 have been described, and aEPEC O26 may be ancestral to STEC O26 [46], [47]. We believe it is unlikely this strain or others lost Shiga toxin genes during passage through enrichment since *stx1/stx2* was not detected in TSB pre-enrichment broths by qPCR, provided the concentration of bacteria was within our level of detection. However, aEPEC strains could potentially re-acquire Shiga toxin genes and become STEC during passage through human or animal hosts [47].

We encountered several methodological challenges during this study related to culture and identification of STEC serogroups. For example, non-specific binding of STEC O-antigens (O103, O145, O157, O26) to IMS beads and method of capture may have caused isolation of multiple O-groups not originally targeted (Table 3). We also found discordant results between O-group specific latex agglutination and confirmatory tests, which has been described previously [48]. The utility of using a commercial latex agglutination screen to identify presumptive STEC from environ-

Table 4. Serotypes and virulence factors of *Escherichia coli* strains isolated from dog and coyote fecal samples, southwestern desert, November 3, 2010 through May 5, 2011.

Serotype[a]	Source			Virulence Factor		
	Coyote	Dog	Total	*stx1/stx2*	*eaeA*	*hlyA*
aEPEC						
O-: H2	1	0	1	-	+	-
O-: H8	0	1	1	-	+	-
O-: H25	0	1	1	-	+	-
O114: H8[b]	0	1	1	-	+	-
O123: H+	0	2	2	-	+	-
O126: H9	0	1	1	-	+	-
O128: H2	1	0	1	-	+	-
O145: H34	0	3	3	-	+	-
O153: H21/36	0	1	1	-	+	-
O157: H+	2	0	2	-	+	-
O167: H9	0	1	1	-	+	-
O26: H11	0	1	1	-	+	+
O26: H8	1	0	1	-	+	-
O64: H19	0	1	1	-	+	-
Subtotal	5	13	18			
OTHER						
O103: H-	1	0	1	-	-	-
O103: H2	1	4	5	-	-	-
O103: H7	0	7	7	-	-	-
O103: H9	0	2	2	-	-	-
O103: H16[b]	0	11	11	-	-	-
O103: H19	1	0	1	-	-	-
O103: H21	1	0	1	-	-	-
O103: H21/36	2	1	3	-	-	-
O103: H40/44	2	1	3	-	-	-
O103: H43	0	1	1	-	-	-
O103: H49[b]	2	26	28	-	-	-
O113: H4[b]	0	3	3	-	-	-
O145: H11[b]	0	2	2	-	-	-
O26: H2	0	1	1	-	-	-
O26: H32[b]	0	3	3	-	-	-
Subtotal	10	62	72			
TOTAL	15	76	91			

[a]O-, O antigen non-determinant.
[b]Twelve dog fecal samples contained two different serotypes including O103:H16/O113:H4 (n = 6); O26:H32/O103:H49 (n = 5); and O114:H8/O145:H11 (n = 1).

mental samples such as feces needs further assessment and comparison with other methods including the serology gold standard. Nevertheless, we concluded that screening for Shiga toxin genes by qPCR would be a more efficient approach to detect STEC colonies in the future, rather than screening for STEC O-groups.

Salmonella detection and characterization

Reported *Salmonella* prevalence in dogs varies greatly between studies, from 5% to over 70% [49], [50], [51], [52]. The predominance of two *Salmonella* serovars in the shelter dog population, but not in the sampled coyote population, is interesting. According to a USDA NVSL 2011 annual report, Typhimurium and Senftenberg, followed by Muenchen, Newport, and Javiana, were the most common serovars isolated nationally from dogs and cats during 2011 [53]. We speculate that the dominant serovars among shelter dog samples could indicate a common source of exposure to *Salmonella* in the environment. For example, contaminated dog food is increasingly recognized as a risk factor for *Salmonella* infections in dogs [54], [55], [56]. The findings could also be indicative of poor sanitation practices and overcrowding. Staffing at the facility in Mexico was noticeably limited and animals were typically kenneled in large groups of ten or more dogs, making the transmission of *Salmonella* between

PFGE-Xba I   PFGE-Xba I

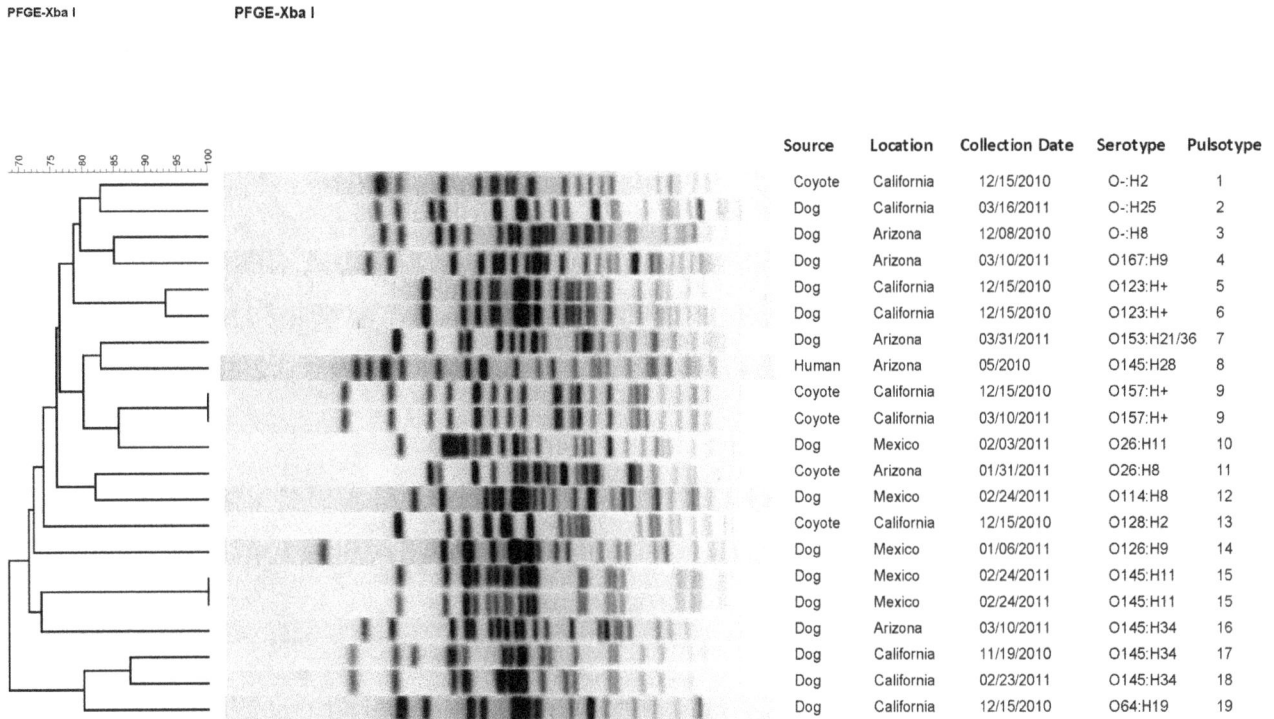

	Source	Location	Collection Date	Serotype	Pulsotype
	Coyote	California	12/15/2010	O-:H2	1
	Dog	California	03/16/2011	O-:H25	2
	Dog	Arizona	12/08/2010	O-:H8	3
	Dog	Arizona	03/10/2011	O167:H9	4
	Dog	California	12/15/2010	O123:H+	5
	Dog	California	12/15/2010	O123:H+	6
	Dog	Arizona	03/31/2011	O153:H21/36	7
	Human	Arizona	05/2010	O145:H28	8
	Coyote	California	12/15/2010	O157:H+	9
	Coyote	California	03/10/2011	O157:H+	9
	Dog	Mexico	02/03/2011	O26:H11	10
	Coyote	Arizona	01/31/2011	O26:H8	11
	Dog	Mexico	02/24/2011	O114:H8	12
	Coyote	California	12/15/2010	O128:H2	13
	Dog	Mexico	01/06/2011	O126:H9	14
	Dog	Mexico	02/24/2011	O145:H11	15
	Dog	Mexico	02/24/2011	O145:H11	15
	Dog	Arizona	03/10/2011	O145:H34	16
	Dog	California	11/19/2010	O145:H34	17
	Dog	California	02/23/2011	O145:H34	18
	Dog	California	12/15/2010	O64:H19	19

Figure 2. *Escherichia coli* (*XbaI* restriction) pulsotypes of 18 aEPEC isolates and 2 non-pathogenicirulent *E. coli* O145:H11 isolates from dog and coyote fecal samples in the southwest desert produce growing areas of Arizona, California, and northern Mexico, November 3, 2010 through May 5, 2011. A human clinical *E. coli* O145:H28 outbreak strain associated with a Romaine lettuce-related outbreak traced to Arizona in May 2010 is also shown.

individuals more likely to occur. We controlled for this variable by limiting fecal collections to fresh feces from individual animals impounded within the last 24 hours. Often, however, it was noted that most of the individual animals sampled on a given day from the Mexican shelter had been collected from the same location and were likely living in groups together. Thus, dogs shedding genetically related strains may have shared common exposures prior to or during impound.

The fact that nearly one in three (32%) coyote fecal samples collected near leafy green fields were positive for *Salmonella* was surprising. In contrast, a survey from the central California coast found that the organism was recovered from only 3 (7.5%) of 40 coyote colonic fecal sample enrichment broths [11]. Variations in sampling and laboratory culture methods between the two studies could, in part, explain these differences. It is also possible that individual coyotes were re-sampled over the course of this study. However, given the wide geographic and temporal distribution of sampling locations in combination with the high diversity of serovars and PFGE subtypes (Figure 3), we do not believe that repeat sampling of individual coyotes contributed significantly to our overall prevalence. Even in the event of re-sampling, the apparent high prevalence of *Salmonella* in coyote fecal material found in or near the production area is of importance to growers given the potential to contaminate the plants directly, or indirectly via agriculture water sources and farm equipment, Additional studies using trapping or hunting techniques are needed to determine actual prevalence of foodborne pathogens in the southwest desert coyote population.

There was some seasonality observed in *Salmonella* recovery from both dog and coyote samples including a significantly higher prevalence in November compared with samples collected in late winter and spring months. More long-term studies, however, are indicated to reveal any true seasonality or temporal patterns of *Salmonella* occurrence. The relatively high prevalence of *Salmonella* shedding in dogs and coyotes may be due in part to the hunting/scavenging behaviors of canids in the region. In desert regions where prey and water resources are limited, both coyotes and stray dogs opportunistically forage on fresh and rotten nutritional sources, including garbage and other refuse, vegetable and fruit matter, and the meat of dead animals. Such scavenging behaviors may put dogs and coyotes at a higher risk of exposure to *Salmonella*. High prevalence of *Salmonella* has previously been found in other scavenging species of the southwest, such as turkey vultures (*Cathartes aura*) [57]. Additionally, these animals may be obtaining water from anthropogenic sources, such as irrigation ditches, sediment basins, and camp-sites in the absence of natural water sources. Water samples, from both static sources and flowing streams, rivers, and creeks, often yield high percentages of *Salmonella* positive samples [10], [58], [59].

We found more antibiotic resistance among aEPEC isolates compared with *Salmonella* isolates (Table 5). Four (22.2%) of 18 aEPEC and 4 (6.1%) of 66 *Salmonella* isolates were resistant to two or more antibiotic classes; two dog aEPEC isolates (O114:H8 and O167:H9), a dog *S.* Senftenberg, and a coyote *S.* Newport displayed resistance to 3 or more antibiotic classes. Of note, Newport and Senftenberg have been identified as emerging multi-drug-resistant serovars worldwide [60], [61]. In a wildlife study conducted during the same time period in the central California coast, a majority of *Salmonella enterica* subspecies Group IIIa and IIIb isolates from wild-caught amphibians and reptiles captured near produce fields were resistant to at least one antibiotic [12]. It appears that antibiotic resistance among *Salmonella* isolates is less

Table 5. Antimicrobial resistance patterns among 18 atypical enteropathogenic *Escherichia coli* (aEPEC) and 66 *Salmonella enterica* isolates from coyote and dog fecal samples, southwestern desert, November 3, 2010 through May 5, 2011.

Drug resistance pattern[a]	Source (No. of isolates)			Serotype or antigenic formula[b]
aEPEC	**Coyote (n = 5)**	**Dog (n = 13)**	**Total (% of all isolates)**	
Pansusceptible	2	4	6 (33.3)	O-:H2
				O-:H25
				O126:H9
				O145:H34
				O26:H8
FIS	2	6	8 (44.4)	O-:H8
				O123:H+
				O128:H2
				O145:H34
				O153:H21/36
				O157:H+
				O26:H11
				O64:H19
FIS-STR	1	0	1 (5.6)	O157:H+
FIS-TET	0	1	1 (5.6)	O123:H+
AMP-AXO-CHL-TET	0	1	1 (5.6)	O167:H9
CHL-FIS-STR-SXT	0	1	1 (5.6)	0114:H8
Salmonella	**Coyote (n = 33)**	**Dog (n = 33)**	**Total (% of all isolates)**	
Pansusceptible	29	29	58 (87.9)	Aqua
				Barranquilla
				Drac
				Duisburg Enteritidis
				Javiana
				Livingstone; Montevideo Muenchen
				Newport
				Oranienburg Sandiego
				Senftenberg Typhimurium Typhimurium var 5-
				II 47:b:1,5
				III 17:z29:-
				III 62:z36:- III_40:z4, z32:- III_48:g, z51:-III_48:i:z; IV
				44:z36:-
				IV 47:l,v:e,n,x
AMP	0	1	1 (1.5)	Enteritidis
STR	1	1	2 (3.0)	Typhimurium
				IV Rough O:autoagglutinate
XNL	1	0	1 (1.5)	Sandiego
AXO-TET	1	1	2 (3.0)	Mbandaka IV 44:z36:-
AMP-STR-SXT	0	1	1 (1.5)	Senftenberg
AMP-AUG2-AXO-CHL-SXT	1	0	1 (1.5)	Newport

[a]AMP, ampicillin; AUG2, amoxicillin/clavulanic acid; AXO, ceftriaxone; AZI, azithromycin; CHL, chloramphenicol; FIS, sulfisoxazole; FOX, cefoxitin; KAN, kanamycin; STR, streptomycin; SXT, trimethoprim/sulfamethoxazole; TET, tetracycline; XNL, ceftiofur.
[b]O-, O antigen non-determinant.

Table 6. Subspecies and serovars of *Salmonella* isolated from dog and coyote fecal samples, southwestern desert, November 3, 2010 through May 5, 2011.

Subspecies (Group)	Serovar or antigenic formula	Source (No. isolates)		
		Coyote	Dog	Total
Enterica (I)	Aqua	1	0	1
	Barranquilla	1	0	1
	Derby	0	1	1
	Drac	1	0	1
	Duisburg	1	0	1
	Ealing	0	1	1
	Enteritidis	0	2	2
	Javiana	2	0	2
	Livingstone	0	1	1
	Mbandaka	0	1	1
	Montevideo	1	0	1
	Muenchen	2	0	2
	Newport	3	0	3
	Oranienburg	0	1	1
	Sandiego	2	0	2
	Senftenberg	0	14	14
	Typhimurium	5	5	10
	Typhimurium var. 5-	0	1	1
Salmae (II)	47:b:1,5	1	0	1
Arizonae (IIIa)	17:z29:-	3	0	3
	35:z29:-	0	1	1
	62:z36:-	1	0	1
	40:z4, z32:-	3	0	3
	48:g, z51:-	0	3	3
Diarizonae (IIIb)	48:i:z	2	0	2
	50:r:z	1	0	1
Houtenae (IV)	44:z36:-	1	1	2
	47:l, v:e,n,x	2	0	2
Unknown	Rough O:autoagglutinate	0	1	1
Total		33	33	66

prominent in canid populations tested in the southwest desert compared with the cold-blooded vertebrates surveyed in coastal California.

Prevention and Control Recommendations

The produce industry currently addresses foodborne pathogen hazards from domestic and wild animal sources through adherence to best practices established by the Arizona and California Leafy Green Marketing Agreements [62], [63]. For example, fecal material in the production area is removed, and a minimal 5-foot radius no-harvest buffer zone (Figure 1D) is used to prevent contaminated plants from entering the food supply. Repeated intrusions into produce fields by dogs and coyotes can be managed by use of fencing, control of strays, and depredation (coyotes). In our study region, animal control officers on both sides of the border work closely with the produce growers to assist with stray animal problems in fields. In the Imperial-Yuma region, loose dogs can be particularly problematic in the fall-winter season—which is also the leafy greens growing season—when the area becomes a popular tourist ("snowbird") destination for recreational vehicle enthusiasts who often travel with their dogs. In Mexico, stray dog control is more challenging because of limited resources, large numbers of un-owned dogs, and cultural barriers to dog population control.

It is worth noting that although intrusions by stray dogs may represent a food safety risk for fresh produce, trained working dogs if used properly can actually be an asset and should not be discouraged. For example, dogs have been used to help disperse and deter large flocks of nuisance birds, and scent detection dogs have been used experimentally to identify in-field fecal contamination of raw produce [64].

Conclusions

In summary, results from this study will assist the produce industry by providing baseline information on the occurrence of zoonotic enteric pathogens found in fecal material from common

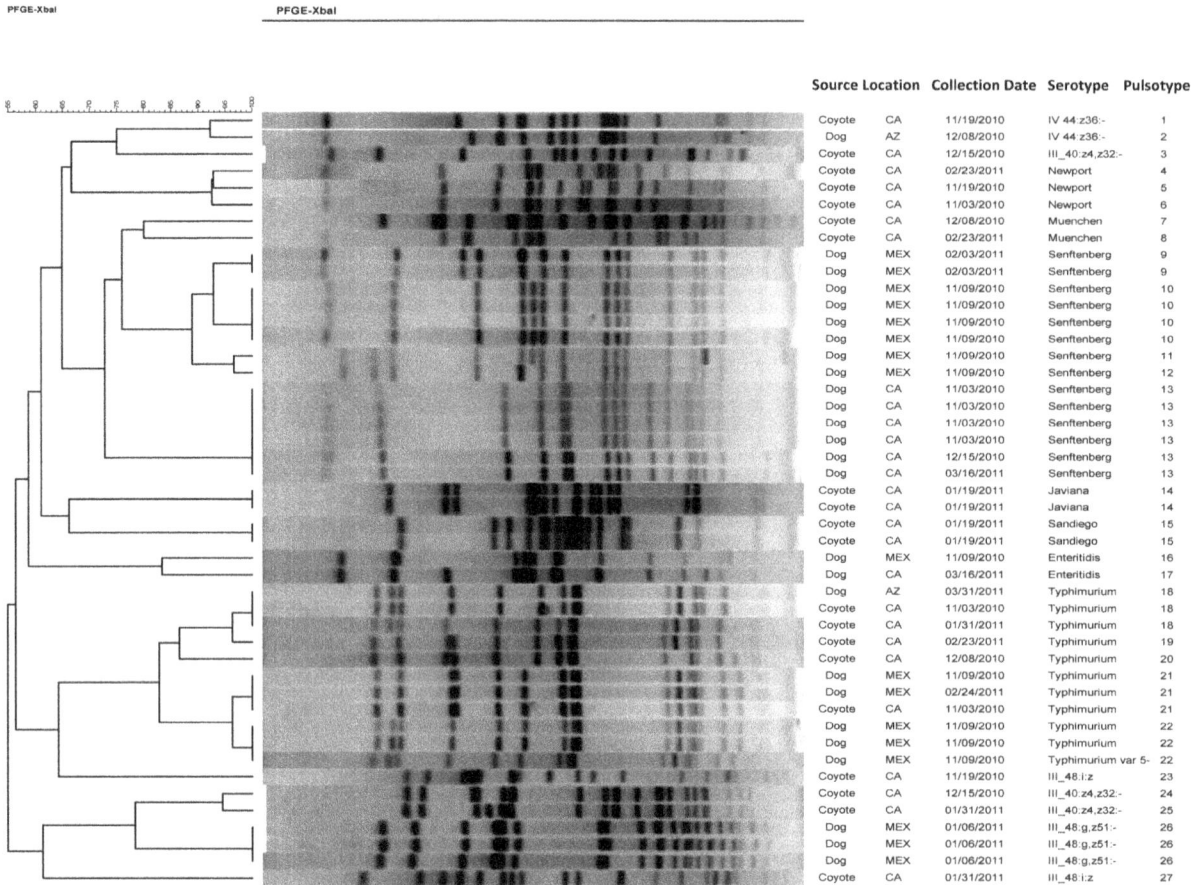

Figure 3. *Salmonella* Pulsotypes (*XbaI* restriction) isolated from dog and coyote fecal samples collected in the southwest desert produce growing areas of Arizona, California, and northern Mexico, November 3, 2010 through May 5, 2011.

domestic and wild canid populations in the desert southwest produce production region. The findings underscore the importance of good agriculture practices for leafy greens and other produce, especially those relating to animal intrusions and pre- and post-harvest environmental assessments. Follow-up surveys are warranted to determine pathogenic *E. coli* and *Salmonella* prevalence in other potential domestic and wildlife reservoirs in this region, and comparison with other possible environmental sources of microbial contamination (e.g., canals, irrigation water, and soil amendments).

Acknowledgments

We are deeply grateful to the growers for their assistance with sample collections and access to study locations. We thank Fatima Corona and Ed Morales for providing information about the leafy greens industry in the southwest desert and on-site coordination. We are also indebted to the local governments and animal shelters that provided us with access to their facilities and data on impounded animals. For technical assistance, we thank Diana Carychao from the USDA ARS Produce Safety and Microbiology Research Unit, Beth Roberts from the Pennsylvania State University E. coli Reference Center, and Peiman Aminabadi at UC Davis. We also thank Robert E. Mandrell for providing the *E. coli* O145:H28 clinical outbreak strain, and Maria D. Pereira for providing the *E. coli* O103, O145, and O26 isolates used as positive controls.

Author Contributions

Conceived and designed the experiments: MTJ AFH YB AT TN. Performed the experiments: MTJ AFH YB AT TN. Analyzed the data: MTJ AFH YB AT TN. Wrote the paper: MTJ AFH.

References

1. Lynch MF, Tauxe RV, Hedberg CW (2009) The growing burden of foodborne outbreaks due to contaminated fresh produce: risks and opportunities. Epidemiol Infect 137: 307–315.
2. Berger CN, Sodha SV, Shaw RK, Griffin PM, Pink D, et al. (2010) Fresh fruit and vegetables as vehicles for the transmission of human pathogens. Environ Microbiol 12: 2385–2397.
3. Lynch M, Painter J, Woodruff R, Braden C (2006) Surveillance for foodborne-disease outbreaks—United States, 1998–2002. MMWR Surveill Summ 55: 1–34.
4. Sivapalasingam S, Friedman CR, Cohen L, Tauxe RV (2004) Fresh produce: a growing cause of outbreaks of foodborne illness in the United States, 1973 through 1997. J Food Prot 67: 2342–2353.
5. Painter JA, Hoekstra RM, Ayers T, Tauxe RV, Braden CR, et al. (2013) Attribution of foodborne illnesses, hospitalizations, and deaths to food commodities by using outbreak data, United States, 1998–2008. Emerg Infect Dis 19: 407–415.
6. Mandrell RE (2011) Tracing pathogens in fruit and vegetable production chains. In: Brul S, Fratamico PM, McMeekin TA, editors. Tracing pathogens in the food chain.Cambridge: Woodhead Publishing Ltd. pp. 548–595.
7. Jay-Russell MT (2013) What is the risk from wild animals in food-borne pathogen contamination of plants? CAB Reviews 8: 040.
8. Boriss H, Brunke H. (2012) Produce Profile. Agricultural Issues Center, University of California. Available: http://www.agmrc.org/commodities__ products/vegetables/lettuce-profile/. Accessed 18 October 2014.

9. Jay MT, Cooley M, Carychao D, Wiscomb GW, Sweitzer RA, et al. (2007) Escherichia coli O157:H7 in feral swine near spinach fields and cattle, central California coast. Emerg Infect Dis 13: 1908–1911.

10. Cooley M, Carychao D, Crawford-Miksza L, Jay MT, Myers C, et al. (2007) Incidence and tracking of Escherichia coli O157:H7 in a major produce production region in California. PLoS ONE 2: e1159.

11. Gorski L, Parker CT, Liang A, Cooley MB, Jay-Russell MT, et al. (2011) Prevalence, distribution, and diversity of Salmonella enterica in a major produce region of California. Appl Environ Microbiol 77: 2734–2748.

12. Gorski L, Jay-Russell MT, Liang AS, Walker S, Bengson Y, et al. (2013) Diversity of pulsed-field gel electrophoresis pulsotypes, serovars and antibiotic resistance among Salmonella isolates from wild amphibians and reptiles in the California central coast. Foodborne Pathog Dis 10: 540–548.

13. Kilonzo C, Li X, Vivas EJ, Jay-Russell MT, Fernandez KL, et al. (2013) Fecal shedding of zoonotic food-borne pathogens by wild rodents in a major agricultural region of the central California coast. Appl Environ Microbiol 79: 6337–6344.

14. Strawn LK, Grohn YT, Warchocki S, Worobo RW, Bihn EA, et al. (2013) Risk factors associated with Salmonella and Listeria monocytogenes contamination of produce fields. Appl Environ Microbiol 79: 7618–7627.

15. Taylor EV, Nguyen TA, Machesky KD, Koch E, Sotir MJ, et al. (2013) Multistate outbreak of Escherichia coli O145 infections associated with Romaine lettuce consumption, 2010. J Food Prot 76: 939–944.

16. Cooper KK, Mandrell RE, Louie JW, Korlach J, Clark TA, et al. (2014) Comparative genomics of enterohemorrhagic Escherichia coli O145:H28 demonstrates a common evolutionary lineage with Escherichia coli O157:H7. BMC Genomics 15: 17.

17. Crawford W, Baloch M, Gerrity K (2010) Environmental assessment: non-O157 Shiga-toxin producing E. coli (STEC). U.S. Food and Drug Administration (FDA). Available: http://www.fda.gov/downloads/Food/RecallsOutbreaksEmergencies/UCM235923.pdf Accessed: 18 October 2014.

18. Kanki M, Seto K, Sakata J, Harada T, Kumeda Y (2009) Simultaneous enrichment of shiga toxin-producing Escherichia coli O157 and O26 and Salmonella in food samples using universal preenrichment broth. J Food Prot 72: 2065–2070.

19. Cooley MB, Jay-Russell M, Atwill ER, Carychao D, Nguyen K, et al. (2013) Development of a robust method for isolation of Shiga toxin-positive Escherichia coli (STEC) from fecal, plant, soil and water samples from a leafy greens production region in California. PLoS ONE 8: e65716.

20. DebRoy C, Roberts E, Fratamico PM (2011) Detection of O antigens in Escherichia coli. Anim Health Res Rev 12: 169–185.

21. Machado J, Grimont F, Grimont PA (2000) Identification of Escherichia coli flagellar types by restriction of the amplified fliC gene. Res Microbiol 151: 535–546.

22. DebRoy C, Maddox C (2001) Assessing virulence of Escherichia coli isolates of veterinary significance. Anim Health Res Rev 1: 129–140.

23. Barkocy-Gallagher GA, Berry ED, Rivera-Betancourt M, Arthur TM, Nou X, et al. (2002). Development of methods for the recovery of Escherichia coli O157:H7 and Salmonella from beef carcass sponge samples and bovine fecal and hide samples. J Food Prot 65: 1527–1534.

24. Grimot PAD, Weill FX (2007) Antigenic formulae of the Salmonella serovars, 9th revision. World Health Organization Collaborating Centre for Reference and Research on Salmonella. Paris: Pasteur Institute.

25. Ribot EM, Fair MA, Gautom R, Cameron DN, Hunter SB, et al. (2006) Standardization of pulsed-field gel electrophoresis protocols for the subtyping of Escherichia coli O157:H7, Salmonella, and Shigella for PulseNet. Foodborne Path Dis 3: 59–67.

26. Tenover FC, Arbeit RD, Goering RV, Mickelsen PA, Murray BE, et al. (1995) Interpreting chromosomal DNA restriction patterns produced by pulsed-field gel electrophoresis: criteria for bacterial strain typing. J Clin Microbiol 33: 2233–2239.

27. Barrett TJ, Gerner-Smidt P, Swaminathan B (2006) Interpretation of pulsed-field gel electrophoresis patterns in foodborne disease investigations and surveillance. Foodborne Pathog Dis 3: 20–31.

28. Anonymous (2008) Performance standards for antimicrobial disk and dilution susceptibility tests for bacteria isolated from animals; Approved Standard—Third Edition. Wayne, PA: Clinical and Laboratory Standards Institute CLSI document M31–A3.

29. Anonymous (2008) Performance standards for antimicrobial susceptibility testing; Twentieth Informational Supplement. Wayne, PA: Clinical and Laboratory Standards Institute CLSI document M100–S20.

30. Anonymous (2012) National Antimicrobial Resistance Monitoring System– Enteric Bacteria (NARMS): 2010 Executive Report. National Antimicrobial Resistance Monitoring System– Enteric Bacteria (NARMS): 2010 Executive Report. U.S. Food and Drug Administration. Available: http://www.fda.gov/AnimalVeterinary/SafetyHealth/AntimicrobialResistance/NationalAntimicrobialResistanceMonitoringSystem/ucm312356.htm. Accessed 18 October 2014.

31. Sjölund-Karlsson M, Joyce K, Blickenstaff K, Ball T, Haro J, et al. (2011) Antimicrobial susceptibility to azithromycin among Salmonella enterica isolates from the United States. Antimicrob Agents Chemother 55: 3985–3989.

32. Trabulsi LR, Keller R, Tardelli Gomes TA (2002) Typical and atypical enteropathogenic Escherichia coli. Emerg Infect Dis 8: 508–513.

33. Bugarel M, Martin A, Fach P, Beutin L (2011) Virulence gene profiling of enterohemorrhagic (EHEC) and enteropathogenic (EPEC) Escherichia coli strains: a basis for molecular risk assessment of typical and atypical EPEC strains. BMC Microbiol 11: 142.

34. Staats JJ, Chengappa MM, DeBey MC, Fickbohm B, Oberst RD (2003) Detection of Escherichia coli Shiga toxin (stx) and enterotoxin (estA and elt) genes in fecal samples from non-diarrheic and diarrheic greyhounds. Vet Microbiol 94: 303–312.

35. Holland RE, Walker RD, Sriranganathan N, Wilson RA, Ruhl DC (1999) Characterization of Escherichia coli isolated from healthy dogs. Vet Microbiol 70: 261–268.

36. Kataoka Y, Irie Y, Sawada T, Nakazawa M (2010) A 3-year epidemiological surveillance of Escherichia coli O157:H7 in dogs and cats in Japan. J Vet Med Sci 72: 791–794.

37. Sasaki Y, Tsujiyama Y, Kusukawa M, Murakami M, Katayama S, et al. (2011) Prevalence and characterization of Shiga toxin-producing Escherichia coli O157 and O26 in beef farms. Vet Microbiol 150: 140–145.

38. Bentancor A, Rumi MV, Gentilini MV, Sardoy C, Irino K, et al. (2007) Shiga toxin-producing and attaching and effacing Escherichia coli in cats and dogs in a high hemolytic uremic syndrome incidence region in Argentina. FEMS Microbiol Lett 267: 251–256.

39. Renter DG, Sargeant JM, Oberst RD, Samadpour M (2003) Diversity, frequency, and persistence of Escherichia coli O157 strains from range cattle environments. Appl Environ Microbiol 69: 542–547.

40. Rice DH, Hancock DD, Besser TE (2003) Faecal culture of wild animals for Escherichia coli O157:H7. Vet Rec 152: 82–83.

41. Bentancor A, Vilte DA, Rumi MV, Carbonari CC, Chinen I, et al. (2010) Characterization of non-Shiga-toxin-producing Escherichia coli O157 strains isolated from dogs. Rev Argent Microbiol 42: 46–48.

42. Goffaux F, China B, Janssen L, Mainil J (2000) Genotypic characterization of enteropathogenic Escherichia coli (EPEC) isolated in Belgium from dogs and cats. Res Microbiol 151: 865–871.

43. Marks SL, Rankin SC, Byrne BA, Weese JS (2011) Enteropathogenic bacteria in dogs and cats: diagnosis, epidemiology, treatment, and control. J Vet Intern Med 25: 1195–1208.

44. Puno-Sarmiento J, Medeiros L, Chiconi C, Martins F, Pelayo J, et al. (2013) Detection of diarrheagenic Escherichia coli strains isolated from dogs and cats in Brazil. Vet Microbiol 166: 676–680.

45. Moura RA, Sircili MP, Leomil L, Matte MH, Trabulsi LR, et al. (2009) Clonal relationship among atypical enteropathogenic Escherichia coli strains isolated from different animal species and humans. Appl Environ Microbiol 75: 7399–7408.

46. Bielaszewska M, Middendorf B, Kock R, Friedrich AW, Fruth A, et al. (2008) Shiga toxin-negative attaching and effacing Escherichia coli: distinct clinical associations with bacterial phylogeny and virulence traits and inferred in-host pathogen evolution. Clin Infect Dis 47: 208–217.

47. Bielaszewska M, Prager R, Kock R, Mellmann A, Zhang W, et al. (2007) Shiga toxin gene loss and transfer in vitro and in vivo during enterohemorrhagic Escherichia coli O26 infection in humans. Appl Environ Microbiol 73: 3144–3150.

48. Kimura R, Mandrell RE, Galland JC, Hyatt D, Riley LW (2000) Restriction-site specific PCR as a rapid test to detect enterohemorrhagic Escherichia coli O157:H7 strains in environmental samples. Appl Environ Microbiol 66: 2513–2519.

49. Joffe DJ, Schlesinger DP (2002) Preliminary assessment of the risk of Salmonella infection in dogs fed raw chicken diets. Can Vet J, 43: 441–442.

50. McKenzie E, Riehl J, Banse H, Kass PH, Nelson S Jr, et al. (2010) Prevalence of diarrhea and enteropathogens in racing sled dogs. J Vet Intern Med 24: 97–103.

51. Lenz J, Joffe D, Kauffman M, Zhang Y, LeJeune J (2009) Perceptions, practices, and consequences associated with foodborne pathogens and the feeding of raw meat to dogs. Can Vet J 50: 637–643.

52. Leonard EK, Pearl DL, Finley RL, Janecko N, Peregrine AS, et al. (2011) Evaluation of pet-related management factors and the risk of Salmonella spp. carriage in pet dogs from volunteer households in Ontario (2005–2006). Zoonoses Public Health 58: 140–149.

53. Lantz K (2012) Salmonella serotypes isolated from animals in the United States: January 1 – December 31, 2011. Report of the Committee on Salmonella, United States Animal Health Association. Available: http://www.usaha.org/Portals/6/Reports/2012/report-sal-2012.pdf. Accessed 18 October 2014.

54. Behravesh CB, Ferraro A, Deasy M 3rd, Dato V, Moll M, et al. (2010) Human Salmonella infections linked to contaminated dry dog and cat food, 2006–2008. Pediatrics 126: 477–483.

55. Anonymous (2008) Multistate outbreak of human Salmonella infections caused by contaminated dry dog food—United States, 2006-2007. MMWR Morbid MortalWeekly Rep 57: 521–524.

56. Selmi M, Stefanelli S, Bilei S, Tolli R, Bertolotti L, et al. (2011) Contaminated commercial dehydrated food as source of multiple Salmonella serotypes outbreak in a municipal kennel in Tuscany Vet Ital 47: 183–190.

57. Winsor DK, Bloebaum AP, Mathewson JJ (1981) Gram-negative, aerobic, enteric pathogens among intestinal microflora of wild turkey vultures (Cathartes aura) in west central Texas. Appl Environ Micrbiol 42: 1123–1124.

58. Economou V, Gousia P, Kansouzidou A, Sakkas H, Karanis P, et al. (2013) Prevalence, antimicrobial resistance and relation to indicator and pathogenic microorganisms of Salmonella enterica isolated from surface waters within an agricultural landscape. Int J Hyg Environ Health 216: 435–444.

59. Wilkes G, Edge TA, Gannon VP, Jokinen C, Lyautey E, et al. (2011) Associations among pathogenic bacteria, parasites, and environmental and land use factors in multiple mixed-use watersheds. Water Res 45: 5807–5825.

60. Whichard JM, Gay K, Stevenson JE, Joyce KJ, Cooper KL, et al. (2007) Human *Salmonella* and concurrent decreased susceptibility to quinolones and extended-spectrum cephalosporins. Emerg Infect Dis 13: 1681–1688.

61. Anonymous (2005) Drug-resistant *Salmonella*. World Health Organization. Available: http://www.who.int/mediacentre/factsheets/fs139/en/ Accessed 18 October 2014.

62. Anonymous (2013) Arizona Leafy Green Marketing Agreement. Available: http://www.arizonaleafygreens.org/. Accessed 18 October 2014.

63. Anonymous (2013) California Leafy Green Marketing Agreement. Available: http://www.caleafygreens.ca.gov/. Accessed 18 October 2014.

64. Partyka ML, Bond RF, Farrar J, Falco A, Cassens B, et al. (2014) Quantifying the sensitivity of scent detection dogs to identify fecal contamination on raw produce. J Food Prot 77: 6–14.

13

Niche-Partitioning of Edaphic Microbial Communities in the Namib Desert Gravel Plain Fairy Circles

Jean-Baptiste Ramond[1], Annelize Pienaar[1], Alacia Armstrong[1], Mary Seely[2], Don A. Cowan[1]*

1 Center for Microbial Ecology and Genomics (CMEG), Genomic Research Institute, University of Pretoria, Pretoria, South Africa, **2** Gobabeb Research and Training Center (GRTC), Walvis Bay, Namibia

Abstract

Endemic to the Namib Desert, Fairy Circles (FCs) are vegetation-free circular patterns surrounded and delineated by grass species. Since first reported the 1970's, many theories have been proposed to explain their appearance, but none provide a fully satisfactory explanation of their origin(s) and/or causative agent(s). In this study, we have evaluated an early hypothesis stating that edaphic microorganisms could be involved in their formation and/or maintenance. Surface soils (0–5cm) from three different zones (FC center, FC margin and external, grass-covered soils) of five independent FCs were collected in April 2013 in the Namib Desert gravel plains. T-RFLP fingerprinting of the bacterial (16S rRNA gene) and fungal (ITS region) communities, in parallel with two-way crossed ANOSIM, showed that FC communities were significantly different to those of external control vegetated soil and that each FC was also characterized by significantly different communities. Intra-FC communities (margin and centre) presented higher variability than the controls. Together, these results provide clear evidence that edaphic microorganisms are involved in the Namib Desert FC phenomenon. However, we are, as yet, unable to confirm whether bacteria and/or fungi communities are responsible for the appearance and development of FCs or are a general consequence of the presence of the grass-free circles.

Editor: Melanie R. Mormile, Missouri University of Science and Technology, United States of America

Funding: This research was funded by the National Research Foundation (http://www.nrf.ac.za/) of South Africa: Grant N °81779, Microbial Ecology of the Namib Desert. The funders had no role in study design, data collection and analysis, decision to publish, or preparation of the manuscript.

Competing Interests: The authors have declared that no competing interests exist.

* Email: don.cowan@up.ac.za

Introduction

The Namib Desert is unique in harboring the enigmatic Fairy Circles (FC) (or Fairy Rings). They occur and have been studied predominantly in the sand dune environment of the eastern Namib but also occur on the gravel plains [1,2]. FCs are circular vegetation-free patterns (generally between 2 to 12 m in diameter), surrounded by grass species of the genus *Stipagrotis* (Figure 1A). Their distribution is restricted to a region in the Pro-Namib zone (60 to 120 km inland), from southern Angola to northern South Africa [2]. The Namib Desert FCs have been described as 'living organisms', as they appear (birth), enlarge (growth) and ultimately disappear (death), with an estimated life span of around 60 years [3].

Since first being reported in the 1970's, many studies and hypotheses have attempted to explain the origins and lifestyles of FCs (most reviewed in [2]): these hypotheses include local radioactivity [4], insect (termite/ant) activities [5–11], the release of volatile chemicals (e.g. allelopathic compounds by dead *Euphorbia damarana* plants [12], semi-volatile products from termite-related activities [8,13] or hydrocarbon-liked compounds of geochemical origins [14]), or plant spatial self-organization [15,16]. The allelopathic compound and radioactivity hypotheses have already been refuted in field studies [2]. The 'abiotic gas leakage' and 'grass harvesting social insect (ant/termite)' hypotheses were contested, based on mathematical modelling (e.g. remote

sensing, spatial pattern analysis and vegetation modelling) [16]. The results obtained by Getzin and colleagues [16] suggested that FC distribution was indicative of self-organized models, i.e. in agreement with the 'spatial self-organization' hypothesis [15]. The authors nevertheless noted that more factors should be implemented (e.g. plant root system and soil moisture data) to fully validate the hypothesis. The current trend to include microbiological data when modeling edaphic ecosystems [17,18] may be even more critical in arid edaphic environments where most ecosystem processes are microbially-mediated [19]. It must also be noted that gravel plain FCs are largely unstudied ([1] and [3] only, to our knowledge) compared to the eastern Namib FCs. The patchiness of gravel plain FCs (as shown in Figure 1A) is not consistent with the self-organization pattern hypothesis [15,16].

Despite the fact that the potential role of microorganisms as contributing agents of FC development has not yet been tested experimentally, it has been previously discussed because of circumstantial support for the involvement of microbial processes. Significantly higher alkene emissions have been observed from FC soils than from the vegetated surroundings soils [14]; and alkene evolution has been shown to be indicative of anaerobic microbially-mediated alkane reduction [20]. Using Namib FC soils, bioassay experiments suggested phytotoxicity [2] and germination studies suggested the involvement of a biologically active abiosis factor leading to plant decay [8]. An '*in situ*' pot trial also

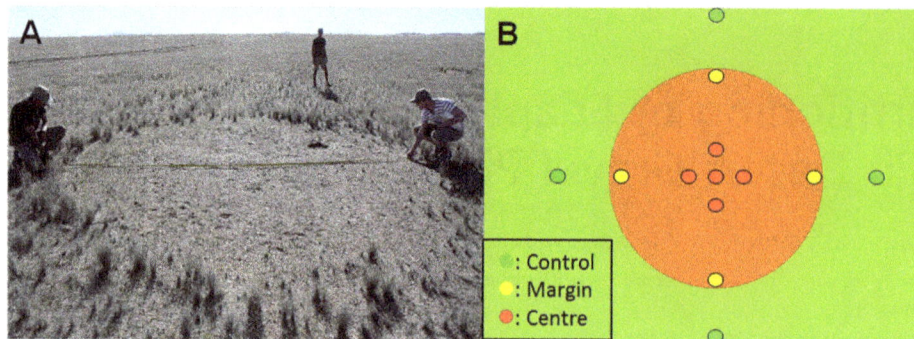

Figure 1. Fairy Circle sampling. Photograph of a Fairy Circle in the gravel plains of the Namib Desert (A) with the schematic of the sampling strategy employed (B).

suggested the possible involvement of either a semi-volatile compound or a (microbial?) toxin in plant-growth inhibition [13].

The current study was initiated with the working hypothesis that edaphic microorganisms could be significant agents in the formation and/or maintenance of Namib Desert FCs. The experimental basis of this study was a comprehensive sampling of surface soils from within FCs (centre and margin) and from external vegetated (controls) sites. Unlike most previous FC studies, which have focused in FC sites situated in the dune sands to the east of the Namib Sand Sea, we selected FCs in the Namib Desert gravel plains. Our analytical approach included T-RFLP fingerprinting of bacterial (16S rRNA gene) and fungal (ITS region) phylogenetic markers and soil chemical analyses (10 different variables), using multivariate statistical analyses to cross-correlate the experimental data.

Materials and Methods

Fairy Circle soil sampling and storage

Namib Desert Fairy Circle soils were sampled under the permits N° P0063476, permit for the importation of controlled goods from the Department of Agriculture, Forestry and Fisheries of the Republic of South Africa, and N° ES 29529, permit for single consignment export of minerals from the Ministry of Mines and Energy of the Republic of Namibia.

Surface soils (0–5 cm; approx. 200g per sample) from five independent Fairy Circles were collected in the gravel plains of Namibia (−23°31′37.09″S/15°11.15″E) near the Gobabeb Research and Training Center (http://www.gobabebtrc.org/) in April 2013, following a strategy shown schematically in Figure 1. A total of thirteen samples per FC were collected: four control soils from outside the FCs and nine within each FCs; 4 at the circle margin and 5 in the center. The samples were stored at 4°C during transport to the CMEG Laboratory (University of Pretoria, South Africa). One gram subsamples were stored at −80°C for subsequent molecular analyses and the residual soil at 4°C for chemical analyses.

Soil Chemistry analyses

Soil chemistry analyses were conducted at the Soil Science Laboratory of the University of Pretoria (South Africa) according to standard quality control procedures [21]. All solutions and reagents used were supplied by Merck Chemicals (South Africa) and, apart for pH measurements, soil samples were sieved (2 mm) prior to analysis.

The slurry technique was used to measure pH recorded with a pH meter (Crison basic 20, Barcelona, Spain) by mixing 4 g of soil with 10 ml of deionized water and allowing it to settle for 1h. The Walkey-Black method [22] was used to determine soil total organic carbon (C) content with minor modifications. 10 mL of 1M potassium dichromate solution was added to 2 g of soil. 10 mL of sulfuric acid (96%) was then added, and the mixture was left to cool at room temperature for 30 min. 150 mL deionized water and 10 mL concentrated (96%) orthophosphoric acid was then added to the mixture which was left to cool at room temperature for 30 min. 1 mL phenylalanine was added and titration was performed by adding iron(II) ammonium sulphate solution until the reaction endpoint was reached; i.e. when the solution changed from purple to green.

Exchangeable ammonium (NH_4^+) and nitrate (NO_3^-) soil content was determined by steam distillation as described [23] with minor modifications. 5 g of soil was mixed with 50 mL of a 2M KCl solution and shaken for 30 min at 220 rpm. Samples were then left to settle for 1 min and the supernatant was filtered through a 110 mm diameter Whatman n°2V filter paper and stored overnight at 4°C. The addition of 0.2 g MgO to the filtrate liberated ammonium, and the residual nitrate was determined by the reduction to nitrite (NO_2) *via* the addition of ∼ 2g Devarda alloy powder [24].

Total organic P was determined by the P Bray method described by [25] with minor modifications. 50 mL of P Bray-1 solution was added to 4 g of soil. The mixture was hand-shaken for exactly 1 min prior to filtration using 110 mm diameter Whatman n°2V filter paper. The phosphorous concentration of the filtrate was then determined by Inductively Coupled Plasma-Optical Emission Spectroscopy (ICP-OES) (Spectro Genesis, Germany).

Total ion concentrations were determined by adding 40 mL of a 0.2M ammonium acetate solution to 4 g of soil. The mixture was shaken for 1 hour and the supernatant filtered through 110 mm diameter Whatman n°2V filter paper. 15 mL of the filtrated were used to determine the concentrations of iron (Fe), calcium (Ca), potassium (K), magnesium (Mg) and sodium (Na) by ICP-OES.

Metagenomic DNA extraction

Total soil DNA was extracted from 0.3 g samples using the Powersoil DNA isolation kit according to the manufacturer's instructions (MOBIO laboratories, San Diego, USA). DNA concentrations were estimated with a NanoDrop spectrophotometer (NanoDrop Technologies, Montchanin, DE, USA).

PCR amplification, purification and restriction digestion

All polymerase chain reactions (PCRs) were carried out in a Bio-Rad Thermocycler (T100 TM Thermal Cycler). Bacterial 16S rRNA encoding genes were amplified using the universal bacterial primers 341F (5′-CCTACGGGAGGCAGCAG-3′)/908R (5′-CCGTCAATTCMTTTGAGTTT -3′) [26] and the Fungal ITS regions amplified using the universal primer set ITS1 (5′-CT-TGGTCATTTAGAGGAAGTAA-3′)/ITS4 (5′-TCCTCCGCT-TATTGATATGC-3′) [27]. PCR was carried out in 50 µl reaction volumes, where each reaction contained 1X PCR buffer, 0.2 U DreamTaq polymerase (Fermentas, USA), 200 µM of each dNTP, 0.5 µM of each primer, 4% bovine serum albumin (BSA), 2% to 6% DMSO to increase PCR specificity (the concentration used was sample-dependent) and between 5 and 20 ng of metagenomic DNA.

16S rRNA gene PCR amplifications were carried out as follows: 5 min at 95°C for denaturation; 25 cycles of 30 s at 94°C, 30 s annealing at 54°C and 105 s at 72°C; and a final elongation step of 7 min at 72°C. Fungal ITS region PCR amplifications were performed as follows: 5 min at 95°C for denaturation; 25 cycles of 1 min at 94°C, 50 s annealing at 55°C and 105 s at 72°C; and a final elongation step of 7 min at 72°C.

To perform T-RFLP analyses, the 341F and ITS1 primers were 5′-end FAM-labelled and the PCR products were purified using the GFX TM PCR DNA and gel band purification kit as directed by the supplier (GE Healthcare, UK). 200 ng of purified PCR products for 16S rRNA gene T-RFLP were digested with the restriction enzymes $MspI$ at 37°C overnight, while 400 ng of purified ITS amplicons was digested overnight with $HaeIII$.

Terminal-Restriction Fragment Length Polymorphism (T-RFLP) analyses

T-RF size was determined by capillary electrophoresis using a Applied Biosystems DNA Sequencer 3130 (Applied Biosystems, Foster City, California, USA) and according to the molecular weight standard GeneScan-600LIZ V2 (Applied Biosystem), with an error of ±1 bp. Individual T-RFs were considered as Operational Taxonomic Units (OTUs), with recognition that each OTU may comprise more than one distinct bacterial ribotype. Peak height was used to identify each T-RF and characterize their relative abundance in the total T-RFLP profiles, which was used as a proxy for OTU abundance in the microbial populations.

Statistical analyses

Multivariate analyses of T-RFLP and environmental data were performed using the software Primer 6 (Primer-E Ltd, UK). Valid T-RF peaks (between 30 and 567 bp for 16S rRNA gene T-RFLP or 30 and 800 bp for ITS region T-RFLP) from triplicate T-RFLP profiles were identified, compiled and aligned to produce large data matrices using the online software T-REX (http://trex. biohpc.org/) [28]. The community structures obtained were analyzed by ordination using non-metric multidimensional scaling (nMDS) of Bray-Curtis similarity matrices of square-root transformed data. Two-way crossed analysis of similarity (ANOSIM) tests were used to assess significant differences in the structure of assemblages among the Fairy Circles sampled and between the different sampling 'zones' (centre/margin/control vegetated soils) [29]. Multivariate dispersion (MVDISP) was used to measure 'within-zone' assemblage dispersion [30]. For each FC, one-way ANOSIM was used to test for differences in microbial community assemblages in the three FC-zones.

Prior to principal component analysis (PCA), the environmental variables were analyzed using a Draftsman plot [31] to evaluate the need for transformations, i.e. any skewness in the dataset. Total Carbon (% C) and ammonium (NH_4^+) were initially $\log(x + 1)$ transformed, and the complete environmental data set was normalized to perform PCA. A resemblance matrix based on Euclidean distances was created using the normalized set prior to two-way crossed ANOSIM.

Results and Discussion

Plant-microbe interactions actively shape plant diversity by various processes (reviewed in [32]), and soil pathogens have been described as active drivers of vegetation-succession [32]. As Namib Desert FCs are circular vegetation-free patterns surrounded by a healthy vegetation-covered matrix, we have hypothesized that a soil-born plant pathogen could be a causative agent of the phenomenon. This would not be unique: a fungus (*Sclerotinia homoeocarpa*) has been shown to be responsible for a plant disease, known as the Dollar Spot, which is morphologically similar to the FC phenomenon. *S. homoeocarpa* infects turfgrass species and, at a much smaller scale (i.e., the size of a dollar coin), creates FC-like circular necrotic patches that grow with time [33]. We therefore investigated the potential involvement of both edaphic bacterial and fungal communities in the origin and/or maintenance of the Namib Desert gravel plain FCs, as members from both phylogenetic groups are known to exhibit phyto-toxicity (e.g., [34-37]).

The measurement of ten edaphic chemical variables (Table 1), analyzed by PCA, provided no clear evidence of an "FC zonation effect" (Figure S1); i.e., the composite soil chemistries of the samples from the three zones tested (FC center/FC margin/ external vegetated control) were randomly scattered on the PCA plot. Two-way crossed ANOSIM confirmed this observation with (marginally significant) global R values <0.2, indicating that the different sampling zones and each FC could not clearly be separated [38]. Van Rooyen and colleagues [2] observed a similar trend with no differences between FC microhabitat soil chemistries in each of the sandy FC sites studied. Contrastingly, clear FC-specific patterns of hydrocarbon-like gas emissions were observed in similar FCs [14] which led to the geochemical origin hypothesis of Namib Desert FCs. Taken together, these results could suggest that gas emission rather than different soil chemistries could play a direct, or indirect, role in FC formation or maintenance.

The nMDS plots showing the ordination of the fungal community structures revealed by T-RFLP suggested a similar trend (Figure 2A); i.e., samples from each FC zones were randomly positioned (2D-stress> 0.2). Contrastingly, the nMDS ordination of the bacterial community T-RFLP fingerprints indicated some discrimination between communities from the FC centre and the control soils (virtually separated by the dashed grey line in Figure 2B), and between communities from different FCs (i.e., FC2 and FC5 communities appeared distinct from those of FC3, FC4 and FC5: virtually separated by the dashed black line in Figure 2B). Two-way crossed ANOSIM (Table 2) showed strongly significant differences in the structure of the bacterial and fungal assemblages between the different FC zones (particularly the control and FC-center bacterial communities: R = 0.702; p = 0.01). The simplest explanation for these results would be habitat filtration due to the absence of rhizosphere and their associated microbial assemblages within the non-vegetated FC soils (e.g., [39,40]). However, highly significant differences between the microbial assemblages of each individual FC were also detected (ANOSIM R>0.9 in 6 of the 10 FC pairwise

Table 1. Characterizations of the Fairy Circles sampled and of their edaphic chemical properties.

Fairy Circle	Dimensions (mxm)	Area (m²)	Zone	% Carbon	NH₄⁺ (mg N/kg)	NO₃⁻ (mg N/kg)	P (mg/kg)	Fe (mg/kg)	Ca (mg/kg)	K (mg/kg)	Mg (mg/kg)	Na (mg/kg)	pH
FC1	4.6×3.7	13.75	Control	0.52 (±0.91)	9.19 (±2.16)	8.47 (±2.54)	13.24 (±1.75)	17.52 (±4.84)	162.95 (±50.98)	17.11 (±0.62)	5.81 (±0.49)	0.95 (±0.39)	8.16 (±0.13)
			Margin	0.88 (±1.54)	13.66 (±4.82)	9.88 (±3.73)	16.79 (±1.26)	14.91 (±2.47)	177.54 (±50.2)	18.42 (±2.53)	6.08 (±0.31)	1.39 (±0.46)	8.07 (±0.08)
			Centre	1.02 (±2.21)	10.3 (±1.71)	10.54 (±4.24)	16.01 (±3.1)	15.72 (±1.32)	138.63 (±11.83)	14.24 (±3.71)	4.85 (±0.46)	2.34 (±1.34)	8.03 (±0.18)
FC2	3.6×3.2	9.05	Control	ND	33.51 (±44.33)	12.86 (±1.38)	13.00 (±1.80)	16.89 (±0.68)	132.54 (±18.56)	14.21 (±2.86)	5.17 (±0.5)	0.83 (±0.57)	8.13 (±0.1)
			Margin	0.64 (±0.79)	18.04 (±10.19)	12.04 (±2.45)	11.90 (±2.97)	13.11 (±6.76)	263.06 (±118.21)	16.72 (±1.78)	6.04 (±0.61)	2.24 (±1.65)	8.15 (±0.09)
			Centre	0.11 (±0.17)	12.29 (±2.19)	12.06 (±2.10)	12.72 (±0.66)	11.35 (±3.97)	193.39 (±42.68)	14.23 (±1.16)	5.68 (±0.27)	1.13 (±0.68)	8.09 (±0.13)
FC3	5×3.7	14.53	Control	0.01 (±0.03)	12.91 (±0.6)	13.76 (±1.80)	13.46 (±1.34)	18.12 (±3.49)	211.71 (±115.24)	15.75 (±1.95)	5.89 (±1.04)	1.43 (±1.70)	8.05 (±0.13)
			Margin	1.84 (±3.67)	9.8 (±2.85)	12.23 (±3.00)	14.62 (±1.56)	17.76 (±3.89)	161.14 (±39.01)	14.13 (±2.26)	5.32 (±0.38)	0.91 (±0.37)	8.19 (±0.05)
			Centre	0.18 (±0.38)	8.91 (±3.04)	9.00 (±1.97)	14.04 (±1.44)	17.53 (±2.05)	158.67 (±20.20)	14.04 (±1.52)	5.38 (±0.14)	0.74 (±0.07)	8.05 (±0.09)
FC4	3.6×4.2	14.7	Control	0.10 (±0.20)	8.50 (±1.30)	6.50 (±1.00)	13.91 (±1.87)	19.59 (±4.38)	131.86 (±29.77)	14.26 (±1.34)	5.69 (±0.25)	0.53 (±0.20)	8.22 (±0.07)
			Margin	0.03 (±0.06)	7.06 (±0.76)	6.63 (±1.22)	14.61 (±1.62)	17.85 (±2.25)	165.94 (±41.3)	13.49 (±1.49)	5.06 (±0.23)	0.63 (±0.18)	8.06 (0.07)
			Centre	ND	8.63 (±2.34)	8.91 (±3.30)	13.01 (±1.1)	15.12 (±3.72)	202.83 (±102.74)	14.02 (±1.46)	5.70 (±0.33)	0.76 (±0.25)	8.09 (±0.09)
FC5	3.8×3.1	9.25	Control	ND	6.16 (±0.87)	6.79 (±0.23)	14.79 (±1.87)	14.2 (±4.62)	191.01 (±80.12)	18.43 (±3.40)	5.43 (±0.17)	1.37 (±0.33)	8.11 (±0.09)
			Margin	ND	11.6 (±6.54)	9.08 (±0.66)	13.65 (±1.33)	17.01 (±6.26)	179.58 (±54.02)	14.34 (±0.70)	5.95 (±0.25)	0.72 (±0.09)	8.13 (±0.14)
			Centre	ND	7.03 (±0.88)	8.58 (±2.30)	12.26 (±2.18)	15.34 (±5.73)	243.54 (±91.70)	14.04 (±0.84)	6.26 (±0.38)	0.88 (±0.15)	8.11 (±0.15)

Values are given as mean ± SD. ND: Not determined as below detection limit.

comparisons, and p consistently <0.05; Table 2). Following this observation, the bacterial and fungal community fingerprint of each FC was analyzed separately by ANOSIM, independently testing for differences in their respective FC zones (Table 3). Excluding the fungal community from FC 2, these analyses showed that both the control and FC-centre communities (pairwise comparisons) and/or the communities from the three zones (global test) were significantly different (p <0.05). These results suggest that each independent FC houses a significantly distinct surface edaphic microbial community which is significantly different from that of the vegetation-covered control soils, and strongly suggest that the absence of rhizosphere/rhizospheric communities cannot satisfactorily explain the structure differences

in edaphic microbial communities observed between FCs and external soils.

Recent satellite data have demonstrated that the Namib Desert Fairy Circles could be considered as 'living organisms', as they appeared (birth), grew and disappeared (death) [3]. It was concluded that any study attempting to unravel the origins of FCs should take into account the fact that they are not static environments but vary over time [3,16]. The underlying corollary from this observation is that individual FCs within a site in the Namib Desert (e.g. such as those in our study which were all sampled within one hectare area) might be at different stages in their 'lifespan' and potentially constitute different biotopes (e.g., re-vegetated 'ghost', newly forming or matured/ing FCs; [3]). If we assume that the surface soil microbial communities are not static

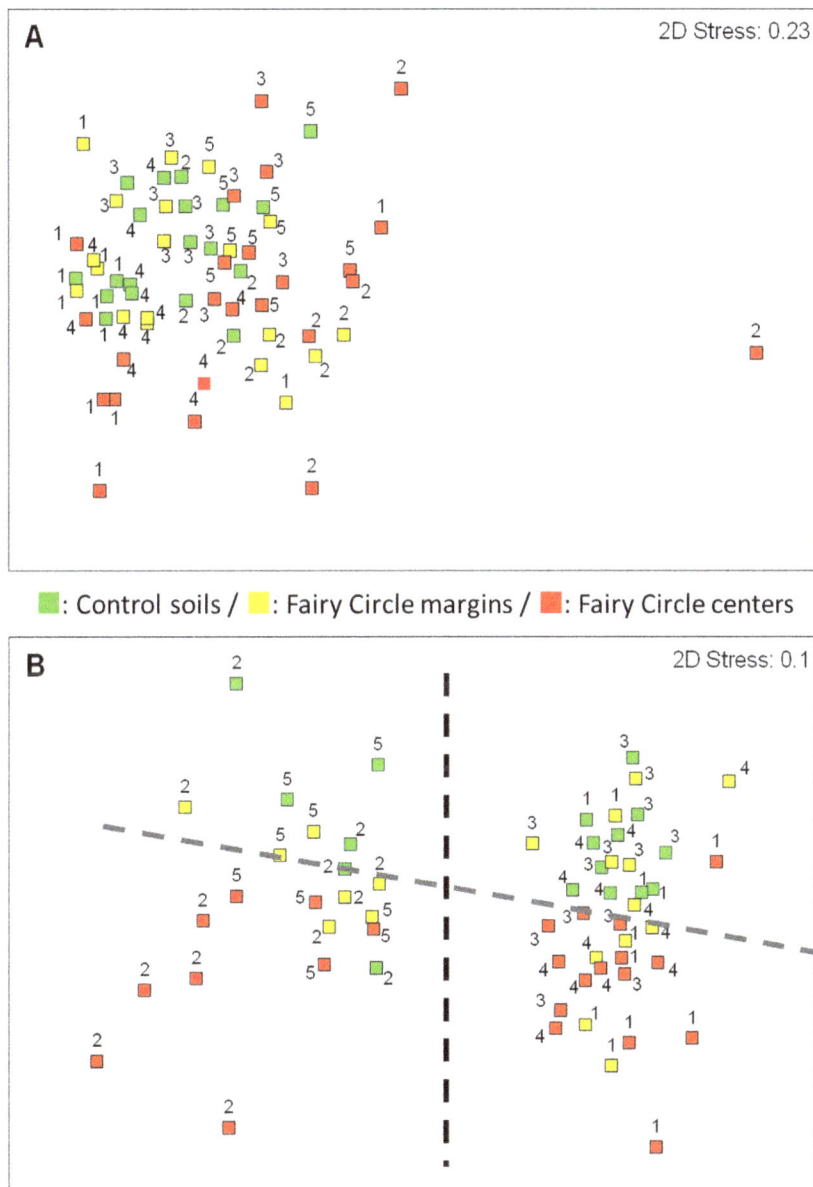

Figure 2. Two dimension nonmetric multidimensiobal scaling (2D-nMDS) plot of Bray–Curtis similarity of fungal (A) and bacterial (B) community structures based on ITS region and 16S rRNA gene square-root transformed T-RFLP profiles respectively. Numbers refer to the respective Fairy Circles. The dashed lines indicate virtual separations between group of samples (Grey: control vs FC centre communities/ Black: FC2 and FC5 vs FC1, FC3 and FC4).

Table 2. Results of two-way crossed ANOSIM tests based on Bray-Curtis similarity matrices from square-root transformed bacterial and fungal T-RFLP profiles.

	Differences among Fairy Circle zones			
	Bacterial communities		Fungal communities	
	R	p	R	p
Global Test	0.418	0.001 *	0.26	0.001 *
Control vs Margin	0.103	0.1	0.26	0.001 *
Control vs Centre	0.702	0.001 *	0.329	0.001 *
Margin vs Centre	0.401	0.001 *	0.234	0.007 *
	Differences among Fairy Circles			
	Bacterial communities		Fungal communities	
	R	p	R	p
Global Test	0.688	0.001 *	0.477	0.001 *
FC 1 vs FC 2	0.994	0.001 *	0.461	0.002 *
FC 1 vs FC 3	0.466	0.001 *	0.599	0.001 *
FC 1 vs FC 4	0.237	0.004 *	0.234	0.023 *
FC 1 vs FC 5	0.993	0.001 *	0.581	0.001 *
FC 2 vs FC 3	0.969	0.001 *	0.429	0.002 *
FC 2 vs FC 4	0.958	0.001 *	0.522	0.001 *
FC 2 vs FC 5	0.214	0.054	0.274	0.004 *
FC 3 vs FC 4	0.454	0.001 *	0.593	0.001 *
FC 3 vs FC 5	1	0.001 *	0.686	0.001 *
FC 4 vs FC 5	1	0.001 *	0.736	0.001 *

R: ANOSIM statistic; p: probability level. *: Significantly different (p <0.05).

throughout the lifespan of a FC (i.e., change in parallel with the evolution of the FC biotope; [39,41]), we are provided with a possible explanation for the differences observed between individual FC microbial communities (Table 2).

Distinct 'drivers' or environmental disturbances could also lead to alternative FC edaphic communities. For example, fluctuating hydrocarbon-like gas seepages [14], the action of different phytopathogenic microorganisms/pathovars [34] or dissimilar toxin(s) infection stages [37] could all lead to different FC biotopes within a single site. The microbial-mediated hypotheses are supported by the fact that two fungal T-RFs (OTUs 246 and 683; data not shown) were consistently observed within the margin and/or center of all five FCs studied, but never in the external control samples and could therefore constitute molecular signatures of "FC-related" fungi [33].

To test whether FC microbial communities were more heterogeneous than those of the vegetated control soils, multivariate dispersion indices were calculated (MVDISP routine in Primer6; [30]). The MVDISP index was lowest for the control communities and highest for the FC-centre communities, for both the bacterial (MVDISP Control 0.854 <MVDISP FC Margin 0.987 <MVDISP FC Centre 1.08) and fungal (MVDISP Control 0.659 <MVDISP FC Margin 0.92 <MVDISP FC Centre 1.261) communities. As a higher MVDISP index is indicative of higher multivariate dispersal, this clearly demonstrates that the vegetated (control) soil microbial communities exhibited lower intra-variability (or higher homogeneity) than the FC communities; the FC centre communities being the more variable, i.e. the more heterogeneous. These intra-variability differences could be explained by the adaptation of the 'intra-FC' (margin and centre)

microbial communities to environmental disturbance leading to their stochastic assembly [41].

Based on the observation that FCs appear to evolve over time [3], we suggest that the control vegetation-covered soils constitute a primary environmental state (or precursor), the FC margins represent a transitional phase, and the centre non-vegetated soils an alternative environment. By applying the three phase community assembly model after disturbance defined by Ferrenberg and colleagues [41] and based on the principles of microbial community dispersal, we propose two models to potentially explain the assembly of Namib Desert gravel plain Fairy Circle microbial communities (Figure 3).

We suggest that an environmental disturbances (e.g., the effect of active microbial phyto-pathogens, (dis)continuous gas leakage or plant spatial self-organization as a response to resource scarcity; [14,33,42]) could lead to plant necrosis which would disrupt the homogenous precursor (control) edaphic communities. Subsequently, FC-margin communities would further develop by stochastic assembly processes (phase 1; Models A and B; Figure 3). Where the disturbance is not continuous in time, the microbial communities of the three FC zones would develop based on niche-partitioning; i.e., vegetation-covered *versus* vegetation-free soils with the FC-margin sites constituting a transitional environment (phase 2, model A). If the disturbance is continuous or multiple sequential disturbances occur over the lifespan of the FC, the FC microbial communities (margin and centre) would initially be determined by stochastic processes. Subsequently, a combination of niche-partitioning (margins constituting a buffered environment separating the well-defined vegetated and not impacted (control) environment and the plant-free FC centers) and stochastic processes (continuous disturbance) would be responsible for

Table 3. One-way ANOSIM statistics comparing the bacterial and fungal community structures the predefined zones of each FC studied.

Fairy Circle	ANOSIM	Bacterial communities		Fungal communities	
		R	p	R	p
FC1	Global test	0.245	0.057	0.247	0.014 *
	Control vs Centre	0.446	0.036 *	0.281	0.048 *
	Margin vs Centre	0.15	0.198	0.244	0.095
	Control vs Margin	0.185	0.229	0.365	0.029 *
FC2	Global test	0.463	0.001 *	0.115	0.104
	Control vs Centre	0.763	0.008 *	0.163	0.103
	Margin vs Centre	0.663	0.008 *	(-)0.094	0.881
	Control vs Margin	(-)0.083	0.771	0.375	0.086
FC3	Global test	0.511	0.002 *	0.364	0.007 *
	Control vs Centre	0.763	0.008 *	0.400	0.048 *
	Margin vs Centre	0.675	0.008 *	0.488	0.024 *
	Control vs Margin	(-)0.104	0.771	0.188	0.143
FC4	Global test	0.463	0.001 *	0.245	0.04 *
	Control vs Centre	0.788	0.008 *	0.313	0.071
	Margin vs Centre	0.244	0.063	0.206	0.103
	Control vs Margin	0.365	0.029 *	0.25	0.2
FC5	Global test	0.408	0.025 *	0.423	0.015 *
	Control vs Centre	0.714	0.067	0.679	0.133
	Margin vs Centre	0.167	0.2	0.407	0.086
	Control vs Margin	0.333	0.2	0	0.6

R: ANOSIM statistic; p: probability level. *: Significantly different ($p < 0.05$).

microbial community assemblies (phase 2, model B). Finally, as FCs die, i.e. disturbance has ceased, a neutral assembly processes would occur, leading to less variable edaphic microbial communities in the newly vegetation-covered soils (phase 3, models A and B). As the structures of the FC-center communities studied presented higher intra-variability than those of the margins and

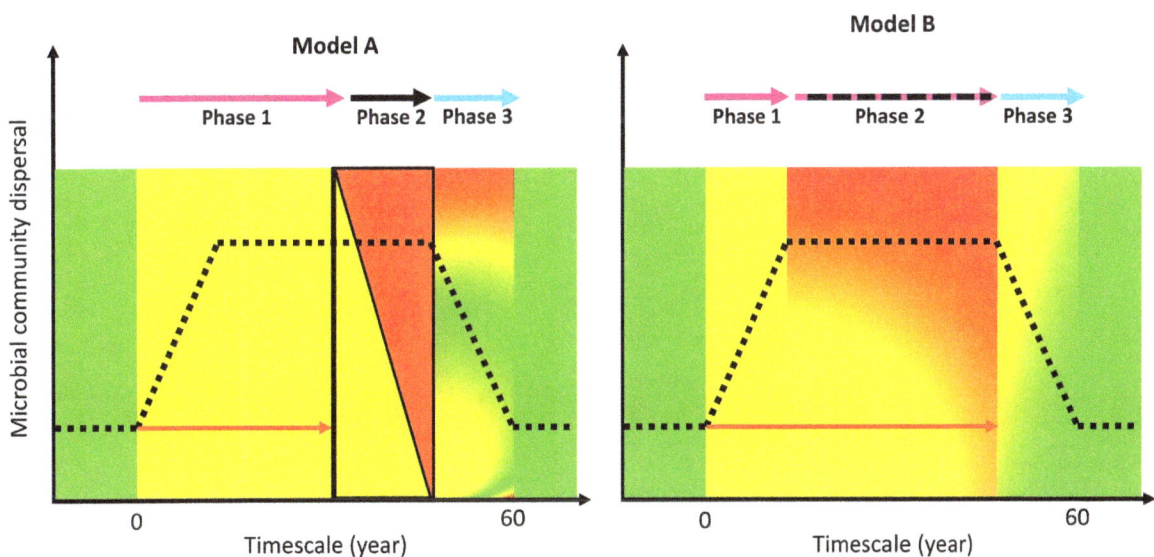

Figure 3. Models hypothesizing microbial community assembly in Namib Desert Fairy Circles. Arrow indicate the assembly processes (purple: stochastic/black: niche-partitioning/blue: neutral). Colors are represent virtually the FC zones (Green: vegetated covered control soils/Yellow: FC margins/Red: FC Centers) where these processes occur. Red arrows indicate the origin in time and length of the environmental disturbance responsible for FC appearance. The x-axis does not reflect proportionally the time scale.

the more homogenous control soils, either the niche-partitioning is still occurring (model A) or FC-centers constitute ever-changing environments (model B).

Conclusion/Perspectives

In ecology, patterns of diversity have been shown to be influenced by four major mechanisms/processes (selection, drift, speciation and dispersal) and by various species interactions (predation, competition and mutualism) [43]. Altogether, our results suggest that FC niche-specific surface edaphic microbial communities are 'selected'. Such selection could originate from (i) habitat/niche-filtration (e.g., rhizospheric communities vs open soil communities; [39]) or (ii) be related to their adaptation to (an) environmental disturbance(s) [41], such as the presence of an active microbial phyto-pathogen [33], a continuous hydrocarbon gas emission [14,20] or vegetation spatial self-organization [15,16,42].

We are yet unable to conclude whether bacteria and/or fungi are actually responsible for the appearance and/or maintenance of Fairy Circles in the Namib Desert. We suggest that more extensive metacommunity/biogeography studies on groups of FCs at different developmental stages [3,44,45] using meta'omic' approaches [46,47] could assist in characterizing the functional basis of FC ecosystems and in potentially identifying the agent(s) responsible for their formation. Such studies should be extended to deep FC soil horizons as it has been shown that (i) sub-surface FC soils are more anoxic than the surrounding vegetation-covered soils [14], (ii) that the abundance of culturable anaerobic bacteria

is higher in FC soils [48] and (iii) that sub-surface FC soils accumulate water [11]. Moreover, to fully address the origin and functioning of Namib FC ecosystems, a comparative chemical (soil and gas) and (micro)biological analysis of 'rocky' and 'sandy' FC formations is recommended.

Supporting Information

Figure S1 Principal component analysis (PCA) plot of the normalized environmental variables measured. Numbers refer to the respective Fairy Circles.

Table S1 Results of two-way crossed ANOSIM tests based on Euclidean distance matrices from normalized soil chemistry measurements. R: ANOSIM statistic; p: probability level. *: Significantly different (p <0.05).

Acknowledgments

The authors gratefully acknowledge the National Research Foundation (NRF, South Africa) and the University of Pretoria for supporting this research.

Author Contributions

Conceived and designed the experiments: JBR DC. Performed the experiments: AP AM JBR. Analyzed the data: JBR. Contributed reagents/materials/analysis tools: DC. Wrote the paper: JBR MS DC. Field knowledge: MS.

References

1. Cox GW (1987) The origin of vegetation circles on stony soils of the Namib Desert near Gobabeb, South West Africa/Namibia. J Arid Environ 13: 237–244.
2. Van Rooyen MW, Theron GK, Van Rooyen N, Jankowitz WJ, Matthews WS (2004) Mysterious circles in the Namib Desert: review of hypotheses on their origin. J Arid Environ 57: 467–485.
3. Tschinkel WR (2012) The life cycle and life span of Namibian fairy circles. PloS one. 7: e38056.
4. Fraley L (1987) Response of short grass plains vegetation to gamma radiation. III. Nine years of chronic irradiation. Environ Exp Bot 27: 193–201.
5. Tinley KL (1971) Etosha and the Kaokoveld. African Wild Life. 25: 1–16.
6. Moll EJ (1994) The origin and distribution of fairy rings in Namibia. In: Proceedings of the 13th Plenary Meeting AETFAT, Malawi. 2: 1203–1209.
7. Becker T, Getzin S (2000) The fairy circles of Kaokoland (North-West Namibia) origin, distribution, and characteristics. Basic Appl Ecol 1: 149–159.
8. Albrecht CF, Joubert JJ, de Rycke PH (2001) Origin of the enigmatic, circular, barren patches ('Fairy Rings') of the pro-Namib. S Afr J Sci 97: 23–27.
9. Becker T (2007) The phenomenon of fairy circles in Kaokoland (NW Namibia). Basic Appl Dryland Res 1: 121–137.
10. Picker MD, Ross-Gillespie VERE, Vlieghe K, Moll E (2012) Ants and the enigmatic Namibian fairy circles-cause and effect? Ecol Entomol 37: 33–42.
11. Juergens N (2013) The biological underpinnings of Namib Desert fairy circles. Science 339: 1618–1621.
12. Theron GK (1979) Die verskynsel van kaal kolle in Kaokoland, SuidWes-Afrika. J S Afr J Biology Soc 20: 43–53.
13. Jankowitz WJ, Van Rooyen MW, Shaw D, Kaumba JS, Rooyen NV (2008) Mysterious circles in the Namib Desert. S Afr J Bot 74: 332–334.
14. Naudé Y, Van Rooyen MW, Rohwer ER (2011) Evidence for a geochemical origin of the mysterious circles in the Pro-Namib desert. J Arid Environ 75: 446–456.
15. Cramer MD, Barger NN (2013) Are Namibian "Fairy Circles" the Consequence of Self-Organizing Spatial Vegetation Patterning? PloS ONE. 8: e70876.
16. Getzin S, Wiegand K, Wiegand T, Yizhaq H, von Hardenberg J, et al. (2014) Adopting a spatially explicit perspective to study the mysterious fairy circles of Namibia. Ecography 37: 001–011.
17. Wieder WR, Bonan GB, Allison SD (2013) Global soil carbon projections are improved by modelling microbial processes. Nat Clim Change 3: 909–912.
18. Graham EB, Wieder WR, Leff JW, Weintraub SR, Townsend AR, et al. (2014) Do we need to understand microbial communities to predict ecosystem function? A comparison of statistical models of nitrogen cycling processes. Soil Biol Biochem 68:279–282.
19. Pointing SB, Belnap J (2012) Microbial colonization and controls in dryland systems. Nat Rev Microbiol 10: 551–562.
20. Grossi V, Cravo-Laureau C, Guyoneaud R, Ranchou-Peyruse A, Hirschler-Réa A (2008) Metabolism of n-alkanes and n-alkenes by anaerobic bacteria: A summary. Org Geochem 39: 1197–1203.
21. SSSA (1996) Methods of Soil Analysis, Part 3. Soil Science Society of America, Madison.
22. Walkley A (1935) An Examination of Methods for Determining Organic Carbon and Nitrogen in Soils. J Agr Sci 25: 598–609.
23. Bremmer JM, Keeney DR (1966) Determination and isotope-ratio analysis of different forms of nitrogen in soils: 3. Exchangeable ammonium, nitrate and nitrite by extraction-distillation methods. Soil Sci Soc Am J 30: 577–582.
24. Keeney DR, Nelson DW (1982) Nitrogen-inorganic forms. In: Page AL, Miller RH, Keeney DR (eds) Methods of soil analysis, part 2. Madison, Wise, Agronomy 9: 643–698.
25. Bray RH, Kurtz LT (1945) Determination of total, organic and available forms of phosphorous in soils. Soil Sci 59: 39–45.
26. Lane DJ, Pace B, Olsen GJ, Stahl DA, Sogin ML, et al. (1985) Rapid determination of 16S ribosomal RNA sequences for phylogenetic analyses. Proc. Natl. Acad. Sci. 82: 6955–6959.
27. White TJ, Bruns T, Lee S, Taylor J (1990) Amplification and direct sequencing of fungal ribosomal RNA genes for phylogenetics. M. Innis, D. Gelfand, J. Sninsky, T. White (Eds.), PCR Protocols: A Guide to Methods and Applications, San Diego, Academic Press, pp. 315–322
28. Culman SW, Bukowski R, Gauch HG, Cadillo-Quiroz H, Buckley DH (2009) T-REX: software for the processing and analysis of T-RFLP data. BMC Bioinformatics 10: 171.
29. Clarke K (1993) Non-parametric multivariate analysis of changes in community structure. Aust J Ecol 18: 117–143.
30. Clarke KR, Warwick RM (2001) A further biodiversity index applicable to species lists: variation in taxonomic distinctness. Mar Ecol-Prog Ser 216: 265–278.
31. Clarke KR, Warwick RM (1994) Change in Marine Communities: An Approach to Statistical Analysis and Interpretation. Plymouth, UK, Plymouth Marine Laboratory, pp. 144.
32. Reynolds HL, Packer A, Bever JD, Clay K (2003) Grassroots ecology: plant-microbe-soil interactions as drivers of plant community structure and dynamics. Ecology 84: 2281–2291.
33. Walsh B, Ikeda SS, Boland GJ (1999) Biology and management of dollar spot (*Sclerotinia homoeocarpa*); an important disease of turfgrass. HortScience. 34: 13–21.
34. Arrebola E, Cazorla FM, Perez-García A, Vicente AD (2011) Chemical and metabolic aspects of antimetabolite toxins produced by *Pseudomonas syringae* pathovars. Toxins. 3: 1089–1110.

35. Dean R, Van Kan JA, Pretorius ZA, Hammond-Kosack KE, Di Pietro A, et al. (2012) The Top 10 fungal pathogens in molecular plant pathology. Mol Plant Pathol 13: 414–430.

36. Hoagland RE (2001) The genus *Streptomyces*: A rich source of novel phytotoxins. In: Ecology of Desert Environments I Parkash (Ed), 139–169 Scientific Publishers, Jodhpur, India.

37. Horbach R, Navarro-Quesada AR, Knogge W, Deising HB (2011) When and how to kill a plant cell: infection strategies of plant pathogenic fungi. J Plant Physiol 168: 51–62.

38. Clarke KR, Gorley RN (2006) PRIMER v6: User Manual. Plymouth Marine Laboratory: Plymouth, UK.

39. Dumbrell AJ, Nelson M, Helgason T, Dytham C, Fitter AH (2009) Relative roles of niche and neutral processes in structuring a soil microbial community. ISME J 4: 337–345.

40. Ramond J-B, Tshabuse F, Bopda CW, Cowan DA, Tuffin MI (2013) Evidence of variability in the structure and recruitment of rhizospheric and endophytic bacterial communities associated with arable sweet sorghum (*Sorghum bicolor* (L) Moench). Plant Soil. 372: 265–278.

41. Ferrenberg S, O'Neill SP, Knelman JE, Todd B, Duggan S, et al. (2013) Changes in assembly processes in soil bacterial communities following a wildfire disturbance. ISME J 7: 1102–1111.

42. Lejeune O, Tlidi M, Couteron P (2002) Localized vegetation patches: a self-organized response to resource scarcity. Phys Rev E 66: 010901.

43. Vellend M (2010) Conceptual synthesis in community ecology. Q Rev Biol 85: 183–206.

44. Hovatter SR, Dejelo C, Case AL, Blackwood CB (2011) Metacommunity organization of soil microorganisms depends on habitat defined by presence of *Lobelia siphilitica* plants. Ecology 92: 57–65.

45. Martiny JBH, Bohannan BJ, Brown JH, Colwell RK, Fuhrman JA, et al. (2006) Microbial biogeography: putting microorganisms on the map. Nat Rev Microbiol 4: 102–112.

46. Jansson JK, Prosser JI (2013) Microbiology: The life beneath our feet. Nature 494: 40–41.

47. Gunnigle E, Ramond JB, Frossard A, Seeley M, Cowan D (2014) A sequential co-extraction method for DNA, RNA and protein recovery from soil for future system-based approaches. J Microbiol Meth 103: 118–123.

48. Eicker A, Theron GK, Grobbelaar N (1982) 'n Mikrobiologiese studie van "kaal kolle" in die Giribesvlakte van Kaokoland, S.W.A.-Namibia. S Afr J Bot 1: 69–74.

14

Intraocular Pressure and Associations in Children. The Gobi Desert Children Eye Study

Da Yong Yang[1,2], Kai Guo[3], Yan Wang[3], Yuan Yuan Guo[3], Xian Rong Yang[3], Xin Xia Jing[3], Hai Ke Guo[2]*, Yong Tao[4]*, Dan Zhu[3]*, Jost B. Jonas[5]

1 Southern Medical University, Guangzhou, Guangdong, China, 2 Department of Ophthalmology, Guangdong General Hospital/Guangdong Academy of Medical Sciences, Guangzhou, Guangdong, China, 3 The Affiliated Hospital of Inner Mongolia Medical University, Hohhot, Inner Mongolia, China, 4 Department of Ophthalmology, People's Hospital, Peking University, & Key Laboratory of Vision Loss and Restoration, Ministry of Education, Beijing, China, 5 Department of Ophthalmology, Medical Faculty Mannheim of the Ruprecht-Karls-University Heidelberg, Mannheim, Germany

Abstract

Purpose: To assess the intraocular pressure (IOP) and its association in children in a population living in an oasis in the Gobi Desert.

Methods: The cross-sectional school-based study included all schools in the Ejina region. The children underwent an ophthalmic examination, non-contact tonometry and measurement of blood pressure and body height and weight.

Results: Out of eligible 1911 children, 1565 (81.9%) children with a mean age of 11.9 ± 3.5 years (range: 6–21 years) participated. Mean spherical refractive error was -1.58 ± 2.00 diopters. In multivariate analysis, higher IOP (right eye) was associated with younger age ($P<0.001$; standardized coefficient beta: -0.13; regression coefficient B: -0.13; 95% Confidence interval (CI):-0.18, -0.07), higher diastolic blood pressure ($P<0.001$;beta:0.13;B:0.05;95%CI:0.03,0.07), higher corneal refractive power ($P<0.001$;beta:0.11;B:0.23;95%CI:0.12,0.34), more myopic refractive error ($P=0.035$;beta:-0.06;B:-0.10;95%CI:-0.19, -0.001), and Han Chinese ethnicity of the father ($P=0.03$;beta:0.06;B:0.42;95%CI:0.04,0.89). If age and diastolic blood pressure were dropped, higher IOP was associated with higher estimated cerebrospinal fluid pressure (CSFP) ($P<0.001$;beta:0.09; B:0.13;95%CI:0.06,0.21) after adjusting for higher corneal refractive power ($P<0.001$) and Han Chinese ethnicity of the father ($P=0.04$). Correspondingly, higher IOP of the left eye was associated with younger age ($P<0.001$;beta:-0.15;B:-0.16;95%CI:-0.21, -0.10), female gender ($P<0.001$;beta:0.09;B:0.65;95%CI:0.30,1.01), higher corneal refractive power ($P<0.001$;beta:0.08;B:0.19;95%CI:0.06,0.32), more myopic refractive error ($P=0.03$;beta:-0.06;B:-0.12;95%CI:-0.22, -0.01), and higher estimated CSFP ($P<0.001$;beta:0.11;B:0.17;95%CI:0.09,0.24).

Conclusions: In school children, higher IOP was associated with steeper corneal curvature and with younger age and higher blood pressure, or alternatively, with higher estimated CSFP. Corneal curvature radius should be included in the correction of IOP measurements. The potential association between IOP and CSFP as also assumed in adults may warrant further research.

Editor: Rajiv R. Mohan, University of Missouri-Columbia, United States of America

Funding: This study was supported by Program for New Century Excellent Talents in University (No: NCET-12-0010) (Beijing, China) and FOK YING TONG education foundation (Hong Kong, China). The funders had no role in study design, data collection and analysis, decision to publish, or preparation of the manuscript.

* Email: guohaike@medmail.com.cn (HKG); drtaoyong@163.com (YT); zhudan1968@163.com (DZ)

Introduction

Intraocular pressure (IOP) is one of an important variable for the physiology and pathophysiology of the eye. Its normal distribution and its associations with other ocular and systemic parameters have been examined in numerous preceding investigations [1–3]. Few studies, however, were focused on the IOP in children [4–18]. These studies had limitations such as a hospital-based study design and inclusion of only a relatively small number of children, and a multivariate analysis was either not performed or did not contain the majority of known factors influencing IOP.

In addition, there was no information available for children in China, in particular not from the vast West China. We therefore performed this study on children in western China, measured the IOP and correlated the measurements with ocular and systemic variables. As study region we chose an oasis city in the mid of the Gobi Desert. This oasis city of Ejinaqi had the advantage that due to its isolated location, the exchange of the population with other regions was limited and that the population was relatively constant.

Table 1. General information of subjects.

Age (Years)	Number (%) Boys	Number (%) Girls	Number (%) Total	Systolic Blood Pressure (mm Hg)	Diastolic Blood Pressure (mm Hg)	Intraocular Pressure (Right Eye) (mmHg)	Intraocular Pressure (Left Eye) (mm Hg)	Body Mass Index (kg/m^2)
6–10	342	335	677	106.5±10.2	66.1±8.4	17.3±3.5	17.4±3.5	17.0±2.9
11–14	270	265	535	116.1±9.8	70.6±8.2	17.1±3.4	17.1±3.6	20.1±3.9
15–21	189	164	353	120.9±10.1	73.0±8.5	16.6±3.3	16.4±3.5	21.6±4.0
Total	801 (51.2)	764 (48.8)	1565 (100.0)	113.0±11.7	69.2±8.8	17.1±3.6	17.1±3.4	19.1±4.0

Methods

The Desert Gobi Children Eye Study was a cross-sectional, school-based study which was performed in city oasis of Ejinaqi, locating in the most western part of the Chinese province of Inner Mongolia. The Ethics Board of the Affiliated Hospital of Inner Mongolia Medical University Hohhot and the local Administration of the Education and School Board of Ejinaqi approved the study and informed written consent was obtained from the parents or guardians of all children. Ejinaqi oasis in the Gobi desert is located in the Ejinaqi region which covers an area of 114,000km^2 in the western part of the Chinese province of Inner Mongolia. The territory of the oasis stretches from 100.90° to 101.42° East longitude and from 41.85° to 42.50° North latitude. With extremely arid conditions, the study area belongs to the north temperature climate zone with a mean annual precipitation of approximately 40 mm and a mean pan evaporation of 3700 and 4000 mm. Average winter temperature minimums are close to −40°C, while summertime temperatures are warm to hot, with highs that range up to 50°C. The main vegetation are poplar trees. The study included all three schools in Ejinaqi which has a total population of 18,030 inhabitants (including 11,301 Han Chinese, 6209 Mongols and 520 individuals from other minorities). The next settlement is located in a distance of approximately 400 km. Ejinaqi can be reached by train (15 hours from Hohhot, the capital of Inner Mongolia) and by road.

The three schools in Ejinaqi (Ejinaqi primary school (911 students), Ejinaqi middle school (765 students), and Minority school (235 students)) included altogether 1911 children with all children from Ejinaqi attending one of the three schools. All children underwent an ophthalmological examination including measurement of presenting visual acuity and uncorrected visual acuity, a slit lamp-based examination of the anterior ocular segment, and tonometry. IOP was measured by a non-contact tonometer (Canon TX-F Full-Auto Tonometer, Canon Co., Tokyo, Japan). Ocular motility, binocularity and presence of strabismus was examined. After instilling 1% cyclopentolate eye drops (Alcon, Ft. Worth, USA) at least three times, cycloplegia was achieved and auto-refractometry was performed (ARK-900, NIDEK, Tokyo, Japan). Each eye was measured at least 3 times. The spherical equivalent of the refractive error was defined as the spherical value of refractive error plus one half of the cylindrical value. After medical mydriasis, ophthalmoscopy was carried out for examination of the fundus.

The non-ophthalmological examination included measurement of body height (using a stadiometer) and body weight, heart rate and blood pressure (using an automatic blood pressure monitor (YE655A, YUYUE, Jiangsu, China). The body mass index was calculated as the ratio of body weight (expressed in kg) divided by the square of body height (expressed in m).

Using the measurements of diastolic blood pressure and body mass index, we estimated the cerebrospinal fluid pressure (CSFP) using the formula of CSFP [mmHg] = 0.44 × Body Mass Index [kg/m^2] + 0.16 × Diastolic Blood Pressure [mmHg] – 0.18 × Age [Years] - 1.91 [19–21]. Previous studies had shown that the higher CSFP was associated with higher body mass index, higher diastolic blood pressure and younger age [22].

Statistical analysis was performed using a commercially available statistical software package (SPSS for Windows, version 21.0, IBM-SPSS, Chicago, IL). The normal distribution of data was tested using the Kolmogorov-Smirnov test. As a first step of the statistical analysis, we calculated the mean and standard deviations of the parameters. As a second step, we search for associations between IOP and other ocular and systemic

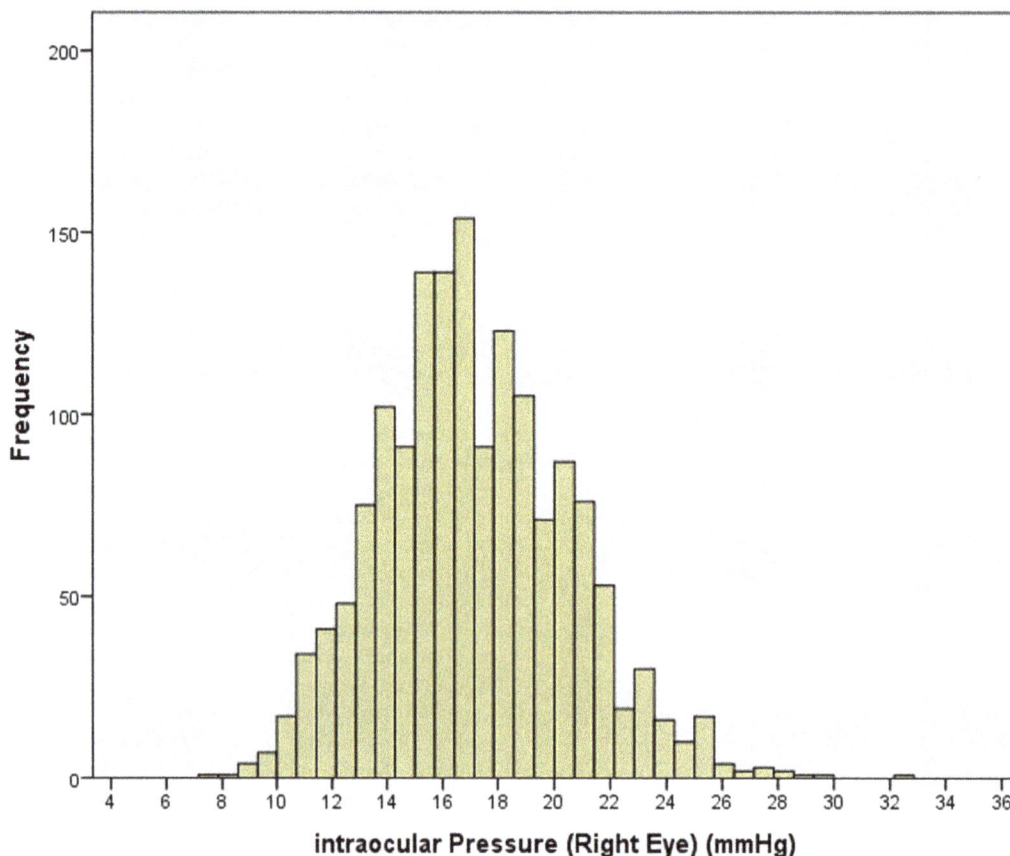

Figure 1. Histogram Showing the Distribution of Intraocular Pressure in the Gobi Desert Children Eye Study.

parameters in a univariate analysis. As a first step, we performed a multivariate regression analysis, with IOP as the dependent variable and all those parameters as independent variables which were significantly associated with IOP in the univariate analysis (indicated by a P-value ≤ 0.10). All P-values were based on two-sided test and were considered statistically significant if less than 0.05.

Results

Out of 1911 children who were primarily eligible for the study, 346 refused the examination, so that the study eventually included 1565 (81.9%) children (801 (51.2%) boys) with a mean age of 11.9 ± 3.5 years (median: 11.7 years; range: 6 to 21 years) (Table 1). Mean spherical refractive error was -1.58 ± 2.00 diopters (median: -1.00 diopters; range: -13.75 to $+5.50$ diopters) in the right eye and -1.54 ± 2.04 diopters (median: -0.75 diopters; range: -25.50 to $+6.25$ diopters) in the left eye. Seventeen (1.0%) of the children had a myopic refractive error exceeding -8 diopters.

Mean IOP was 17.1 ± 3.6 mm Hg (median: 16.8 mm Hg; range: 5.6 to 31.5 mm Hg) in the right eye and 17.1 ± 3.4 mm Hg (median: 16.9 mm Hg; range: 7.8 to 32.3 mm Hg) in the left eye. For both eyes, the distribution of IOP showed approximately a Gaussian distribution curve with a minor skew to the right (Fig. 1). For both eyes, IOP was not normally distributed (Kolmogorov-Smirnov test; $P<0.001$). Mean body mass index (BMI) was 19.1 ± 4.0 kg/m^2, with 6.2% of the children having a BMI higher

than 25 and ≤ 30, and 1.5% of the children having a BMI higher than 30.

When divided into three age groups, mean systolic and diastolic blood pressure and mean body mass index increased with older age group while the mean IOP decreased (Table 1). Within all three age groups, higher IOP was significantly associated with higher blood pressure (all $P<0.05$).

In univariate linear analysis, IOP decreased significantly with older age ($P<0.001$; standardized coefficient beta: -0.07; regression coefficient B: -0.07; 95%CI: -0.11, -0.02) (Fig. 2). In univariate analysis, higher IOP was significantly associated with female gender ($P = 0.02$), higher blood pressure ($P<0.001$), higher estimated CSFP ($P<0.001$) (Fig. 3), Han Chinese ethnicity of the father ($P = 0.04$), and higher keratometric readings ($P<0.001$) (Table 2).

We then performed a multivariate linear regression analysis, which included intraocular pressure as the dependent variable and all those parameters as independent variables which were associated with intraocular pressure in the univariate analysis with a P-value ≤ 0.10. In a first step, we dropped the parameters of mean blood pressure, systolic blood pressure and body weight from the list of independent variables due to the collinearity with diastolic blood pressure and with body mass index. We then dropped in a combination of stepwise, forward and backward regression analysis, all those parameters which were no longer significantly associated with IOP in the multivariate analysis. In the final model, higher IOP of the right eye was significantly

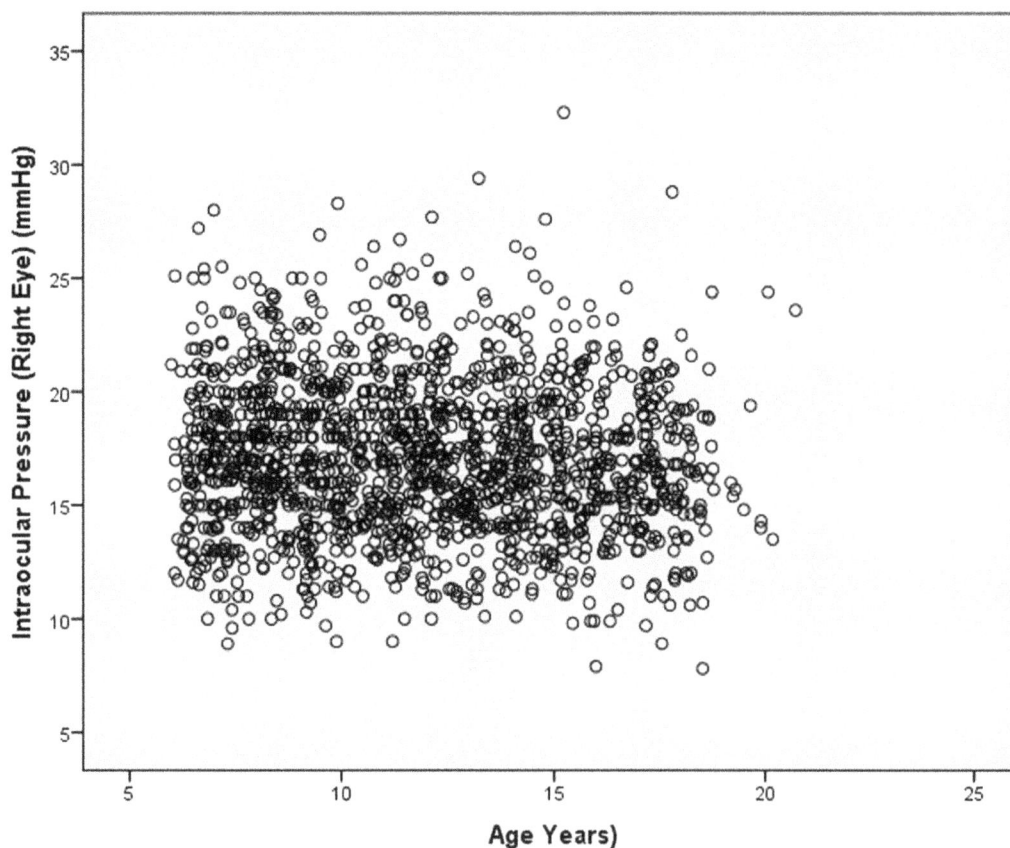

Figure 2. Scattergram Showing the Distribution of Intraocular Pressure and Age in the Gobi Desert Children Eye Study.

associated with the non-ocular parameters of younger age ($P<$ 0.001), higher diastolic blood pressure ($P<0.001$), higher corneal refractive power ($P<0.001$), more myopic refractive error ($P = 0.035$), and Han Chinese ethnicity of the father ($P = 0.03$) (Table 3). If age and diastolic blood pressure were dropped and replaced by the estimated CSFP, higher IOP was associated with higher estimated CSFP ($P<0.001$; beta: 0.09; B: 0.13; 95%CI: 0.06, 0.21), higher corneal refractive power ($P<0.001$; beta: 0.12; B: 0.26; 95%CI: 0.15, 0.37), and Han Chinese ethnicity of the father ($P = 0.04$; beta: 0.05; B: 0.44; 95%CI: 0.02, 0.87).

If the IOP of the left eye was taken as dependent variable in the multivariate analysis, similar results were obtained: higher IOP of the left eyes was associated with younger age ($P<0.001$), higher diastolic blood pressure ($P<0.001$), higher corneal refractive power ($P<0.001$), and additionally, with female gender ($P<$ 0.001) (Table 4). If diastolic blood pressure was dropped and replaced by estimated CSFP, higher IOP was associated with younger age ($P<0.001$; beta: -0.13; B: -0.13; 95%CI: -0.18, -0.08), female gender ($P<0.001$; beta: 0.10; B: 0.70; 95%CI: 0.34, 1.06), higher corneal refractive power ($P<0.001$; beta: 0.08; B: 0.21; 95%CI: 0.08, 0.34), and higher estimated CSFP ($P<$ 0.001; beta: 0.12; B: 0.18; 95%CI: 0.10, 0.25).

Discussion

In our population-based study on school children in an oasis in the Gobi Desert, higher IOP was significantly associated with younger age, higher diastolic blood pressure, steeper cornea and more myopic refractive error. If diastolic blood pressure were dropped from the analysis in the otherwise unchanged statistical model, higher IOP was significantly ($P<0.001$) associated with higher estimated CSFP. In the multivariate model, IOP was not significantly associated with BMI.

The association between higher IOP and higher blood pressure as found in our children study was in agreement with previous population based studies on adults, such as the Rotterdam Study, the Singaporean Tanjong Pagar Study, the Blue Mountains Eye Study, the Beaver Dam Eye Study, and the Los Angeles Latino Eye Study [19–24]. In univariate analysis, IOP increased by 0.4 mm Hg for each increase in diastolic blood pressure by 10 mm Hg (Table 2). In the multivariate model, IOP increased by 0.5 mm Hg for ach increase in diastolic blood pressure by 10 mm Hg (Table 4). It confirms the results of the preceding studies and extends their findings onto children. It shows that independently of age, IOP and blood pressure are connected to each other. It is in agreement with the experimental study by Samuels and colleagues who stereotaxically microinjected the gamma-aminobutyric acid receptor antagonist bicuculline methiodide into the dorsomedial and perifornical hypothalamus of rats and a significant increases in heart rate, mean arterial blood pressure, IOP and intracranial pressure [29].

The association between IOP and steeper cornea (i.e., higher corneal refractive power) again agrees with preceding studies on adults in which similar correlation have been reported [28,30]. The higher the corneal refractive power was, i.e., the steeper the cornea was, the higher were the intraocular pressure readings. The

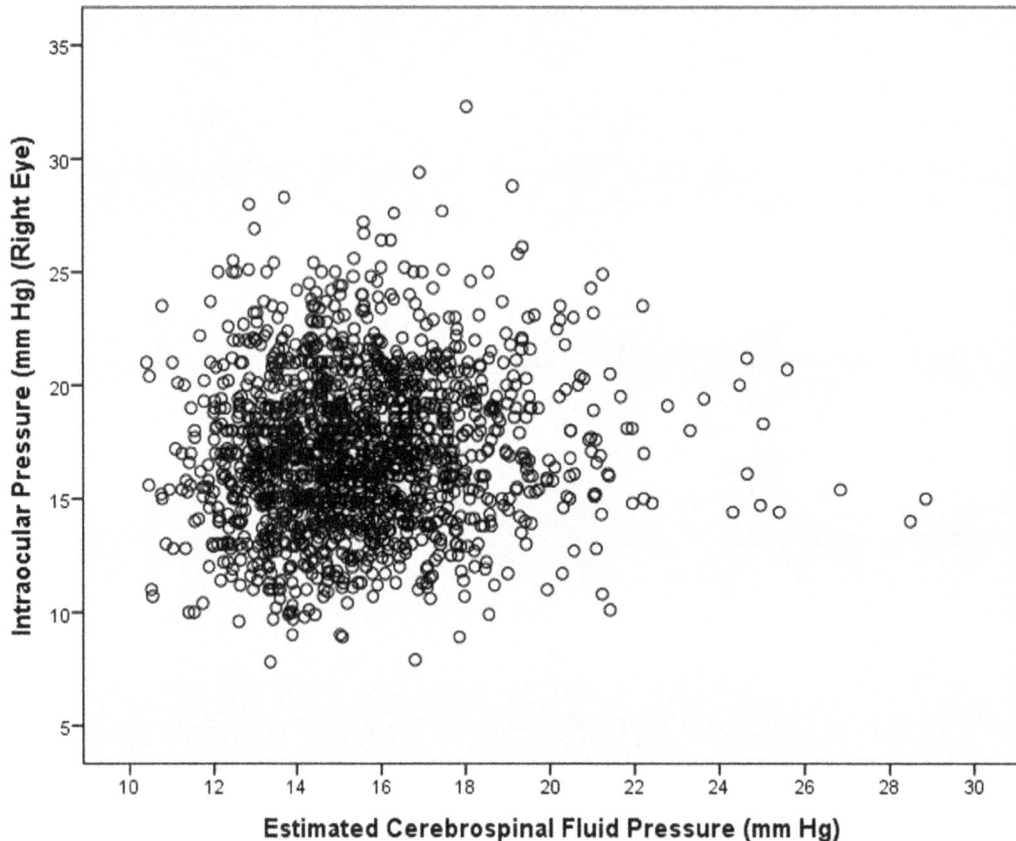

Figure 3. Scattergram Showing the Distribution of Intraocular Pressure and Estimated Cerebrospinal Fluid Pressure Age in the Gobi Desert Children Eye Study.

finding of our study may be due to geometrical reasons, since a flat structure as compared with a steep structure needs less external pressure to be further flattened up to a standardized applanation area. The clinical importance of the finding is that IOP measurements should be corrected for central corneal thickness and corneal curvature, in after corneal refractive surgery for the correction of myopia.

In contrast to previous studies on adults, IOP was not significantly associated with BMI in our study population [31–34]. In the previous studies, higher body mass index was associated with higher blood pressure, higher IOP and higher CSFP, which in turn was correlated with IOP and blood pressure [33,34]. The reason for the discrepancy between the previous studies on adults and our study on children may that most of the children in our study had a normal body mass index, with only approximately 8% of the children having overweight or being obese. Also in contrast to previous studies on adults [15,28], myopic refractive error was only weakly associated with higher IOP in the multivariate analysis in our study. In our study population, only 1.0% of the children had a myopic refractive error exceeding -8 diopters, and only 2.9% of the children had a myopic refractive error exceeding -6 diopters. These figures are markedly lower than those obtained from school children in metropolitan regions at the East coast of China. The relatively low prevalence of high myopia in our study population may have prevented a statistically stronger influence on

IOP in the multivariate models. Interestingly, as in our study, the Los Angeles Latino Eye Study neither found a strong association between IOP and myopia [26].

IOP decreased with older age in our study population (Tables 2–4). These results agree with findings of some studies, and are contradictory to results other investigations. An increase in IOP with older age for 405 children up to an age of 12 years was reported by Sihota and colleagues [13]. An increase in IOP with older age for children aged less than 10 years was also reported by Duckman and colleagues [5]. In a similar manner, studies reported on different association between higher age and IOP in adults, with increased IOP in Westerners and decreasing IOP with older age in Japanese [26,35].

Higher IOP was associated with higher estimated CSFP in the multivariate analysis in our study, if diastolic blood pressure or additionally age (which were included in the formula to calculate the CSFP). With all limitations of the formula to estimate the CSFP, the finding may point to an association between IOP and CSFP. A similar relationship has been measured in adults using lumbar pressure measurements of CSFP, and has also been postulated in other population-based studies of various ethnicities [20–22].

Potential limitations of our study should be mentioned. First, the oasis city of Ejinaqi in West China is not representative for whole China. For the vast Western region of China, however, our study

Table 2. Associations (Univariate Analysis) between Intraocular Pressure (Right Eyes) and Systemic and Ocular Parameters in the Gobi Desert Children Eye Study.

Parameter	P-Value	Standardized Coefficient Beta	Regression Coefficient B	95% Confidence Interval
Age (Years)	<0.001	−0.07	−0.07	−0.11, −0.02
Gender	0.02	0.06	0.42	0.08, 0.76
Body Height (cm)	0.05	−0.05	−0.01	−0.02, 0.00
Body Weight (kg)	0.60	−0.01	−0.003	−0.013, 0.007
Body Mass Index (kg/m^2)	0.46	0.02	0.02	−0.03, 0.06
Systolic Blood Pressure (mm Hg)	<0.001	0.10	0.03	0.01, 0.04
Diastolic Blood Pressure (mm Hg)	<0.001	0.10	0.04	0.02, 0.06
Mean Blood Pressure (m Hg)	<0.001	0.11	0.04	0.02, 0.06
Estimated Cerebrospinal Fluid Pressure (mm Hg)	<0.001	0.09	0.13	0.06, 0.20
Fathers Ethnicity (Han Chinese Versus Mongolian)	0.049	0.05	0.43	0.001, 0.86
Mothers Ethnicity (Han Chinese Versus Mongolian)	0.16	0.04	0.29	−0.11, 0.70
Gestational Time	0.37	0.02	0.01	−0.01, 0.02
Oxygen Supply in the Neonatal Phase	0.40	−0.02	−0.29	−0.95, 0.38
Number of Hours Spent Indoors (Hours)	0.91	−0.003	−0.01	−0.14, 0.12
Number of Hours Spent Outdoors (Hours)	0.66	0.01	0.03	−0.10, 0.16
Keratometric Reading (Diopters)	<0.001	0.12	0.26	0.15, 0.36
Refractive Error (Spherical Equivalent) (Diopters)	0.13	−0.04	−0.07	−0.15, 0.02
Intraocular Pressure Contralateral Eye (mm Hg)	<0.001	0.74	0.71	0.68, 0.74
Best Corrected Visual Acuity	0.78	−0.01	−0.01	−0.05, 0.04

was the first one to report on IOP in children. Second, as in any population-based study, the participation rate is of concern. In our study, the response rate of 81.9% of eligible children participating may be sufficient as basis for a population-based statistical analysis. Third, we did not measure central corneal thickness which without doubt influences the IOP measurements. Previous studies on adults showed however, that with or without inclusion of central corneal thickness into the multivariate model, higher IOP was associated with higher blood pressure and flatter cornea shape [23–28,30]. Fourth, we did not asses the educational level of the parents which could also have influenced the IOP. Strengths of our study were that we included almost all children of the region in

contrast to previous school-based studies, in which usually schools were randomly selected and their children were asked to participate in the study.

In conclusion, higher IOP in school children in West China was associated with steeper corneal curvature and with younger age and higher blood pressure, or alternatively, with higher estimated CSFP. Corneal curvature radius should be included for the correction of IOP measurements. The potential association between IOP and CSFP as also assumed in adults may warrant further research.

Table 3. Associations (Multivariate Analysis) between Intraocular Pressure (Right Eyes) and Systemic and Ocular Parameters in the Gobi Desert Children Eye Study.

Parameter	P-Value	Standardized Coefficient Beta	Regression Coefficient B	95% Confidence Interval
Age (Years)	<0.001	−0.13	−0.13	−0.18, −0.07
Diastolic Blood Pressure (mm Hg)	<0.001	0.13	0.05	0.03, 0.07
Corneal Refractive Power (Diopters)	<0.001	0.11	0.23	0.12, 0.34
Refractive Error (Spher. Equiv.) (Diopters)	0.035	−0.06	−0.10	−0.19, −0.001
Fathers Ethnicity (Han Chinese/Non-Han Chinese)	0.03	0.06	0.42	0.04, 0.89

Table 4. Associations (Multivariate Analysis) between Intraocular Pressure (Left Eyes) and Systemic and Ocular Parameters in the Gobi Desert Children Eye Study.

Parameter	*P*-Value	Standardized Coefficient Beta	Regression Coefficient B	95% Confidence Interval
Age (Years)	<0.001	−0.14	−0.13	−0.19, −0.08
Gender (Boys/Girls)	<0.001	0.09	0.67	0.32,1.03
Diastolic Blood Pressure (mm Hg)	<0.001	0.13	0.05	0.03, 0.07
Corneal Refractive Power (Diopters)	0.002	0.08	0.21	0.08, 0.33

Author Contributions

Conceived and designed the experiments: DYY DZ KG YW YYG XRY XXJ HKG YT JBJ. Performed the experiments: DYY DZ KG YW YYG XRY XXJ HKG YT. Analyzed the data: DYY DZ KG YW YYG XRY XXJ HKG YT JBJ. Contributed reagents/materials/analysis tools: DYY DZ KG YW YYG XRY XXJ HKG YT JBJ. Wrote the paper: DYY DZ KG YW YYG XRY XXJ HKG YT JBJ.

References

1. Klein BE, Klein R, Linton KL (1992) Intraocular pressure in an American community. The Beaver Dam Study. Invest Ophthalmol Vis Sci 33:2224–2228.
2. Dielemans I, Vingerling JR, Algra D, Hofman A, Grobbee DE, et al. (1995) Primary open-angle glaucoma, intraocular pressure, and systemic blood pressure in the general elderly population. The Rotterdam Study. Ophthalmology 102:54–60.
3. Wu SY, Leske MC (1997) Associations with intraocular pressure in the Barbados Eye Study. Arch Ophthalmol 115:1572–1576.
4. Youn DH, Yu YS, Park IW (1990) Intraocular pressure and axial length in children. Korean J Ophthalmol 4:26–29.
5. Duckman RH, Fitzgerald DE (1992) Evaluation of intraocular pressure in a pediatric population. Optom Vis Sci 69:705–709.
6. Pensiero S, Da Pozzo S, Perissutti P, Cavallini GM, Guerra R (1992) Normal intraocular pressure in children. J Pediatr Ophthalmol Strabismus 29:79–84.
7. Edwards MH, Chun CY, Leung SS (1993) Intraocular pressure in an unselected sample of 6- to 7-year-old Chinese children. Optom Vis Sci 70:198–200.
8. Edwards MH, Brown B (1993) Intraocular pressure in a selected sample of myopic and nonmyopic Chinese children. Optom Vis Sci 70:15–17.
9. Jaafar MS, Kazi GA (1993) Normal intraocular pressure in children: a comparative study of the Perkins applanation tonometer and the pneumatonometer. J Pediatr Ophthalmol Strabismus 30:284–287.
10. Broman AT, Congdon NG, Bandeen-Roche K, Quigley HA (2007) Influence of corneal structure, corneal responsiveness, and other ocular parameters on tonometric measurement of intraocular pressure. J Glaucoma 16:581–588.
11. Yildirim N, Sahin A, Basmak H, Bal C (2007) Effect of central corneal thickness and radius of the corneal curvature on intraocular pressure measured with the Tono-Pen and noncontact tonometer in healthy schoolchildren. J Pediatr Ophthalmol Strabismus 44:216–222.
12. Sahin A, Basmak H, Yildirim N (2008) The influence of central corneal thickness and corneal curvature on intraocular pressure measured by tono-pen and reboundtonometer in children. J Glaucoma 17:57–61.
13. Sihota R, Tuli D, Dada T, Gupta V, Sachdeva MM (2006) Distribution and determinants of intraocular pressure in a normal pediatric population. J Pediatr Ophthalmol Strabismus 43:14–18.
14. Sahin A, Basmak H, Yildirim N (2008) The influence of central corneal thickness and corneal curvature on intraocular pressure measured by tono-pen and reboundtonometer in children. J Glaucoma 17:57–61.
15. Kawase K, Tomidokoro A, Araie M, Iwase A, Yamamoto T, et al. (2008) Ocular and systemic factors related to intraocular pressure in Japanese adults: the Tajimi study. Br J Ophthalmol 92:1175–1179.
16. Wong TY, Klein BE, Klein R, Knudtson M, Lee KE (2003) Refractive errors, intraocular pressure, and glaucoma in a white population. Ophthalmology 110:211–217.
17. Edwards MH, Brown B (1996) IOP in myopic children: the relationship between increases in IOP and the development of myopia. Ophthalmic Physiol Opt 16:243–246.
18. Goss DA, Caffey TW (1999) Clinical findings before the onset of myopia in youth: 5. Intraocular pressure. Optom Vis Sci 76:286–291.
19. Xie XB, Zhang XJ, Fu J, Wang H, Jonas JB, et al. (2013) Intracranial pressure estimation by orbital subarachnoid space measurement. Crit Care 17:R162.
20. Jonas JB, Nangia V, Wang N, Bhate K, Nangia P, et al. (2013) Trans-lamina cribrosa pressure difference and open-angle glaucoma: The Central India Eye and Medical Study. PLoS One 8:e82284.
21. Jonas JB, Wang N, Wang YX, You QS, Xie X, et al. (2014) Body height, estimated cerebrospinal fluid pressure, and open-angle glaucoma. The Beijing Eye Study 2011. PLoS One 9:e86678.
22. Ren R, Jonas JB, Tian G, Zhen Y, Ma K, et al. (2010) Cerebrospinal fluid pressure in glaucoma. A prospective study. Ophthalmology 117:259–266.
23. Foster PJ, Machin D, Wong TY, Ng TP, Kirwan JF, et al. (2003) Determinants of intraocular pressure and its association with glaucomatous optic neuropathy in Chinese Singaporeans: the Tanjong Pagar Study. Invest Ophthalmol Vis Sci 44:3885–3891.
24. Mitchell P, Lee AJ, Wang JJ, Rochtchina E (2005) Intraocular pressure over the clinical range of blood pressure: blue mountains eye study findings. Am J Ophthalmol 140:131–132.
25. Klein BE, Klein R, Knudtson MD (2005) Intraocular pressure and systemic blood pressure: longitudinal perspective: the Beaver Dam Eye Study. Br J Ophthalmol 89:284–287.
26. Memarzadeh F, Ying-Lai M, Azen SP, Varma R; Los Angeles Latino Eye Study Group (2008) Associations with intraocular pressure in Latinos: the Los Angeles Latino Eye Study. Am J Ophthalmol 146:69–76.
27. Xu L, Wang H, Wang Y, Jonas JB (2007) Intraocular pressure correlated with arterial blood pressure. The Beijing Eye Study. Am J Ophthalmol 144: 461–462.
28. Jonas JB, Nangia V, Matin A, Sinha A, Kulkarni M, et al. (2011) Intraocular pressure and associated factors. The Central India Eye and Medical Study. J Glaucoma 20:405–409.
29. Samuels BC, Hammes NM, Johnson PL, Shekhar A, McKinnon SJ, et al. (2012) Dorsomedial/Perifornical hypothalamic stimulation increases intraocular pressure, intracranial pressure, and the translaminar pressure gradient. Invest Ophthalmol Vis Sci 53:7328–7335.
30. Mark HH, Mark TL (2003) Corneal astigmatism in applanation tonometry. Eye 17:617–618.
31. Mori K, Ando F, Nomura H, Sato Y, Shimokata H (2000) Relationship between intraocular pressure and obesity in Japan. Int J Epidemiol 29:661–666.
32. Shiose Y, Kawase Y (1986) A new approach to stratified normal intraocular pressure in a general population. Am J Ophthalmol 101:714–721.
33. Zhou Q, Liang YB, Wong TY, Yang XH, Lian L, et al. (2012) Intraocular pressure and its relationship to ocular and systemic factors in a healthy Chinese rural population: the Handan Eye Study. Ophthalmic Epidemiol 19:278–284.
34. Wang YX, Xu L, Zhang XH, You QS, Zhao L, et al. (2013) Five-year change in intraocular pressure associated with changes in arterial blood pressure and body mass index. The Beijing Eye Study. PLoS One 8:e77180.
35. Tomoyose E, Higa A, Sakai H, Sawaguchi S, Iwase A, et al. (2010) Intraocular pressure and related systemic and ocular biometric factors in a population-based study in Japan: the Kumejima study. Am J Ophthalmol 150:279–286.

15

Saharan Dust Deposition May Affect Phytoplankton Growth in the Mediterranean Sea at Ecological Time Scales

Rachele Gallisai[1]*, Francesc Peters[1], Gianluca Volpe[2], Sara Basart[3], José Maria Baldasano[3,4]

1 Departament de Biologia Marina i Oceanografia, Institut de Ciències del Mar, CSIC, Barcelona, Spain, 2 Istituto di Scienze dell'Atmosfera e del Clima, Roma, Italy, 3 Earth Sciences Department, Barcelona Supercomputing Center-Centro Nacional de Supercomputación, BSC-CNS, Barcelona, Spain, 4 Environmental Modelling Laboratory, Technical University of Catalonia, Barcelona, Spain

Abstract

The surface waters of the Mediterranean Sea are extremely poor in the nutrients necessary for plankton growth. At the same time, the Mediterranean Sea borders with the largest and most active desert areas in the world and the atmosphere over the basin is subject to frequent injections of mineral dust particles. We describe statistical correlations between dust deposition over the Mediterranean Sea and surface chlorophyll concentrations at ecological time scales. Aerosol deposition of Saharan origin may explain 1 to 10% (average 5%) of seasonally detrended chlorophyll variability in the low nutrient-low chlorophyll Mediterranean. Most of the statistically significant correlations are positive with main effects in spring over the Eastern and Central Mediterranean, conforming to a view of dust events fueling needed nutrients to the planktonic community. Some areas show negative effects of dust deposition on chlorophyll, coinciding with regions under a large influence of aerosols from European origin. The influence of dust deposition on chlorophyll dynamics may become larger in future scenarios of increased aridity and shallowing of the mixed layer.

Editor: Tomoya Iwata, University of Yamanashi, Japan

Funding: The work was supported by the following: JAE-Predoc fellowship from the Spanish Scientific Research Council (CSIC), RG; Aerosol deposition and ocean plankton dynamics (CTM2011-23458), Spanish Ministry of Science and Innovation project, http://www.idi.mineco.gob.es/, RG and FP; Estructura de la materia orgánica en el océano costero: implicaciones biogeoquímicas y ecológicas (CTM2009-09352), Spanish Ministry of Science and Innovation project, http://www.idi.mineco.gob.es/, FP; Grup d'Oceanografia Mediterrània (2009/SGR/588), Generalitat de Catalunya project, http://www10.gencat.cat/agaur_web/AppJava/catala/index.jsp, RG and FP; La ricerca italiana per il mare, Italian Ministry of Education, University and Research, http://www.ritmare.it/, GV; Acoplamiento on-line de un modulo completo de aerosoles multicomponente al modelo atmosferico global regional NMMB (CGL2010-19652), Spanish Ministry of Science and Innovation project, http://www.idi.mineco.gob.es/, SB and JMB; Supercomputación and e-ciencia project (CSD2007-0050) from the Consolider-Ingenio 2010 program of the Spanish Ministry of Economy and Competitivity, http://www.idi.mineco.gob.es/, SB and JMB; and Severo Ochoa program (SEV-2011-00067) from Spanish Ministry of Economy and Competitivity, http://www.idi.mineco.gob.es/, SB and JMB. The funders had no role in study design, data collection and analysis, decision to publish, or preparation of the manuscript.

Competing Interests: The authors have declared that no competing interests exist.

* Email: gallisai@icm.csic.es

Introduction

Aerosols have major impacts on weather and climate regulations [1,2] and even on crop production [3]. Atmospheric desert dust may travel large distances from its source and has been proposed to have ocean production regulation effects over geological times scales [4]. The Mediterranean Sea (hereafter Med) atmosphere is subject to the continuous injection of Saharan and Middle East mineral dust particles [5]. The deposition of these mineral particles supply numerous macro and micro- nutrients to the ocean surface [6–14] and some authors consider it as the major source of "new" nutrients [15] for system production.

Calculations show that the atmospheric input of nutrients in the Med is of the same magnitude as riverine inputs [16–18], thus playing a significant role in the regulation of the nutrient balance of the basin at decadal or longer time scales [19,20]. The contribution of atmospheric deposition can be especially important and efficient in oligotrophic environments such as the Med, which has a marked stratification period and a pronounced nutrient limitation [21]. The deposition of some of these soluble compounds on surface waters may influence biological production, at least during certain events [9,22,23]. Dust deposition spreads over vast areas and dilutes into the water column often preventing the potential effects on system production to be unequivocally detected at ecological time scales. Experiments and observations in low nutrient – low chlorophyll areas have so far shown mixed results [24–26]. Reasons may include a tremendous variability in dust nutrient bioavailability content [27–29] and a relatively small increase of the background nutrient concentration when vertical mixing is active and represents the major source of nutrients [21] as well as a rapid transfer of increased primary production to other trophic levels and a variety of plankton community structures and physiological states.

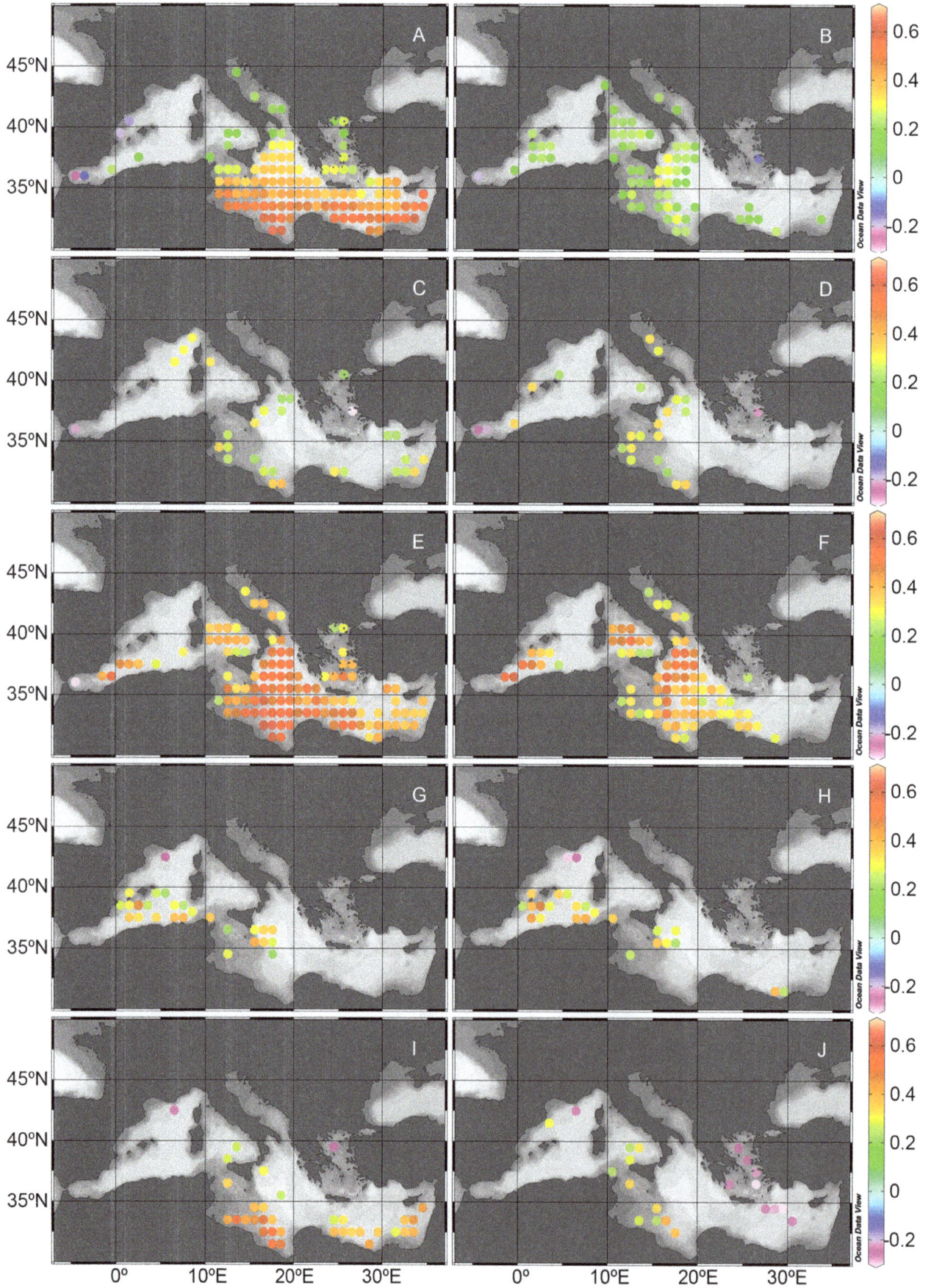

Figure 1. Correlation between chlorophyll concentration and dust deposition. Statistically significant (p<0.05) correlation coefficient (r) between chlorophyll concentration and dust deposition (left panels) and between the seasonally detrended chlorophyll concentration and the seasonally detrended dust deposition (right panels) for the whole time series and for different seasons. Panels: a, b) annual; c, d) winter (January to March); e, f) spring (April to June); g, h) summer (July to September) and i, j) autumn (October to December).

Given the episodic nature of dust events, an additional complication may reside in the human capacity of detecting the dust event deposition with sufficient space-time resolution in order to build a statistically significant dust event database. Previous attempts used satellite-derived aerosol optical thickness (AOT) as a proxy of dust in the atmosphere to infer the deposition events [26,30,31]. Dust generally travels from several hundreds to thousands of meters high in the atmosphere, making this approach not quantitatively adequate for discerning between transport and deposition. Deposition is measured *in situ* at a few terrestrial (mainland and islands) sites, which are extremely valuable for ground truth validation but are dependent on local conditions, making generalizations hard to draw especially towards the open ocean. Here, we employ a state-of-art atmospheric transport and deposition model, the BSC-DREAM8b model [32], which has been validated [33–35] and gives the power of having aerosol deposition data over the whole Med basin with daily temporal resolution. A previous study showed the potential positive effects of dust deposition on SeaWiFS-derived chlorophyll (Chl) in the Med [32]. However, in the Med, the used NASA OC4v4 algorithm falls far short to retrieve Chl with accuracy smaller than 100%, casting doubts on the relationships found. Here we extend this approach by relating deposition to SeaWiFS Chl using the Med-specific algorithm MedOC4 [36]. When we think of dust deposition, we tend to think about very large events, those that are obvious in true color images or that we recognize because we find our cars covered with red dust, but the truth is that, to some extent, there is Saharan dust in the atmosphere over the Mediterranean almost continuously and deposition does not occur only during large events but also when atmospheric aerosol concentrations are not so high. Thus, rather than focusing on single events or experiments, we take a correlational approach using an 8-year data time series in order to find relationships between Chl dynamics and dust deposition over the Mediterranean Sea.

Methods

Chlorophyll data

SeaWiFS HRPT Level-1A data (2000–2007) were collected at the Istituto di Scienze dell'Atmosfera e del Clima of Rome, Italy, and processed up to Level-3 using the MedOC4 regional algorithm (http://www.myocean.eu/web/69-myocean-interactive-catalogue.php) [36]. This algorithm takes into account the peculiar blue-green ratio of Med waters. Level-3 Chl data, with a native 1 km resolution, were $\log_{10}$-transformed averaged, over a period of eight days, and regridded over the $1°$ resolution grid of the basin (179 cells, see Table S1). A previous study showed that it is recommended not to use 8-d averages when computing correlation analysis between Chl and dust events [26]. To account for the possible contamination by atmospheric dust mimicking chlorophyll, here, before averaging over the period of eight days, the quality of the entire Chl dataset was carefully checked by i) applying all the SeaDAS Level-2 processing masks and flags (http://oceancolor.gsfc.nasa.gov/VALIDATION/flags.html), ii) removing all isolated pixels, iii) removing all pixels exceeding 3 standard deviations within a moving box of 3×3 pixels, and iv) by applying a median filter over all remaining good pixels. This procedure increases the confidence level on data quality, with the only shortcoming of reducing the

number of observations with respect to the NASA standard processing. The time series of daily observations was temporally binned into periods of 8 days. This results into 45 bins up to the 360th day of the year. The last bin was computed with the remaining 5 days, and in the case of leap years, with the remaining 6 days. The climatic mean is then calculated across years for each of the natural 8-d time periods.

Dust deposition

For the present study, a dust deposition simulation from the BSC-DREAM8b model (http://www.bsc.es/earth-sciences/mineral-dust/catalogo-datos-dust) model [33,37] was used for the period between 1 January 2000 and 31 December 2007, over the Med basin. BSC-DREAM8b tracks mineral dust particles from their sources in the Sahara and Middle East regions. Output, after being $\log_{10}$-transformed, was provided for the same space and time resolution as for chlorophyll. A low cut-off threshold (10^{-8} Kg m^{-2} d^{-1}) is applied to the numerical deposition output from BSC-DREAM8b since the dataset showed numerically correct but physically unrealistic low value spikes [32]. The model main features were described in detail in Pérez et al. [37] and Basart et al. [35]. It has been used for dust forecasting and as a dust research tool in North Africa and the Med [32,38–40]. Several studies have checked its performance [33,41], concerning both the horizontal and vertical extent of the dust plumes in the Med Basin. The model daily evaluation with near-real time observations is conducted at the Barcelona Supercomputing Center, and includes satellite data (MODIS and MSG) and AERONET sun photometers. BSC-DREAM8b has also been validated and tested over longer time periods in the European region [34,42,43] and against measurements at source regions [44].

Aerosol Optical Thickness

AOT at 865 nm data were derived from SeaWiFS radiometer measurements and they were downloaded from the Giovanni database (http://gdata1.sci.gsfc.nasa.gov/daac-bin/G3/gui.cgi?instance_id=ocean_8day). We acquired 8-d averaged, 9 km resolution product from 2000 to 2007. Similarly to Chl and deposition data, AOT data were $\log_{10}$-transformed and regridded over the $1°$ resolution grid of the basin, with the same temporal binning. This was done for the same 179 $1°×1°$ cells as for chlorophyll. It should be noted that AOT contains information of total aerosol particles in the atmosphere, not only of particles from Saharan origin. However, over much of the Mediterranean Sea most particles are indeed of Saharan origin [45].

Statistical analyses

Pearson's correlation coefficient (r) was calculated between chlorophyll concentration, modeled dust deposition and AOT time series for each grid cell. Significance was considered at p< 0.05 using Student's t-test. In addition, the degrees of freedom used for significance testing were adjusted to take into account the possible presence of autocorrelation in the time series. The number of effective independent observations, N*, were calculated as described in Pyper el al. [46]. Correlations were computed both for the entire series and for each season. The same analyses were performed after seasonally detrending the data by subtracting the

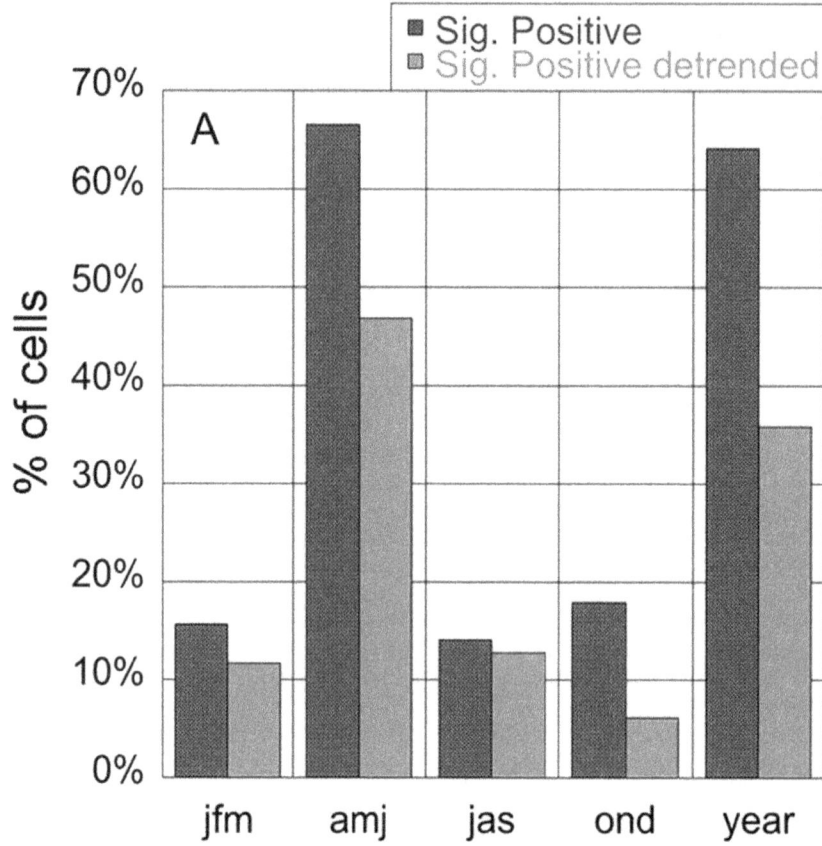

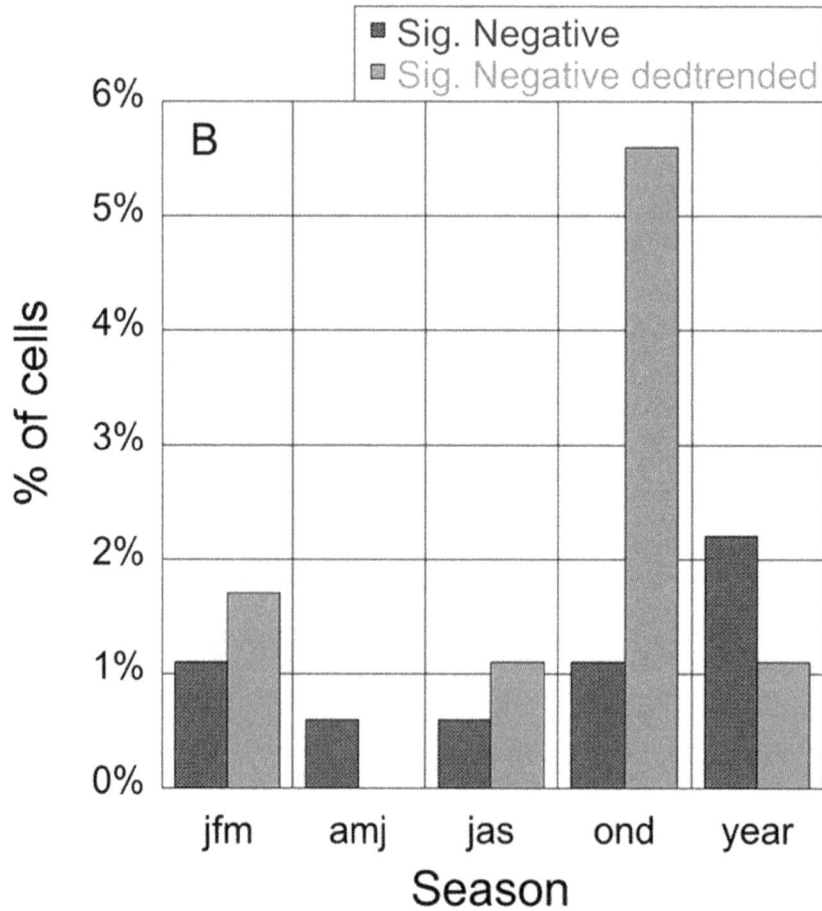

Figure 2. Percentage of cells showing significant correlations between chlorophyll and deposition. Left panel: positive correlations. Right panel: negative correlations. Red bars represent non seasonally detrended data and blue bars seasonally detrended data.

climatic mean at each time series data point. The r^2 of the correlation in a cell is the variance explained by the correlation in that cell. The minimum, maximum and average variance-explained values (expressed in %variability) were calculated for the population of cells with a $p<0.05$.

Results and Discussion

We have found statistically significant positive correlations between surface chlorophyll and mineral dust deposition in large areas of the Med, covering 64% of the analyzed surface and located mainly in the Central and Eastern basins (Fig. 1) and with a clear south to north gradient in correlation intensity from 0.63 to 0.12. Significant negative correlations (r from −0.15 to −0.25) are observed in only 4 cells located in the Alboran Sea and in the eastern coast of Spain. Positive correlations can be found during all seasons, although it is in spring when we see the largest effects with correlations ranging from 0.22 to 0.65 mainly in the Central, Eastern and Southwestern Med. The Western and Central Med also show regions with positive correlations in summer, while in autumn there are some areas affected in the Central and Eastern Med. Most of the Med phytoplankton variability (>80%) is well explained by the variability of the mixed layer depth [47], and especially the winter-spring mixing bringing nutrient-rich deep waters to the surface. Thus, at least part of the explained variability between our deposition and chlorophyll time series must be due to the partial matching of the annual cycles of both variables.

The relationship between the seasonally detrended data of chlorophyll and dust deposition, that represents more of a response of short-term chlorophyll peaks to dust outbreaks, is somewhat weaker in intensity and in area covered. Largest positive correlations are found in the Central Med (from 0.13 to 0.32) (Fig. 1). Again, it is in spring where the largest impacted area is found, mainly in the Central Med and extending into the Eastern Med and Southwestern Med with r ranging from 0.24 to 0.58. The Western Med shows the largest area affected in summer. This is not surprising given that the seasonal dust event frequency peaks during spring in the Central-Eastern Med, and during summer over the Central-Western basin [26]. Seasonally detrended data tend to slightly increase the number of cells showing significant negative correlations and decrease the number of significantly positive correlated cells (Fig. 2). It is in autumn when we see the largest number of negatively correlated cells (6% of analyzed surface) and located mainly in the Aegean Sea and extending southeasterly of Crete.

For the seasonally detrended data, we checked that the correlation values were not caused by chance. We generated synthetic seasonally detrended chlorophyll time series with the observed mean and standard deviation for each cell. Correlations were computed with dust deposition model outputs, and the process repeated 100 times (Fig. 3). The observed significant correlations were compared to the distribution of the synthetic correlations for each cell, and in all cases they were statistically different with an $\alpha<0.001$ and a power $(1-\beta)$ undistinguishable from 1. This confirmed the non-spurious nature of the relationships between dust deposition and non-seasonal chlorophyll time series.

Bulk Saharan dust deposition over the Med is not straightforwardly related to dust travelling in the atmosphere (Fig. S1). Meteorological conditions and wind patterns at different times of the year often have large amounts of dust (AOT) travelling at altitude with little deposition [48–50]. AOT and dust deposition

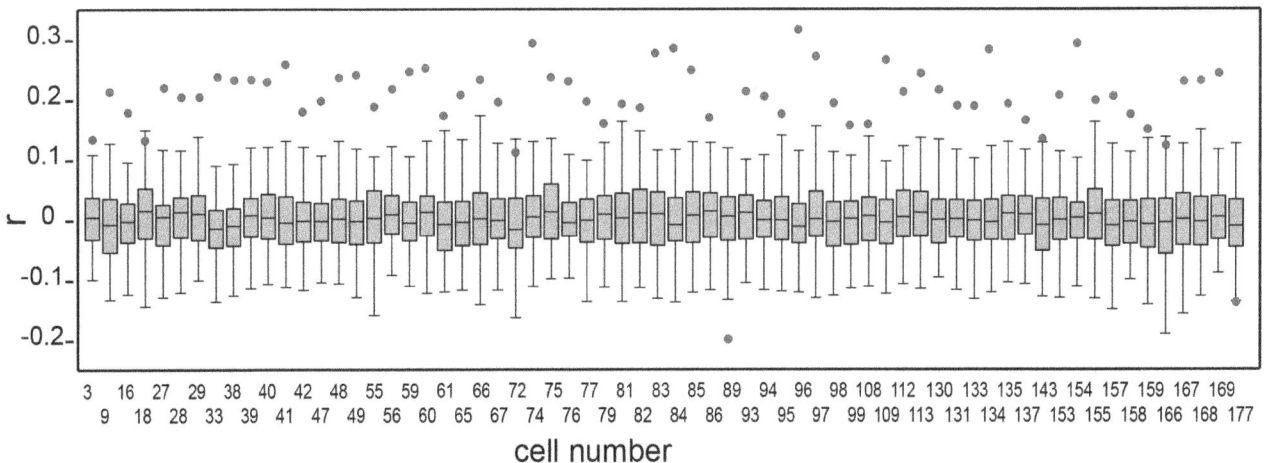

Figure 3. Analysis of the chance of significant correlations being spurious. Comparison between the box plots of the distribution of correlation coefficients between synthetic seasonally detrended chlorophyll time series and seasonally detrended dust deposition model outputs (N = 100) and the actual observed correlation between the seasonally detrended chlorophyll and the seasonally detrended dust deposition model outputs (dots). Data is shown only for those cells showing significant (p<0.05) observed correlations. Dots in blue represent significantly positive correlations and red significantly negative correlations. Box plots show the median, the grey box englobing all data between the 25 and 75 percentiles, and the range between the smallest and largest values that are not outliers. Starting from the detrended data of the cells that show statistically significant correlations between detrended chlorophyll and deposition (Fig. 1b), synthetic Chl time series with the same mean and standard deviation (normal distribution) as the original detrended chlorophyll time series, were computed for each cell. The correlation between these synthetic Chl time series and the modelled dust deposition were computed. For each cell, this process was repeated 100 times, and the probability distribution functions (PDFs) of the correlations were then obtained and presented as box plots.

Table 1. Percentage of observed chlorophyll variability explained by modeled dust deposition.

Season	Non-detrended data				Seasonally detrended data			
	N	Min	Max	Average	n	Min	Max	Average
Annual	115	1.4	40.2	16.4	64	1.3	10.1	4.7
Winter	28	4.3	15.1	7.8	21	4.4	12.6	8.5
Spring	119	5.0	41.6	19.1	84	5.7	33.3	15.0
Summer	25	4.2	19.6	10.2	23	4.7	21.3	11.1
Autumn	32	6.5	26.9	15.6	11	5.4	16.5	9.6

Number of cells (n) with significantly (p<0.05) positive correlations. Minimum (min), maximum (max) and average percentage of chlorophyll variability explained in the significantly positive cells.

show positive correlation in the Western Med (Fig. S2.) especially in spring and summer with correlated areas shifting depending on the season. The Eastern sub-basin presents the highest correlations in spring (from 0.23 to 0.51) and the Central Med (Tyrrhenian Sea, Sicily channel and Dardanelle strait) in autumn. Once the data are seasonally detrended, deposition events are more related to AOT events, both when the whole series is considered and when the data are analyzed for the different times of the year (Fig. S2). With respect to non-detrended data, seasonally detrended data show main increases in correlation and correlated area in the Central Med for most of the year, as well as in the Eastern Med in autumn. Some overall hotspots appeared in the Alboran and in the Tyrrhenian Sea and around Crete, where the correlations ranged from 0.36 to 0.48.

The annual cycles of chlorophyll and AOT do not match (Fig. S1). The maximum chlorophyll concentrations occur in winter and minima coincide with the summer months. On the contrary, the highest AOT is found in summer and the minimum in autumn. Overall, AOT and chlorophyll (Fig. S3) show no significant correlations in the Med, except for some areas near the African coasts, where the correlation is negative (from -0.28 to -0.36). While no correlations are evident between AOT and chlorophyll there are significant correlations between seasonally detrended AOT and chlorophyll data (Fig. S3). A plume of higher correlation, with r-values between 0.33 and 0.39, appear in the northern part of Cyrenaica region with an extension up to the south of Italy. The best match between both series was found in summer (Fig. S3). Volpe et al. [36] ground truthed the chlorophyll satellite estimates with in situ measurements and concluded that the atmospheric correction was appropriate. In addition, we compared the data from the chlorophyll measurements at the DYFAMED station (1998–2007) with SeaWiFS estimates corrected with a regional algorithm giving a slope of ~ 1 (logDYF = 0.0129+1.0497·logSW; Adjusted $R^2 = 0.68$; N = 91; p<0.001). Moreover DYFAMED chlorophyll was unrelated to AOT, providing further evidence of the independence between satellite measurements of chlorophyll and AOT. Aerosols travelling over a certain area are not necessarily depositing. When a deposition event is occurring, it should coincide with high aerosol content in the air (AOT), thus if we find relationships between dust deposition events and non-seasonal chlorophyll peaks it is also logical to expect that chlorophyll is related to AOT, while the non-detrended AOT data show little or no relationship.

As mentioned before, the largest positive correlations between dust deposition and chlorophyll occur around the Central and Eastern Med. Calculations [8,20,51] and experiments [11,20,24] tell us that aerosol deposition effects on primary production should be small in most situations and thus we do not expect African dust deposition in general to explain a large portion of chlorophyll variability. Accordingly, positive significant correlations between mineral dust deposition of Saharan origin and chlorophyll do explain only a 1 to 10% (average 5%) of chlorophyll variability for seasonally detrended and a 1 to 40% (average 16%) for non-detrended data although it may be higher for certain seasons (Table 1). It should be noted that the explained variability does not provide direct information of the magnitude of chlorophyll impacted.

Winter shows overall the lowest significantly positive correlations, while spring presents the highest. This is to be expected since the entrance of new nutrients should be mostly due to seasonal winter overturning and mixing of nutrient-rich deep waters with upper ocean surface waters, through a number of physical processes that increase vertical diffusion at certain moments. But even at times when nutrient concentrations are expected to be relatively high in the water, low concentrations and strong imbalances between N and P are often observed [52,53], opening windows of opportunity for the nutrients from atmospheric deposition to have an impact in the sustainment of phytoplankton production. We can only speculate on the positive cause-effect relationship between aerosol deposition and chlorophyll in the Med at certain times. Terrestrial inputs through major rivers occur mainly in the Western Med [54], and atmospheric inputs may dominate nutrient supply at certain times [10,55]. Phosphorus (P) limitation alleviation has often been invoked [8,56] as the surface waters of the Med are among the most P-limited in the world [57]. Although aerosols show a disproportionally large ratio of nitrogen to phosphorus [18], potentially only exacerbating P-limitation, they do carry an amount of P that could be used by phytoplankton and bacteria, especially in spring and summer when the concentrations of this element in surface waters of the open Med are at their lowest. Guieu et al. [10] calculated that, if P is considered the limiting element for phytoplankton growth, atmospheric deposition could account for chlorophyll increases of ca. 0.2 $\mu g\ L^{-1}$ in the upper mixed layer for a single large deposition event or for the average total deposition during the summer-stratified period. The Central and Eastern Med do not show the typical spring phytoplankton bloom and have been defined as no blooming areas [58]. The ultra-oligotrophic conditions [59] found in these areas should make them most responsive to external nutrient supplies. As is the case for high nutrient – low chlorophyll areas, micronutrients such as iron from aerosols have also been proposed to stimulate Med phytoplankton production under certain situations [60], albeit addition experiments have not shown a direct increase in dissolved iron (Fe) [61]. Fe in the mixed layer of the Mediterranean is found at concentrations from 0.13 to 2.7 nM [62,63]. It seems though that Fe is, relative to the needs of plankton, in excess with respect to P in the Mediterranean [64]. Nevertheless, in a system where all elements are relatively scarce, responses to the combination of elements arriving through aerosol deposition, may be very complex, with elements becoming successively limiting in a chained reaction. Ridame et al. [65] found stimulation of nitrogen (N) fixation in dust pulse experiments, in general related to a primary alleviation of P-limitation. In their Central Med experiment though, they found high N-fixation stimulation unrelated to P- or Fe-limitation, further showing the complexity of the processes involved and the potential spatial and temporal variability. An initial stimulation of heterotrophic bacteria [25,66] should not be discarded since these organisms have a potential advantage at low nutrient concentrations owing to their high surface to volume ratio. Secondarily, released nutrients from recycling could then stimulate phytoplankton processes. Contrary to the Eastern Med showing the lowest nutrient concentrations in the Mediterranean [59], the Central Med was found somewhat more responsive to dust deposition in the present study. Pey et al. [5] mention the Central Med as a transitional area, receiving a higher frequency of dust outbreaks than similar latitudes in the Western and Eastern Med. Additionally, the dust source areas are not homogeneous. The Libyan Desert is the main source of dust for the Central Med while the Eastern Med receives dust from Libya and from the Middle East [67]. Thus, positive correlations between dust deposition and surface chlorophyll seem to arise from the combination of areas of low nutrient concentrations with the right nature, timing and frequency of dust outbreaks.

Negative relationships between dust deposition and chlorophyll have been related to metal (mainly Cu but also Al) inhibition of phytoplankton growth [13,68]. The toxicity of Cu in reducing phytoplankton growth rate has been shown in laboratory experiments (see [68] and references therein). A recent correlation study between chlorophyll and metals from onshore-measured aerosols in the Northwest Med shows negative relationships in the area under northerly wind (Tramontane) conditions [13]. These winds favor the transport of anthropogenic aerosols from Europe to the Med. Although most Cu pulses are anthropogenically derived, pulses originating in Africa showed effects on chlorophyll undistinguishable from those originating locally [13]. A reduction in chlorophyll growth of up to 20% can be seen along the French and Spanish coasts. In addition, Jordi et al. [13] argue that since Cu toxicity seems to be taxon specific, the summer phytoplankton community with a predominance of nanoflagellates over the less sensitive diatoms, is more vulnerable to atmospheric deposition. This is an area where we also see some negative correlations between the modeled deposition and chlorophyll. We only track Saharan mineral dust, while some of the high load of metals may be more related to local anthropogenic sources. Most of the large deposition events in the Northwest Med come in the form of wet deposition [69]. In our model, the deposition field only originates from Saharan and Middle East dust transport and does not account for local anthropogenic aerosol sources, but rain washes out the entire atmospheric column aerosol loading, no matter the origin. Results from our correlation analysis agree with previous more detailed local studies [13]. We also see a negative relationship between deposition and chlorophyll, both seasonally detrended, mainly in autumn in the Aegean region (Fig. 1). This area is affected by long-range transport of air pollutants from Eastern Europe [70] but it is also heavily impacted by anthropogenic emissions generated in Athens and Istanbul [71,72]]. A high-density population together with a massive number of vehicles, many of them still using non-catalytic or old technology diesel engines, contributed to exceed the EU annual aerosol limit. The amount of Cu in these aerosols is high with an annual mean concentration between 0.013 and 0.22 $\mu g\ m^{-3}$ ([73] and references therein). An estimated dry deposition flux of Cu over the sea ranges then between 22 and 380 $\mu g\ Cu\ m^{-2}\ d^{-1}$ surpassing the threshold limit for Cu to inhibit phytoplankton growth rate according to [13] and [68].

Conclusion

Desert dust storm events seem to be increasing in frequency and intensity [14,69,72–76] in the last decades, due to human activities and climate forcing. This means that the presence of aerosols over the Med is likely to increase with future aridity. Thus, it is important to understand basin level patterns in the response of Med biogeochemistry to aerosol deposition. Only a few studies [26,30] have tried to analyze the potential links between aerosols in the air column and chlorophyll for the entire Med basin, with non definitive results. In this study, we use a modeled actual aerosol deposition product and show both positive and negative significant correlations with chlorophyll dynamics in certain areas and times of the year. Mineral dust from North Africa and the Middle East correlates to chlorophyll in large areas of the Med Sea. This is especially true for the Central and Eastern Med sub-basins, where Saharan dust deposition dynamics matches that of chlorophyll, particularly during spring. Here the atmospheric input may be an intrinsic part of the annual ecosystem dynamics. In terms of large dust outbreaks, chlorophyll best relates to aerosol deposition in the Central Med, extending both into the Eastern and Southwestern Med (Fig. S3). Some areas of the Western Med and Aegean Sea show negative correlations between chlorophyll and deposition in accordance with some recent findings of toxicity brought by metals in aerosols. As expected, dust deposition does not explain an overall large amount of chlorophyll variability since the main ecosystem production driver in the Med is the vertical mixing of nutrients from deep waters. Variability related to carbon to chlorophyll ratios, the consumption of biomass with a varying degree of coupling and the variable settling of primary production, are all additional sources of surface chlorophyll variability that we could not account for in our correlations and thus add to the noise. No matter how small significant correlations are, they are not distributed randomly in space and coincide with independent estimates that follow the same trend. Thus, albeit the mechanisms that affect chlorophyll through aerosol deposition cannot be pinpointed and may be indirect, non-unique, and dependent on local spatio-temporal conditions, our study shows a clear potential for effects at ecological scales. These effects should become more important in a future scenario with increased aerosols over the Med and a shallower mixed layer depth, owing to increased temperatures, over which aerosols may leach out nutrients.

Supporting Information

Figure S1   Seasonal average values of chlorophyll concentration, dust deposition and aerosol optical thickness. Average chlorophyll concentration (left panels). Average dust deposition (central panels) and average aerosol optical thickness (right panels) for different seasons. Winter (a, b, c), spring (d, e, f), summer (g, h, i) and autumn (j, k, l).

Figure S2   Correlation between dust deposition and aerosol optical thickness. Statistically significant ($p < 0.05$) correlation coefficient (r) between dust deposition and aerosol optical thickness (left panels) and between seasonally detrended dust deposition and seasonally detrended aerosol optical thickness

(right panels) for the whole time series and for different seasons. Panels: a, b) annual; c, d) winter (January to March); e, f) spring (April to June); g, h) summer (July to September) and i, j) autumn (October to December).

Figure S3   Correlation between chlorophyll concentration and aerosol optical thickness. Same as Fig. S2 but for chlorophyll concentration versus aerosol optical thickness.

Table S1   Geographical coordinates for the 1° ×1° grid cells analyzed in this study. Coordinates refer to the central point of the cell. Latitudes are all North. Positive longitudes are East and negative longitudes West.

Acknowledgments

The Goddard Earth Sciences Data and Information Services Center (GES DISC), the SeaWiFS mission scientists and associated NASA personnel for the production of the data used in this study and for the development and maintenance of SeaDAS software. The Barcelona Supercomputing Center (BSC) for hosting the BSC-DREAM8b simulations on the Mare Nostrum Supercomputer. The DYFAMED (CNRS-INSU) Observatoire Océanologique de Villefranche-sur-mer (http://www.obs-vlfr.fr/dyfBase/) for providing *in situ* chlorophyll measurements. The developers of Ocean Data View (http://odv.awi.de) for their useful tool.

Author Contributions

Conceived and designed the experiments: RG FP. Performed the experiments: RG FP GV SB JMB. Analyzed the data: RG FP GV SB JMB. Contributed reagents/materials/analysis tools: RG FP GV SB JMB. Contributed to the writing of the manuscript: RG FP GV SB JMB.

References

1. Booth BBB, Dunstone NJ, Halloran PR, Andrews T, Bellouin N (2012) Aerosols implicated as a prime driver of twentieth-century North Atlantic climate variability. Nature 484: 228–232.

2. Creamean JM, Suski KJ, Rosenfeld D, Cazorla A, DeMott PJ, et al. (2013) Dust and Biological Aerosols from the Sahara and Asia Influence Precipitation in the Western U.S. Science 339: 1572–1578.

3. Liu X, Zhang Y, Han W, Tang A, Shen J, et al. (2013) Enhanced nitrogen deposition over China. Nature 494: 459–462.

4. Jaccard SL, Hayes CT, Martinez-Garcia A, Hodell DA, Anderson RF, et al. (2013) Two Modes of Change in Southern Ocean Productivity Over the Past Million Years. Science 339: 1419–1423.

5. Pey J, Querol X, Alastuey A, Forastiere F, Stafoggia M (2013) African dust outbreaks over the Mediterranean Basin during 2001–2011: PM10 concentrations, phenomenology and trends, and its relation with synoptic and mesoscale meteorology. Atmospheric Chemistry and Physics 13: 1395–1410.

6. Bonnet S, Guieu C, Chiaverini J, Ras J, Stock A (2005) Effect of atmospheric nutrients on the autotrophic communities in a low nutrient, low chlorophyll system. Limnology and Oceanography 50: 1810–1819.

7. Bergametti G, Remoudaki E, Losno R, Steiner E, Chatenet B, et al. (1992) Source, transport and deposition of atmospheric phosphorus over the northwestern Mediterranean. Journal of Atmospheric Chemistry 14: 501–513.

8. Ridame C, Guieu C (2002) Saharan input of phosphate to the oligotrophic water of the open western Mediterranean sea. Limnology and Oceanography 47: 856–869.

9. Markaki Z, Oikonomou K, Kocak M, Kouvarakis G, Chaniotaki A, et al. (2003) Atmospheric deposition of inorganic phosphorus in the Levantine Basin, eastern Mediterranean: Spatial and temporal variability and its role in seawater productivity. Limnology and Oceanography 48: 1557–1568.

10. Guieu C, Loÿe-Pilot MD, Benyahya L, Dufour A (2010) Spatial variability of atmospheric fluxes of metals (Al, Fe, Cd, Zn and Pb) and phosphorus over the whole Mediterranean from a one-year monitoring experiment: Biogeochemical implications. Marine Chemistry 120: 164–178.

11. Pulido-Villena E, Rerolle V, Guieu C (2010) Transient fertilizing effect of dust in P-deficient LNLC surface ocean. Geophysical Research Letters 37: L01603.

12. Herut B, Krom MD, Pan G, Mortimer R (1999) Atmospheric input of nitrogen and phosphorus to the Southeast Mediterranean: Sources, fluxes, and possible impact. Limnology and Oceanography 44: 1683–1692.

13. Jordi A, Basterretxea G, Tovar-Sanchez A, Alastuey A, Querol X (2012) Copper aerosols inhibit phytoplankton growth in the Mediterranean Sea. Proceedings of the National Academy of Sciences of the United States of America 109: 21246–21249.

14. Goudie AS, Middleton NJ (2001) Saharan dust storms: nature and consequences. Earth-Science Reviews 56: 179–204.

15. Ternon E, Guieu C, Ridame C, L'Helguen S, Catala P (2011) Longitudinal variability of the biogeochemical role of Mediterranean aerosols in the Mediterranean Sea. Biogeosciences 8: 1067–1080.

16. Guieu C, Martin JM, Thomas AJ, Elbazpoulichet F (1991) Atmospheric versus river inputs of metals to the Gulf of Lions: Total concentrations, partitioning and fluxes. Marine Pollution Bulletin 22: 176–183.

17. Ludwig W, Bouwman AF, Dumont E, Lespinas F (2010) Water and nutrient fluxes from major Mediterranean and Black Sea rivers: Past and future trends and their implications for the basin-scale budgets. Global Biogeochemical Cycles 24: GB0A13.

18. Markaki Z, Loÿe-Pilot MD, Violaki K, Benyahya L, Mihalopoulos N (2010) Variability of atmospheric deposition of dissolved nitrogen and phosphorus in the Mediterranean and possible link to the anomalous seawater N/P ratio. Marine Chemistry 120: 187–194.

19. Bethoux JP, Morin P, Ruiz-Pino DP (2002) Temporal trends in nutrient ratios: chemical evidence of Mediterranean ecosystem changes driven by human activity. Deep-Sea Research Part Ii-Topical Studies in Oceanography 49: 2007–2016.

20. Herut B, Zohary T, Krom MD, Mantoura RFC, Pitta P, et al. (2005) Response of East Mediterranean surface water to Saharan dust: On-board microcosm experiment and field observations. Deep-Sea Research Part Ii-Topical Studies in Oceanography 52: 3024–3040.

21. Estrada M (1996) Primary production in the northwestern Mediterranean. Scientia Marina 60: 55–64.

22. Guerzoni S, Chester R, Dulac F, Herut B, Loÿe-Pilot MD, et al. (1999) The role of atmospheric deposition in the biogeochemistry of the Mediterranean Sea. Progress in Oceanography 44: 147–190.

23. Morales-Baquero R, Pulido-Villena E, Reche I (2006) Atmospheric Inputs of Phosphorus and Nitrogen to the Southwest Mediterranean Region: Biogeochemical Responses of High Mountain Lakes. Limnology and Oceanography 51: 830–837.

24. Romero E, Peters F, Marrasé C, Guadayol T, Gasol JM, et al. (2011) Coastal Mediterranean plankton stimulation dynamics through a dust storm event: An experimental simulation. Estuarine, Coastal and Shelf Science 93: 27–39.

25. Lekunberri I, Lefort T, Romero E, Vázquez-Domínguez E, Romera-Castillo C, et al. (2010) Effects of a dust deposition event on coastal marine microbial abundance and activity, bacterial community structure and ecosystem function. Journal of Plankton Research 32: 381–396.

26. Volpe G, Banzon VF, Evans RH, Santoleri R, Mariano AJ, et al. (2009) Satellite observations of the impact of dust in a low-nutrient, low-chlorophyll region: Fertilization or artifact? Global Biogeochem Cycles 23: GB3007.

27. Carbo P, Krom MD, Homoky WB, Benning LG, Herut B (2005) Impact of atmospheric deposition on N and P geochemistry in the southeastern Levantine basin. Deep-Sea Research Part II: Topical Studies in Oceanography 52: 3041–3053.

28. Herut B, Collier R, Krom MD (2002) The role of dust in supplying nitrogen and phosphorus to the Southeast Mediterranean. Limnology and Oceanography 47: 870–878.

29. Baker AR, French M, Linge KL (2006) Trends in aerosol nutrient solubility along a west-east transect of the Saharan dust plume. Geophysical Research Letters 33: L07805.

30. Cropp RA, Gabric AJ, McTainsh GH, Braddock RD, Tindale N (2005) Coupling between ocean biota and atmospheric aerosols: Dust, dimethylsulphide, or artifact? Global Biogeochemical Cycles 19: GB4002.

31. Gabric AJ, Cropp R, Ayers GP, McTainsh G, Braddock R (2002) Coupling between cycles of phytoplankton biomass and aerosol optical depth as derived from SeaWiFS time series in the Subantarctic Southern Ocean. Geophysical Research Letters 29: 1112.

32. Gallisai R, Peters F, Basart S, Baldasano JM (2012) Mediterranean basin-wide correlations between Saharan dust deposition and ocean chlorophyll concentration. Biogeosciences Discussion 9: 28.

33. Pérez C, Nickovic S, Baldasano JM, Sicard M, Rocadenbosch F, et al. (2006) A long Saharan dust event over the western Mediterranean: Lidar, Sun photometer observations, and regional dust modeling. Journal of Geophysical Research-Atmospheres 111: D15214.

34. Basart S, Pay MT, Jorba O, Pérez C, Jinez-Guerrero P, et al. (2012) Aerosols in the CALIOPE air quality modelling system: Evaluation and analysis of PM levels, optical depths and chemical composition over Europe. Atmospheric Chemistry and Physics 12: 3363–3392.

35. Basart S, Perez C, Nickovic S, Cuevas E, Baldasano JM (2012) Development and evaluation of the BSC-DREAM8b dust regional model over Northern Africa, the Mediterranean and the Middle East. Tellus Series B-Chemical and Physical Meteorology 64: 18539.

36. Volpe G, Santoleri R, Vellucci V, d'Alcala MR, Marullo S, et al. (2007) The colour of the Mediterranean Sea: Global versus regional bio-optical algorithms evaluation and implication for satellite chlorophyll estimates. Remote Sensing of Environment 107: 625–638.

37. Pérez C, Nickovic S, Pejanovic G, Baldasano JM, Özsoy E (2006) Interactive dust-radiation modeling: A step to improve weather forecasts. Journal of Geophysical Research-Atmospheres 111: D16206.

38. Amiridis V, Kafatos M, Perez C, Kazadzis S, Gerasopoulos E, et al. (2009) The potential of the synergistic use of passive and active remote sensing measurements for the validation of a regional dust model. Annales Geophysicae 27: 3155–3164.

39. Alonso-Perez S, Cuevas E, Perez C, Querol X, Baldasano JM, et al. (2011) Trend changes of African airmass intrusions in the marine boundary layer over the subtropical Eastern North Atlantic region in winter. Tellus, Series B: Chemical and Physical Meteorology 63: 255–265.

40. Pay MT, Jiménez-Guerrero P, Jorba O, Basart S, Querol X, et al. (2012) Spatio-temporal variability of concentrations and speciation of particulate matter across Spain in the CALIOPE modeling system. Atmospheric Environment 46: 376–396.

41. Papanastasiou DK, Poupkou A, Katragkou E, Amiridis V, Melas D, et al. (2010) An Assessment of the Efficiency of Dust Regional Modelling to Predict Saharan Dust Transport Episodes. Advances in Meteorology 2010: 154368.

42. Jiménez-Guerrero P, Pérez C, Jorba O, Baldasano JM (2008) Contribution of Saharan dust in an integrated air quality system and its on-line assessment. Geophysical Research Letters 35: L03814.

43. Pay MT, Piot M, Jorba O, Gassó S, Gonçalves M, et al. (2010) A full year evaluation of the CALIOPE-EU air quality modeling system over Europe for 2004. Atmospheric Environment 44: 3322–3342.

44. Haustein K, Pérez C, Baldasano JM, Müller D, Tesche M, et al. (2009) Regional dust model performance during SAMUM 2006. Geophysical Research Letters 36: L03812.

45. Barnaba F, Gobbi GP (2004) Aerosol seasonal variability over the Mediterranean region and relative impact of maritime, continental and Saharan dust particles over the basin from MODIS data in the year 2001. Atmos. Chem. Phys. 4: 2367–2391.

46. Pyper BJ, Peterman RM (1998) Comparison of methods to account for autocorrelation in correlation analyses of fish data. Can. J. Fish. Aquat. Sci. 55: 2127–2140.

47. Volpe G, Nardelli BB, Cipollini P, Santoleri R, Robinson IS (2012) Seasonal to interannual phytoplankton response to physical processes in the Mediterranean Sea from satellite observations. Remote Sens. Environ. 117: 223–235.

48. Papayannis A, Amiridis V, Mona L, Tsaknakis G, Balis D, et al. (2008) Systematic lidar observations of Saharan dust over Europe in the frame of EARLINET (2000–2002). J. Geophys. Res. (Atmos.) 113: D10204.

49. Mona L, Amodeo A, Pandolfi M, Pappalardo G (2006) Saharan dust intrusions in the Mediterranean area: Three years of Raman lidar measurements. J. Geophys. Res. (Atmos.) 111: D16203.

50. Gobbi GP, Angelini F, Barnaba F, Costabile F, Baldasano JM, et al. (2013) Changes in particulate matter physical properties during Saharan advections over Rome (Italy): a four-year study, 2001–2004. Atmos. Chem. Phys. 13: 7395–7404.

51. Eker-Develi E, Kideys AE, Tugrul S (2006) Role of Saharan dust on phytoplankton dynamics in the northeastern Mediterranean. Mar. Ecol. Prog. Ser. 314: 61–75.

52. Diaz F, Raimbault P, Boudjellal B, Garcia N, Moutin T (2001) Early spring phosphorus limitation of primary productivity in a NW Mediterranean coastal zone (Gulf of Lions). Mar. Ecol. Prog. Ser. 211: 51–62.

53. Rahav E, Herut B, Levi A, Mulholland MR, Berman-Frank I (2013) Springtime contribution of dinitrogen fixation to primary production across the Mediterranean Sea. Ocean Sci. 9: 489–498.

54. Struglia MV, Mariotti A, Filograsso A (2004) River discharge into the Mediterranean Sea: Climatology and aspects of the observed variability. J. Clim. 17: 4740–4751.

55. Durrieu de Madron X, Guieu C, Sempéré R, Conan P, Cossa D, et al. (2011) Marine ecosystems' responses to climatic and anthropogenic forcings in the Mediterranean. Prog. Oceanogr. 91: 97–166.

56. Izquierdo R, Benítez-Nelson CR, Masqué P, Castillo S, Alastuey A, et al. (2012) Atmospheric phosphorus deposition in a near-coastal rural site in the NE Iberian Peninsula and its role in marine productivity. Atmos. Environ. 49: 361–370.

57. Marty JC, Chiaverini J, Pizay MD, Avril B (2002) Seasonal and interannual dynamics of nutrients and phytoplankton pigments in the western Mediterranean Sea at the DYFAMED time-series station (1991–1999). Deep-Sea Res. Part II. 49: 1965–1985.

58. D'Ortenzio F, Ribera D'Alcalà M (2009) On the trophic regimes of the Mediterranean Sea: A satellite analysis. Biogeosciences 6: 139–148.

59. Pujo-Pay M, Conan P, Oriol L, Cornet-Barthaux V, Falco C, et al. (2011) Integrated survey of elemental stoichiometry (C, N, P) from the western to eastern Mediterranean Sea. Biogeosciences 8: 883–899.

60. Bonnet S, Guieu C (2006) Atmospheric forcing on the annual iron cycle in the western Mediterranean Sea: A 1-year survey. J. Geophys. Res. (Oceans) 111: C09010.

61. Wagener T, Guieu C, Leblond N (2010) Effects of dust deposition on iron cycle in the surface Mediterranean Sea: results from a mesocosm seeding experiment. Biogeosciences 7: 3769–3781.

62. Sarthou G, Jeandel C (2001) Seasonal variations of iron concentrations in the Ligurian Sea and iron budget in the Western Mediterranean Sea. Mar. Chem. 74: 115–129.

63. Guieu C, Bozec Y, Blain S, Ridame C, Sarthou G, et al. (2002) Impact of high Saharan dust inputs on dissolved iron concentrations in the Mediterranean Sea. Geophys. Res. Lett. 29: 1911.

64. Guieu C, Dulac F, Desboeufs K, Wagener T, Pulido-Villena E, et al. (2010) Large clean mesocosms and simulated dust deposition: a new methodology to investigate responses of marine oligotrophic ecosystems to atmospheric inputs. Biogeosciences 7: 2765–2784.

65. Ridame C, Le Moal M, Guieu C, Ternon E, Biegala IC, et al. (2011) Nutrient control of N-2 fixation in the oligotrophic Mediterranean Sea and the impact of Saharan dust events. Biogeosciences 8: 2773–2783.

66. Pulido-Villena E, Wagener T, Guieu C (2008) Bacterial response to dust pulses in the western Mediterranean: Implications for carbon cycling in the oligotropic ocean. Global. Biogeochem. Cy. 22: GB1020.

67. Gaetani M, Pasqui M (2012) Synoptic patterns associated with extreme dust events in the Mediterranean Basin. Reg. Environ. Change.: 1–14.

68. Paytan A, Mackey KRM, Chen Y, Lima ID, Doney SC, et al. (2009) Toxicity of atmospheric aerosols on marine phytoplankton. Proc. Natl. Acad. Sci. USA. 106: 4601–4605.

69. Avila A, Peñuelas J (1999) Increasing frequency of Saharan rains over northeastern Spain and its ecological consequences. Sci. Total Environ. 228: 153–156.

70. Lelieveld J, Berresheim H, Borrmann S, Crutzen PJ, Dentener FJ, et al. (2002) Global air pollution crossroads over the Mediterranean. Science 298: 794–799.

71. Kanakidou M, Mihalopoulos N, Kindap T, Im U, Vrekoussis M, et al. (2011) Megacities as hot spots of air pollution in the East Mediterranean. Atmos. Environ. 45: 1223–1235.

72. Querol X, Alastuey A, Pey J, Cusack M, Perez N, et al. (2009) Variability in regional background aerosols within the Mediterranean. Atmos. Chem. Phys. 9: 4575–4591.

73. Theodosi C, Grivas G, Zarmpas P, Chaloulakou A, Mihalopoulos N (2011) Mass and chemical composition of size-segregated aerosols (PM1, PM2.5,

PM10) over Athens, Greece: local versus regional sources. Atmos. Chem. Phys. 11: 11895–11911.

74. Ganor E, Osetinsky I, Stupp A, Alpert P (2010) Increasing trend of African dust, over 49 years, in the eastern Mediterranean. J. Geophys. Res. (Atmos.) 115: D07201.

75. Goudie AS (2009) Dust storms: Recent developments. J. Environ. Manage. 90: 89–94.

76. Mahowald NM, Kloster S, Engelstaedter S, Moore JK, Mukhopadhyay S, et al. (2010) Observed 20th century desert dust variability: impact on climate and biogeochemistry. Atmos. Chem. Phys. 10: 10875–10893.

16

Alpha, Beta and Gamma Diversity Differ in Response to Precipitation in the Inner Mongolia Grassland

Qing Zhang[1,3], Xiangyang Hou[2], Frank Yonghong Li[4], Jianming Niu[1,3]*, **Yanlin Zhou[1], Yong Ding[2], Liqing Zhao[1], Xin Li[3], Wenjing Ma[1], Sarula Kang[1]**

1 School of Life Sciences, Inner Mongolia University, Hohhot, China, **2** Grassland Research Institute of Chinese Academic of Agricultural Science, Hohhot, China, **3** Sino-US Center for Conservation, Energy and Sustainability Science, Inner Mongolia University, Hohhot, China, **4** AgResearch Grasslands Research Centre, Palmerston North, New Zealand

Abstract

Understanding the distribution pattern and maintenance mechanism of species diversity along environmental gradients is essential for developing biodiversity conservation strategies under environmental change. We have surveyed the species diversity at 192 vegetation sites across different steppe zones in Inner Mongolia, China. We analysed the total species diversity (γ diversity) and its composition (α diversity and β diversity) of different steppe types, and their changes along a precipitation gradient. Our results showed that (i) β diversity contributed more than α diversity to the total (γ) diversity in the Inner Mongolia grassland; the contribution of β diversity increased with precipitation, thus the species-rich (meadow steppe) grassland had greater contribution of β diversity than species-poor (desert steppe) grassland. (ii) All α, β and γ species diversity increased significantly ($P<0.05$) with precipitation, but their sensitivity to precipitation (diversity change per mm precipitation increase) was different between the steppe types. The sensitivity of α diversity of different steppe community types was negatively ($P<0.05$) correlated with mean annual precipitation, whereas the sensitivity of β and γ diversity showed no trend along the precipitation gradient ($P>0.10$). (iii) The α diversity increased logarithmically, while β diversity increased exponentially, with γ diversity. Our results suggest that for local species diversity patterns, the site species pool is more important in lower precipitation areas, while local ecological processes are more important in high precipitation areas. In addition, for β diversity maintenance niche processes and diffusion processes are more important in low and high precipitation areas, respectively. Our results imply that a policy of "multiple small reserves" is better than one of a "single large reserve" for conserving species diversity of a steppe ecosystem, and indicate an urgent need to develop management strategies for climate-sensitive desert steppe ecosystem.

Editor: Dafeng Hui, Tennessee State University, United States of America

Funding: This study was supported by the State Key Basic Research Development Programme of China (Grant nos. 2012CB722201 and 2014CB138805), The National Basic Research Programme of China (Grant no. 31200414), The Specialised Research Fund for the Doctoral Programme of Higher Education of China (Grant no. 20121501120006), and the Start Research Funding Project of Inner Mongolia University (Grant no. 125106). The funders had no role in study design, data collection and analysis, decision to publish, or preparation of the manuscript.

Competing Interests: The authors have declared that no competing interests exist.

* E-mail: jmniu2005@163.com

Introduction

Half a century ago, Whittaker first defined species diversity from three different perspectives: alpha (α) diversity, beta (β) diversity, and gamma (γ) diversity [1]. Alpha and γ diversity are grouped as inventory diversity [2], sharing the same characteristics and differentiated only by scale. Beta diversity is defined as the difference in species composition between communities and is closely related to many facets of ecology and evolutionary biology [3,4,5]. Most research on species diversity has focused on inventory diversity, but research on β diversity has recently increased [3,6,7]. The distribution patterns and maintenance mechanisms of species diversity along environmental gradients have been core topics in ecological research [4,8,9]. It is essential to understand such patterns and mechanisms for the development of strategies and measures for conserving species diversity under environmental change. Whether or not a single large or several small (SLOSS) reserves are superior means of conserving biodiversity [10] depends on the dominant type of diversity

present. For example, a high β species diversity within a community type may theoretically imply that the community occupies a heterogeneous environment. In that situation, the use of the 'several small' strategy will be superior to the "single large" strategy in reserve design for species diversity conservation [11].

Studies on the geographical patterns of species diversity generally showed a decreasing trend of α and γ diversity with increasing latitude from tropics to poles, which was primarily driven by temperature [8,12]. However, the correlations between β diversity and latitude were inconsistent; they could be positively [13,14], negatively [15,16], or not correlated [9,17]. Numerous studies have also reported species diversity patterns along other environmental gradients [18,19,20], and in many cases the species diversity-precipitation relationships have been studied [21,22,23]. Precipitation is the most important factor affecting species diversity and ecosystem functioning in arid and semiarid grasslands, such as in the Eurasian steppe, North American prairie and African savanna [23,24,25]. However, reported relationships between species diversity and precipitation have been inconsistent for

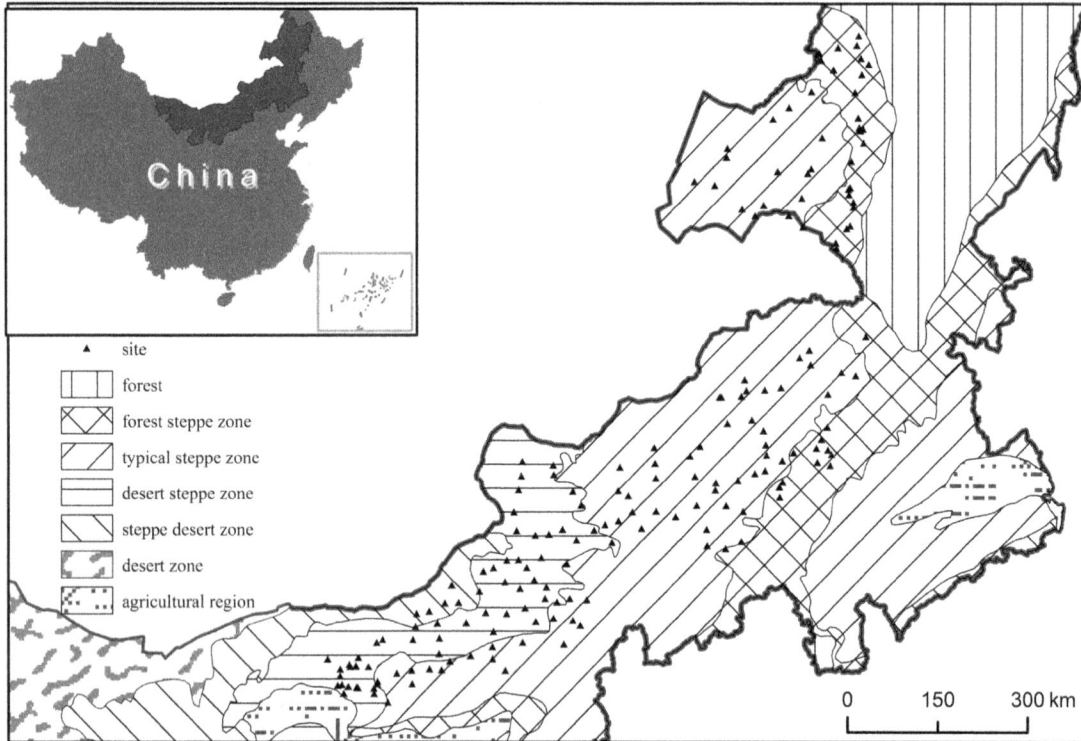

Figure 1. Map of the study region, showing the vegetation zones and sampling sites.

grassland biomes and scales; in some cases increasing precipitation promoted species diversity [21,22,26], but in other cases it had little effect [21,25].

Species evolution and diffusion, inter-specific competition, and environmental changes commonly influence the α, β and γ diversity of plant communities, but the response of species diversity pattern to these biological and environmental changes and the mechanisms for the response differ among the three species diversity measures [27,28,29,30]. Climate, geological history and ecological randomness are considered to be the main factors affecting species diversity pattern at large scales [31,32], whereas the local ecological processes and regional species pool are considered to be important at a small scale [33,34]. Environmental heterogeneity and species diffusion are closely related to β diversity [9,35]. The relative importance of different components of species diversity in maintaining the total species diversity differs with the spatial and temporal scales or across different regions [15,36].

The Inner Mongolia steppe grassland is an essential part of the Eurasian steppe [37]. Owing to the heterogeneous environment and the particular geological and evolutionary history, the grassland is rich in plant species diversity. It has more than 2,300 vascular plant species [38], and is recognised as the second most species-rich grassland biome in the world in terms of indigenous plant biodiversity after the African savanna [39]. The Inner Mongolia grassland is one of the key areas for biodiversity conservation in the world. The grassland types in Inner Mongolia show clear zones along a climatic gradient, from the desert steppe in dry areas, to the typical steppe, and through to the meadow steppe [40,41]. There have been several studies on the effects of precipitation on the species diversity and grassland productivity in the Inner Mongolia [24,42], but the composition of species diversity (α, β, γ) and their changes across different steppe grassland types on environmental gradients has not been studied.

In the present study, we measured the components of species diversity (α, β, γ) in each of the major steppe community types. The aim of the study was to present the composition of, and the relationships between α, β, and γ diversity of main steppes types in Inner Mongolia, and analyse the pattern of variation along the precipitation gradient. Then, based on these diversity patterns and relationships, we discuss the mechanism of species diversity maintenance in the Inner Mongolia grassland, and suggest the best strategies for species diversity conservation under environmental change.

Methods

Ethics statement

All the survey sites were owned and/or managed by pastoral farmers, who gave permission to the survey. The field studies did not involve any endangered or protected species.

Study area

We surveyed the grassland species diversity in the whole region of the Inner Mongolia grassland in northern China. The region stretches from latitudes 41.31°N to 50.78°N and longitudes 108.16°E to 120.39°E with elevation ranging from 532 to 1725 m above sea level (Fig. 1). The typical landforms in this region include gently rolling plains, tablelands, and hills. Mean annual temperature (MAT) ranges from −3.0 to 6.7°C, and mean annual precipitation (MAP) varies from about 150 to 450 mm, decreasing from southeast to northwest [43]. Distributed along the climate gradient from the relatively humid southeast to relatively dry northwest are soil types that very from chernozems, to chestnut- and then calcic brown soils; and grassland biomes that vary from meadow steppe, to typical steppe and then desert steppe grassland (Fig. 1).

Data collection

We investigated species diversity on 192 sites in the grassland region from late July to mid-August in 2012, when the grassland community biomass was at its peak. Seven dominant steppe community types were investigated: *Leymus chinensis* meadow steppe, *Stipa baicalensis* meadow steppe, *S. grandis* typical steppe, *L. chinensis* typical steppe, *S. krylovii* typical steppe, *S. breviflora* desert steppe, and *S. klemenzii* desert steppe. Nineteen to 39 sites were surveyed for each community type; the number of sites for each type was approximately determined by the relative size of its distribution area. The position of each site was located using GPS (Fig. 1). To focus on the relationships between species diversity and climate, and to minimize the influence of domestic animal grazing, all the surveyed sites were selected either in fenced grassland under protection, or in mowed grassland (surveyed before haymaking at the end of August). Grassland sites with obvious grazing effects (recognised by species composition) were excluded from the survey. At each site an area of 10×10 m was delineated, and ten 1×1 m plots were randomly placed in the delineated area to record all the plant species.

Meteorological records from the 156 meteorological stations in the study area were used in the analysis of the diversity-precipitation relations [44].

Data analyses

Calculation of species diversity. Since the concept of β diversity was introduced in 1972 [45], more than 40 different methods have been proposed for its calculation [6,46]. Jurasinski [2] suggested that these calculation methods could be split into two groups: the first group, designated differentiation diversity, includes a similarity coefficient, similarity attenuation slope with distance, gradient length in ordination space, and total variance of community composition [9,47]. The second group, designated proportional diversity, includes additive partition diversity and multiplicative partitioning diversity [28,45]. Additive diversity partition expresses α and β diversity in the same unit so that their relative importance can be easily quantified and interpreted, and can be directly compared across spatial and temporal scales or land-use practices [28,34,48]. We used the additive partition approach to calculate total species diversity of the studied grassland and its components. The methods were described in [27,28], and are briefly described below:

- α diversity, also called community diversity, is defined as the mean of the species richness (number) in the surveyed ten plots at a site:

$$\alpha = \frac{1}{10} \sum_{i=1}^{10} S_i \qquad (Eq.1)$$

where S_i represents the species richness in each plot.

- γ diversity, also called site diversity, is the total species richness at a site.

- β diversity, defined as the difference in species composition among the ten plots at a site, is calculated by subtracting α diversity from γ diversity:

$$\beta = \gamma - \alpha \qquad (Eq.2)$$

The β diversity has two components, one quantifying the degree of nestedness of species richness (β_N), i.e. the degree to which species richness differs between plots within one site from the most species-rich plot, and the other component

reflecting the difference in species composition among the plots (β_R). With S_{max} representing the species number in the richest plot, β_N and β_R are calculated as follows:

$$\beta_N = \frac{1}{10} \sum_{i=1}^{10} (S_{max} = S_i) \qquad (Eq.3)$$

$$\beta_R = \beta - \beta_N \qquad (Eq.4)$$

Consequently, the total species richness at a site (γ) is the sum of the mean species richness of all the plots at the site (α), the differences in species richness due to the nestedness (β_N) and species composition (β_R) among the plots:

$$\gamma = \alpha + \beta_R + \beta_N \qquad (Eq.5)$$

The components of species diversity (α, β and γ) were assessed for each grassland site. The species diversity at all of the sites for each steppe community type was averaged to represent the species diversity of the community type. The species diversity composition of three steppe vegetation types (i.e., the desert steppe, typical steppe and meadow steppe) were also aggregated in the same way to represent the species diversity at all grassland sites of a steppe vegetation type.

Calculation of precipitation at each site and for each grassland type. We calculated mean annual precipitation (MAP) at each vegetation site using the approach of Thornthwaite [49] based on precipitation records of the 156 meteorological stations in the region. Each site MAP was derived according to the latitude (LAT), longitude (LNG) and altitude (ALT) of the site, using a previously developed model on the relationship between MAP and the geographical coordinates of each meteorological station [44]:

$$\begin{aligned} MAP = {} & 13872.1241 - 0.8941 ALT - 198.2731 LAT \\ & + 2.2360 LAT^2 - 0.0313 LAT \times ALT \\ & - 176.0567 LNG + 0.8296 LNG^2 + 0.0203 LNG \\ & \times ALT (r^2 = 0.84) \end{aligned} \qquad (Eq.6)$$

The MAP of each community type was calculated as the mean of the MAP at all the sites that belongs to the community type.

Relationships among species diversity components. First, the proportions of α and β to γ diversity were used to quantify the relative contribution of these two kinds of diversity to γ diversity. The composition of γ diversity (i.e., the proportions of α, β, β_R and β_N) of the seven steppe community types, and of the three steppe vegetation types, was ordered according to the precipitation of their distribution areas in order to examine the species diversity composition changes across these steppe types in relation to precipitation.

Second, we used regression analysis to model relationships between γ diversity and its components (α and β) across all 192 sites. The relationships between γ diversity and the 'occasional species' (species recorded in only one or two of the ten plots at a site) were also analysed. These relationships were used to interpret the mechanism of species diversity maintenance in the grasslands.

Table 1. The environmental characteristics (mean annual precipitation MAP, mean annual temperature MAT, and major soil types) and species diversity composition (α, β and γ) of the seven steppe community types and of the three vegetation types in the Inner Mongolia grassland.

Type	No. of sites	MAP	MAT	Major soil types	α	β	γ	β_N	β_R	$\alpha\%$	$\beta\%$
Stipa klemenzii desert steppe	36	196	3.52	calcic brown soil	9.5±0.3	9.0±0.4	18±0.7	3.6±0.2	5.4±0.4	51.9±0.9	48.1±0.9
Stipa breviflora desert steppe	39	212	3.18	Calcic brown and light chestnut soil	10.3±0.3	10.4±0.5	20.7±0.6	3.9±0.2	6.5±0.4	50.4±1.2	49.6±1.2
Stipa krylovii typical steppe	31	263	0.85	chestnut and light chestnut soil	12.8±0.4	13.7±0.6	26.4±0.9	4.7±0.2	8.9±0.5	48.7±0.6	51.3±0.6
Leymus chinensis typical steppe	28	305	0.15	chestnut and dark chestnut soil	12.6±0.4	13.5±0.7	26.1±1.0	3.7±0.2	9.8±0.6	48.9±0.9	51.1±0.9
Stipa grandis typical steppe	21	329	−0.40	chestnut soil	15.1±0.5	17.8±0.8	32.8±1.3	5.3±0.2	12.4±0.7	46.5±0.8	53.5±0.8
Stipa baicalensis meadow steppe	18	348	−2.02	chernozem and light chernozem soil	20.7±0.8	32.3±1.7	53.0±2.5	7.8±0.4	24.5±1.5	39.4±0.9	60.6±0.9
Leymus chinensis meadow steppe	19	354	−1.85	chernozem soil	18.2±0.7	22.6±0.9	40.8±1.3	6.0±0.4	16.5±0.8	44.8±1.1	55.2±1.1
Desert steppe	75	205	3.34	mainly on calcic brown soil	9.9±0.2	9.7±0.3	19.6±0.5	3.7±0.2	5.97±0.3	51.1±0.7	48.9±0.7
Typical steppe	80	302	0.27	mainly on chestnut soil	13.3±0.3	14.7±0.5	28.0±0.7	4.5±0.2	10.2±0.4	48.2±0.5	51.8±0.5
Meadow steppe	37	351	−1.93	mainly on chernozem soil	19.4±0.5	27.3±0.2	46.7±1.5	6.9±0.4	20.4±1.0	42.2±0.8	57.9±0.8
Inner Mongolia grassland	192	273	1.05	all the soils above	13.1±0.3	15.2±0.6	28.3±0.8	4.7±0.1	10.5±0.5	48.2±0.4	51.8±0.4

The β diversity has two components of β nestedness diversity (β_N) and β replacement diversity (β_R). The diversity values are the mean ± s.e.m of species number recorded in ten plots at each site. The percentage of α and β diversity in γ diversity are also shown as $\alpha\%$ and $\beta\%$.

Patterns of species diversity along a precipitation gradient. With α, β and γ diversity and the precipitation (MAP) calculated for all the vegetation sites, we plotted each diversity measure against precipitation, and tested for a significant correlation between species diversity and MAP using linear regression analysis. We also did a linear regression analysis between the species diversity and precipitation for each of the seven steppe community types, and used the slope of the regression (i.e. diversity change per mm precipitation change) to represent the sensitivity of the species diversity of each community type to precipitation. A steep slope or high sensitivity means a greater effect of precipitation on community species diversity. The changes in sensitivity across seven community types along a precipitation gradient were tested by examining if the sensitivity and precipitation of these community types were significantly correlated.

All statistical analyses were performed using Excel 2010 and SPSS 17.0.

Results

Species diversity and diversity composition of the Inner Mongolia grassland

The species diversity (γ) and its components (α, β_R and β_N) were consistently low in desert steppe and high in meadow steppe, and in-between in the typical steppe (Table 1). The β diversity had a slightly greater contribution (51.83%) to γ diversity than α diversity (48.17%) in the Inner Mongolia grassland (Table 1 and Fig. 2A). The contribution of α and β diversity to γ diversity differs among the seven steppe community types and among the three steppe vegetation types (Fig. 2A); the contribution of α diversity was higher in desert steppe than in meadow steppe, and conversely the contribution of β diversity was lower in desert steppe than in

meadow steppe. The change in the contribution of β diversity among the steppe community types was mainly due to changes in β replacement diversity (β_R), while β nestedness diversity (β_N) showed little change. When these steppe types were ordered according to the MAP of their distribution areas, a trend of decreasing α and increasing β with increasing MAP was shown (Fig. 2B).

Species diversity pattern and their components along a precipitation gradient

The α diversity increased significantly (P<0.001) with increasing precipitation (Fig. 3A), but the sensitivity of α diversity to precipitation across the seven steppe community types significantly decreased (P = 0.028) with increasing precipitation (Fig. 3B).

Both β_N diversity and β_R diversity increased with precipitation, and consequently total β diversity also increased (Fig. 3C, E, G). However, there was no correlation between the sensitivity of β, β_N or β_R diversity to precipitation across the seven community types along the precipitation gradient (P>0.05), though the sensitivity of β_R diversity showed a non-significant increasing trend (P = 0.10) (Fig. 3D, F, H).

The γ diversity also significantly increased with precipitation (Fig. 3I), but the sensitivity of γ diversity to precipitation showed no correlation with MAP across the seven community types (Fig. 3J).

Relations among the γ diversity and its components

The α diversity increased with γ diversity, and the increase was gradually saturates and could be described by a logarithmic curve (P<0.001) (Fig. 4A). The β diversity also increased with γ diversity, but the increase was, complementarily to α diversity, accelerated with the increase of γ diversity (Fig. 4B). There was a significant linear correlation between γ diversity and occasional species

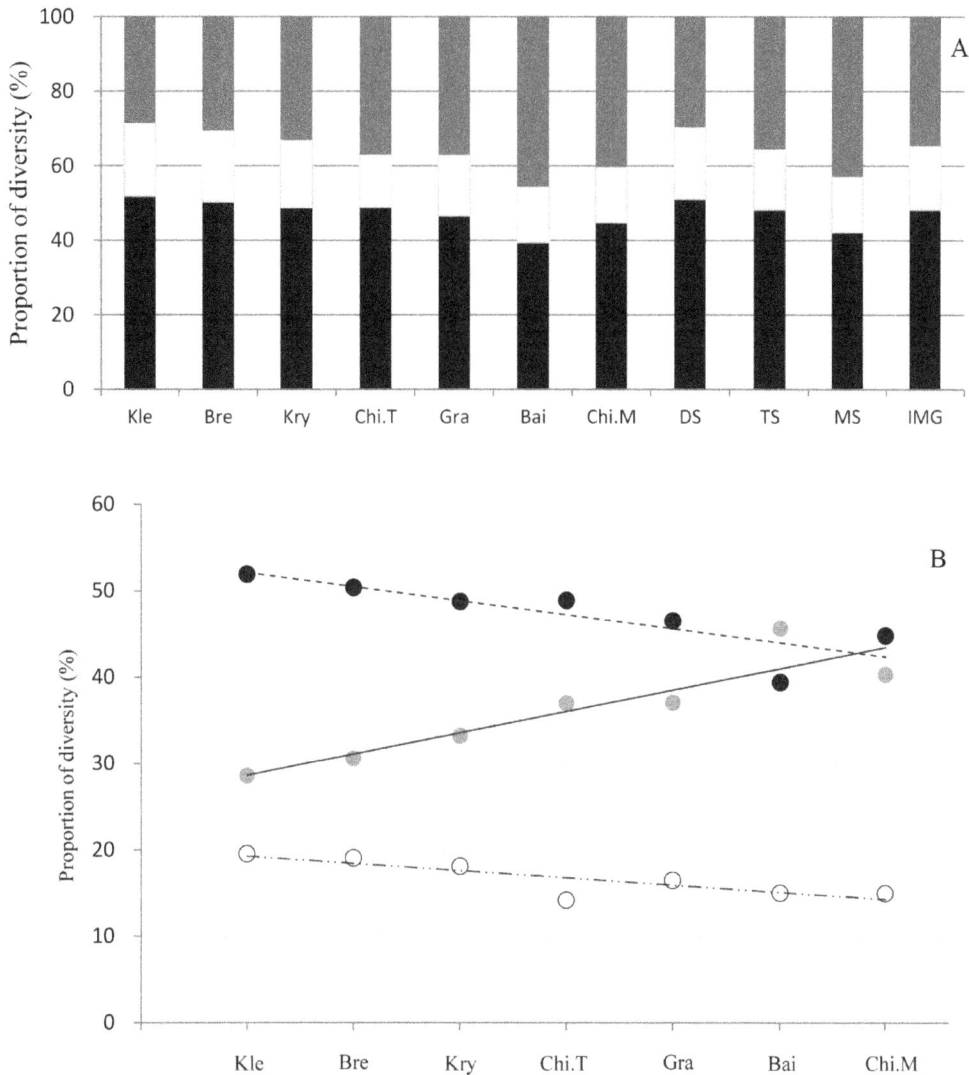

Figure 2. Species diversity of the steppe grassland in Inner Mongolia. The data are presented for seven major steppe community types: Kle (*Stipa klemenzii* desert steppe), Bre (*Stipa breviflora* desert steppe), Kry (*Stipa krylovii* typical steppe), Chi.T (*Leymus chinensis* typical steppe), Gra (*Stipa grandis* typical steppe), Bai (*Stipa baicalensis* meadow steppe), Chi.M (*Leymus chinensis* meadow steppe); three steppe vegetation types: DS (desert steppe), TS (typical steppe), MS (meadow steppe); and the Inner Mongolia grassland as a whole (IMG). A: Proportion of α diversity (black), β nestedness diversity (β_N) (white) and β replacement diversity (β_R) (grey) in γ diversity; B: The trend of the proportion of α diversity (black dot), β_N diversity (white dot) and β_R diversity (grey dot) in γ diversity in the seven grassland types with the types ordered according to the annual mean precipitation of their distribution areas, with precipitation increase from left to right.

diversity (Fig. 4C). The occasional species diversity also increased significantly (P<0.001) with increasing precipitation (Fig. 4D).

Discussion

Contribution of α and β to γ diversity changes across different steppe types along a precipitation gradient

The contributions of α and β diversity to γ diversity form the basis for understanding the biodiversity components [4,46]. Controversial opinions exist on the relative importance of α and β diversity to γ diversity: some believe that α diversity is more important, while the others contend that β diversity is more important. A third group suggest that α and β diversity work together [4,27,46]. We have found that β diversity contributes slightly more (51.83%) than α diversity (48.17%) to γ diversity in the Inner Mongolia grassland (Fig. 2), and that contribution of β

diversity is greater in species-rich grassland (meadow steppe) in high precipitation areas than in species-poor grassland (desert steppe) in low precipitation areas. In other words, the contribution of β diversity has an increasing trend with precipitation (Fig. 2B).

The relative contribution of α and β to γ diversity in biological communities depends on the ecological heterogeneity and capability of species diffusion [27,28]. The α diversity is more important in communities with an homogeneous environment and strong-diffusion species, whereas β diversity is, on the contrary, more important in communities with an heterogeneous environment and weak-diffusion species. The increase of ecological heterogeneity with increasing precipitation in the studied grassland region [22,42] may be attributable to the increase in the contribution of β diversity in γ diversity with precipitation increase. This is supported by the high occasional species diversity in the species-rich grasslands (Fig. 4C). High occasional species

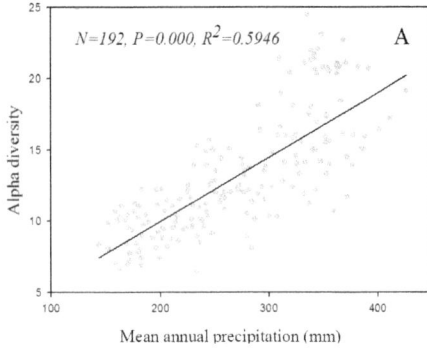

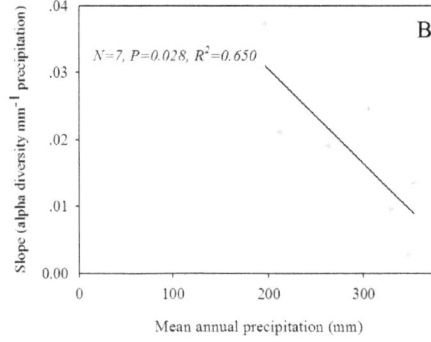

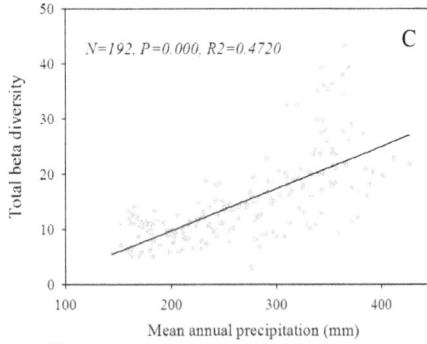

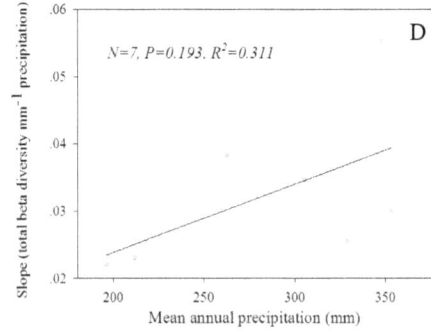

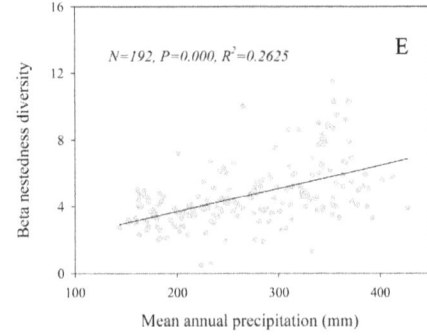

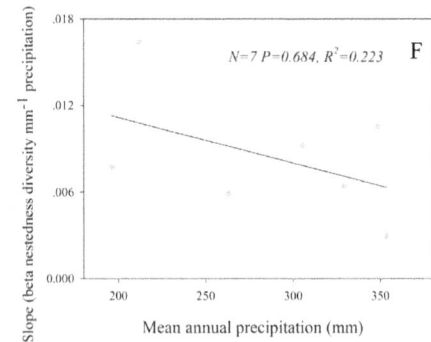

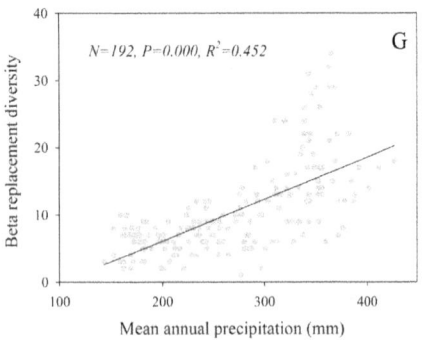

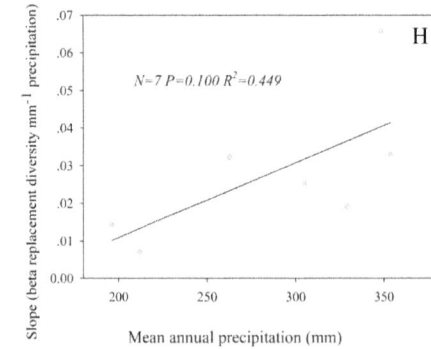

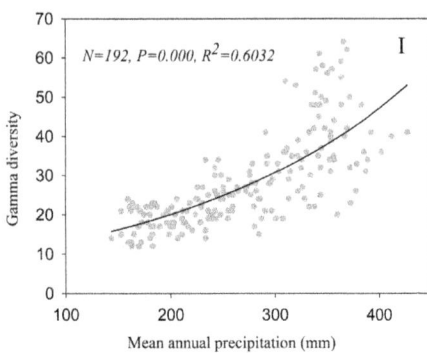

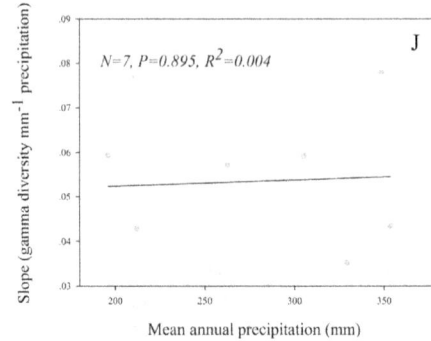

Figure 3. Response of species diversity to precipitation (left column), and the sensitivity of species diversity of seven steppe community types to precipitation change (right column) in the Inner Mongolia grassland. The sensitivity is represented by the regression slope of the linear regression of species diversity to precipitation within each steppe type. A: response of α diversity; B: sensitivity of α diversity; C: response of β diversity; D: sensitivity of β diversity; E: response of β_N diversity; F: sensitivity of β_N diversity; G: response of β_R diversity; H: sensitivity of β_R diversity; I: response of γ diversity; J: sensitivity of γ diversity.

diversity is also in accordance with high species replacement diversity (β_R) (Fig. 2). Thus, the increase of the contribution of β to γ diversity along the gradient of increasing precipitation is most likely associated with the increase in ecological heterogeneity and the decrease in species diffusion.

Maintenance mechanism of species diversity along a precipitation gradient changes from regional species pool to the effect of local ecological processes

The maintenance mechanism of local species diversity has always been an important topic in ecology. Local ecological processes (such as predation, competition, resource supply and diffusion) [50,51] and the regional species pool [33,34] have been considered as the mechanisms for diversity maintenance. However, there is insufficient understanding on which of these two mechanisms is more important for local species diversity patterns

[36]. Examining the relationship between regional and local species diversity can help quantify the importance of these two mechanisms [34]. A linear correlation between γ diversity and α diversity would suggest that the regional species pool was the main limiting factor. Alternatively, a saturated curve between α and γ diversity would suggest ecological processes are more important in maintaining diversity [34].

In the studied grassland, both α and γ diversity increased as precipitation increased (Fig. 3), but the increase of α diversity was gradually saturated (a logarithmical increase with γ diversity) (Fig. 4A). That is, the increase of α diversity with increasing γ increasing was rapid and approximately linear when γ diversity was low, but the increase slowed down when γ diversity was high in high precipitation areas. In species-poor grasslands (low γ) with low precipitation, the inter-specific competition is relatively weak resulting in much spare niche capacity. Niche theory indicates that

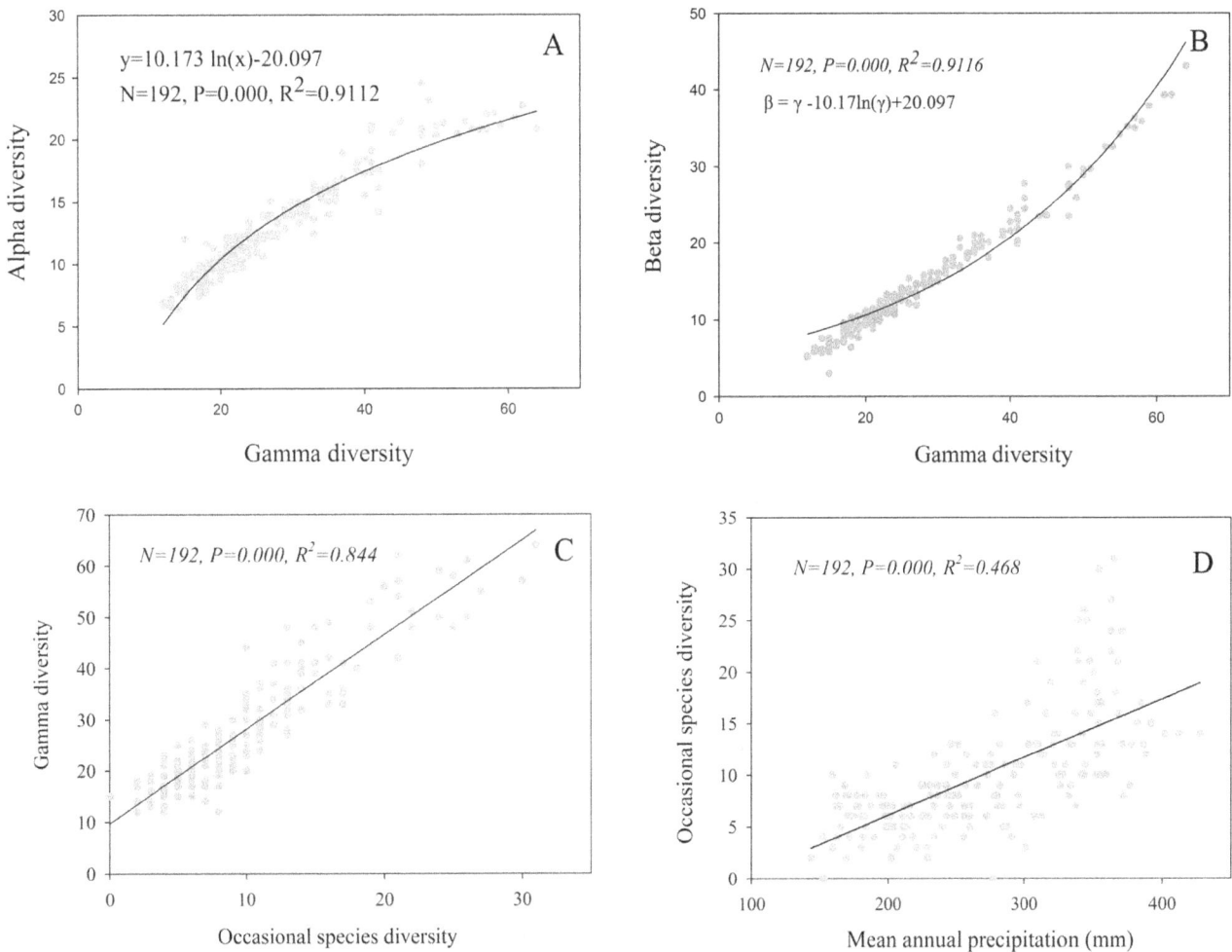

Figure 4. Relations of α diversity (A), β diversity (B) and occasional diversity (C) with γ diversity in the Inner Mongolia grassland. The β diversity is complementary to α diversity in γ diversity, and the curve in (B) is derived as $\beta = \gamma - \alpha$. D: response of occasional species diversity to mean annual precipitation.

every species occupies its unique corresponding niche [52], thus an increasing regional species pool provides the possibility for species to occupy more niches in the community. The near linear increase of γ diversity with α diversity (Fig. 4A) in low precipitation areas may indicate the importance of the regional species pool in determining local species diversity. In the high precipitation grassland region, more species appeared in the community, and inter-specific competition was relatively strong. In the case of full use of resources, the ecological niche was occupied to its fullest extent [53]. This means there was very little spare niche capacity, and no new species could be present in the community. Therefore, the regional species pool was still increasing, but local species diversity was saturated. In other words, with an increase in mean annual precipitation, the dominant maintenance mechanism of community diversity changed from the regional species pool to the effect of local ecological processes.

The maintenance mechanism of β diversity changes from niche processes to diffusion processes with increasing precipitation

The β diversity reflects different degrees of species composition. The environment heterogeneity (or the niche process [1,45]) and species diffusion processes [54] have been recognised as the main mechanisms that combine to maintain β diversity [9,29]. The relative importance of the two processes varies across regions and scales [15,35]. The nestedness species diversity (β_N) represents the degree to which species richness in each plot differs from the richest plot at a site, and reflects the extent of the variation in species number within each community. It is closely related to resource heterogeneity or the niche processes [27]. In low precipitation areas (*i.e.*, desert steppe), strong winds may erode and move the soil from grass-dominant areas to areas occupied by shrub clamps to create fertile "islands", thus increasing the environmental heterogeneity [55]. In high precipitation areas (*i.e.*, meadow steppe), β_N diversity increased with an increasing species pool. However, since α diversity is almost saturated with respect to γ diversity (Fig. 4A), β_N diversity does not increase at the same rate as γ diversity, resulting in a slight reduction in the proportion of β_N in species-rich grasslands in high precipitation areas (Fig. 2B).

On the contrary, the strong increase of β_R with increasing precipitation (Fig. 4G) indicates a strong species-replacement effect in high precipitation areas [27,28]. Communities with more weak-diffusing species, such as those recorded as occasional species, tend to form high β_R diversity [27,35]. Considering the increase of occasional species diversity with precipitation (Fig. 4D) and with γ diversity (Fig. 4C), β_R is much more important in the diversity of species-rich than species-poor grassland communities. In summa-

ry, the dominant mechanism for β diversity maintenance changes from niche processes to diffusion processes across steppe community types from species-poor desert steppe to species-rich meadow steppe along precipitation gradient.

Management implications

Changes in regional precipitation patterns under global climate change will undoubtedly affect species diversity and ecosystem function and stability [56,57], and the effects will be especially profound in arid and semiarid grassland regions [24,58]. Our results have important implications for understanding the potential effects of climate change on the semiarid grassland, and for developing biodiversity conservation strategies. First, the generally greater contribution of β diversity than α diversity to γ diversity (Fig. 2) implies it is better to construct several small reserves than a single large reserve for protecting species diversity of a steppe community in the region. The much greater contribution of β diversity in species-rich (meadow steppe) than in species-poor (desert steppe) communities suggests it more useful to apply multiple small reserves for protecting the meadow steppe grassland in relatively humid areas. Second, since species diversity provides a mechanism for maintaining ecosystem stability through compensatory interactions among species [59], the greater sensitivity of species diversity (mainly α diversity) to precipitation in desert steppe than in meadow steppe (Fig. 3B) suggests that more efforts are urgently needed to understand the effects of climate change on desert steppe for developing adaptive ecosystem management strategies. Our study has focused on species diversity composition changes along climatic gradients by excluding the grassland sites under heavy animal grazing. Human activities, mainly through animal grazing, have profound impacts on grassland species diversity [60,61,62]. The effects of animal grazing on species diversity along precipitation gradients in the Inner Mongolia grassland also need future studies.

Acknowledgments

The authors are grateful to Dr Dafeng Hui, Dr Rafael B. de Andrade and an anonymous reviewer for the comments and suggestions on the early version of this manuscript. We also thank Dr Adrian Walcroft for improving the English readability of this manuscript.

Author Contributions

Conceived and designed the experiments: QZ JN XH. Performed the experiments: QZ JN YZ LZ YD WM SK. Analyzed the data: QZ FL JN. Wrote the paper: QZ FL XL.

References

1. Whittaker RH (1960) Vegetation of the siskiyou mountains, oregon and california. Ecological Monographs 30: 280–338.

2. Jurasinski G, Retzer V, Beierkuhnlein C (2009) Inventory, differentiation, and proportional diversity: a consistent terminology for quantifying species diversity. Oecologia 159: 15–26. doi:10.1007/s00442-008-1190-z.

3. Sfenthourakis S, Panitsa M (2012) From plots to islands: species diversity at different scales. Journal of Biogeography 39: 750–759. doi:10.1111/j.1365-2699.2011.02639.x.

4. Meynard CN, Devictor V, Mouillot D, Thuiller W, Jiguet F, et al. (2011) Beyond taxonomic diversity patterns: how do alpha, beta and gamma components of bird functional and phylogenetic diversity respond to environmental gradients across France? Global Ecology And Biogeography 20: 893–903. doi:10.1111/j.1466-8238.2010.00647.x.

5. Leprieur F, Albouy C, De Bortoli J, Cowman PF, Bellwood DR, et al. (2012) Quantifying Phylogenetic Beta Diversity: Distinguishing between 'True' Turnover of Lineages and Phylogenetic Diversity Gradients. Plos One 7. doi:e4276010.1371/journal.pone.0042760.

6. Anderson MJ, Crist TO, Chase JM, Vellend M, Inouye BD, et al. (2011) Navigating the multiple meanings of beta diversity: a roadmap for the practicing ecologist. Ecology Letters 14: 19–28. doi:10.1111/j.1461-0248.2010.01552.x.

7. de Juan S, Thrush SF, Hewitt JE (2013) Counting on beta-Diversity to Safeguard the Resilience of Estuaries. Plos One 8. doi:e6557510.1371/journal.pone.0065575.

8. Wang Z, Brown JH, Tang Z, Fang J (2009) Temperature dependence, spatial scale, and tree species diversity in eastern Asia and North America. Proceedings of the National Academy of Sciences 106: 13388. doi:10.1073/pnas.0905030106.

9. Melo AS, Rangel TFLVB, Diniz-Filho JAF (2009) Environmental drivers of beta-diversity patterns in New-World birds and mammals. Ecography 32: 226–236. doi:10.1111/j.1600-0587.2008.05502.x.

10. Patterson BD, Atmar W (1986) Nested subsets and the structure of insular mammalian faunas and archipelagos. Biological Journal of the Linnean Society 28: 65–82.

11. Wiersma YF, Urban DL (2005) Beta diversity and nature reserve system design in the Yukon, Canada. Conservation Biology 19: 1262–1272.

12. Hillebrand H (2004) On the generality of the latitudinal diversity gradient. American Naturalist 163: 192–211. doi:10.1086/381004.

13. Novotny V, Miller SE, Hulcr J, Drew RAI, Basset Y, et al. (2007) Low beta diversity of herbivorous insects in tropical forests. Nature 448: 692–U698. doi:10.1038/nature06021.

14. Koleff P, Lennon JJ, Gaston KJ (2003) Are there latitudinal gradients in species turnover? Global Ecology And Biogeography 12: 483–498. doi:10.1046/j.1466-822X.2003.00056.x.

15. Qian H, Ricklefs RE (2007) A latitudinal gradient in large-scale beta diversity for vascular plants in North America. Ecology Letters 10: 737–744. doi:10.1111/j.1461-0248.2007.01066.x.

16. Dyer LA, Singer MS, Lill JT, Stireman JO, Gentry GL, et al. (2007) Host specificity of Lepidoptera in tropical and temperate forests. Nature 448: 696–U699. doi:10.1038/nature05884.

17. McKnight MW, White PS, McDonald RI, Lamoreux JF, Sechrest W, et al. (2007) Putting beta-diversity on the map: Broad-scale congruence and coincidence in the extremes. Plos Biology 5: 2424–2432. doi:10.1371/journal.pbio.0050272.

18. Tang Z, Fang J, Chi X, Feng J, Liu Y, et al. (2012) Patterns of plant beta-diversity along elevational and latitudinal gradients in mountain forests of China. Ecography 35: 1083–1091. doi:10.1111/j.1600-0587.2012.06882.x.

19. Kluge J, Kessler M, Dunn RR (2006) What drives elevational patterns of diversity? A test of geometric constraints, climate and species pool effects for pteridophytes on an elevational gradient in Costa Rica. Global Ecology And Biogeography 15: 358–371. doi:10.1111/j.1466-822x.2006.00223.x.

20. de Andrade RB, Barlow J, Louzada J, Vaz-de-Mello FZ, Souza M, et al. (2011) Quantifying Responses of Dung Beetles to Fire Disturbance in Tropical Forests: The Importance of Trapping Method and Seasonality. Plos One 6. doi:e2620810.1371/journal.pone.0026208.

21. Adler PB, Levine JM (2007) Contrasting relationships between precipitation and species richness in space and time. Oikos 116: 221–232. doi:10.1111/j.2006.0030-1299.15327.x.

22. Bai YF, Wu JG, Pan QM, Huang JH, Wang QB, et al. (2007) Positive linear relationship between productivity and diversity: evidence from the Eurasian Steppe. Journal of Applied Ecology 44: 1023–1034. doi:10.1111/j.1365-2664.2007.01351.x.

23. Volder A, Briske DD, Tjoelker MG (2013) Climate warming and precipitation redistribution modify tree-grass interactions and tree species establishment in a warm-temperate savanna. Global Change Biology 19: 843–857. doi:10.1111/gcb.12068.

24. Bai YF, Wu JG, Xing Q, Pan QM, Huang JH, et al. (2008) Primary production and rain use efficiency across a precipitation gradient on the Mongolia plateau. Ecology 89: 2140–2153.

25. Collins SL, Koerner SE, Plaut JA, Okie JG, Brese D, et al. (2012) Stability of tallgrass prairie during a 19-year increase in growing season precipitation. Functional Ecology 26: 1450–1459. doi:10.1111/j.1365-2435.2012.01995.x.

26. Cleland EE, Collins SL, Dickson TL, Farrer EC, Gross KL, et al. (2013) Sensitivity of grassland plant community composition to spatial vs. temporal variation in precipitation. Ecology 94: 1687–1696. doi:10.1890/12-1006.1.

27. Chiarucci A, Bacaro G, Arevalo JR, Delgado JD, Fernandez-Palacios JM (2010) Additive partitioning as a tool for investigating the flora diversity in oceanic archipelagos. Perspectives in Plant Ecology Evolution and Systematics 12: 83–91. doi:10.1016/j.ppees.2010.01.001.

28. Crist TO, Veech JA (2006) Additive partitioning of rarefaction curves and species-area relationships: unifying alpha-, beta- and gamma-diversity with sample size and habitat area. Ecology Letters 9: 923–932. doi:10.1111/j.1461-0248.2006.00941.x.

29. Legendre P, Borcard D, Peres-Neto PR (2005) Analyzing beta diversity: Partitioning the spatial variation of community composition data. Ecological Monographs 75: 435–450. doi:10.1890/05-0549.

30. Boieiro M, Carvalho JC, Cardoso P, Aguiar CAS, Rego C, et al. (2013) Spatial Factors Play a Major Role as Determinants of Endemic Ground Beetle Beta Diversity of Madeira Island Laurisilva. Plos One 8. doi:e6459110.1371/journal.pone.0064591.

31. Field R, Hawkins BA, Cornell HV, Currie DJ, Diniz-Filho JAF, et al. (2009) Spatial species-richness gradients across scales: a meta-analysis. Journal of Biogeography 36: 132–147. doi:10.1111/j.1365-2699.2008.01963.x.

32. Colwell RK, Rahbek C, Gotelli NJ (2004) The mid-domain effect and species richness patterns: what have we learned so far. The American Naturalist 163: E1–E23. doi:10.1086/382056.

33. Zobel M (1992) Plant species coexistence: the role of historical, evolutionary and ecological factors. Oikos: 314–320.

34. Gering JC, Crist TO (2002) The alpha-beta-regional relationship: providing new insights into local-regional patterns of species richness and scale dependence of diversity components. Ecology Letters 5: 433–444. doi:10.1046/j.1461-0248.2002.00335.x.

35. Green JL, Ostling A (2003) Endemics-area relationships: The influence of species dominance and spatial aggregation. Ecology 84: 3090–3097. doi:10.1890/02-3096.

36. Russell R, Wood SA, Allison G, Menge BA (2006) Scale, environment, and trophic status: The context dependency of community saturation in rocky intertidal communities. American Naturalist 167: E158–E170. doi:10.1086/504603.

37. Suttie JM, Reynolds SG, Batello C (2005) Grasslands of the world. Rome: Food and Agriculture Organization of the United Nations.

38. Ma YQ (1995–1998) Flora of Inner Mongolia. Hohhot: Inner Mongolia People's Publishing House.

39. World Conservation Monitoring Centre (1992) Global biodiversity: status of the earth's living resources: Chapman & Hall.

40. Li B (1962) Basic types and eco-geographic distribution of zonal vegetation in Inner Mongolia. Acta Scientiarum Naturalium Universitatis Neimongol 4: 42–72.

41. Li Y (1996) Ecological vicariance of steppe species on Mongolian Plateau and its indication to vegetation change under climate change. Acta Phytoecologica Sinica 22: 1–12.

42. Zhang Q, Niu J, Buyantuyev A, Zhang J, Ding Y, et al. (2011) Productivity-species richness relationship changes from unimodal to positive linear with increasing spatial scale in the Inner Mongolia steppe. Ecological Research 26: 649–658. doi:10.1007/s11284-011-0825-4.

43. Inner Mongolia-Ningxia Joint Inspection Group of Chinese Sciences of Academy (1985) Vegetation of Inner Mongolia. Beijing: Science Publishing House.

44. Niu JM (2001) Climate-based digital simulation on spatial distribution of vegetation-A case study in Inner Mongolia. Acta Ecologica Sinica 21: 1064–1071.

45. Whittaker RH (1972) Evolution and measurement of species diversity. Taxon 21: 213–251.

46. Jost L (2007) Partitioning diversity into independent alpha and beta components. Ecology 88: 2427–2439. doi:10.1890/06-1736.1.

47. Chao A, Chazdon RL, Colwell RK, Shen TJ (2005) A new statistical approach for assessing similarity of species composition with incidence and abundance data. Ecology Letters 8: 148–159. doi:10.1111/j.1461-0248.2004.00707.x.

48. Veech JA, Crist TO (2010) Diversity partitioning without statistical independence of alpha and beta. Ecology 91: 1964–U1998. doi:10.1890/08-1727.1.

49. Fang JY, Yoda K (1990) Climate and Vegetation in China III water balance and distribution of vegetation. Ecological Research 5: 9–23.

50. Huston M (1979) A general hypothesis of species diversity. American Naturalist: 81–101.

51. Wimp GM, Whitham TG (2001) Biodiversity consequences of predation and host plant hybridization on an aphid-ant mutualism. Ecology 82: 440–452. doi:10.1890/0012-9658(2001)082[0440:bcopah]2.0.co;2.

52. Case TJ, Gilpin ME (1974) Interference competition and niche theory. Proceedings of the National Academy of Sciences of the United States of America 71: 3073–3077. doi:10.1073/pnas.71.8.3073.

53. Loreau M (2000) Are communities saturated? On the relationship between alpha, beta and gamma diversity. Ecology Letters 3: 73–76. doi:10.1046/j.1461-0248.2000.00127.x.

54. Cody ML (1970) Chilean bird distribution. Ecology: 455–464.

55. Zhang P, Yang J, Zhao L, Bao S, Song B (2011) Effect of Caragana tibetica nebkhas on sand entrapment and fertile islands in steppe-desert ecotones on the Inner Mongolia Plateau, China. Plant and Soil 347: 79–90. doi:10.1007/s11104-011-0813-z.

56. Hooper DU, Chapin FS, Ewel JJ, Hector A, Inchausti P, et al. (2005) Effects of biodiversity on ecosystem functioning: A consensus of current knowledge. Ecological Monographs 75: 3–35. doi:10.1890/04-0922.

57. Turnbull LA, Levine JM, Loreau M, Hector A (2013) Coexistence, niches and biodiversity effects on ecosystem functioning. Ecology Letters 16: 116–127. doi:10.1111/ele.12056.

58. Sala OE, Parton W, Joyce L, Lauenroth W (1988) Primary production of the central grassland region of the United States. Ecology 69: 40–45.

59. Bai YF, Han XG, Wu JG, Chen ZZ, Li LH (2004) Ecosystem stability and compensatory effects in the Inner Mongolia grassland. Nature 431: 181–184. doi:10.1038/nature02850.

60. Chapin III FS, Zavaleta ES, Eviner VT, Naylor RL, Vitousek PM, et al. (2000) Consequences of changing biodiversity. Nature 405: 234–242. doi:10.1038/35012241.

61. La Sorte FA, McKinney ML, Pyšek P, Klotz S, Rapson G, et al. (2008) Distance decay of similarity among European urban floras: the impact of anthropogenic activities on β diversity. Global Ecology And Biogeography 17: 363–371. doi:10.1111/j.1466-8238.2007.00369.x.

62. Li Y, Wang W, Liu Z, Jiang S (2008) Grazing gradient versus restoration succession of Leymus chinensis (Trin.) Tzvel. grassland in Inner Mongolia. Restoration Ecology 16: 572–583.

17

Discovery of a Rare Pterosaur Bone Bed in a Cretaceous Desert with Insights on Ontogeny and Behavior of Flying Reptiles

Paulo C. Manzig[1,2], **Alexander W. A. Kellner**[3]*, **Luiz C. Weinschütz**[1], **Carlos E. Fragoso**[4], **Cristina S. Vega**[5], **Gilson B. Guimarães**[6], **Luiz C. Godoy**[6], **Antonio Liccardo**[6], **João H. Z. Ricetti**[1], **Camila C. de Moura**[1]

1 Centro Paleontológico da UnC (CENPÁLEO), Universidade do Contestado, Mafra, Santa Catarina, Brazil, 2 Programa de Pós-Graduação IEL-Labjor, Universidade Estadual de Campinas (UNICAMP), Campinas, São Paulo, Brazil, 3 Laboratory of Systematics and Taphonomy of Fossil Vertebrates, Departamento de Geologia e Paleontologia, Museu Nacional/Universidade Federal do Rio de Janeiro, Rio de Janeiro, Brazil, 4 Universidade Estadual de Ponta Grossa, Ponta Grossa, Paraná, Brazil, 5 Departamento de Geologia, Universidade Federal do Paraná, Curitiba, Paraná, Brazil, 6 Departamento de Geociências, Universidade Estadual de Ponta Grossa, Ponta Grossa, Paraná, Brazil

Abstract

A pterosaur bone bed with at least 47 individuals (wing spans: 0.65–2.35 m) of a new species is reported from southern Brazil from an interdunal lake deposit of a Cretaceous desert, shedding new light on several biological aspects of those flying reptiles. The material represents a new pterosaur, *Caiuajara dobruskii* gen. et sp. nov., that is the southermost occurrence of the edentulous clade Tapejaridae (Tapejarinae, Pterodactyloidea) recovered so far. *Caiuajara dobruskii* differs from all other members of this clade in several cranial features, including the presence of a ventral sagittal bony expansion projected inside the nasoantorbital fenestra, which is formed by the premaxillae; and features of the lower jaw, like a marked rounded depression in the occlusal concavity of the dentary. Ontogenetic variation of *Caiuajara dobruskii* is mainly reflected in the size and inclination of the premaxillary crest, changing from small and inclined (~115°) in juveniles to large and steep (~90°) in adults. No particular ontogenetic features are observed in postcranial elements. The available information suggests that this species was gregarious, living in colonies, and most likely precocial, being able to fly at a very young age, which might have been a general trend for at least derived pterosaurs.

Editor: Andrew A. Farke, Raymond M. Alf Museum of Paleontology, United States of America

Funding: AWAK acknowledges funding from the Fundação Carlos Chagas Filho de Amparo à pesquisa do Rio de Janeiro (FAPERJ # E-26/102.737/2012) and the Conselho Nacional de Desenvolvimento Científico e Tecnológico (CNPq # 307276/2009-9). The funders had no role in study design, data collection and analysis, decision to publish, or preparation of the manuscript.

Competing Interests: The authors have declared that no competing interests exist.

* Email: kellner@mn.ufrj.br

Introduction

Pterosaurs comprise an extinct group of flying reptiles that have been recovered on all continents [1]. Notwithstanding their distribution, their record is rather patchy, with most occurrences limited to fragmentary remains that in several cases were only briefly reported in the literature [2]. Most pterosaurs are known from ancient coastal or shallow marine deposits and the number of species that lived deep inside the continents is limited [3,4], particularly from desert environments [5]. Most species are based on one incomplete individual, and aside from one potential exception of a collection of flattened specimens [6], no pterosaur accumulation can be regarded as a bone bed preserving several individuals that can confidently be assigned to the same species and at least potentially be regarded as representing the same or successive populations [7]. This has hampered the discussion of several biological questions regarding those animals, such as ontogenetic growth, development of cranial crests, and behavior.

Here we describe a rare pterosaur bone bed composed of hundreds of bones from the outskirts of Cruzeiro do Oeste, southern Brazil. The deposits correspond to the Caiuá Group [8]

that represents a sand sea formed in an interior paleodesert whose paleontological content was up to know limited to infrequent tetrapod ichnofossils [9,10] (Figure 1). This exceptional occurrence, combined with the large number of three-dimensionally preserved individuals, sheds new light on the biology of those rather enigmatic volant animals.

Materials and Methods

Phylogenetic Analysis

In order to determine the phylogenetic position of *Caiuajara dobruskii* gen. et sp. nov., we performed a phylogenetic analysis using PAUP 4.0b10 for Microsoft Windows [11] using the TBR heuristic searches performed using maximum parsimony. Characters were given equal weight and treated as unordered (ACCTRAN setting). This analysis is based on previous cladistic studies (List S3 in File S1).

Nomenclatural Acts

The electronic edition of this article conforms to the requirements of the amended International Code of Zoological

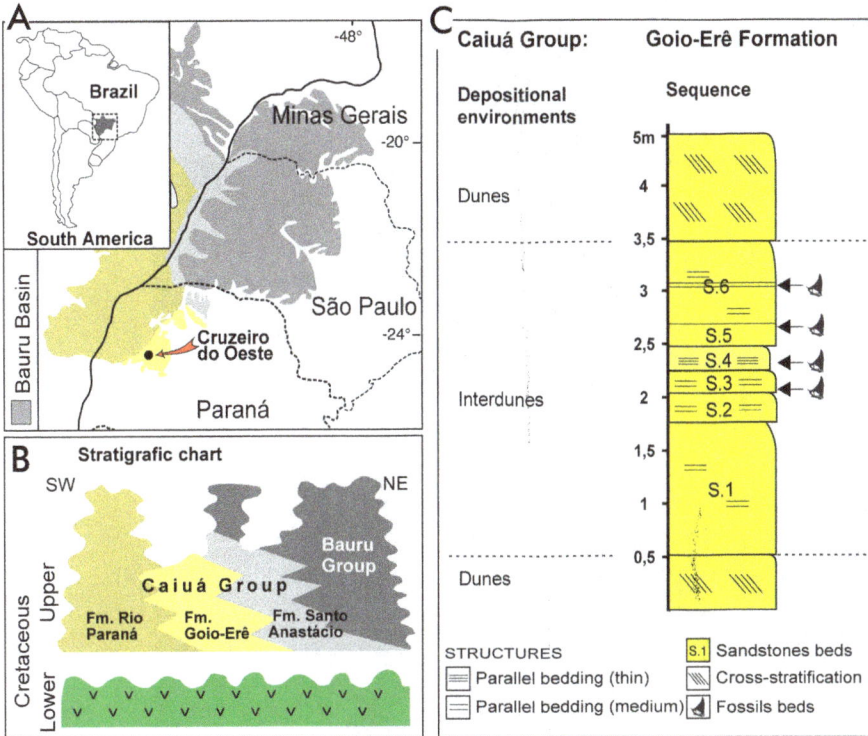

Figure 1. Localization and stratigraphic framework of the new pterosaur locality. (A) Map of South America and the geographic position of Cruzeiro do Oeste. **(B)** Stratigraphic chart showing the relation between the distinct stratigraphic units of the Bauru Basin [10]. **(C)** Detailed stratigraphic section of the quarried beds of the Goio-Erê Formation, showing the location where the fossils were recovered.

Figure 2. Holotype (CP.V 1449 left) and one paratype (CP.V 2003, right) of *Caiuajara dobruskii* **gen. et sp. nov. separated by a red line, showing skull and postcranial elements.** Scale bar equals 100 mm. Abbreviations: cor, coracoid; cv, cervical vertebra; d, dentary; dca, distal carpal series; dcr, dentary crest; f, frontal; hu, humerus; hy, hyoid bone; l, left; man, mandible; mcfo, meckelian fossa; mcI-III, metacarpal I–III; mcIV, metacarpal IV; oc, occipital condyle; p, parietal; pmcr, premaxillary crest; ph1d4, first phalanx of manual digit IV; ph2d4, second phalanx of manual digit IV; ph4d4, forth phalanx of manual digit IV; pof, postfrontal; ptd, pteroid; q, quadrate; r, right; ra, radius; ri, rib; un, ungueal.

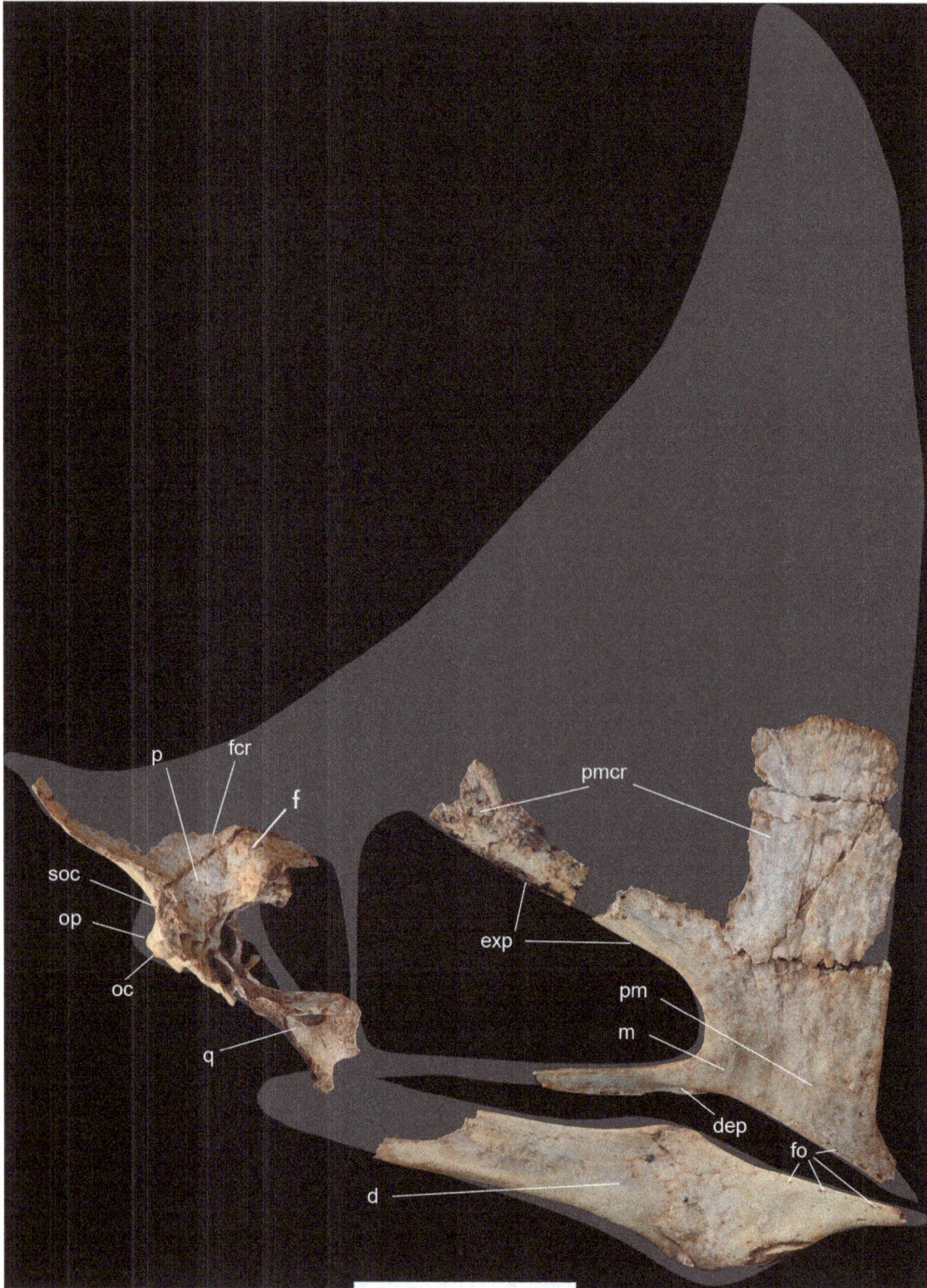

Figure 3. Skull of *Caiuajara dobruskii* **gen. et sp. nov. (holotype, CP.V 1449) with the shape of an adult individual.** Scale bar equals 50 mm. Abbreviations: d, dentary; dcr, dentary crest; dep, depression; exp, ventral expansion of the premaxilla; f, frontal; fcr, frontal crest; fo, foraminae; m, maxilla; oc, occipital condyle; op, opisthotic; p, parietal; pm, premaxilla; pmcr, premaxillary crest; q, quadrate; soc, supraoccipital. The quadrate is inverted.

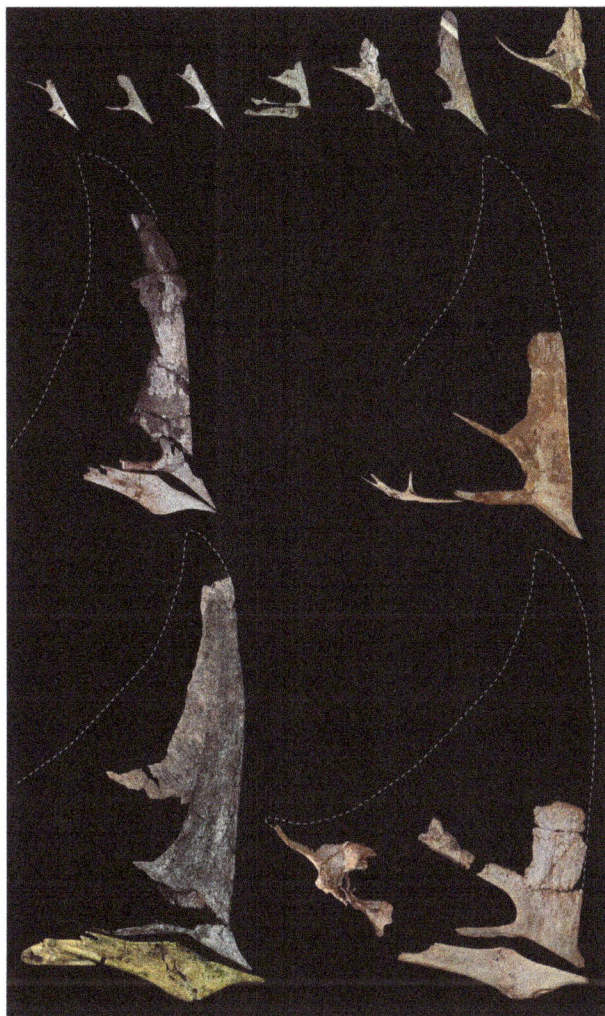

Figure 4. Selected cranial material of *Caiuajara dobruskii* showing anatomical changes during ontogeny. Note that the cranial crest gets gradually larger in older individuals. From top left to bottom right: CP.V 1050-1 (inverted), CP.V 1050-2 (inverted), CP.V 1003, CP.V 866 (inverted), UEPG/DEGEO/MP-4151, CP.V 1023 (inverted), UEPG/ DEGEO/MP-4151 (second skull), CP.V 1001, CP.V 1447, CP.V 1005 (with posterior part of lower jaw reconstructed), CP.V 1449 (holotype). Scale bar, equals 50 mm.

Nomenclature, and hence the new names contained herein are available under the Code from the electronic edition of this article. This published work and the nomenclatural acts it contains have been registered in ZooBank, the online registration system for the ICZN. The ZooBank LSIDs (Life Science Identifiers) can be resolved and the associated information viewed through any standard web browser by appending the LSID to the previx "http://zoobank.org/". The LSID for this publication is: urn:lsid:zoobank.org:pub:E6A57D0A-3F3A-4F56-9279-B12CFA222337. The electronic edition of this work was published in a journal with an ISSN, and has been archieved and is available from the following digital repositories: PubMed Central, LOCKSS.

All permits were obtained for the described study, which complied with all relevant regulations. The permit for collecting the specimens was issued by the Departamento Nacional de Produção Mineral (DNPM, Brasília), under the number DNPM n° 48400-000807/2012-94. See appropriate section of Systematic

Paleontology for locality, stratigraphy and repository, and specimen numbers.

Results

Systematic Paleontology

Pterosauria Kaup, 1834
Pterodactyloidea Plieninger, 1901
Azhdarchoidea Nessov, 1984
Tapejaridae Kellner, 1989
Tapejarinae Kellner, 1989 *sensu* Kellner & Campos [12]
Caiuajara dobruskii gen. et sp. nov.
ZooBank Life Science Identifier (LSID) for genus. urn:lsid:zoobank.org:act:9E5919F7-7A2A-4065-9FC1-11EB1960BF5C
ZooBank LSID for species. urn:lsid:zoobank.org:act: CF251616-A7AA-4C25-A6BE-B69AB448D93B
Etymology. Combination of Caiuá and *Tapejara*, the internal specifier of the Tapejarinae [13]; species honors Alexandre Dobruski, who with his son, João Dobruski, found the new site back in 1971.
Holotype. Partial skeleton including skull and lower jaw, cervical vertebrae and wing elements (CP.V 1449), housed at the Centro Paleontológico (CENPALEO) of the Universidade do Contestado, Mafra, Santa Catarina, Brazil (Figures 2, 3).
Paratypes. CP.V 865, consisting of the anterior portion of a skull, the posterior portion of the lower jaw, the right jugal, vertebrae, ribs and metatarsals; CP.V 867, rostral end of a skull and long bones; CP.V 868, rostral end of a skull, wing elements and other postcranial bones; CP.V 869, incomplete skeleton with a partial vertebral column (posterior cervicals vertebrae, dorsal elements to the first five caudal vertebrae), right humerus, radius and ulna, carpal elements, coracoid, sternum, some wing phalanges, gastralia, pelvic elements and the right femur; CP.V 870, incomplete postcranial elements, with humeri and pectoral girdle; CP.V 871, fused right scapulocoracoid and incomplete long bones; CP.V 872, partial skeleton including a fragmentary skull, lower jaw, right humerus, radius, ulna, carpals, cervical vertebrae and other long bones; CP.V 873, rostral end of a skull and manual phalanges; CP.V 999, partial skull; CP.V 1001, one incomplete skull with elongated premaxillary crest and lower jaw, and postcranial elements of at least three individuals, wing bones (with humeri), cervical vertebrae and pelvic elements; CP.V 1003, incomplete skull and the rostral tip of the lower jaw; CP.V 1004, rostral end of a skull; CP.V 1005, incomplete skull with an elongated premaxillary crest and a complete lower jaw; CP.V 1006, partial skull with anterior rostral end missing, with large premaxillary crest and several postcranial bones; CP.V 1023, anterior portion of a skull and several postcranial elements; CP.V 1024, skull and several postcranial bones of at least three small individuals; CP.V 1025, isolated femur; CP.V 1026, isolated femur; CP.V 1450, several small individuals (at least 14); CP.V 2003, skull and lower jaw associated with wing bones; UEPG/ DEGEO/MP-4151, two skulls on one slab and postcranial elements; and UEPG/DEGEO/MP-4152, a rostrum and several postcranial elements (Figures 2, 4–8). For referred specimens, see List S1 in File S1.
Type locality, horizon and age. Cruzeiro do Oeste, Paraná State, Brazil; Bauru Basin, Caiuá Group, Goio-Erê Formation, Upper Cretaceous [8,12,14].
Diagnosis. Tapejarine tapejarid with the following autapomorphies: anterior end of the premaxillary strongly deflected ventrally (~142–149°) relative to the ventral margin of the upper jaw; premaxillae with ventral sagittal bony expansion projected inside the nasoantorbital fenestra; rounded depression in the

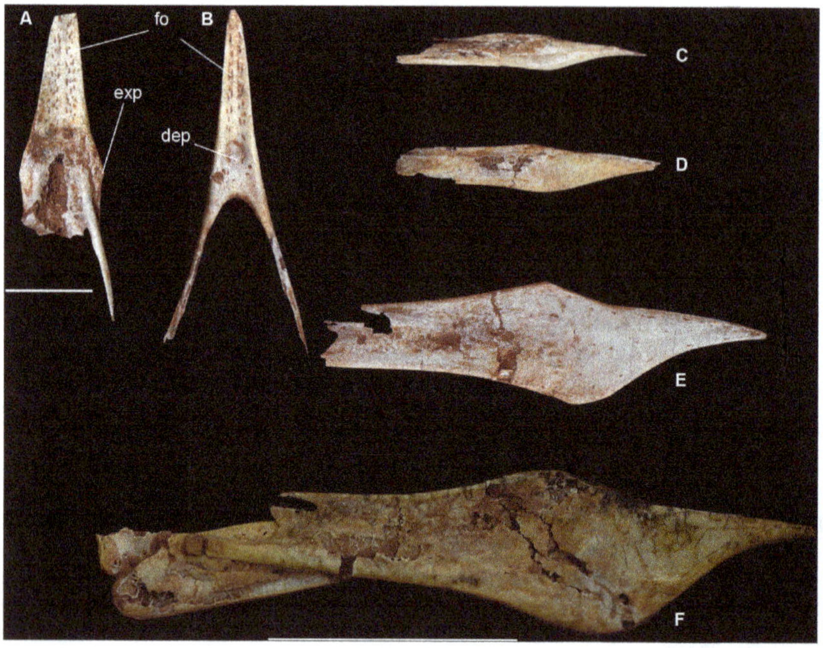

Figure 5. *Caiuajara dobruskii* **gen. et sp. nov., occlusal view of upper jaw (A) and mandible (B) (CP.V 1449).** Scale bar equals 10 mm; different lower jaws, from from top to bottom: juvenile (CP.V 1450-2) (C); older juvenile (CP.V. 1450-1) (D); young/subadult (CP.V. 1001a-1) (E); and adult (CP.V 1005a) (F) specimen. Scale bar equals 50 mm. Abbreviations: dep, depression; exp, lateral expansion; fo, foraminae.

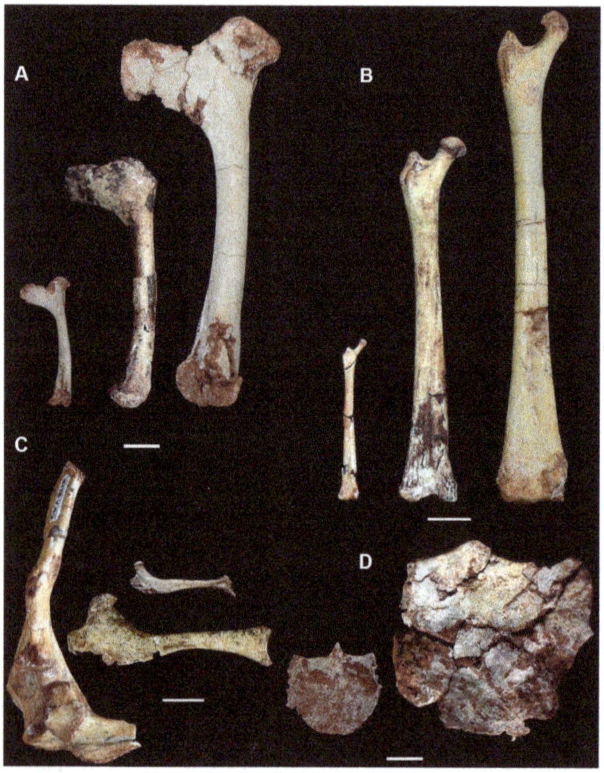

Figure 6. Selected post-cranial elements of *Caiuajara dobruskii* showing the anatomical changes during ontogeny. A, humeri (CP.V 1450 - inverted; CP.V 1009; CP.V 1013); B, femura (CP.V 1883-1; CPV 872a-1; CP.V 1025), scapulocoracoid (CP.V 871b - inverted), coracoids (CP.V 1006-1; CP.V 866b - inverted); and sterna (CP.V 1000; CP.V 1001a-1). Scale bar equals 10 mm.

occlusal concavity of the dentary; elongates groove on the anterolateral margin of the quadrate; and marked lateral depression on maxilla ventral to anterior part of the nasoantorbital fenestra. The new species can be further distinguished from other tapejarine pterosaurs by the following combination of characters: ventral margin of the orbit rounded; gap between upper and lower jaws during occlusion wider; and marked depression on ventral side of the pteroid lacking a pneumatic foramen.

Description and Comparisons

Several anatomical features show that *Caiuajara* belongs to the toothless pterodactyloid clade Tapejaridae [13], such as a premaxillary crest from the anterior rostral end extended above the occipital region, nasoantorbital fenestra elongated comprising more than 40% of the cranial length (Figure 3), and a well-developed tubercle on anterior surface of the coracoid. It further has all synapomorphies of the Tapejarinae, such as the down-turned anterior part of the rostrum [12,13,15–17], with the inclination varying from about 138° to 150°; most species average 142°. The orbit is piriform, with the ventral margin more rounded than in other tapejarids [15,18]. The nasoantorbital fenestra anterior margin is wide, similar to other tapejarines [15,16,19] but differing from the narrower condition of thalassodromines [12,20].

The premaxilla is perforated by a large number of foraminae on the lateral and palatal surface, similar to *Tapejara*, suggesting that the beak was covered by a horny covering analogous to the rhamphoteca in birds. A developed premaxillary sagittal crest is present in the smallest and the largest individuals (Figure 4), casting doubts on previous interpretations that the presence and absence of cranial crests might be sexually dimorphic [21]. The anterior part of the crest is very high similar to *Tupandactylus* [13,22], but differs by being more expanded. The occipital portion of the crest, formed by the supraoccipital and parietal, is dorsally curved, differing from the much longer and straighter structure found in *Tupandactylus* [22]. Starting close to the anterior margin of

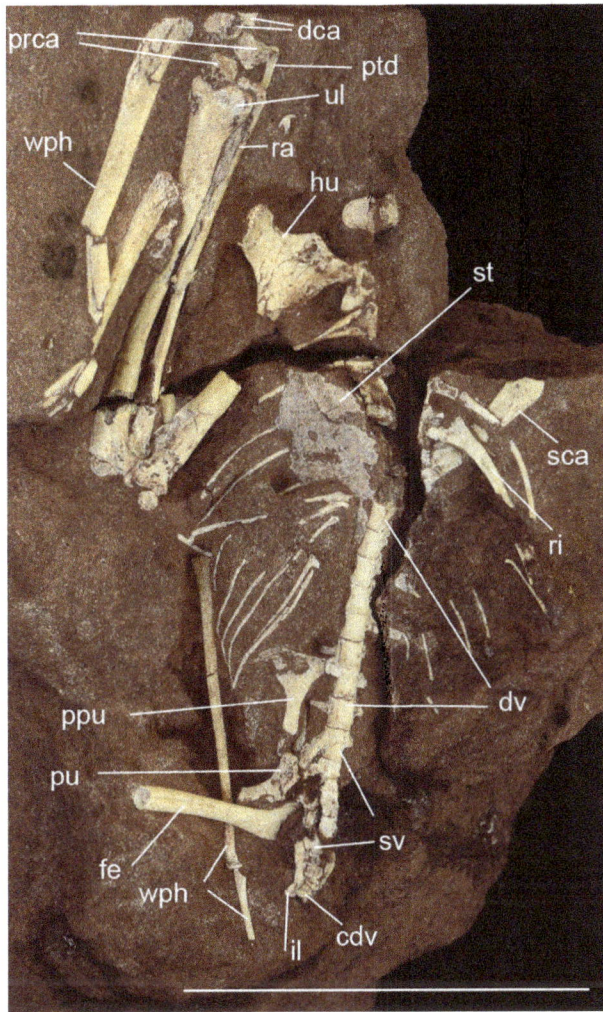

Figure 7. *Caiuajara dobruskii* **gen. et sp. nov., (CP.V 869), partial articulated skeleton.** Scale bar equals 50 mm. Abbreviations: cdv, caudal vertebrae; dca, distal carpal series; dv, dorsal vertebrae; fe, femur; hu, humerus; il, ilium; prca, proximal carpal series; ptd, pteroid; ppu, prepubis; pu, pubis; ra, radius; ri, ribs; sca, scapula; st, sternum; sv, sacral vertebrae; ul, ulna; wph, wing phalanx.

the nasoantorbital fenestra, the premaxilla has a sagittal bony expansion that extends posteriorly, where it merges with the lateral margin of this opening (Figure 3). This structure, whose function is unknown, is present in all specimens, from the smallest to the largest, and has not been reported in any other pterosaur before (Figure 4). As in *Tapejara*, *Caiuajara* shows a deep concavity in the palate restricted to the anterior part, followed by a posterior convexity. Differing from all other tapejarids where the occlusal surface can be observed, some specimens of *Caiuajara* have a faint longitudinal crest inside the most concave portion of the palate that does not form a palatal ridge as in some thalassodromines [12] and some pteranodontoids [17,23]. As in other tapejarines, the upper jaw shows a small lateral expansion close to the anterior margin of the nasoantorbital fenestra. The occlusal surface of the dentary also displays a deep concavity as in *Tapejara*, but *Caiuajara* differs by showing a distinctive rounded depression (Figure 5). The dentary shows the typical tapejarine step-like dorsal margin and a blunt dentary sagittal crest that is more developed in larger individuals, similar to *Tapejara* [18] and *Europejara* [24], but

differing from the Chinese tapejarines [15,16]. There is no helical jaw joint, differing from *Caupedactylus* [25].

The cervical vertebrae (Figure 2) are slightly elongated, more so than in pteranodontoids [26], but not to the same degree as in archaeopterodactyloids [17,23] or azhdarchids [27,28]. The neural spine is blade-like and the centrum pierced laterally by small pneumatic foraminae. One lateral pneumatic opening occurs on each side of the neural canal on the anterior articulation surface. No notarium is developed. The sacrum is formed by five sacrals. The scapula is longer than coracoid, and where complete, the coracoid shows a developed tubercle on the anterior surface (Figure 6). The sternum is semicircular in shape. The humerus displays an elongated deltopectoral crest that is rectangular and slightly curved medially, particularly at the most posterodistal end, but the crest is not warped as in pteranodontoids [17,23,26,29]. About 35 humeri were identified so far (20 right, 14 left and one unidentifiable) with lengths ranging from 31 mm to 115.6 mm. Overall, the radius is thinner than the ulna, but not to the same degree as observed in istiodactylids and anhanguerids [23]. Distal sincarpals show a rectangular shape. The pteroid clearly articulates with the proximal carpal series, showing a developed ventral depression but no pneumatic foramen (Figure 7). Wing metacarpal IV is similar to that seen in other tapejarids, being proportionally longer relative to other wing elements when compared to anhanguerids (Figure 2), but does not approach the extreme elongation reported in nyctosaurids [23,26]. The femur is bowed and about the same size as the humerus (Figure 6).

A phylogenetic analysis based on a previous study of tapejarid phylogenetic relationships [24] shows that *Caiuajara* is a member of the Tapejarinae, falling in a polytomy with other tapejarine tapejarids (Figure 9). If *Eopteranodon*, a poorly described taxon, and *Europejara* (unfortunately very incomplete) are removed, *Caiuajara* falls in a sister group relationship with *Tupandactylus*, in a trichotomy with *Tapejara* and *Sinopterus+Huaxiapterus*, indicating that the known tapejarines from China form a monophyletic entity. This exercise also shows that much more has to be done to resolve the relationships of the Tapejaridae, particularly the Tapejarinae.

Previous studies of pterosaur ontogeny were based on isolated specimens mostly recovered without stratigraphic control that, despite important contributions [30–39], have fostered some controversy, particularly over whether or not the studied specimens represent the same species [40,41]. *Caiuajara* is the first case where a pterosaur ontogenetic series is provided based on specimens from a pterosaur bone bed that can be confidently assigned to the same species. The sample also has the advantage of having most elements preserved three-dimensionally and not flattened, avoiding the problems related to change of morphology due to distortion [6,7,36]. Regarding postcranial elements, there are few differences from smaller to larger individuals except for size and the tendency for ontogenetically more developed individuals to show more ossified bones, particularly the sternum (Figure 6). The humerus, for example, shows the same proportion in smaller and larger individuals, including the development of the deltopectoral crest that corresponds to about 38–40% of the humerus length. This indicates that the general shape of most postcranial elements is formed at a juvenile stage and does not change significantly, as the animal grows older. The most conspicuous exceptions are the prepubis, with older individuals showing a more developed and larger distal plate, and the coracoid, where ontogenetically more developed individuals display a slightly larger ventral expansion. Furthermore, as reported in other pterodactyloids, the scapula and coracoid are fused in adult individuals but unfused in younger individuals, with

Figure 8. Hundreds of bones, including at least 14 partial skulls of *Caiuajara dobruskii* (CP.V. 1450). Scale bar equals 200 mm. Abbreviations: cra - skulls, man - mandible.

the same trend happening in the epiphyses of the humerus and the carpal series [32,35].

Regarding the skull, the main ontogenetic differences can be found in the rostrum and the cranial crest (Figures 4, 10). Younger individuals display a reduced rostrum that grows, becoming more massive in older individuals. The inclination of the occlusal margin relative to the horizontal plane does not vary significantly, mostly being around 142°. The premaxillary crest, on the contrary, shows marked variation, being reduced and inclined posteriorly for about ~115° relative to the horizontal plane in small individuals. As the animal grew, the crest got rapidly larger and steeper (up to ~90°). Similar changes are observed in the dentary crest, which is almost absent in young individuals and gets more developed in older ones (Figures 5, 10).

Discussion and Conclusions

There are several interesting aspects of this discovery. So far, all other pterosaur material recorded from Brazil comes from the northeastern part of the country [1], and this is the first in the southern part. Besides the Crato and Romualdo formations of Brazil [12], tapejarid pterosaurs have also been recorded in China [15,16], Morocco [42], and Spain [24], all in deposits that range from the Barremian to the Cenomanian [24]. Based on stratigraphic correlations, the age of the Goio-Erê Formation is regarded as Turonian to Campanian [14], or even having a Coniacian basal limit [10]. Therefore, either *Caiuajara* is the youngest member of this group known to date, or this deposit is older than previously thought. In any case, Cruzeiro do Oeste is the southernmost occurrence of Tapejarinae recorded so far, suggesting that those pterosaurs, which are regarded to be frugivorous [18,24], had a cosmopolitan distribution.

The discovery of this pterosaur bone bed also allows inference of some aspects regarding the behavior of the new species that might be applied for other flying reptiles. *Caiuajara dobruskii* is known from hundreds of bones representing individuals of different sizes that were collected in an area of less than 20 m². Based on the premaxillae, a minimum of 47 individuals can be established (List S2 in File S1), but the actual number present in this site must be well in the hundreds.

All parts of the skeleton are represented. The skeletons of a few specimens were found articulated (Figure 7) or closely associated (Figure 2), but most are mixed together, making it difficult establish which elements belong to the same individual. In one extreme case, at least 14 individuals could be identified based on the premaxillae in one small block of sandstone (40 cm by 60 cm; more was left in the outcrop) with hundreds of bones, including 11 lower jaws, all belonging to small individuals (Figure 8), indicating that this pterosaur accumulation was partially submitted to hydraulic selection. In several instances, there is indication that bones were broken prior to fossilization and that some might have been exposed more than others before being buried.

Regarding the ontogenetic stage of the recovered specimens, the fact that it is very hard to associated elements to the same individual, makes is difficult to establish with certainty the number of juveniles, sub-adults and adults. However, most of the bones from *Caiuajara dobruskii* recovered are predominatly small, as can be exemplified by the humerus (Table S1 in File S1) and also the sizes of the skull (Figure 5). Therefore we can observe that most recovered specimens are predominantly juveniles or very young animals, with adults being quite rare, represented by only two skulls and three humeri.

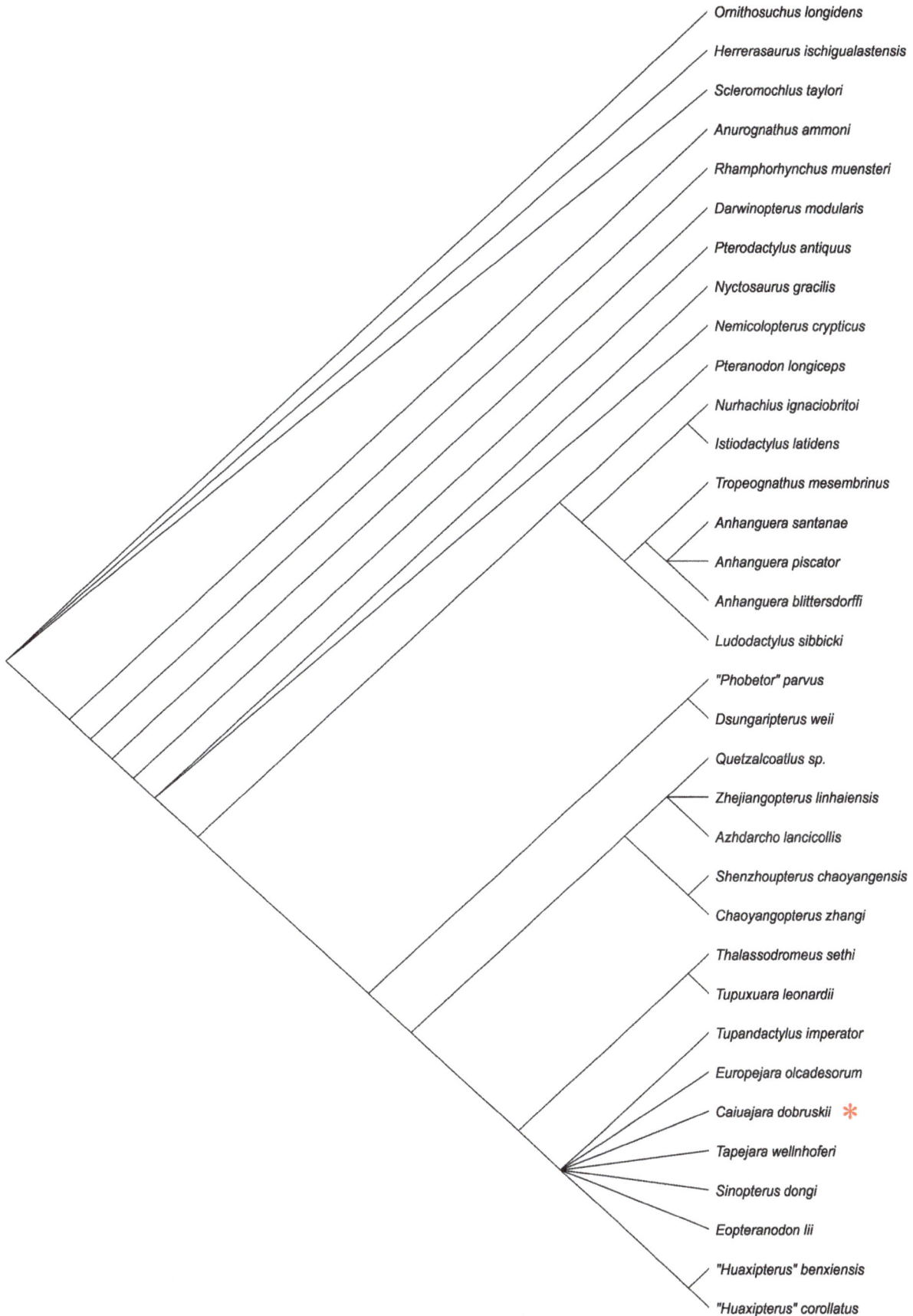

Figure 9. Phylogenetic analysis showing the relationships of *Caiuajara dobruskii* **gen. et sp. nov.** Based on Vullo et al. 2012 [23].

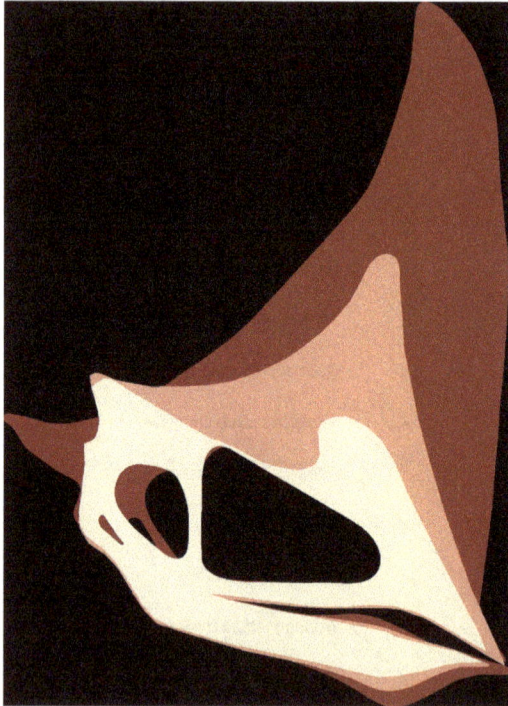

Figure 10. *Caiuajara dobruskii* **gen. et sp. nov., reconstruction of shapes from juveniles (bright color) to adults (darker color).** Outlines not to scale.

It is also very difficult to establish precisely the wing span variation of the sample collected so far. Comparing the size of several postcranial elements with other tapejarids [1,15,16], particularly the humerus, allows us to estimate the wingspan variation of what is presently known of *Caiuajara dobruskii* between 0.65 and 2.35 m.

Most specimens were collected in two different levels less than 0.5 m apart vertically. A third level of accumulation with hundreds of bones of small individuals in a more restricted area is less than 0.5 m above the last one (Figure 1). A fourth one yielded only isolated elements, indicating distinct events that generated this pterosaur accumulation. Sedimentological data supports the interpretation that the Goio-Erê Formation was formed in a desert environment with interdunal wetland [9,10]. So far there is no evidence of invertebrate or plant material.

The fact that several pterosaur individuals were found in such close association is compelling evidence that *Caiuajara dobruskii* was gregarious, as has been suggested based on similar evidences for other extinct reptiles, including dinosaurs [43,44]. Previous evidences of this kind of social behavior were restricted to some specimens of *Quetzalcoatlus* sp. found in close proximity [45] and a concentration of the archaeopterodactyloid *Pterodaustro* in Argentina [6]. Besides those, close associations of pterosaur individuals are exceedingly rare, limited to fragmentary remains of unknown

affinity from Chile [5], one duplicate bone in one nodule from the Romualdo Formation [46] and two pterosaur specimens from Kazakhstan [47], making the pterosaur bone bed in Cruzeiro do Oeste particularly important.

Based on the available information, we conclude that *Caiuajara dobruskii* lived in colonies around an inland lake situated in a desert. Although some parental care might have been possible, the fact that the postcranial skeleton does not differ among juveniles and adults suggests that the new species was precocial and most likely could fly at a very young age. Other researcher have also pointed out to this possibility (e.g., [48,49]). The taphonomic and geological conditions suggest that individuals died around an oasis over the years, being exposed and gradually disarticulated. The degree of disarticulation was dependent on the exposure time. Episodic events (e.g., desert storms) likely carried the disarticulated and partially articulated skeletons to the bottom of the lake where they got eventually preserved. The presence of three main levels of accumulation in a section of less than one meter suggests that this region was home to pterosaur populations for an extended period of time. It is also plausible that *Caiuajara* was a migratory pterosaur that visited this area from times to time, although the first possibility is favored here. The causes of death remain unknown, although similarities with dinosaur drought-related mortality are striking [42]. However, it is also possible that desert storms could have been responsible for the occasional demise of these pterosaurs.

Supporting Information

File S1   Supporting information. List S1, Specimens referred to *Caiuajara dobruskii* gen. et sp. nov. List S2, Minimum number of individuals of *Caiuajara dobruskii* gen. et sp. nov. List S3, Phylogenetic analysis, characters and character matrix. Table S1, Measurements of wing elements of *Caiuajara dobruskii* gen. et sp. nov. Table S2, Measurements of hindlimb elements and the pteroid of *Caiuajara dobruskii* gen. et sp. nov.

Acknowledgments

We thank Valter Pereira da Rocha (Mayor of Municipality of Cruzeiro do Oeste) and João Gustavo Dobruski, Neurides Oliveira Martins and Maristela Sanches Morcelli, all residents of Cruzeiro do Oeste, for their help in the fieldwork. José Alceu Valério, Solange Salete Sprandel da Silva, Ademir Flores, and Itaíra Susko, all from the Universidade do Contestado, are thanked for supporting the research at the CENPALEO. Vilson Greinert is acknowledged for the preparation of several specimens. CEF thanks Conselho Nacional de Desenvolvimento Científico e Tecnológico (CNPq) for a scholarship. Juliana Manso Sayão (Universidade Federal de Pernambuco, Pernambuco) and an anonymous reviewer are thanked for several suggestions in earlier versions of the manuscript.

Author Contributions

Conceived and designed the experiments: AWAK PCM LCW CEF. Analyzed the data: AWAK PCM LCW CEF CSV GBG LCG AL JHZR CCM. Wrote the paper: AWAK PCM LCW CEF CSV GBG LCG AL JHZR CCM.

References

1. Kellner AWA (2006) Pterossauros - os senhores do céu do Brasil. Vieira & Lent, Rio de Janeiro 176p.
2. Barrett PM, Butler RJ, Edwards NP, Milner (2008). Pterosaur distribution in time and space: an atlas. Zitteliana 28: 61–107.
3. Wang X, Kellner AWA, Zhou Z, Campos DA (2005) Pterosaur diversity and faunal turnover in Cretaceous terrestrial ecosystems in China. Nature 437: 875–879.

4. Witton MP, Naish D (2008) A reappraisal of Azhdarchid pterosaur functional morphology and paleoecology. PLoS ONE 3: e2271
5. Bell CM, Padian K (1995) Pterosaur fossils from the Cretaceous of Chile: evidence for a pterosaur colony on an inland desert plain. Geol Mag 132: 31–38.
6. Chiappe LM, Rivarola D, Romero E, Davila S, Codorniu L (1998) Recent Advances in the Paleontology of the Lower Cretaceous Lagarcito Formation (Parque Nacional Sierra de Las Quijadas, San Luis, Argentina). In Lucas SG,

Kirkland JI, Estep JW editors. Lower and Middle Cretaceous Terrestrial Ecosystems, New Mexico: Museum of Natural History and Science Bulletin 14: 187–192.

7. Kellner AWA, Campos DA, Sayão JM, Saraiva AAF, Rodrigues T, et al. (2013) The largest flying reptile from Gondwana: a new specimen of *Tropeognathus* cf. *T. mesembrinus* Wellnhofer, 1987 (Pterodactyloidea, Anhangueridae) and other large pterosaurs from the Romualdo Formation, Lower Cretaceous, Brazil. An Acad Bras Cienc 85:113–135.

8. Manzig PC, Weinschütz LC (2012) Museus e Fósseis da Região Sul do Brasil: Uma Experiência Visual com a Paleontologia. Curitiba. Ed. Germânica.

9. Fernandes LA, Sedor FA, Silva RC, Silva LR, Azevedo AA, et al. (2009) Icnofósseis da Usina Porto Primavera, SP: rastros de dinossauros e de mamíferos em rochas do deserto neocretáceo. In: Winge M, Schobbenhaus C, Souza CRG, Fernandes ACS, Bebert-Born M, Queiroz ET, Campo DA, editors. Sítios Geológicos e Paleontológicos do Brasil. Brasília: CPRM - Serviço Geológico do Brasil p. 479–488.

10. Milani EJ, Melo JH, Souza PA, Fernandes LA, França AB (2007) Bacia do Paraná. Bol Geoc Petrobras 15: 265–287.

11. Swofford DL (2000) Paup: Phylogenetic Analysis Using Parsimony, Version 4.0B10 (for Microsoft Windows) Massachusetts, Sinauer Associates, Inc. Sunderland.

12. Kellner AWA, Campos DA (2007) Short note on the ingroup relationships of the Tapejaridae (Pterosauria, Pterodactyloidea). Boletim do Museu Nacional – Geologia 75: 1–14.

13. Kellner AWA (2004) New information on the Tapejaridae (Pterosauria, Pterodactyloidea) and discussion of the relationships of this clade. Ameghiniana 41: 521–534.

14. Basilici G, Sgarbi GN, Dal'Bó PFF (2012) A Sub-bacia Bauru: um sistema continental entre deserto e cerrado. In: Y. Hasui et al., Eds. Geologia do Brasil Beca 520–543.

15. Wang X, Zhou Z (2003) A new pterosaur (Pterodactyloidea, Tapejaridae) from the Early Cretaceous Jiufotang Formation of western Liaoning, China and its implications for biostratigraphy. Chin Sci Bull 48: 16–23.

16. Lü J, Jin S, Unwin D, Zhao L, Azuma Y, et al (2006) A new species of *Huaxiapterus* (Pterosauria: Pterodactyloidea) from the Lower Cretaceous of western Liaoning, China with comments on the systematics of tapejarid pterosaurs. Acta Geol Sin 80: 315–326.

17. Andres B, Ji Q (2008) A new pterosaur from the Liaoning Province of China, the phylogeny of the Pterodactyloidea, and the convergence in their cervical vertebrae. Palaeontology 51: 453–469.

18. Wellnhofer P, Kellner AWA (1991) The skull of *Tapejara wellnhoferi* Kellner (Reptilia, Pterosauria) from the Lower Cretaceous Santana Formation of the Araripe Basin, Northeastern Brazil. Mitt Bayer Staatsslg Paläont hist Geol 31: 89–106.

19. Frey E, Martill DM, Buchy M-C (2003) A new species of tapejarid pterosaur with soft-tissue head crest. In: E. Buffetaut J, Mazin M, editors. Evolution and palaeobiology of pterosaurs. Geol. Soc. London Spec. Pub 217: 65–72.

20. Kellner AWA, Campos DA (2002) The function of the cranial crest and jaws of a unique pterosaur from the Early Cretaceous of Brazil. Science 297: 389–392.

21. Lu JC, Unwin DM, Deeming DC, Jin X, Liu Y, et al. (2011) An egg-adult association, gender, and reproduction in pterosaurs. Science 331: 321–324.

22. Pinheiro FL, Fortier DC, Schultz CL, Andrade JAFG, Bantim RAM (2011) New information of the pterosaur *Tupandactylus imperator*, with comments on the relationships of Tapejaridae. Acta Palaeontologica Polonica 56: 567–580.

23. Kellner AWA (2003) Pterosaur phylogeny and comments on the evolutionary history of the group. In: Buffetaut E, Mazin J-M, editors. Evolution and palaeobiology of pterosaurs. Geol. Soc. London Spec. Pub. 217: 105–137.

24. Vullo R, Marugán-Lobón J, Kellner AWA, Buscalioni AD, Gomez B, et al. (2012) A new crested pterosaur from the Early Cretaceous of Spain: the first European tapejarid (Pterodactyloidea: Azhdarchoidea). PLoS ONE 7, e38900.

25. Kellner AWA (2013) A new unusual tapejarid (Pterosauria, Pterodactyloidea) from the Early Cretaceous Romualdo Formation, Araripe Basin, Brazil. Earth and Environmental Science Transaction of the Royal Society of Edinburgh 103: 1–14.

26. Bennett SC (2001) The osteology and functional morphology of the Late Cretaceous pterosaur *Pteranodon* part 1 - General description and osteology. Palaeontographica 260: 1–112.

27. Nessov LA (1994) Pterosaurs and birds of the Late Cretaceous of Central Asia. Paläontologische Zeitschrift 1: 47–57.

28. Averianov AO (2010) The osteology of *Azhdarcho lancicollis* Nessov, 1984 (Pterosauria, Azhdarchidae) from the Late Cretaceous of Uzbekistan. Proceedings of the Zoological Institute RAS 314: 264–317.

29. Bennett SC (1989) A pteranodontid pterosaur from the Early Cretaceous of Peru, with comments on the relationships of Cretaceous pterosaurs. J Paleont 63: 669–677.

30. Wellnhofer P (1970) Die Pterodactyloidea (Pterosauria) der Oberjura-Plattenkalke Süddeutschlands. Abhandlungen der Bayerischen Akademie der Wissenschqften, N F 141: 1–13.

31. Mateer NJ (1976) A statistical study of the genus *Pterodactylus*. Bul Geol Inst Univ Uppsala 6: 97–105.

32. Bennett SC (1993) The ontogeny of *Pteranodon* and other pterosaurs. Paleobiology 19: 92–106.

33. Bennett SC (1995) A statistical study of *Rhamphorhynchus* from the Solnhofen Limestone of Germany: Year-classes of a single large species. J Paleont 69: 569–580.

34. Bennett SC (1996) Year-classes of pterosaurs from the Solnhofen limestones of Germany: taxonomic and systematic implications. J Vert Paleont 16: 432–444.

35. Kellner AWA, Tomida Y (2000) Description of a new species of Anhangueridae (Pterodactyloidea) with comments on the pterosaur fauna from the Santana Formation (Aptian-Albian), Northeastern Brazil. Nat Sci Museum Monogr 17: 1–135.

36. Codorniú L, Chiappe LM (2004) Early juvenile pterosaurs (Pterodactyloidea: *Pterodaustro guinazui*) from the Lower Cretaceous of central Argentina. Can J Earth Sci 41: 9–18

37. Jouve S (2004) Description of the skull of *Ctenochasma* (Pterosauria) from the latest Jurassic of eastern France, with a taxonomic revision of European Tithonian Pterodactyloidea. J Vert Paleont 24: 542–554.

38. Bennett SC (2007) A review of the pterosaur *Ctenochasma*: taxonomy and ontogeny. N Jb Geol Paläont Abh, 245, 23–31.

39. Chinsamy A, Codorniú L, Chiappe L (2008) Developmental growth patterns of the filter-feeder pterosaur, *Pterodaustro guinazui*. Biol Lett 4: 282–285.

40. Kellner AWA (2010) Comments on the Pteranodontidae (Pterosauria, Pterodactyloidea) with the description of two new species. An Acad Bras Cienc 82: 1063–1084.

41. Peters D (2011) A catalog of pterosaur pedes for trackmaker identification, Ichnos 18: 114–141.

42. Wellnhofer P, Buffetaut E (1999) Pterosaur remains from the Cretaceous of Morocco. Paläontol Ƶ 73: 133–142.

43. Rogers RR (1990) Taphonomy of three dinosaur bone beds in the Upper Cretaceous Two Medicine Formation of Northwestern Montana: evidence for drought-related mortality. Palaios 5: 394–413.

44. Dodson P, Forster CA, Sampson SD (2004) Ceratopsidae. In Weishampel DB, Dodson P, Osmólska, H, editors. The Dinosauria (second edition). University of California Press: Berkeley pp. 494–513.

45. Kellner AWA (1994) Remarks on pterosaur taphonomy and paleoecology. Acta Geologica Leopoldensia 39: 175–189.

46. Eck K, Elgin RA, Frey E (2011) On the osteology of *Tapejara wellnhoferi* Kellner 1989 and the first occurrence of multiple specimen assemblage from the Santana Formation, Araripe Basin, NE-Brazil. Swiss Journal of Palaeontology 130: 277–96.

47. Costa FR, Alifanov V, Dalla Vecchia FM, Kellner AWA (2013) On the presence of an elongated tail in an undescribed specimen of *Batrachognathus volans* (Pterosauria: Anurognathidae: Batrachognathinae). In: Rio Ptero 2013 - International Symposium on Pterosaurs, Short Communications: 54–56.

48. Wellnhofer P (1991) The illustrated encyclopedia of pterosaurs. Salamander Books, London 192 p.

49. Unwin DM, Deming CV (2008) Pterosaur eggshell structure and its implications for pterosaur reproductive biology. Zitteliana 28: 199–207.

18

Impact of Environmental Factors and Biological Soil Crust Types on Soil Respiration in a Desert Ecosystem

Wei Feng, Yuqing Zhang*, Xin Jia, Bin Wu*, Tianshan Zha, Shugao Qin, Ben Wang, Chenxi Shao, Jiabin Liu, Keyu Fa

Yanchi Research Station, College of Soil and Water Conservation, Beijing Forestry University, Beijing, China

Abstract

The responses of soil respiration to environmental conditions have been studied extensively in various ecosystems. However, little is known about the impacts of temperature and moisture on soils respiration under biological soil crusts. In this study, CO_2 efflux from biologically-crusted soils was measured continuously with an automated chamber system in Ningxia, northwest China, from June to October 2012. The highest soil respiration was observed in lichen-crusted soil (0.93 ± 0.43 µmol m^{-2} s^{-1}) and the lowest values in algae-crusted soil (0.73 ± 0.31 µmol m^{-2} s^{-1}). Over the diurnal scale, soil respiration was highest in the morning whereas soil temperature was highest in the midday, which resulted in diurnal hysteresis between the two variables. In addition, the lag time between soil respiration and soil temperature was negatively correlated with the soil volumetric water content and was reduced as soil water content increased. Over the seasonal scale, daily mean nighttime soil respiration was positively correlated with soil temperature when moisture exceeded 0.075 and 0.085 m^3 m^{-3} in lichen- and moss-crusted soil, respectively. However, moisture did not affect on soil respiration in algae-crusted soil during the study period. Daily mean nighttime soil respiration normalized by soil temperature increased with water content in lichen- and moss-crusted soil. Our results indicated that different types of biological soil crusts could affect response of soil respiration to environmental factors. There is a need to consider the spatial distribution of different types of biological soil crusts and their relative contributions to the total C budgets at the ecosystem or landscape level.

Editor: Xiujun Wang, University of Maryland, United States of America

Funding: This research was fund by the National Key Technology Research and Development Program of China for 12th Five-year Plan (2012BAD16B02) and the National Natural Science Foundation of China (31170666). The funders had no role in study design, data collection and analysis, decision to publish, or preparation of the manuscript.

Competing Interests: The authors have declared that no competing interests exist.

* Email: zhangyqbjfu@gmail.com (YZ); wubin@bjfu.edu.cn (BW)

Introduction

Soil respiration (R_s) accounts for the second largest carbon flux between terrestrial ecosystems and atmosphere, after gross primary productivity. Physical (e.g., soil temperature, moisture) and biological factors (e.g., microbial community) affecting Rs should be taken into consideration in order to accurately estimate global carbon balance [1]. However, we have limited knowledge on the biophysical controls of R_s in dryland ecosystems. Drylands cover 41–47% of the terrestrial surface [2]. Biological soil crusts (BSCs) as a biological factor commonly cover 70% of the inter-canopy earth in dryland and are found in all ecosystems around the world [3]. BSCs consist of algae, lichen, moss, fungi, cyanobacteria, and bacteria and cover the top few millimeters of the soil surface [3,4]. However, knowledge about the role of BSCs as a modulator of R_s is still lacking [5–7]. It is important to study the effects of environmental factors, such as temperature and moisture, on R_s under BSCs. This knowledge can reduce bias in ecosystem-level estimation of R_s and can help us predict how climate changes will affect CO_2 flux in desert ecosystems.

BSCs are an integral part of the soil system in arid regions worldwide [4]. R_s studies in relation to BSCs have drawn much attention in the past decade [4]. In the Gurbantunggute desert, the mean R_s of cyanobacteria/lichen-crusted soil is significantly higher than that of bare land after 15 mm rainfall [8]. In Kalahari sand, the CO_2 flux of cyanobacteria-crusted soil is lower than that of disturbed crusted soil [6]. In the Iberian Peninsula, lichen-crusted soils are the main contributor to R_s [9]. In the Mu Us desert, R_s does not differ between BSC-dominated areas and bare land [10]. However, the limited knowledge about the role of BSCs as a modulator of R_s on C cycle merely focused on particular species or communities. Although those have provided valuable insights on the effects of BSCs on C fluxes, in-situ data remain rare and we have incomplete understanding of the impact of different types of BSCs on R_s.

Soil temperature (T_s) and soil water content (VWC) are the key environmental factors responsible for variation in R_s [11]. T_s is the major control of R_s through its influence on the kinetics of microbial decomposition, root respiration, and the diffusion of enzymes and substrate [12]. VWC controls the decomposition of soil organic matter, root respiration, and microbial actively [3,4,12,13]. T_s and VWC were been predicted to increase at global scales in the following decades [2]. In order to assess the impact of the changing climate on ecosystem C flux, quantification of the effects of T_s and VWC on R_s is needed. Recent studies have shown that diurnal variations in R_s are usually highly correlated with temperature of the surface soil layers [14,15]. However, a few studies have reported a hysteresis effect and a decoupling between R_s and soil surface temperature during drought conditions in boreal forests [16], tropical forests [17], Mediterranean ecosystems [18], and desert ecosystems [19]. Low water content may increase the degree of hysteresis between R_s and T_s [17,18,19] or, in some cases, may reduce it [20]. At the seasonal scale, R_s is also highly correlated with changes in T_s when water content is not limited [19,21,22]. Strong inhibition of R_s has often been observed when soil water content is low [23]. All those are mainly focused on shrub soils or bare-land soils. However, our ability to capture the effects of environmental factors on R_s in biologically-crusted soil is still lacking.

Understanding of how biologically-crusted soil types and environmental factors influence R_s in a desert ecosystem, we measured R_s in algae-, lichen-, and moss-crusted soil in the Mu Us Desert, northwestern China. The specific objectives of this study were: (1) to examine and compare the temporal variability of R_s in three crusted soils; (2) to determine seasonal and diurnal patterns of R_s; and (3) to assess the contributions of the three crusted soils to the amount of C released by R_s at the ecosystem level.

Materials and Methods

2.1 Ethics Statement
The study site is owned by Beijing Forestry University. The field work did not involve any endangered or protected species, and did not involve destructive sampling. Specific permits were required for the described study.

2.2 Site description
The research was conducted at the Yanchi Research Station (37°04′ to 38°10′ N and 106°30′ to 107°41′ E, 1550 m a.s.l.), Ningxia, northwest China. The area is located in the mid-temperate zone and characterized by a semiarid continental monsoon climate. The mean annual temperature is 8.1°C, the mean annual rainfall is 292 mm, 62% of which falls between July and September. The mean annual potential evaporation is 2100–2500 mm. All meteorological data were provided by the meteorological station of Yanchi County and represent 51 year averages (1954–2004). The vegetation in the area is dominated by *Artemisia ordosica*. The soil surface of inter-canopy is commonly covered by algae, lichen, and moss crusts, which are mainly composed of *Microcoleus vaginatus*, *Oscillatoria chlorine*, *Collema tenax*, and *Byumargenteum*, respectively [10,24]. The physical and chemical characteristics of the three crusted soils are shown in Table 1. The soil of the area is aripsamment with 1.61 g cm^{-3} in soil bulk density.

2.3 Soil respiration measurements
Continuous measurements of soil surface CO_2 efflux (R_s) were made in an open area at *Artemisia ordosica* shrub land with intact algae, lichen and moss crusts between June and October in 2012. An automated soil respiration system (Model LI 8100A fitted with a LI-8150 multiplexer, LI-COR, Nebraska, USA) was used to measure R_s. Three permanent PVC collars (20.3 cm in diameter, 10 cm in height, inserted ~7 cm) were separately installed in intact algae-, lichen- and moss-crusted soil in March 2012, three months before the start of measurements. A permanent opaque chamber (model LI-104, LI-COR, Nebraska, USA) was set on each collar. The measurement time for each chamber was 3 min and 15 s, including a 30 s pre-purge, a 45 s post-purge, and a 2 min observation period. Hourly T_s and VWC at 5-cm depth were measured near the chamber using an 8150-203 temperature sensor and an EC_{H2O} soil moisture sensor (Li-COR, Nebraska USA), respectively. During observation, any plants re-growing within collars were manually removed. Rainfall was measured near the chamber by a manual rain gauge and a tipping-bucket rain gauge (model TE525MM, Campbell Scientific, UT, USA). Half-hourly incident photosynthetically active radiation (PAR) was measured using a quantum sensor (PAR-LITE, Kipp & Zonen, The Netherlands) near the chambers.

2.4 Data treatment and analysis
The CO_2 efflux values greater than 15 μmol m^2 s^{-1} or less than -1 μmol m^2 s^{-1} were considered abnormal and removed from the dataset. Instrument failure, sensor calibration, and poor-quality measurements together resulted in the loss of 4% to 5.4% of the values for three chambers from June to October 2012 (Fig. 1).

To avoid including the impacts of photosynthesis and Birch effects on the seasonal responses of R_s to T_s and VWC, certain observations were removed from the dataset. (1) Daytime (photosynthetically active radiation, PAR >5 μmol m^{-2} s^{-1}) CO_2 efflux values were removed to ensure that no photosynthesis effects were included. (2) Measurements recorded immediately (within 30 min) after a rain event were excluded because they were potentially affected by the rewetting of the upper soil layers, which could stimulate respiration [25,26]. The daily mean nighttime value (R_s, T_s, and VWC) was computed as the average of the hourly values when PAR was below 5 μmol m^{-2} s^{-1}. Daily mean nighttime values were used to examine the seasonal responses of R_s to T_s and VWC. The seasonal relationships between R_s and T_s were estimated using four common models: Exponential (Q_{10}), Arrhenius, Quadratic, and Logistic (see Table 2). The four models were fitted separately for each crusted soil. Root mean square error (RMSE) and the coefficient of determination (R^2) were used to evaluate model performance. Temperature-normalized daily mean nighttime R_s (R_{sN}), calculated as the ratio of the observed nighttime R_s to the value predicted by the Q_{10} model, was used to analyze the seasonal dependence of daily mean nighttime R_s on VWC. Three bivariate models with T_s and VWC as independent variables were developed to show the combined effect of both variables (Table 3).

To ensure that the measurements of diurnal responses of R_s to T_s and VWC were not affected by photosynthesis, CO_2 flux measurements taken within two days after a significant rain event (>10 mm) were removed from the dataset. Field observation

Table 1. Physical and chemical characteristics of BSC layer in the study sites [41,42].

Soil type	SOC (%)	TNC (%)	SBD (g·cm^{-3})	pH	Particle content (<0.05 mm)(%)
Algae-crusted soil	0.34±0.13	0.02±0.01	1.69±0.10	8.81±1.40	6.16±1.14
Lichen-crusted soil	1.33±0.09	0.07±0.01	1.60±0.03	8.62±1.10	8.43±1.41
Moss-crusted soil	2.14±0.19	0.10±0.02	1.70±0.45	7.84±1.60	11.07±0.81

SOC: soil organic carbon; TNC: total nitrogen content; SBD: soil bulk density.

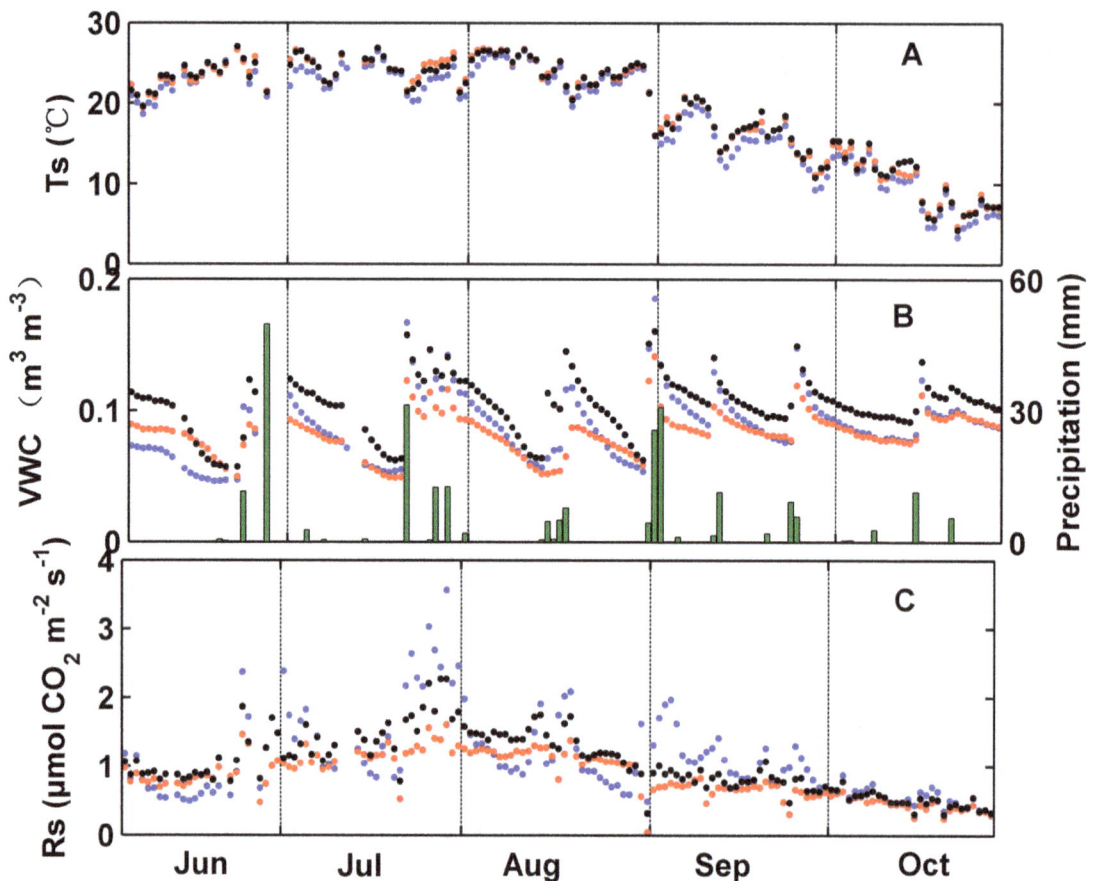

Figure 1. Daily mean of soil respiration (R_s), soil temperature (T_s), and soil volumetric water content (VWC) in soil crusted with algae (red), lichen (black), and moss (blue).

revealed that the water content of BSCs layers decreased to the water compensation point of photosynthesis within two days after the last significant rain event (>10 mm) in all three crusted soils [24,27]. The mean diurnal courses of R_s, T_s, and VWC were computed for each month by averaging the hourly means for each time of day. Cross-correlation analysis was used to detect hysteresis between R_s and T_s at the diurnal scale. Correlation analysis was used to evaluate the relationship between R_s and T_s (Table 4). All analyses were processed in Matlab 7.11.1 (R2010b, the Mathworks Inc., Natick, MA, USA).

To examine whether daily mean nighttime R_s, T_s, and VWC differed among biologically-crusted soils, we used a two-way (biologically-crusted soil types and time) ANOVA, with repeated measures of one of the factors (time). The environmental factors show relatively small variation within three days. Thus, we selected consecutive three-day periods as the three replication for statistical requirements. When significant biologically-crusted soils effects were found ($P<0.05$), the Tukey HSD post hoc test was employed to evaluate differences between biologically-crusted soil types. Prior to these analyses, data were tested for assumptions of normality and homogeneity of variances and were log-transformed when necessary. All the ANOVA analyses were performed using the SPSS 15.0 statistical software (SPSS Inc., Chicago, Illinois, USA).

Results

3.1. Hysteresis between R_s and T_s

Over the course of the diurnal period, R_s (μmol m^{-2} s^{-1}) reached its minimum at 6:00 and peaked at around 10:00–11:00 (Fig. 2), and T_s arrived at its minimum at 7:00–8:00 and peaked at 16:00 in the three crusted soils. The diurnal variation of R_s was out of phase with T_s, causing hysteresis between R_s and T_s. The maximum mean lag time between R_s and T_s was 5 h in June in moss-crusted soil, and the minimum mean lag time was 1 h in August in lichen-crusted soil, with R_s peaking earlier than T_s (Table 4). The degree of hysteresis was small in lichen-crusted soil, and large in moss-crusted soil (Table 4). The lag time between R_s and T_s was negatively and linearly correlated with VWC in crusted soil (Fig. 3). The lag time was reduced as VWC increased. The r values, derived from the data set with synchronized R_s and T_s, were higher than that without synchronization (Table 4).

3.2. Seasonal variation in R_s, T_s, and VWC

Similar changes in daily mean T_s, VWC, and CO_2 flux (including both daytime and nighttime data) were detected in algae-, lichen-, and moss-crusted soils (Fig. 1). Daily mean T_s was high from June to August, after which it gradually declined (Fig. 1A). No differences were observed in the daily mean nighttime T_s between algae- ($18.15\pm5.61°$C, mean $\pm$ standard deviation, SD) and lichen-crusted soil ($18.14\pm7.13°$C). However, daily mean nighttime T_s in moss-crusted soil ($17.45\pm5.56°$C) was

Table 2. Parameters and statistics for the analysis of the dependence of daily mean nighttime R_s (μmol m^{-2} s^{-1}) on daily mean nighttime T_s (°C) at 5-cm depth when daily mean nighttime VWC (m³ m^{-3}) was above and below 0.075 m³ m^{-3} in algae-and moss-crusted soil, and 0.085 m³ m^{-3} in lichen-crusted soil.

Soil Type	Model	VWC >0.075 m³ m^{-3}					VWC <0.075 m³ m^{-3}				
		a	b	c	Adj.R²	RMSE	a	b	c	Adj.R²	RMSE
Algae-crusted soil	Q$_{10}$	0.38	2.01		**0.82**	0.1254	0.55	1.52		0.57	0.1482
	Quadratic	0.0014	−0.002	0.25	**0.82**	0.1262	−0.01	0.68	−7.67	0.52	0.1511
	Logistic	32.2	0.07	71.79	**0.82**	0.1262	1.10	0.51	19.28	0.57	0.1513
	Arrhenius	0.38	0.0005		**0.82**	0.1255	0.54	0.00031		0.57	0.1482
Moss-crusted soil	Q$_{10}$	0.55	1.97		**0.53**	0.377	0.32	1.81		0.08	0.2446
	Quadratic	0.00053	0.06	−0.01	**0.53**	0.3708	0.01	−0.45	5.158	0.10	0.2419
	Logistic	1.52	0.19	13.58	**0.53**	0.3639	3.60	0.026	79.02	0.0006	0.2545
	Arrhenius	0.55	0.00047		**0.53**	0.3759	0.32	0.00043		0.076	0.244

Type	Model	VWC>0.085 m³ m^{-3}					VWC<0.085 m³ m^{-3}				
		a	b	c	Adj.R²	RMSE	a	b	c	Adj.R²	RMSE
Lichen-crusted soil	Q$_{10}$	0.46	2.13		**0.74**	0.2196	0.43	2.00		0.062	0.2849
	Quadratic	0.002	0.007	0.22	**0.74**	0.2198	−0.07	3.30	−39.6	0.12	0.2759
	Logistic	3.31	0.10	27.95	**0.74**	0.2198	1.26	1.33	21.65	0.092	0.2803
	Arrhenius	0.46	0.00053		**0.74**	0.2191	0.41	0.00051		0.063	0.2848

Q_{10}: $R_s = ab^{(T_s-10)/10}$; Arrhenius: $R_s = ae \exp(b/283.15\ 8.314)(1-283.15/T_s)$; Quadratic: $R_s = a \cdot T_s^2 + b \cdot T_s + c$; Logistic: $R_s = a/(1+\exp(b(c-T_s))$; Q_{10}: relative increase in R_s for a 10°C increase in T_s; $Adj.R^2$ is the adjusted coefficient of determination; RMSE is the root-mean-square error; a, b, and c are fitted parameters; values in bold indicate best fits according to $Adj.R^2$ and RMSE.

Table 3. Parameters, statistics, and predicted values from temperature-only and bivariate models of soil respiration on the basis of daily mean values.

Soil Type	Model	a	b	c	d	$Adj.R^2$	RMSE	Predicted R_s (g C m^{-2})
Algae-crusted soil	Q_{10}	0.38	1.98			0.82	0.1293	123.22
	Q_{10}-power	0.53	2.03	0.13		0.82	0.1268	127.46
	Q_{10}-linear	1.76	1.99	0.13	0.21	0.82	0.1262	126.65
	Q_{10}-hyperbolic	2.042	−0.13	3.83	0.015	**0.83**	0.1268	**126.21**
Lichen-crusted soil	Q_{10}	0.48	1.98			0.68	0.2393	155.92
	Q_{10}-power	1.34	2.20	0.53		0.74	0.2151	132.43
	Q_{10}-linear	1.19	2.07	0.68	0.32	0.72	0.7129	158.84
	Q_{10}-hyperbolic	2.167	−0.62	6.86	0.036	**0.76**	0.2066	**165.39**
Moss-crusted soil	Q_{10}	0.56	1.54			0.22	0.4127	130.43
	Q_{10}-power	12.61	2.07	1.36		0.67	0.2699	146.92
	Q_{10}-linear	1.43	1.82	1.72	0.19	0.43	0.3522	134.57
	Q_{10}-hyperbolic	2.08	−0.51	9.31	0.013	**0.67**	0.2691	**147.08**

Q_{10}-power: $R_s = a \cdot b^{((T-10)/10)} VWC^c$. Q_{10}-linear: $R_s = a \cdot b^{((T-10)/10)} (c\ VWC+d)$. Q_{10}-hyperbolic: $R_s = a^{((T-10)/10)} (b+c \cdot VWC+d/VWC)$; $Adj.R^2$ is the adjusted coefficient of determination; RMSE is the root-mean-square error; a, b, and c are fitted parameters; values in bold indicate best fits according to $Adj.R^2$ and RMSE.

Table 4. Correlation and hysteresis analysis of monthly diurnal courses of soil respiration (R_s) and soil temperature (T_s) at 5-cm depth.

Soil Type		Jun		Jul		Aug		Sep		Oct	
		r	P	r	P	r	P	r	P	r	P
T_s-R_s						Non-Synchronized					
	Algae-crusted soil	0.231	0.279	0.521**	0.009	0.441*	0.031	0.571**	0.001	0.438*	0.032
	Lichen-crusted soil	0.435*	0.034	0.728**	0.001	0.591**	0.002	0.625**	0.01	0.405*	0.05
	Moss-crusted soil	−0.210	0.362	0.531**	0.008	0.208	0.331	0.668**	0.001	0.212	0.781
						Synchronized					
T_s-R_s		r	P	r	P	r	P	r	P	r	P
	Algae-crusted soil	0.844	0.001	0.943	0.001	0.914	0.001	0.962	0.001	0.971	0.001
	Lichen-crusted soil	0.701	0.001	0.875	0.001	0.658	0.001	0.952	0.001	0.917	0.001
	Moss-crusted soil	0.932	0.001	0.903	0.001	0.851	0.001	0.966	0.001	0.781	0.0001
Lag time	Algae-crusted soil	2		3		2		3		4	
	Lichen-crusted soil	3		2		1		3		3	
	Moss-crusted soil	5		3		3		3		4	

r is the Pearson correlation coefficient; P is the significance level.

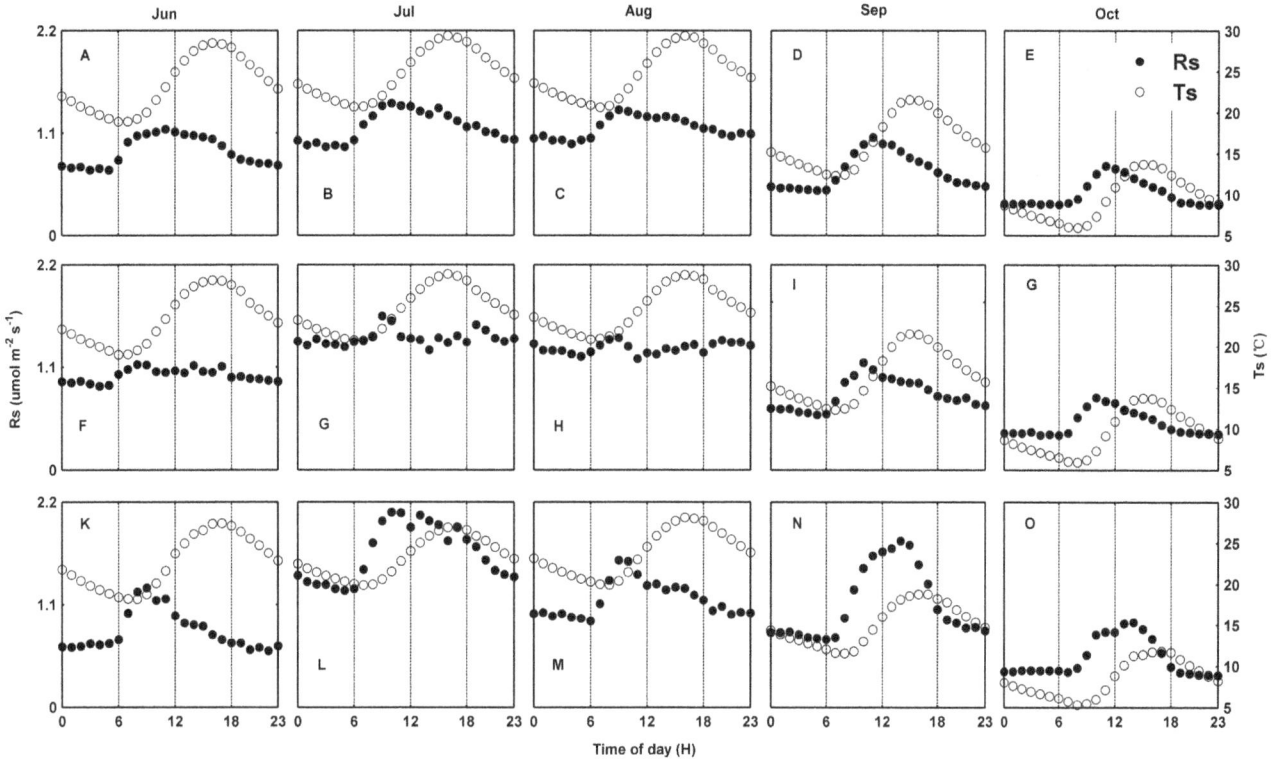

Figure 2. Monthly diurnal courses of soil respiration (R_s) and soil temperature (T_s) in soil crusted with algae (A-E), lichen (F-J), and moss (K-O). Each point is the monthly mean for a particular time of day.

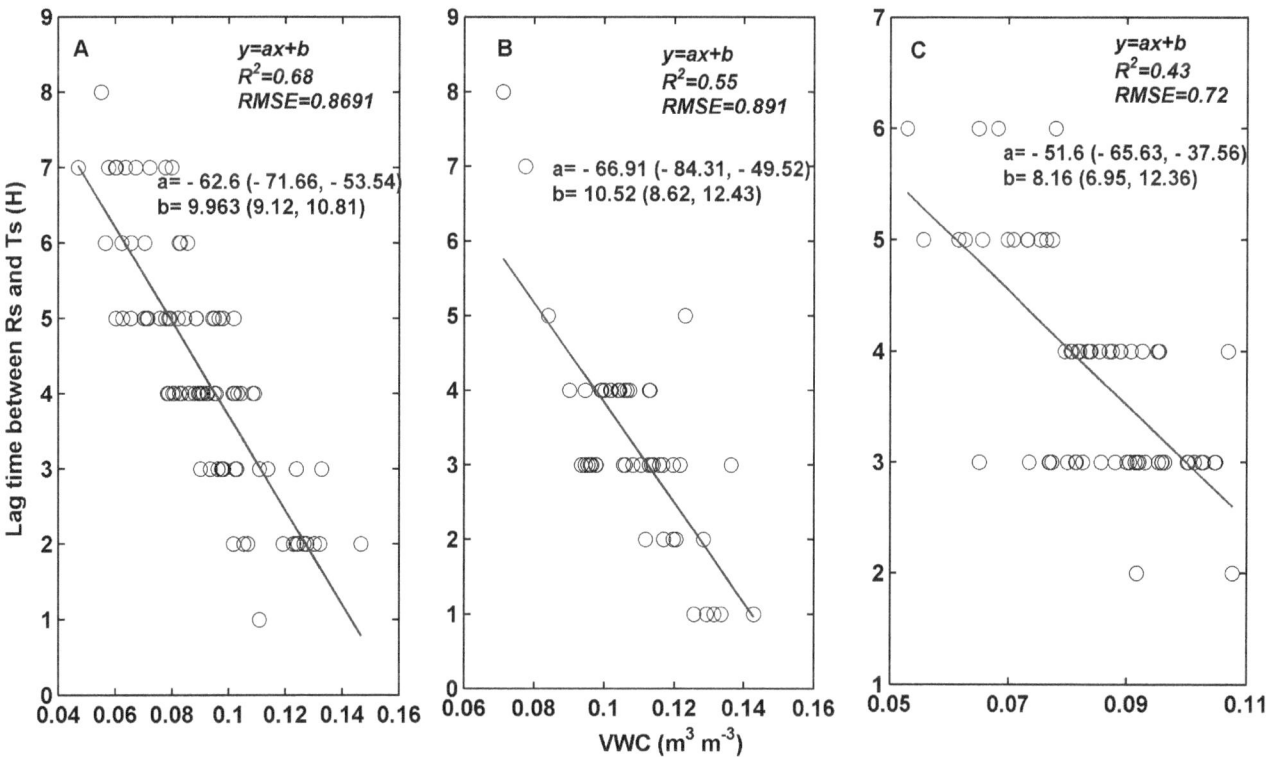

Figure 3. Lag time between soil respiration (R_s) and soil temperature (T_s) over diurnal courses, in relation to soil volumetric water content (VWC) in soil crusted with moss (A), lichen (B), and algae (C). The solid line is fitted using linear regression.

significantly lower than that in algae- and lichen-crusted soil ($df = 2$, $F = 11.92$, $P = 0.013$). Daily mean VWC sharply increased after each precipitation pulse (Fig. 1B). Daily mean nighttime VWC ranged from 0.049 to 0.14 m^3 m^{-3}, 0.057 to 0.16 m^3 m^{-3}, and 0.046 to 0.19 m^3 m^{-3} in algae-, lichen-, and moss-crusted soil, respectively. Daily mean nighttime VWC was significantly higher in lichen-crusted soil (0.104 ± 0.026 m^3 m^{-3}) than in algae- and moss-crusted soils (0.083 ± 0.015 m^3 m^{-3} and 0.089 ± 0.026 m^3 m^{-3}, respectively) ($df = 2$, $F = 251.91$, $P<0.001$). Daily mean CO_2 flux varied markedly following the changes in T_s and VWC, especially after a rain pulse. Daily mean CO_2 flux peaked in late July and then generally declined following the decrease in T_s (Fig. 1C). The limiting effect of VWC on CO_2 flux was clear as CO_2 flux reached its highest value in a quick, sharp response to each rain event and then decreased to pre-rain values (Fig. 1B, C). Daily mean nighttime R_s was significantly different in three crusted soils ($df = 2$, $F = 56.69$, $P<0.001$) with the highest values in lichen-crusted soil (0.93 ± 0.43 μmol m^{-2} s^{-1}) and lowest values in algae-crusted soil (0.73 ± 0.31 μmol m^{-2} s^{-1}).

Daily mean nighttime R_s was positively related to T_s when VWC was higher than 0.075 m^3 m^{-3} in moss-crusted soil and 0.085 m^3 m^{-3} in lichen-crusted soil (Fig. 4). There were no differences among the four temperature-response models examined (Table 2). T_s at the 5-cm depth explained 82%, 74%, and 51% of the seasonal variation of daily mean nighttime R_s when VWC was not a limiting factor in algae-, lichen-, and moss-crusted soil, respectively (Table 2). In algae-crusted soil, however, R_s was controlled by T_s below the VWC threshold value (Table 2). As no differences were observed among the temperature-response models, the remainder of the analysis was performed using the Q_{10} model. Over the study period, daily mean nighttime R_s normalized using the Q_{10} model with T_s at 5 cm depth (R_{sN}) increased with VWC, except in algae-crusted soil (Fig. 5).

The seasonal sensitivity of R_s to T_s (parameter b from the Q_{10} model in Table 2) were 2.01, 2.13, and 1.97 in algae-, lichen-, and

moss-crusted soil, respectively. The long-term basal respiration rate at 10°C (R_{s10}, parameter a from the Q_{10} model in Table 2) for these same soils was 0.38, 0.46, and 0.55 μmol m^{-2} s^{-1}.

The bivariate model Q_{10}-hyperbolic with T_s and VWC as independent variables produced higher R^2 and lower RMSE values than the other models in lichen- and moss-crusted soil (Table 3). There was no significant difference observed between the temperature-only and the bivariate model in algae-crusted soil (Table 3), and the estimated total C release calculated with the Q_{10} model and gap-filled T_s was 123.22 g C m^{-2} in algae-crusted soil (Table 3). The estimated total R_s, as computed using the Q_{10}-hyperbolic model and gap-filled T_s and VWC, was 165.39 and 147.08 g C m^{-2} over the study period in lichen- and moss-crusted soils, respectively. Lichen-crusted soil was the main contributor to this flux among crusted soils during the study period.

Discussion

4.1. Interactive effects of T_s and VWC on R_s

Over the course of the diurnal cycle, there was a significant hysteresis between R_s and T_s (Table 4, Fig. 2). Diurnal hysteresis has been observed in many other ecosystems [16–19,28,29] and is affected by many physical and biological processes, such as mismatch between the depth of temperature measurement and the depth of CO_2 production, photosynthetic carbon supply for diurnal R_s [30], wind-induced pressure pumping [31], and different responses of autotrophic and heterotrophic respiration to environmental factors [20]. We observed that the lag time between R_s and T_s was negatively related to VWC in the three crusted soils, which is consistent with the finding from the Mu Us desert [19]. The increased lag time at low VWC in crusted soils was mainly due to the decoupling of R_s from T_s when VWC is low, and which indicate the sensitivity of root and microbial activity to soil moisture. The timing of the diurnal R_s peak is highly sensitive to VWC, with progressively earlier peaks as the VWC reduces. At

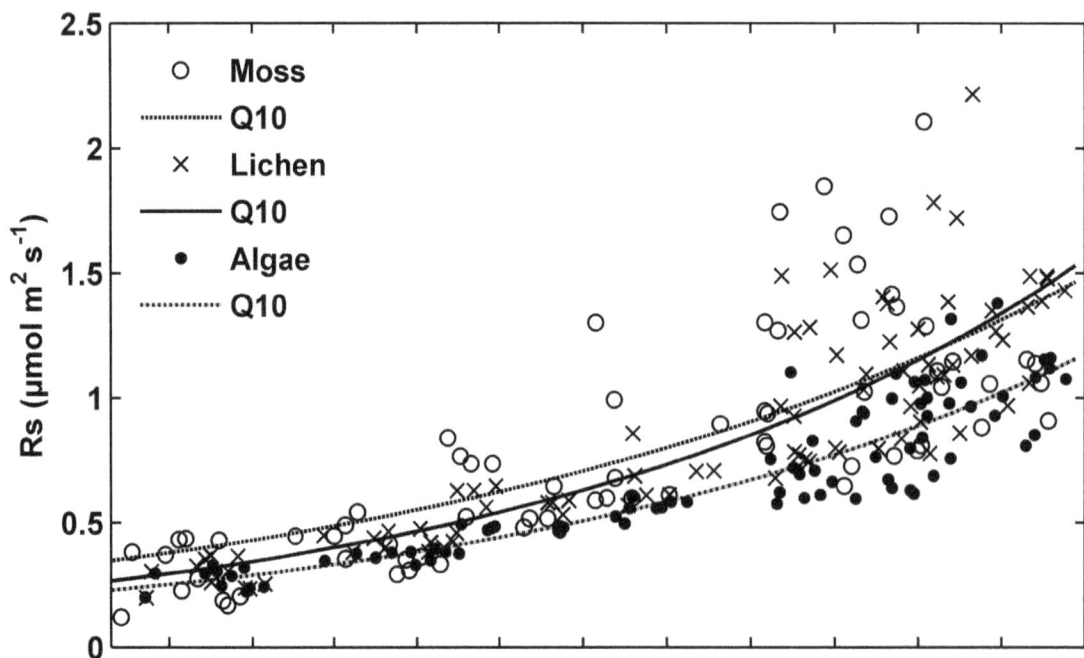

Figure 4. Relationships between daily mean nighttime soil respiration (R_s) and soil temperature (T_s) in algae-, lichen-, and moss-crusted soil.

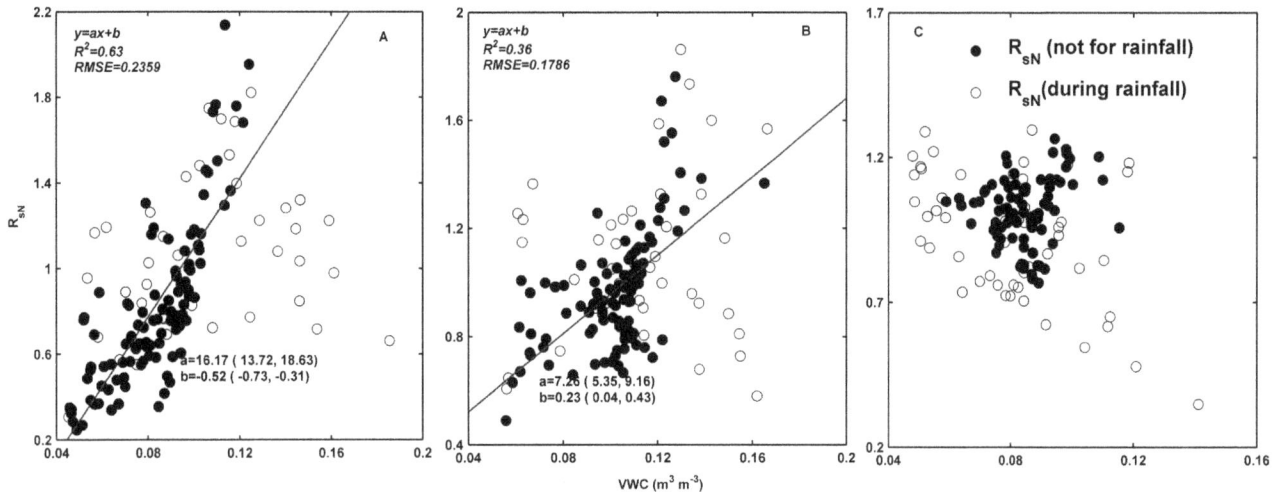

Figure 5. Relationship between daily temperature-normalized mean nighttime soil respiration (R_{sN}) and soil volumetric water content (*VWC*) at 5-cm depth in moss- (A), lichen- (B), and algae- (C) crusted soil, respectively. R_{sN} is the ration of the observed soil respiration (R_s) value to the value predicted by the Q_{10} function. The solid line is fitted using linear regression.

low *VWC*, R_s peaks in the early morning due to root and microbial activity may strongly increased with condensation water, resulting to significant hysteresis between T_s and R_s (Fig. 2, Table 4) [19].

The seasonal changes in daily mean nighttime R_s were mainly controlled by T_s (Table 2, Fig. 4). The four temperature-only models performed well with the same R^2. T_s explained 74% and 53% of the variation in R_s when *VWC* was above 0.085 and 0.075 m^3 m^{-3} in lichen- and moss-crusted soil, respectively, but it was uncorrelated with R_s when *VWC* fell below those thresholds (Table 2). Our observations are in line with those of previous studies in many other ecosystems [8,21,28,32]. Wang et al. [19] reported that T_s explained 76% of the variation in R_s for *VWC* values above 0.08 m^3 m^{-3}, but it was uncorrelated with R_s when *VWC* fell below 0.08 m^3 m^{-3}. Castillo-Monroy et al. [9] found that R_s was controlled by T_s when soil moisture was higher than 11% in microsites dominated by BSCs. Below this level, R_s was driven by soil moisture alone. The decreased R_s under low *VWC* was limited by reduced microbial contact with the available substrate, dormancy and/or death of microorganisms, and substrate supply, which was affected by reduced photosynthesis and drying out of the litter in the surface layer [15,33].

R_{sN} increased with *VWC* and did not show a threshold value in moss- and lichen-crusted soils during the seasonal cycle. Our observation contrasts with the results of previous studies that found a distinct *VWC* threshold [16]. The difference mainly resulted from low *VWC* (0.04–0.16 m^3 m^{-3}) and high soil porosity did not limit CO_2 transport out of soil and CO_2 production due to a lack of O_2.

4.2. Differences in R_s among biologically-crusted soil types

Daily mean nighttime R_s was significantly different in three types of crusted soils (algae-, lichen- and moss-crusted soil) ($df = 2$, $F = 56.69$, $P<0.001$) with the highest values in lichen-crusted soil and lowest values in algae-crusted soil. This result contrasts with those of other studies in desert ecosystems. Su et al.'s [8] study of Gurbantunggute Desert reported no differences in carbon flux between moss- and lichen/cyanobacteria-crusted soil. The differences in the present study can be explained by the following aspects. It is possible that the lowest R_s in algae-crusted soil

resulted from the differences in soil fertility induced by BSCs, total N was significantly lower in algae-crusted soil (0.17±0.09 g kg^{-1}) than in lichen- (0.23±0.08 g kg^{-1}) and moss-crusted soil (0.28±0.13 g kg^{-1}) (unpublished data). In addition, the assemblage of microbial and microfaunal organisms varied in the three crusted soils [10,24,34–36]. The observation of the highest values occurred in lichen-crusted soil was in line with the result conducted in dry condition in the Mu Us desert. The highest values in lichen-crusted soil is mainly due to highest water content and total porosity of lichen layer [10]. T_s was significantly lower in moss-crusted soil than in algae- and lichen-crusted soil (Fig. 1). This result is attributed to the darkening of the surface by cyanobacteria and lichens, resulting in greater absorption of solar radiation and a higher surface temperature [39]. *VWC* in lichen-crusted soil was consistently significantly higher than in moss- and algae-crusted soils (Fig. 1). The difference may be attributed to higher dew deposition (soil moisture input by dewfall can be an important mechanism in dryland environment) and water infiltration in lichen-crusted soil than in moss- and algae-crusted soil [37].

The lag time between R_s and T_s differed depending on the type of crusted soil, suggesting that the response of species in biologically-crusted soils to *VWC* was different among crusted types. The timing of the diurnal R_s peak is highly sensitive to *VWC*, with progressively earlier peaks as the soil *VWC* declines [19]. Moss crusts need more *VWC* than lichen and algae crust to achieve metabolic activity [24]. In water stressed ecosystems, algae and lichen can utilize dew and light rainfall that moss are unable to use [24,27]. Thus the diurnal R_s in moss-crusted soil peaks earlier than algae- and lichen-crusted soils, which lead to significant hysteresis between R_s and T_s in moss-crusted soil. Hysteresis had a smaller impact on lichen-crusted soil than on algae-crusted soil. The result may be partly attributed to the higher water level in lichen- than in algae-crusted soil.

The average Q_{10} of 1.83 from three biologically-crusted soil types from June to October is at the lower end of the range of 1.28 to 4.75 from alpine, temperate, and tropical ecosystems across China [38]. The low Q_{10} value is attributed to their low levels of soil organic matter, small microbial community, and dry soil conditions [19,39,40]. The Q_{10} of algae-, lichen-, and moss-crusted soil was 1.98, 1.98, and 1.54, respectively. The majority of C associated with BSCs, in the forms of microbial biomass or their secretions [31,32], is close to or at the soil surface and is directly in contact with small precipitation or dew captured by algae and lichen crusts. However, small amounts of hydration cannot directly reach the soil surface because the soil is covered with moss. The relatively small amounts of hydration in moss-crusted soils result in the lower Q_{10} [16,21,22,32].

The effects of VWC and T_s on R_s should be considered in carbon cycle models in moss- and lichen-crusted soils. However, we did not find any effect of VWC on daily mean nighttime R_s in algae-crusted soil from June to October 2012 (Tables 2, 3). This observation coincided with the result that R_{sN} was independent of VWC in algae-crusted soil. The independence of VWC from R_s in algae-crusted soil may be attributed to the low water requirement of algae for active metabolism [24,27]. Even a very small hydration event, such as water vapor and dew in the early morning, is sufficient to allow algae to achieve microbial metabolism. Further examination is needed to justify our conclusion regarding the role of VWC on algae-crusted soil due to the dew data gap. We used the Q_{10}-hyperbolic model, with T_s and VWC as independent variables, to predict changes in R_s. Using Q_{10}-hyperbolic model to predict R_s was also reported in a boreal trembling aspen stand [16].

Using temperature-only and Q_{10}-hyperbolic model, we obtained an approximate estimate of the total amount of C released at each crusted soil via soil respiration of 123.2, 165.4, and 147.1 g C m^{-2} over 5 months studied in algae-, lichen- and moss-crusted soils, respectively. Lichen-crusted soil was the main contributor to the total C released by R_s. We found that total C released by R_s in lichen-crusted soil was 2.5% higher than the mean total C released by R_s (161.4 g C m^{-2}, unpublished data) over 5 months, whereas total C released by R_s in algae- and moss-crusted soil were 23.65% and 8.87% smaller than the mean total C released by R_s, respectively. Our results show the importance of BSCs as modulators of R_s in the C release and indicate that we should

not ignore their relative contributions to the total C budgets in desert ecosystems.

Conclusions

Our study showed that R_s was significantly different in three crusted soils with highest values in lichen-crusted soil and lowest values in algae-crusted soil. Lichen-crusted soil was the main contributor to the total C released by R_s. Over the diurnal cycle, T_s exerted dominant control over R_s in the three crusted soils. There was a significant lag between T_s and R_s over the diurnal cycle, and that the lag time increased as VWC decreased. Over the seasonal scale, the response of R_s to T_s was regulated by VWC, and R_s was uncorrelated with T_s when VWC dropped below 0.075 and 0.085 m^3 m^{-3} in lichen- and moss-crusted soils, respectively. However, VWC was not a limiting factor on R_s in algae-crusted soil. Our results indicated that different types of BSCs may affect response of R_s to environmental factors. There is a need to consider the spatial distribution of different types of BSCs and their relative contributions to the total C budgets at the ecosystem or landscape level.

Acknowledgments

We thank Su Lu, Huishu Shi, Yuming Zhang, Xuewu Yang for their assistance with the field measurements and instrumentation maintenance. We are grateful to the anonymous reviewers and the Academic Editor for providing insightful comments and suggestions. We also thank language service company for their help with language revision, and valuable comments to the manuscript.

Author Contributions

Conceived and designed the experiments: WF YZ BW TZ SQ XJ CS. Performed the experiments: SQ WF BW KF. Analyzed the data: WF SQ XJ BW. Contributed reagents/materials/analysis tools: YZ BW TZ XJ BW JL KF. Wrote the paper: WF YZ BW XJ SQ. Designed the software used in analysis: XJ.

References

1. Schimel DS (1995) Terrestrial ecosystems and the carbon cycle. Glob Change Biol 1: 77–91.
2. Le Houérou HN (1996) Climate change, drought and desertification. J Arid Environ 34: 133–185.
3. Belnap J (2003a) Comparative structure of physical and biological soil crusts. In: Belnap J, Lange OL, editors. Biological soil crusts: Structure, function, and management. Berlin: Springer-Verlag. pp. 177–191.
4. Belnap J (2003) The world at your feet: desert biological soil crusts. Front Ecol Environ 1: 181–189.
5. Maestre FT, Cortina J (2003) Small-scale spatial variation in soil CO$_2$ efflux in a Mediterranean semiarid steppe. Appl Soil Ecol 23: 199–209.
6. Thomas AD, Hoon SR, Dougill AJ (2011) soil respiration at five sites along the Kalahari Transect: effects of temperature, precipitation pulse and biological soil crust cover. Geoderma 167: 284–294.
7. Thomas AD, Hoon SR (2010) Carbon dioxide fluxes from biologically-crusted Kalahari Sands after simulated wetting. J Arid Environ 74: 131–139.
8. Su YG, Wu L, Zhou ZB, Liu YB, Zhang YM (2013) Carbon flux in deserts depends on soil cover type: A case study in the Gurbantunggute desert, North China. Soil Biol Biochem 58: 332–340.
9. Castillo-Monroy AP, Maestre FT, Rey A, Soliveres S, García-Palacios P (2011) Biological soil crust microsites are the main contributor to soil respiration in a semiarid ecosystem. Ecosystems 14: 835–847.
10. Feng W, Zhang YQ, Wu B, Zha TS, Jia X, et al. (2013) Influence of disturbance on soil respiration in biologically-crusted soil during the dry season. The Scientific World J 2013.
11. Fang C, Moncrieff JB (2001) The dependence of soil CO$_2$ efflux on temperature. Soil Biol Biochem 33: 155–165.
12. Jassal RS, Black TA, Novak MD, Gaumont-Guay D, Nesic Z (2008) Effects of soil water stress on soil respiration and its temperature sensitivity in an 18-year-old temperate Douglas-fir stand. Glob Change Biol 14: 1305–1318.
13. Bouma TJ, Nielsen KL, Eissenstat DM, Lynch JP (1997) Estimating respiration of roots in soil: interactions with soil CO$_2$, soil temperature and soil water content. Plant Soil 195: 221–232.
14. Drewitt GB, Black TA, Nesic Z, Humphreys ER, Jork EM, et al. (2002) Measuring forest floor CO$_2$ fluxes in a Douglas-fir forest. Agr Forest Meteorol 110: 299–317.
15. Jassal R, Black A, Novak M, Morgenstern K, Nesic Z, et al. (2005) Relationship between soil CO$_2$ concentrations and forest-floor CO$_2$ effluxes. Agr Forest Meteorol 130: 176–192.
16. Gaumont-Guay D, Black TA, Griffis TJ, Barr AG, Jassal RS, et al. (2006) Interpreting the dependence of soil respiration on soil temperature and water content in a boreal aspen stand. Agr Forest Meteorol 140: 220–235.
17. Vargas R, Allen MF (2008) Environmental controls and the influence of vegetation type, fine roots and rhizomorphs on diel and seasonal variation in soil respiration. New Phytol 179: 460–471.
18. Tang J, Baldocchi DD, Xu L (2005) Tree photosynthesis modulates soil respiration on a diurnal time scale. Glob Change Biol 11: 1298–1304.
19. Wang B, Zha TX, Jia X, Wu B, Zhang YQ, et al. (2014) Soil moisture modifies the response of soil respiration to temperature in a desert shrub ecosystem. Biogeosciences 11: 259–268.
20. Riveros-Iregui DA, Emanuel RE, Muth DJ, McGlynn BL, Epstein HE, et al. (2007) Diurnal hysteresis between soil CO$_2$ and soil temperature is controlled by soil water content. Geophys Res Lett 34.
21. Yuste JC, Janssens IA, Carrara A, Meiresonne L, Ceulemans R (2003) Interactive effects of temperature and precipitation on soil respiration in a temperate maritime pine forest. Tree Physiol 23: 1263–1270.
22. Jassal RS, Black TA, Novak MD, Gaumont-Guay D, Nesic Z (2008) Effects of soil water stress on soil respiration and its temperature sensitivityin an 18-year-old temperate Douglas-fir stand. Glob Change Biol 14: 1305–1318.
23. Harper CW, Blair JM, Fay PA, Knapp AK, Carlisle JD (2005) Increased rainfall variability and reduced rainfall amount decreases soil CO$_2$ flux in a grassland ecosystem. Glob Change Biol 11: 322–334.
24. Feng W, Zhang YQ, Wu B, Qin SG, Lai ZR (2014) Influence of environmental factors on carbon dioxide exchange in biological soil crusts in desert areas. Arid Land Res Manage 28: 186–196.

25. Birch H (1958) The effect of soil drying on humus decomposition and nitrogen availability. Plant Soil 10: 9–31.

26. Rey A, Pegoraro E, Tedeschi V, De Parri I, Jarvis PG, et al. (2002) Annual variation in soil respiration and its components in a coppice oak forest in Central Italy. Glob Change Biol 8: 851–866.

27. Lange OL (2003) Photosynthesis of soil-crust biota as dependent on environmental factors. In: Belnap J, Lange OL, editors. Biological soil crusts: Structure, function, and management. Berlin: Springer-Verlag. pp. 217–240.

28. Vargas R, Baldocchi DD, Allen MF, Bahn M, Black TA, et al. (2010) Looking deeper into the soil: biophysical controls and seasonal lags of soil CO_2 production and efflux. Ecol Appl 20: 1569–1582.

29. Jia X, Zha TS, Wu B, Zhang YQ, Chen WJ, et al. (2013) Temperature response of soil respiration in a Chinese pine plantation: hysteresis and seasonal vs. diel Q_{10}. PLoS one 8: e57858.

30. Stoy PC, Palmroth S, Oishi AC, Siqueira MB, Juang JY, et al. (2007) Are ecosystem carbon inputs and outputs coupled at short time scales? A case study from adjacent pine and hardwood forests using impulse-response analysis. Plant Cell Environ 30: 700–710.

31. Flechard CR, Neftel A, Jocher M, Ammann C, Leifeld J, et al. (2007) Temporal changes in soil pore space CO_2 concentration and storage under permanent grassland. Agr Forest Meteorol 142: 66–84.

32. Xu M, Qi Y (2001) Soil-surface CO_2 efflux and its spatial and temporal variations in a young ponderosa pine plantation in northern California. Glob Change Biol 7: 667–677.

33. Högberg P, Nordgren A, Buchmann N, Taylor AFS, Ekblad A, et al. (2001) Large-scale forest girdling shows that current photosynthesis drives soil respiration. Nature 411: 789–792.

34. Housman DC, Yeager CM, Darby BJ, Sanford RL, Kuske CR, et al. (2007) Heterogeneity of soil nutrients and subsurface biota in a dryland ecosystem. Soil Biol Biochem 39: 2138–2149.

35. CastilloMonroy AP, Bowker MA, Maestre FT, Rodríguez-Echeverría S, Martinez I, et al. (2011) Relationships between biological soil crusts, bacterial diversity and abundance, and ecosystem functioning: Insights from a semi-arid Mediterranean environment. J Veg Sci 22: 165–174.

36. Warren S (2003) Biological soil crusts and hydrology in North American deserts. In: Belnap J, Lange OL, editors. Biological soil crusts: Structure, function, and management. Berlin: Springer-Verlag.pp. 327–337.

37. Liu LC, Li SZ, Duan ZH, Wang T, Zhang ZS, et al. (2006) Effects of microbiotic crusts on dew deposition in the restored vegetation area at Shapotou, northwest China. J Hydrol 328: 331–337.

38. Zheng ZM, Yu GR, Fu YL, Wang YS, Sun XM, et al. (2009) Temperature sensitivity of soil respiration is affected by prevailing climatic conditions and soil organic carbon content: a trans-China based case study. Soil Biol Biochem 41: 1531–1540.

39. Gershenson A, Bader NE, Cheng W (2009) Effects of substrate availability on the temperature sensitivity of soil organic matter decomposition. Glob Change Biol 15: 176–183.

40. Cable JM, Ogle K, Lucas RW, Huxman TE, Loik ME, et al. (2011) The temperature responses of soil respiration in deserts: a seven desert synthesis. Biogeochemistry 103: 71–90.

41. Gao GL, Ding GD, Wu B, Zhang YQ, Qin SG, et al. (2014) Fractal scaling of particle size distribution and relationships with topsoil properties affected by biological soil crusts. PloSone 9: e88559.

42. Bao YF, Ding GD, Wu B, Zhang YQ, Liang WJ, et al. (2013) Study on the wind-sand flow structure of windward slope in the Mu Us Desert, China. J Food Agric Environ 11: 1449–1454.

19

The Formation of the Patterns of Desert Shrub Communities on the Western Ordos Plateau, China: The Roles of Seed Dispersal and Sand Burial

Yange Wang, Xiaohui Yang*, Zhongjie Shi*

Institute of Desertification Studies, Chinese Academy of Forestry, Beijing, China

Abstract

The western Ordos Plateau is a key area of shrub diversity and a National Nature Reserve of endangered shrub species in north-west China. Desert expansion is becoming the most important threat to these endangered species. However, little is known about the effects of sand burial on the dynamics of the shrub community. This study aims to investigate how the shrubs as a community and as different individual shrubs respond to the disturbances caused by the desert expansion. The approach used by this study is to separate the seed-dispersal strategy from the sand-burial forces that are involved in structuring the shrub communities at different disturbance stages. Four communities for different disturbance stages were surveyed by using 50×50 m plots. The individual shrubs were classified into coloniser and successor groups at the seed-dispersal stage and strong and weak sand-burial tolerance groups at the sand-expansion stage. We employed spatial point pattern analysis with null models for each community to examine the seed-dispersal strategy and sand-burial forces affecting community distribution patterns. At the seed-dispersal stage, the interactions between the colonisers and the successors showed significant positive correlation at a scale of 0–1 m and significant negative correlation at a scale of 2 m; significant negative correlations between the groups with strong and weak sand-burial tolerance in the early stage of sand expansion at scales of 3–6 m, and significant positive correlation in the later stage of sand expansion at a scale of 13 m, were found. Seed-dispersal strategy is a reasonable mechanism to explain the shrub community pattern formation in the earlier stages, whereas sand burial is the primary reason for the disappearance of shrubs with weak sand-burial tolerance, this irreversible disturbance causes homogenisation of the community structure and produces aging populations of shrub species. This has an important influence on the succession direction of desert shrub communities.

Editor: Christopher J. Lortie, York U, Canada

Funding: This research was funded in part by State Forestry Administration Public Welfare Research Foundation (201204203), the National Natural Science Foundation of China (41271033) and State Sci. & Tech. Supporting Project (2006BAD26B05). We thank Prof. Longjun Ci for her comments on this research. The funders had no role in study design, data collection and analysis, decision to publish, or preparation of the manuscript.

Competing Interests: The authors have declared that no competing interests exist.

* E-mail: yangxh@caf.ac.cn (XHY); shizj@caf.ac.cn (ZJS)

Introduction

The ecosystems of arid areas are among those in which plant facilitation is most frequently studied [1–4], especially in the seedling growth stage. In this stage, the 'nursing' of seedlings by shrubs is a pivotal feature of the shrub community [5–9]. Evidence from other arid ecosystems shows that in the community-formation stage, small-seeded species colonise bare ground, whereas large, wind-blown seeds are trapped by and become established under these existing small-seeded species initially [10]. Additionally, it is often assumed that low nutrient availability, harsh conditions and, in particular, low water availability are the predominant abiotic forces structuring plant communities in desert ecosystems [11–12]. Relatively few studies have assessed the impacts of biotic and abiotic forces on the abundance and distribution of shrubs and the community dynamics at different disturbance stages [13].

The western Ordos Plateau is the distribution center for northwest China's endemic genera and the key area of shrub diversity in northern China. More than 100 of the shrub and sub-shrub species found in arid and semi-arid areas of China are found in this area [14–16]. It is also a central area for endangered shrub species [17]. A National Nature Reserve focused on shrub protection has been established in the area. Meanwhile, human activities in the Ordos over the last 30 years have led to an expansion of the desert – arguably greater than the natural amount of expansion that occurred over the previous 2000 years. Desert expansion is becoming the most important threat to endangered shrub species [17]. Owing to irrational land reclamation, mining and grazing, a large area of fixed sand has changed to shifting sand. Driven eastward by the wind, the sandy areas have expanded. The shrub communities on the western Ordos Plateau have been affected by vegetation degradation and soil desertification over a long period [18]. However, little is known about how sand burial affects the dynamics of the shrub community.

This study aims to investigate how shrub-communities and different individual shrubs respond to desert expansion during escalating sand disturbances by separating the seed-dispersal stages from the sand-burial stages that eventually consume this ecosystem. We hypothesised that seed dispersal would determine shrub establishment success and that sand burial would affect plant health and growth, with important consequences for community structure and dynamics. Shrubs may be divided into two categories as coloniser and successor, according to their thousand-seed kernel weights (Table 1), which can be used to reflect the seed-disperal strategy [19]. Shrubs may also fall into two functional groups with respect to sand burial according to the sand thickness in the area in which their distribution is concentrated. We identified these two groups as the strong sand-burial tolerance group and the weak sand-burial tolerance group. We compared the abundance of the shrub species and the functional groups at each stage of disturbance, and we described the spatial distribution pattern of the members of the same or different groups at different scales to gain an understanding of these ecological systems and the dynamics of arid shrub communities.

Materials and Methods

Study Site and Species

The study region is located in the western part of the Ordos Plateau, east of the Yellow River. It is administered from Wuhai City (106° 36′ – 107° 08′ E, 39° 13′ – 39° 54′ N, 1,150 m a.s.l.), Inner Mongolia. The Yellow River flows from south to north on the piedmont plain formed between the Zhuozi Mountains to the east and the Helan Mountains and Ulan Buh Desert to the west (Figure 1). The local climate is arid and temperate, with an annual precipitation of 150–200 mm, more than 60% of which occurs in July and August. The annual evaporation is 2,470.5–3,481 mm, the annual mean temperature is 7.8–8.1°C, the annual mean relative humidity is 42%, the annual mean wind speed is 2.7 m s^{-1}, the sand-blown period is 41–67 days, and the total annual amount of solar radiation is 155.9 kcal cm^{-2}. The main soil types are grey desert soil, aeolian soil, and meadow soil, with pH values of 9.0–10.0. Wuhai is located in an ecotone ranging from desert steppe to steppe desert. The average vegetation cover is 25%, and the desert shrubs and semi-shrubs are *Tetraena mongolica* Maxim. and *Ammopiptanthus mongolicus* (Maxim.) S. H. Cheng.

Data Collection

Experiments to study pattern-generating processes in arid and semi-arid plant communities, such as the shrub communities investigated here, are difficult. These communities are characterized by slowly changing dynamics and by a disturbance process defined over long time scales because of the difficulty of seed germination and the presence of species that exhibit relatively slow growth and long life spans [20]. Therefore, this study used the

spatial distribution of shrubs as a proxy for the progress of time to identify the drivers and trends of disturbance.

In a preliminary survey, we found that the shrub communities surveyed were located in the leeward region of a large sand-source area to the east of the city of Wuhai. This area is expanding toward the east. The sand source was a sand mound devoid of vegetation. Most of the plant species that we studied are wind dispersed, whereas few are gravity dispersed. Because both seed dispersal and sand transmission are primarily wind driven, both the disturbance stages and the intensity of sand burial as well as seed dispersal should follow gradients along the wind direction. In August 2006, we selected four typical sample plots as the representative of a gradient to represent different disturbance stages of the shrub communities along the main wind direction. Sample plot 1 was located at a distance of 2,000 m from the sand source on the west slope of a low hill (Figure 1) covered by gravel at the eastern edge of the concentrated shrub distribution area (furthest from the sand source). This plot was considered to represent an earlier stage of community pattern formation. Sample plot 2 was located at a distance of 1,500 m from the sand source and was covered by gravel and a few shallow patches of sand. It was considered to represent a later stage of community pattern formation. Sample plot 3 was located at a distance of 1,000 m from the sand source. It was covered by a 5-cm deep layer of sand and was considered to represent an earlier stage of sand expansion. Sample plot 4 was located 500 m from the sand source and was covered by sand approximately 12 cm deep. This plot was considered to represent a later stage of sand expansion. Both the shrub density and the proportion of seedlings decreased systematically from east to west (sample plot 1 to sample plot 4). Many sand dunes were present around the shrubs in sample plot 4. Some of the dunes reached 1.5 m high and 3 m wide along the wind direction. The plots had not been disturbed by humans or grazed by livestock over the 15 years preceding the study. Rare natural granivores and herbivores were found in the area.

One 50 m×50 m subplot was established for each of the four sample plots, which is suitable size to document shrub diversity (see species-area curve by Yang [21]). All individual shrubs in the four subplots were mapped using a total station transit (model GTS-3B, Topcon, Paramus, New Jersey) and classified into different groups according to their biological, physiological and ecological characteristics (Table 1, Figure 2). Because some shrubs are capable of clonal reproduction, we considered the ramets as distinct individuals if they grew more than 10 cm apart.

Table 1. Classification of shrubs and classification criteria.

Species	Seed		Sand burial tolerance
	Thousand-seed kernel weight	Classification	
Ammopiptanthus mongolicus (Maxim.) S. H.Cheng	>10 g	successor	Strong
Convolvulus fruticosus Pall.	<10 g	colonizer	Weak
Nitraria tangutorum Bobrov.	>10 g	successor	Strong
Potaninia mongolica Maxim.	<10 g	colonizer	Weak
Reaumuria songarica (Pall.) Maxim.	<10 g	colonizer	Weak
Reaumuria trigyna Maxim.	<10 g	colonizer	Weak
Sarcozygium xanthoxylon Bunge	>10 g	successor	Weak
Tetraena mongolica Maxim.	<10 g	colonizer	Weak

(Ma, 1989; Zeng et al., 2000; Ma et al., 2003; Zhou et al., 2006).

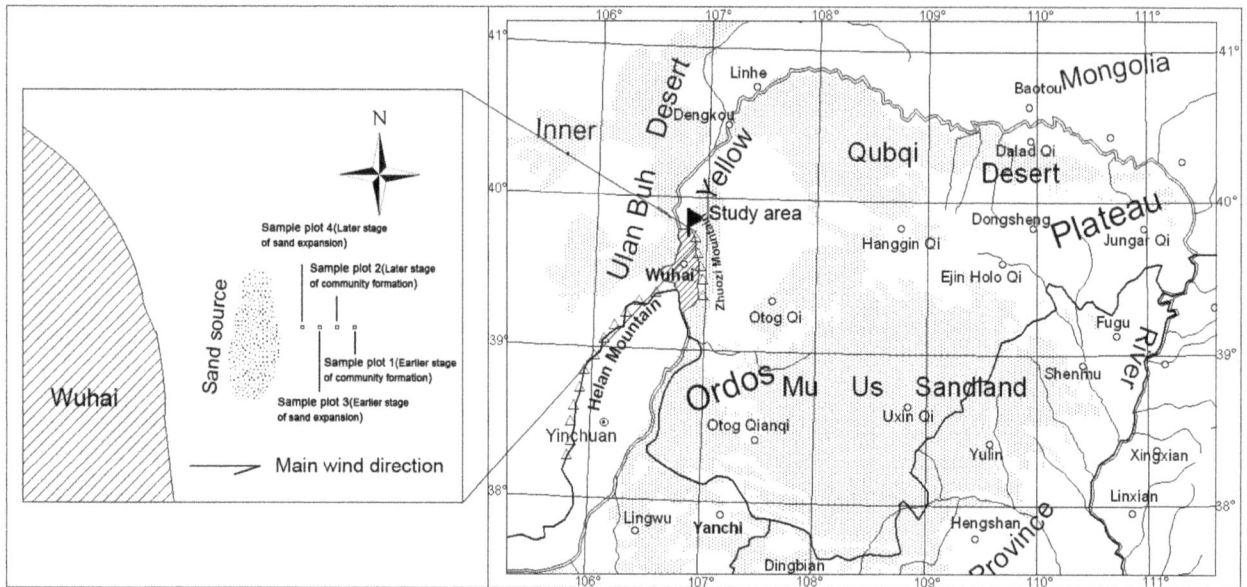

Figure 1. Map of Ordos Plateau and study region.

Spatial Statistics

Ripley's K-function and the pair-correlation g-function are common techniques for univariate and bivariate point-pattern analysis [22–23]. They are used together by ecologists because they have different sensitivities. In some situations, it is difficult to precisely reflect the pattern of certain scale features using Ripley's K-function because of the cumulative 'memory'. However, the g-function, which has been developed from Ripley's K-function and uses rings instead of circles, allows specific distance classes to be isolated so that it can detect aggregation or dispersion accurately at a given distance, t [24]. The g-function can also reflect spatial pattern types more effectively [25]. The g-function has been shown to be preferable to the commonly used Ripley's K-function in research similar to the current study [19]. Consequently, in this study we only used the g-function in spatial point pattern analysis.

The univariate g-function is determined by Equation 1 [24]:

$$\hat{g}(t) = \frac{1}{2\pi t} \frac{A^2}{n^2} \sum_{i=1}^{n} \sum_{\substack{j=1 \\ j \neq i}}^{n} w_{ij}^{-1} k_h\left(t - |x_i - x_j|\right) \qquad (1)$$

Where A is the plot area, n is the total number of plants and w_{ij} is a weighting factor correcting for edge effects. k_h is a kernel function – a weight function applying maximum weight to point pairs with a distance exactly equal to t but incorporating point pairs at an approximate distance t with reduced weight. This weight falls to zero if the actual distance between the points differs from t more than h, the so-called bandwidth parameter.

Values of $\hat{g}(t) > 1$ indicate that inter-point distances of t are more frequent and that samples are aggregated, whereas values of $\hat{g}(t) < 1$ indicate that they are less frequent and that samples are dispersed compared to spatial randomness.

The bivariate point patterns concern the spatial distribution of the individuals of one species in relation to those of another species. The corresponding bivariate spatial point pattern is estimated by

$$\hat{g}_{12}(t) = \frac{1}{2\pi t} \frac{A^2}{n_1 n_2} \sum_{i=1}^{n_1} \sum_{j=1}^{n_2} w_{ij}^{-1} k_h\left(t - |x_i - y_j|\right) \qquad (2)$$

Where x_i, $i = 1, \ldots, n_1$, and y_j, $j = 1, \ldots, n_2$ are the points of groups 1 and 2, respectively, with the same weights w_{ij} and kernel function k_h as Equation 1.

Univariate and Bivariate Null Models

The homogeneity test showed that the colonisers and successors as well as all of the shrubs in sample plot 1 were homogeneously distributed, whereas on sample plots 2, 3, and 4, heterogeneity was found for each functional group and for all the shrubs.

We applied the complete spatial randomness model (CSR) [26] to sample plot 1 because the CSR is the simplest and most widely used null model for homogeneous patterns. For sample plots 2, 3 and 4, we applied the heterogeneous Poisson process null model, the simplest alternative to CSR for heterogeneous patterns [27].

We applied an antecedent-condition null model to test the hypothesis that small-seeded species colonise bare ground and that large, wind-blown seeds are trapped by and become established under the existing small-seeded species [10] for the communities on sample plot 1 and sample plot 2. This null model is commonly used when two types of points are created in sequence rather than at the same time. We kept the location of the colonisers fixed and randomised the successors according to the CSR model and the heterogeneous Poisson process in simulations for sample plots 1 and 2, respectively [27].

To test the sand-burial hypothesis on sample plots 3 and 4, we applied a random-labelling null model. The model was used to investigate whether the labels 'weak sand-burial tolerance' and 'strong sand-burial tolerance' have a random structure within the given spatial structure of the joint pattern [27].

We calculated the 99% Monte Carlo confidence envelopes for each null model by running 99 Monte Carlo simulations. The Programita software, written by Wiegand and Moloney [25], was used for the calculations.

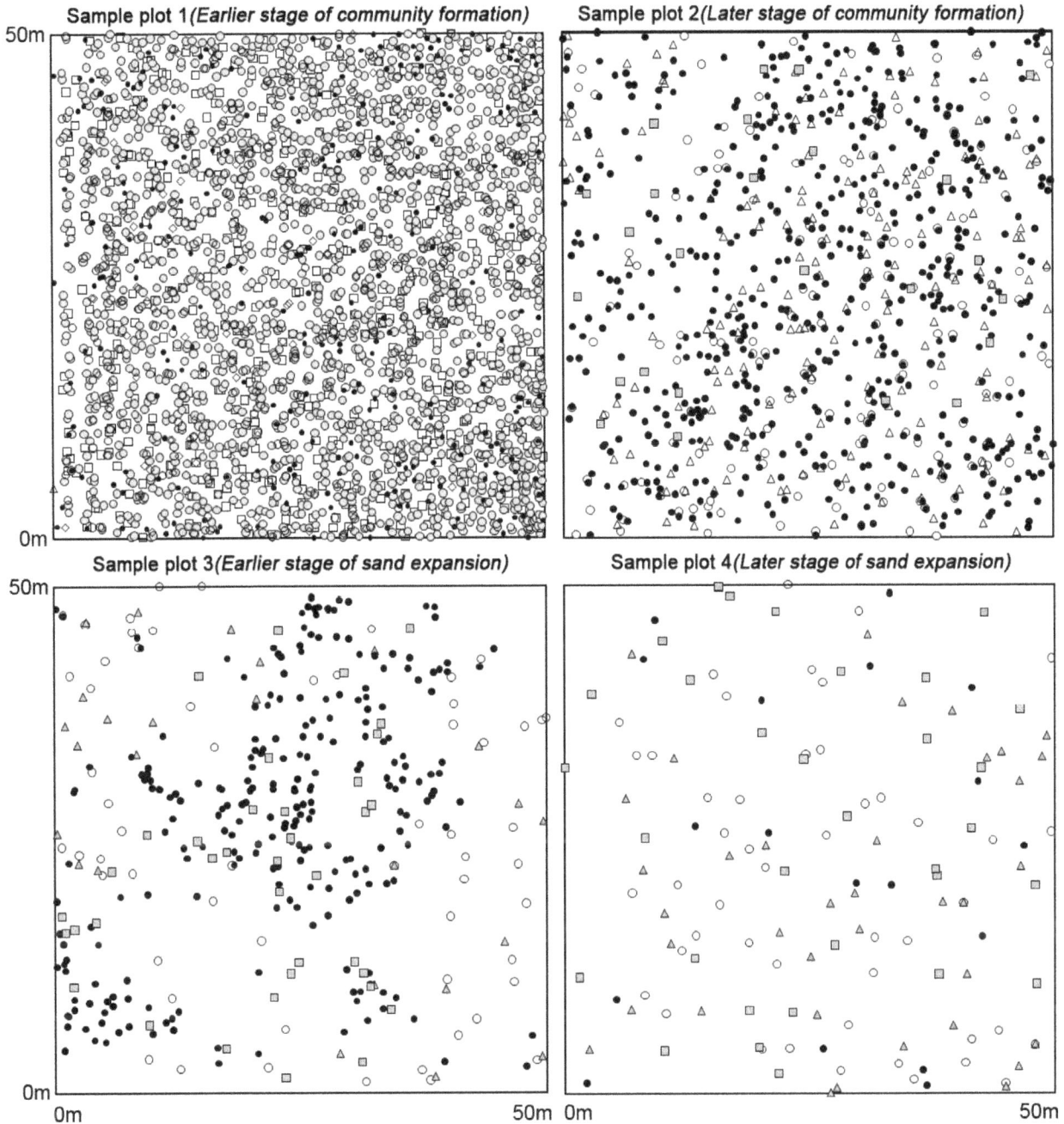

Figure 2. Point map of shrubs.

Results

Changes of Species in Different Sand-Burial Gradients

In sample plot 1, which represents the early stages of community pattern formation, four out of the five species were colonisers. The abundance of *Tetraena mongolica*, a weak sand-burial tolerant coloniser, reached a peak in sample plot 2 and then decreased gradually after sand burial. The abundance of *Nitraria tangutorum*, a strong sand-burial tolerant successor, reached its peak in sample plot 3 and then decreased. Shrubs with weak sand-burial tolerance disappeared gradually during the sand expansion stage. Plot 4 primarily included strongly sand-burial tolerant shrubs. The

total number of shrubs decreased gradually from 2,893 to 135 during the process of disturbance produced by the sand expansion (Figure 2, Table 2).

Spatial Point Pattern Analysis Based on the Hypothesis of Seed Dispersal Strategy

Based on the seed dispersal strategy, we classified the shrubs into colonisers and successors. The spatial point pattern analysis showed that in the early stage of community pattern formation (sample plot 1), the colonisers had a uniform distribution at a scale of 0–1 m and an aggregated distribution at scales of 3–4 m. The successor *Sarcozygium xanthoxylon* had a uniform distribution at

Table 2. Shrub numbers in four sample plots.

Species	Sample plot 1	Sample plot 2	Sample plot 3	Sample plot 4
Ammopiptanthus mongolicus			25	37
Convolvulus fruticosus	135			
Nitraria tangutorum		25	117	32
Potaninia mongolica	1 784			
Reaumuria songarica		164		
Reaumuria trigyna	494			
Sarcozygium xanthoxylon	209	162	59	48
Tetraena mongolica	271	555	246	18

scales of 1 m and 18 m and an aggregated distribution at a scale of 11 m. In the later stage of community pattern formation (sample plot 2), the colonisers had a uniform distribution at a scale of 0–1 m, whereas the successors *Nitraria tangutorum* and *Sarcozygium xanthoxylon* had a uniform distribution at scales of 0–1 m, 3 m and 12 m (Figure 3).

The interactions between the colonisers and the successors showed a significant positive correlation at a scale of 0–1 m and a significant negative correlation at a scale of 2 m in sample plot 1 and plot 2 (Figure 4).

Spatial Point Pattern Analysis Based on the Hypothesis of Sand Burial Effects

In the early stage of sand expansion (sample plot 3), the weak sand-burial tolerance group had an aggregated distribution at a scale of 1–2 m, and the strong sand-burial tolerance group had a uniform distribution at a scale of 7 m (Figure 5). In the later stage of sand expansion (sample plot 4), the weak sand-burial tolerance group had a random distribution at all scales that we studied, and the strong sand- burial tolerance group had a uniform distribution at the scales of 9 m and 13 m (Figure 5).

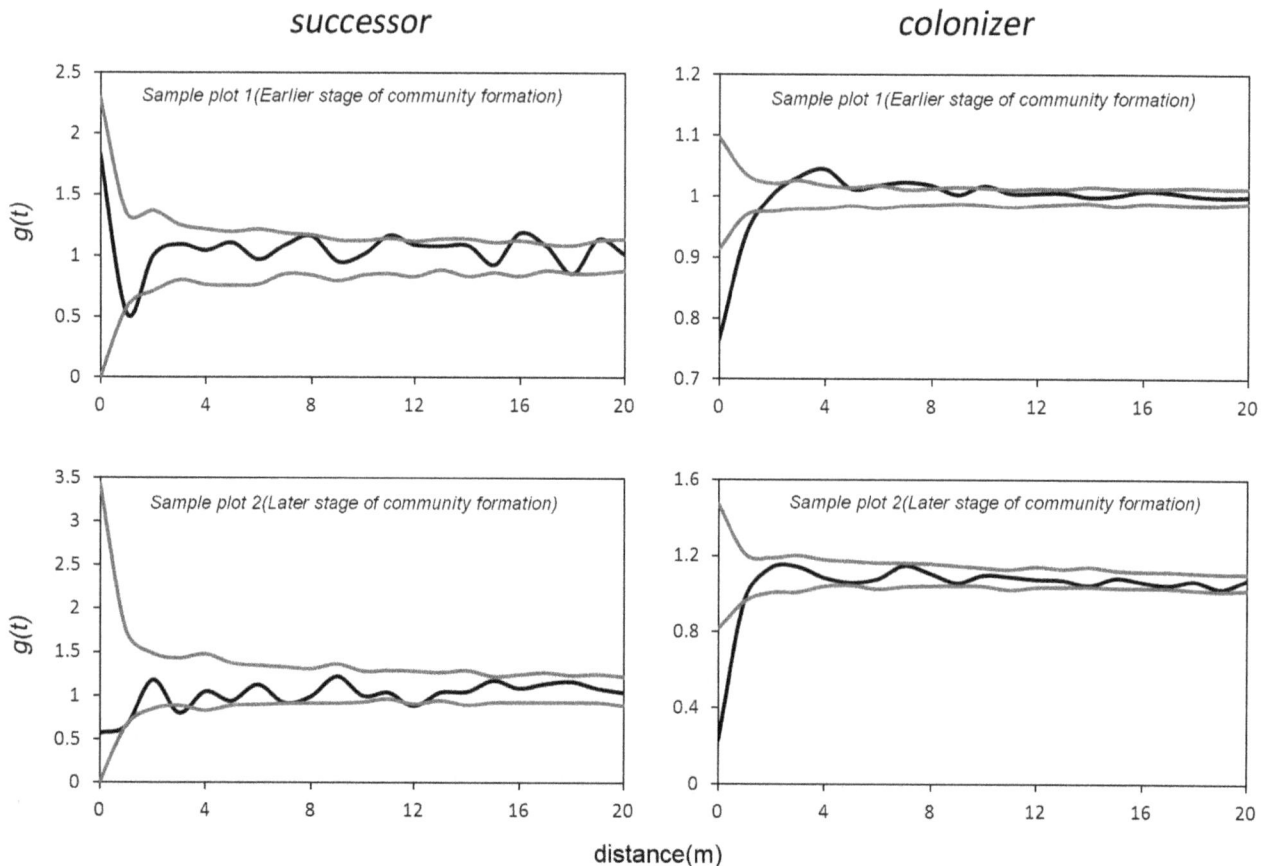

Figure 3. Univariate g(t) function of colonizers and successors in the community formation stage.

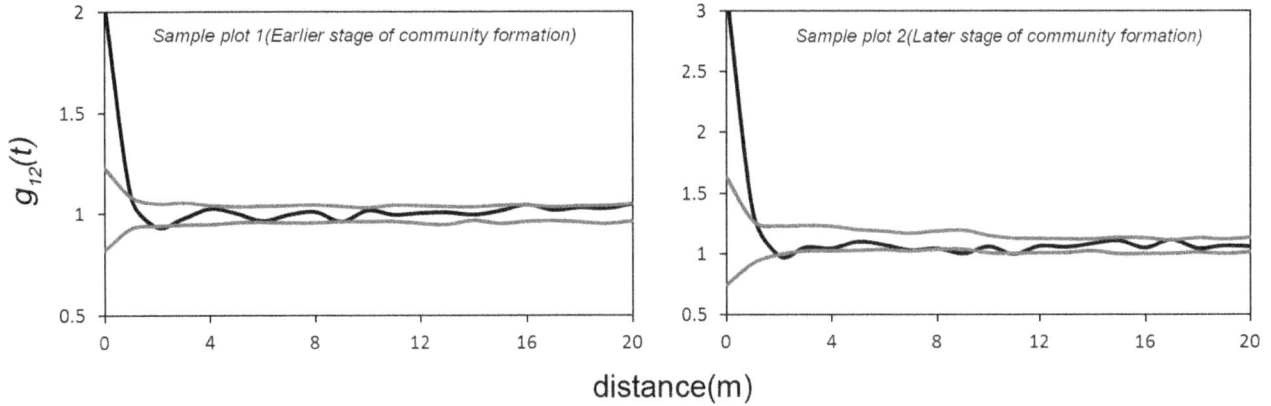

Figure 4. Bivariate g12(t) function between colonizers and successors in the community formation stage.

Significant negative correlations were found between the groups with strong and weak sand-burial tolerance in the early stage of sand expansion (sample plot 3) at scales of 3–6 m, whereas in the later stage of sand expansion (sample plot 4) a significant positive correlation was found at a scale of 13 m (Figure 6).

Discussion

Effect of Sand Burial on Community Succession

Sand burial, as an important natural selection pressure of plant distribution in desert areas, play an important role on the community succession [28–29]. Some studies showed that sand burial can enhance the seed germination and seedling emergence of the sand-burial tolerant shrubs [30–31] (Zhang et al., 2010; Li and Zhao, 2006). During the process of disturbance in the western Ordos Plateau shrub communities, small-seeded shrubs play a

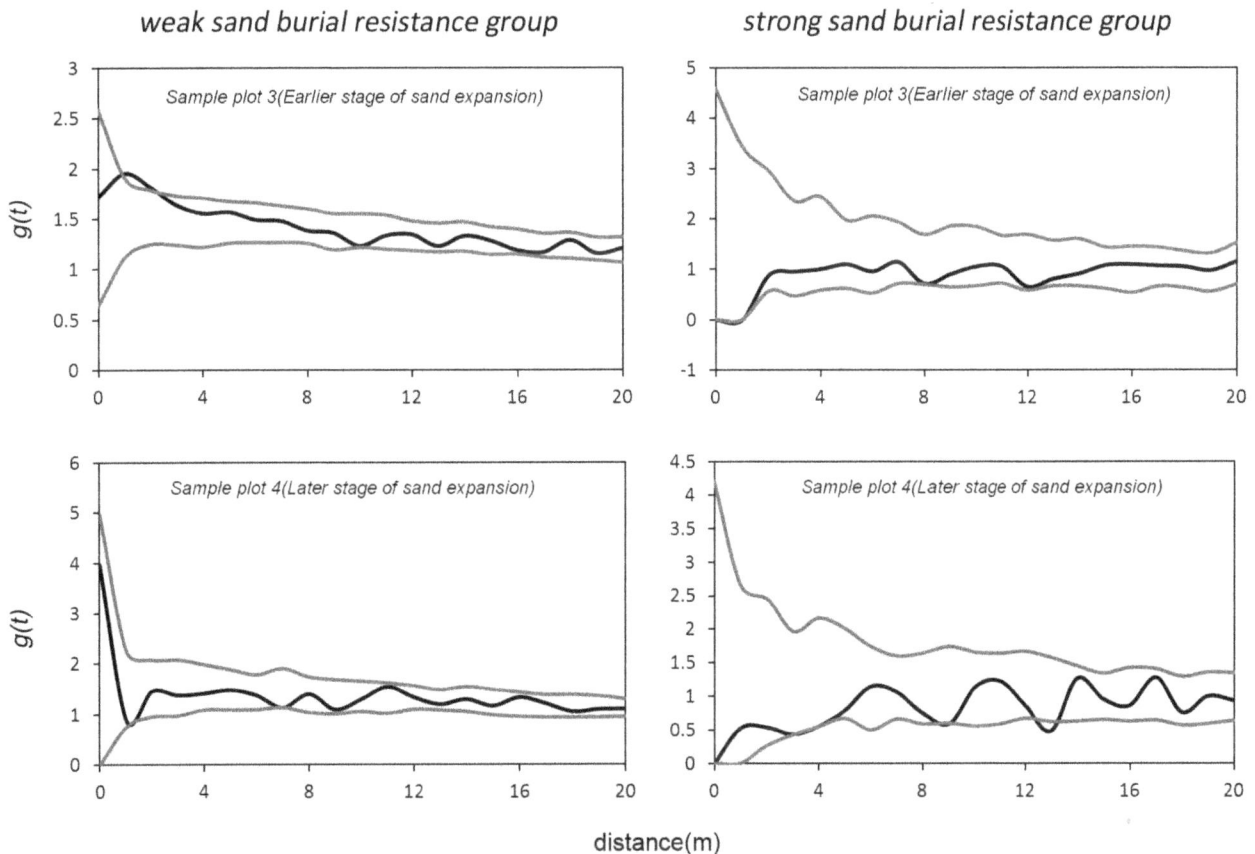

Figure 5. Univariate g(t) function of strong and weak sand burial resistance groups in the sand burial stage.

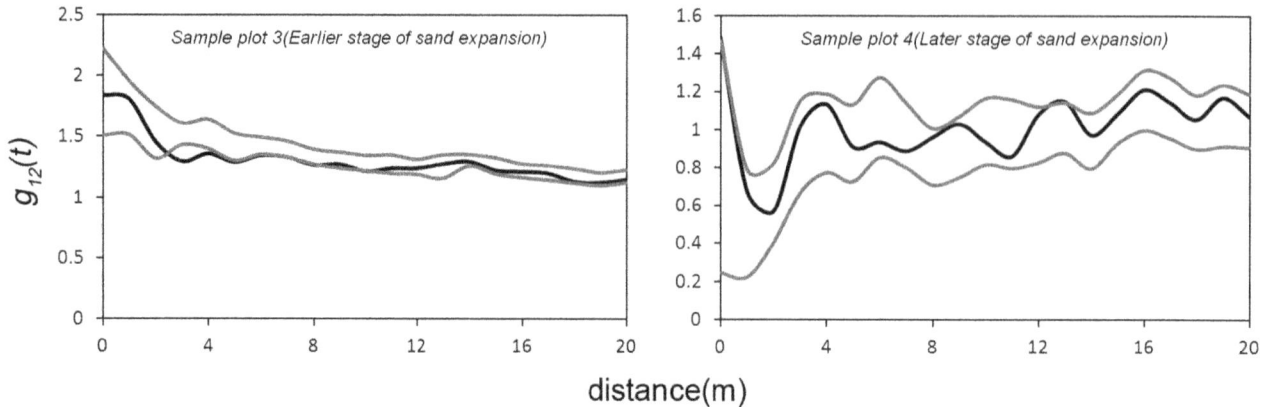

Figure 6. Bivariate g12(t) function between strong and weak sand burial tolerance groups in the sand burial stage.

dominant role in the early stage of community development, whereas strongly sand-burial tolerant shrubs play a dominant role after sand invasion. The total number of individuals in the shrub communities decreased gradually as the sand invaded. We assumed that each sample plot corresponded to a different stage of community disturbance. During disturbance, as successors invade, the occurrence of the colonisers is gradually reduced and the colonisers may even disappear. In the first disturbance stage, the only successor was *Sarcozygium xanthoxylon*; in the second disturbance stage, the coloniser *Reaumuria songarica* and the successor *Nitraria tangutorum* had invaded, whereas the colonisers *Reaumuria trigyna*, *Convolvulus fruticosus* and *Potaninia mongolica* had disappeared; and in the third and fourth stages, as a result of sand burial, the only remaining shrub species were the strongly sand-burial tolerant shrubs *Nitraria tangutorum* and *Ammopiptanthus mongolicus* and the weakly sand-burial tolerant shrubs *Sarcozygium xanthoxylon* and *Tetraena mongolica*. The thick sand covering was only suitable for the seed germination of a few large-seeded and strongly sand-burial tolerant shrubs that have a stronger emergence capability [32]. Consequently, few seedlings occurred in the sand-burial communities, and the community composition tended to be simplified and tended to include older-aged shrubs. Thus, following sand invasion, the abundant shrubs and the high diversity of the shrub communities in study area, even on the whole western Ordos Plateau [16], gradually declined.

Small-Scale Aggregated Distributions in the Early Stage of Community Pattern Formation

In the early stages of community pattern formation, the colonisers had an aggregated distribution at a scale of 3–4 m, which may result from accumulation of small seeds in the lowlands by wind transport when they invade the communities [33–36]. However, both sample plots 1 and 2 showed a uniform distribution on a scale of less than 1 m. This pattern was probably a result of shrub competition [19]. Moreover, in these two stages all the coloniser shrubs had their own crown sizes but were represented by points in the statistical analysis. This simplification may obscure an aggregated distribution at small scales [37]. The successors were distributed almost randomly in the early stage of community pattern formation. This result indicates that large seeds do not accumulate easily in one place.

The bivariate spatial point pattern analysis indicated that a significant positive correlation occurred between the coloniser and successor groups in the early stages of community pattern formation. This result indicates that the seeds of the successors

may be trapped by the colonisers that may facilitate the establishment of successors there [10].

The univariate and bivariate pattern analysis above showed that the colonisers had an aggregated distribution at small scales in the early stage of community pattern formation. Such a pattern may occur because shrubs with small seeds, such as *P. mongolica*, colonise bare ground first [10] by accumulating in small depressions, such as those formed by wind [33–34,38–39]. Both colonisers and successors (for example, *N. tangutorum*) may co-occur at a small scale in the early stage because new seeds can be trapped by the existing colonisers and become established there [1,40]. Other factors causing small-scale aggregated distribution may include soil heterogeneity resulting from the presence of micro-environments in which seedlings can grow under adult plants, the so-called 'nurse-plant syndrome' [40–44]. In these situations, adult plants alter the environment. For example, the surface temperatures in the open space between shrubs and under the canopy are different, and the adult plants may regulate the temperature [1,45–47]. The seed-dispersal strategy appears to represent a reasonable mechanism for explaining the dominant shrub community distribution pattern that occurs during the early stages of disturbance. This explanation is similar to the conclusions of a previous analysis of spatial patterns in a semi-arid Karoo shrubland [19].

Sand-Burial Effects

Plants encounter different levels of sand burial in sand dune habitats [48–50]. Sand burial can change the physiology, morphology, growth, and survival of plants by altering the biotic and abiotic environment [51–53] through parameters such as available photosynthetically active radiation [54–55], humidity, and temperature [29,56]. Different species have different degrees of sand-burial tolerance, and sand burial, as an environmental sieve, has driven the development of sand-burial tolerance mechanisms over the course of evolution [51]. Only those species with a certain degree of tolerance to sand burial can survive and grow in areas with drifting sand.

Moderate sand burial can improve the environment by increasing the humidity around plant roots and promoting the activities of soil microorganisms. Shrub species with root-suckering ability are prone to grow adventitious roots and continuously increase in size buried by sand [57–58]. In turn, this process increases the height of the plant buried by sand, promoting its biomass accumulation and the formation of a new branch [54,59]. However, sand burial gradually changes from a positive effect to a negative effect if the intensity of the sand burial is greater than a certain threshold. Plants cannot photosynthesise in the dark following sand burial [54]. This process weakens plant growth and can affect plant survival [48,51,56]. Sand burial results in the formation of a large, high dune around the bush, changing the horizontal distribution of soil material [60–61]. Nutrients and moisture within the communities can be displaced horizontally, causing marked differences between the soil on the sand dunes and in deflation hollows between the dunes and thereby affecting the distribution patterns of plants.

At the beginning of sand accumulation, the weak sand-burial tolerance group had an aggregated distribution at small scales. This distribution may have resulted from the primary reproductive mode of weakly sand-burial tolerant shrubs changing to sprouting from seed propagation and by small shrubs growing around the maternal shrub [16]. At the later stage of sand burial, only a few weakly sand-burial tolerant maternal shrubs were left in the community with random distribution, while most of seedlings or juvenile shrubs with weak tolerance around them disappeared. The number of strongly sand-burial tolerant shrubs also decreased under sand-burial pressure, compared with the earlier stage. They were randomly distributed at most of the scales studied and showed a uniform distribution at only a few scales (i.e. 9 m and 13 m).

Species with strong and weak sand-burial tolerance exhibited a significant negative correlation at the early stage of sand-burial expansion (sample plot 3). A thick layer of sand covered the ground, and sand dunes formed around the shrubs with strong sand-burial tolerance. As a result, these shrubs had few of the weakly sand-burial tolerant shrubs in their vicinity. At the later stage of sand burial, weakly sand-burial tolerant species tended to become extinct, and the number of strongly sand-burial tolerant species also decreased. The correlation between the groups was weak or even absent because shrubs were sparsely distributed in the community.

Sand burial and wind erosion are two important sources of selection pressure for plant distributions in arid areas [62] and are also the most common source of perturbation? in grasslands, desertified grasslands, and desert areas in China [63]. These two factors play important roles in the formation of plant communities in habitats with drifting sand [48] and represent important driving forces of community disturbance [64]. Our analysis shows that sand burial may cuase the disappearance of species with weak sand-burial tolerance which are dominant species in undegraded shrub communities. Sand burial has an important influence on the direction of development in shrub communities on the western Ordos Plateau and is also an important cause of decreasing shrub diversity in this region. Following the decline and death of the shrubs, large numbers of sand dunes become active and mobile. This change initiates a vicious cycle that affects all downwind areas.

In this paper, we hypothesized two primary driving forces for shrub-community spatial patterns in different disturbance stages on the western Ordos Plateau. Spatial analyses showed that the patterns of shrub communities may be formed by the seed-dispersal strategy. However, this mechanism breaks down in the later sand-expansion stage because seed germination is difficult in the changed environment. Sand burial seems to be the main force driving community degradation. This process causes the species with weak sand-burial tolerance to disappear. The disturbance is irreversible, causes homogenisation of the community structure and produces ageing populations of shrub species. We should be aware that this study is a preliminary attempt using a typical community investigation, which may over- or underestimate influence of the two primary driving forces on the western Ordos Plateau shrub pattern formation, and using spatial pattern models, which can only reflect underlying ecological processes with apparent patterns, hence, further works should put emphasis on more experimental studies of more shrub communities in this region in order to have deeper insight into underlying ecological processes of shrub community pattern formation.

Acknowledgments

We thank Prof. Longjun Ci for her comments on the research.

Author Contributions

Conceived and designed the experiments: XHY ZJS. Performed the experiments: YGW. Analyzed the data: YGW XHY ZJS. Contributed reagents/materials/analysis tools: XHY ZJS. Wrote the paper: YGW XHY ZJS.

References

1. Flores J, Jurado E (2003) Are nurse-protégé interactions more common among plants from arid environments? Journal of Vegetation Science 14: 911–916.
2. Miriti MN (2007) Twenty years of changes in spatial association and community structure among desert perennials. Ecology 88: 1177–1190.
3. Montane' F, Casals P, Dale MRT (2011) How Spatial Heterogeneity of Cover Affects Patterns of Shrub Encroachment into Mesic Grasslands. PLoS ONE 6(12): e28652. doi:10.1371/journal.pone.0028652.
4. Taylor S, Kumar L, Reid N, Kriticos DJ (2012) Climate Change and the Potential Distribution of an Invasive Shrub, Lantana camara L. PLoS ONE 7(4): e35565. doi:10.1371/journal.pone.0035565.
5. Valiente-Banuet A, Bolongaro-Crevenna A, Briones O, Ezcurra E, Rosas M, et al (1991) Spatial relationships between cacti and nurse shrubs in a semi-arid environment in central Mexico. Journal of Vegetation Science 2: 15–20.
6. Callaway RM (1995) Positive interactions among plants. The Botany Review 61: 306–349.
7. Armas C, Pugnaire FI (2005) Plant interactions govern population dynamics in a semi-arid plant community. Journal of Ecology 93: 978–989.
8. Padilla FM, Pugnaire FI (2006) The role of nurse plants in the restoration of degraded environments. Frontiers in Ecology and the Environment 4: 196–202.
9. Yang X, Baskin CC, Baskin JM, Liu G, Huang Z (2012) Seed Mucilage Improves Seedling Emergence of a Sand Desert Shrub. PLoS ONE 7(4): e34597. doi:10.1371/journal.pone.0034597.
10. Yeaton RI, Esler KJ (1990) The dynamics of a succulent karoo vegetation. Vegetation 88: 103–113.
11. Noy-Meir I (1973) Desert ecosystems: environment and producers. Annual Review of Ecology, Evolution, and Systematics 4: 25–51.
12. Sperry JS, Hacke UG (2002) Desert shrub water relations with respect to soil characteristics and plant functional type. Functional Ecology 16: 367–378.
13. Armas C, Pugnaire FI (2009) Ontogenetic shifts in interactions of two dominant shrub species in a semi-arid coastal sand dune system. Journal of Vegetation Science 20: 535–546.
14. West NE (1983) Temperate deserts and semi-deserts. Amsterdam: Elsevier Scientific Publishing Company. 522p.
15. Zhang XS (1994) Principles and optimal models for development of Maowusu (Mu Us) sandy grassland. Acta Phytoecologica Sinica 18: 1–16. (in Chinese with English abstract).
16. Li XR (2001) Study on shrub community diversity of Ordos Plateau, Inner Mongolia, Northern China. Journal of Arid Environments 47: 271–279.
17. Wang GH (2005) The western Ordos Plateau as a biodiversity center of relic shrubs in arid areas of China. Biodiversity and Conservation 14: 3187–3200.
18. Yang MX (1997) The brief introduction to the West Ordos natural reserve. Inner Mongolia Environmental Protection 9: 25–26. (in Chinese with English abstract).
19. Schurr FM, Bossdorf O, Milton SJ, Schumacher J (2004) Spatial pattern formation in semi-arid shrubland: a priori predicted versus observed pattern characteristics. Plant Ecology 173: 271–282.
20. Cody ML (2000) Slow-motion population dynamics in Mojave Desert perennial plants. Journal of Vegetation Science 11: 351–358.
21. Yang C (2002) Conservation biology of Tetraena mongolica Maxim. Beijing:Scientific Press. 160p (in Chinese).
22. Ripley BD (1981) Spatial statistics. New York: Wiley. 272p.
23. Stoyan D, Stoyan H (2000) Improving ratio estimators of second order point process characteristics. Scandinavian Journal of Statistics 27: 641–656.

24. Stoyan D, Stoyan H (1994) Fractals, random shapes and point fields: Methods of geometrical statistics. Chichester: Wiley. 406p.

25. Wiegand T, Moloney KA (2004) Rings, circles, and null-models for point pattern analysis in ecology. Oikos 104: 209–229.

26. Diggle PJ (2003) Statistical analysis of spatial point patterns, 2nd edition. New York: Hodder Education Publishers. 159p.

27. Wiegand T (2004) Introduction to point pattern analysis with Ripley's L and the O-ring statistic using the Programita software (Second draft version). User Manual for PROGRAMITA.

28. Peng F, Wang T, Liu LC, Huang CH (2012) Evolution phases and spatial pattern of Nebkhas in Minqin desert-oasis ecotone. Journal of desert research, 32(3): 593–599.

29. Maun MA (1998) Adaptations of plants to burial in coastal sand dunes. Canadian Journal of Botany 76: 713–738.

30. Li QY, Zhao WZ (2006) Seedling emergence and growth responses of five desert species to sand burial depth. Acta Ecologica Sinica, 26(6): 1802–1808.

31. Zhang YJ, Wang YS (2010) Effects of sand burial on seed germination and seedling emergence of four rare species in West Ordos area. Acta Bot. Boreal.-Occident. Sin. 30(1): 126–130.

32. Zeng YJ, Wang YR, Zhuang GH, Yang ZS (2004) Seed germination responses of Reaumuria soongorica and Zygophyllum xanthoxylum to drought stress and sowing depth. Acta Ecologica Sinica 24: 1629–1634. (in Chinese with English abstract).

33. Howe HF, Smallwood J (1982) Ecology of seed dispersal. Annual Review of Ecology, Evolution, and Systematics 13: 201–228.

34. Marone L, Rossi BE, Horno ME (1998) Timing and spatial patterning of seed dispersal and redistribution in a South American warm desert. Plant Ecology 137: 143–150.

35. Toft C A, Fraizer T (2003) Spatial dispersion and density dependence in a perennial desert shrub (Chrysothamnus nauseosus : Asteraceae). Ecological Monographs 73: 605–624.

36. Vander-Wall SB (2003) Effects of seed size of wind-dispersed pines (Pinus) on secondary seed dispersal and the caching behavior of rodents. Oikos 100: 25–34.

37. Prentice IC, Werger MJA (1985) Clump spacing in a desert dwarf shrub community. Vegetation 63: 133–139.

38. Dean WRJ, Milton SJ (1991) Disturbances in semi-arid shrubland and arid grassland in the Karoo, South Africa: mammal diggings as germination sites. African Journal of Ecology 29: 11–16.

39. Milton SJ, Dean WRJ (1993) Selection of seeds by harvester ants (Messor capensis) in relation to condition of arid rangeland. Journal of Arid Environments 24: 63–74.

40. Day TA, Wright RG (1989) Positive plant spatial association with Eriogonum ovalifolium in primary succession on cinder cones: seed-trapping nurse plants. Vegetation 80: 37–45.

41. Franco AC, Nobel PS (1988) Interactions between seedlings of Agave deserti and the nurse plant Hilaria rigida. Ecology 69: 1731–1740.

42. Callaway RM, Walker LR (1997) Competition and facilitation: a synthetic approach to interactions in plant communities. Ecology 78: 1958–1965.

43. Holmgren M, Scheffer M, Huston MA (1997) The interplay of facilitation and competition in plant communities. Ecology 78: 1966–1975.

44. Gomez-Aparicio L, Zamora R, Gomez JM, Hodar JA, Castro J, et al. (2004) Applying plant facilitation to forest restoration: a meta-analysis of the use of shrubs as nurse plants. Ecological Applications 14: 1128–1138.

45. Fulbright TE, Kuti JO, Tipton AR (1995) Effects of nurse-plant canopy temperatures on shrub seed germination and seedling growth. Acta Oecologica 16: 621–632.

46. Breshears DD, Nyhan JW, Heil CE, Wilcox BP (1998) Effects of woody plants on microclimate in a semiarid woodland: Soil temperature and evaporation in canopy and intercanopy patches. International Journal Plant Science 159: 1010–1017.

47. Esler KJ (1999) Plant reproductive ecology. In: Dean WRJ, Milton SJ. eds. The Karoo: ecological patterns and processes. Cambridge: Cambridge University Press. 123–144.

48. Maun MA, Lapierre J (1986) Effects of burial by sand on seed germination and seedling emergence of four dune species. American Journal of Botany 73: 450–455.

49. Kent M, Owen NW, Dale P, Newnham RM, Giles TM (2001) Studies of vegetation burial: a focus for biogeography and biogeomorphology? Progress in Physical Geography 25: 455–482.

50. Benard RB, Toft CA (2008). Fine-scale spatial heterogeneity and seed size determine early seedling survival in a desert perennial shrub (Ericameria nauseosa: Asteraceae). Plant Ecology 194: 195–205.

51. Yu FH, Chen YF, Dong M (2001) Clonal integration enhances survival and performance of Potentilla anserina, suffering from partial sand burial on Ordos plateau, China. Evolutionary Ecology 15: 303–318.

52. Liu F, Ye X, Yu F, Dong M (2007) Responses of Hedysarum Laeve, a guerrilla clonal semi-shrub in the Mu Us sandland, to local sand burial. Frontiers of Biology in China 2: 431–436.

53. Jia RL, Li XR, Liu LC, Gao YH, Li XR (2008) Responses of biological soil crusts to sand burial in a revegetated area of the Tengger Desert, Northern China. Soil Biology and Biochemistry 40: 2827–2834.

54. Brown JF (1997) Effects of experimental burial on survival, growth, and resource allocation of three species of dune plants. Journal of Ecology 85: 151–158.

55. Terrados J (1997) Is light involved in the vertical growth response of seagrasses when buried by sand? Marine Ecology Progress Series 152: 295–299.

56. Zhang JH, Maun MA (1990) Effects of sand burial on seed germination, seedling emergence, survival, and growth of Agropyron psammophilum. Canadian Journal of Botany 68: 304–310.

57. Jia BQ, Cai TJ, Gao ZH, Ding F, Zhang GZ (2002) Biomass forecast models of Nitraria tangutorum shrub in sand dune. Journal of Arid Land Resources and Environment 16: 96–99. (in Chinese with English abstract).

58. Yu QS, Wang JH, Li CL, Zhuang GH, Chen SK (2005) A preliminary study on the distribution patterns and characteristics of Ammopiptanthus mongolicus populations in different desert environment. Acta Phytoecologica Sinica 29: 591–598. (in Chinese with English abstract).

59. Disraeli DJ (1984) The effect of sand deposits on the growth and morphology of Ammophila Breviligulata. Journal of Ecology 72: 145–154.

60. Zhang H, Shi PJ Zheng QH (2001) Research progress in relationship between shrub invasion and soil heterogeneity in a natural semi-arid grassland. Acta Phytoecologica Sinica 25: 366–370. (in Chinese with English abstract).

61. Zhao WZ, He ZB, Li ZG (2003) Biological mechanism of sandy desertification in grassland reclamation area in north China. Advance in Earth Science 18: 257–262. (in Chinese with English abstract).

62. Li RP, Jiang DM, Liu ZM, Li XH, Li XL (2004) Effects of sand-burying on seed germination and seedling emergence of six psammophytes species. Chinese Journal of Applied Ecology 15: 1865–1868. (in Chinese with English abstract).

63. Liu YX (1988) Phytoreclamation of sand dunes in the northwest, north, and northeast of China. Journal of Desert Research 8: 11–17. (in Chinese with English abstract).

64. Zhao WZ (1998) A preliminary study on the arenaceous adaptability of Sophora moorcrftiana. Acta Phytoecologica Sinica 22: 379–384. (in Chinese with English abstract).

20

Getting "Just Deserts" or Seeing the "Silver Lining": The Relation between Judgments of Immanent and Ultimate Justice

Annelie J. Harvey*, Mitchell J. Callan*

Department of Psychology, University of Essex, Essex, United Kingdom

Abstract

People can perceive misfortunes as caused by previous bad deeds (immanent justice reasoning) or resulting in ultimate compensation (ultimate justice reasoning). Across two studies, we investigated the relation between these types of justice reasoning and identified the processes (perceptions of deservingness) that underlie them for both others (Study 1) and the self (Study 2). Study 1 demonstrated that observers engaged in more ultimate (vs. immanent) justice reasoning for a "good" victim and greater immanent (vs. ultimate) justice reasoning for a "bad" victim. In Study 2, participants' construals of their bad breaks varied as a function of their self-worth, with greater ultimate (immanent) justice reasoning for participants with higher (lower) self-esteem. Across both studies, perceived deservingness of bad breaks or perceived deservingness of ultimate compensation mediated immanent and ultimate justice reasoning respectively.

Editor: Cheryl McCormick, Brock University, Canada

Funding: The authors received no specific funding for this work.

Competing Interests: The authors have declared that no competing interests exist.

* Email: aharve@essex.ac.uk (AJH); mcallan@essex.ac.uk (MJC)

Introduction

A long history of research into the psychology of justice and deservingness has demonstrated that people are motivated to make sense of and find meaning in their own and others' experiences of suffering and misfortune [1], [2], [3], and they do so in a variety of ways [4], [5], [6]. For example, on the one hand, people may attempt to perceive a "silver lining" in someone's undeserved suffering by adopting the belief that although a victim is currently suffering, she will ultimately be compensated for her misfortune [3]. In other words, through *ultimate justice reasoning*, people are able to extend the temporal framework of an injustice, such that any negative outcome previously endured will be ultimately compensated with a positive outcome. Research has confirmed that perceiving benefits in the later lives of victims of misfortunes is one way observers cognitively manage the threat imposed when observing undeserved suffering [7], [8], [9], [10]. For example, Anderson and colleagues found that participants, whose belief in a just world had been previously threatened, displayed a tendency to see a teenager's later life as more enjoyable and meaningful if he had been badly injured than if he suffered only a mild injury [7].

On the other hand, people may try to make sense of suffering and misfortune by engaging in *immanent justice reasoning* [11], [12], [13], for a review see [14], which involves causally attributing a negative outcome to someone's prior misdeeds, even if such a causal connection is illogical. For example, Callan and colleagues found that participants causally related a freak car accident to a man's prior behavior to a greater extent when they learned he stole from children than when he did not steal [15]. Immanent

justice reasoning, then, allows an observer to maintain a perception of deservingness by locating the *cause* of a random misfortune in the prior misdeeds of the victim [11], [15], [14]. Indeed, research has shown that people engage in greater immanent justice reasoning when their justice concerns are heightened by first focusing on their long-term goals [15], cf. [16] or after being exposed to an unrelated instance of injustice [11].

Although research has shown that people readily engage in immanent and ultimate justice reasoning in response to suffering and misfortune, much less is known about how these responses interact and how they operate. Indeed, only a handful of studies have thus far examined ultimate and immanent justice reasoning simultaneously [17], [18], [19], and have primarily done so in the context of assessing individual differences in these justice beliefs. Understanding how these different reactions to misfortune operate not only informs future theorizing see [1], but also carries practical implications in predicting how people will react to victims in different circumstances. Thus, we sought to extend the literature on immanent and ultimate justice reasoning in three important ways: (1) by investigating whether there is a relation between immanent and ultimate justice reasoning, (2) by identifying the underlying processes that give rise to this relation, and (3) by examining whether immanent and ultimate justice reasoning operate the same way when people consider their own misfortune as when they consider the misfortunes of others (Study 2).

The relation between immanent and ultimate justice reasoning

Maes and colleagues [18], [19] identified that people's individual endorsement of immanent and ultimate justice reasoning resulted in opposite reactions to victims. That is, people who believe strongly in ultimate justice reasoning are more likely to positively evaluate victims of misfortune, whereas people scoring highly in immanent justice beliefs blamed and derogated a victim for their plight. As immanent and ultimate justice reasoning are associated with conflicting victim reactions, these reactions to injustice may have a negative relation, such that the adoption of one form of justice reasoning reduces the extent to which people engage in the other. In Study 1, we sought to test this negative relation between these two types of justice reasoning empirically by assessing how people make sense out of misfortunes. We predicted that when people are given to ultimate justice reasoning (i.e., when the victim is a good person; see [7]), they would be less likely to engage in immanent justice reasoning. When people are given to immanent justice reasoning (i.e., when the victim is a bad person; see [14]), however, they would be less likely to perceive ultimate justice. We propose that the relation between the worth of the victim and justice reasoning is at least partly due to people's perceptions of what is considered as deserved.

Perceived deservingness and immanent and ultimate justice reasoning

Responding to instances of suffering and misfortune with ultimate and immanent justice reasoning can be considered seemingly irrational. Although there may be logical reasons why good and bad people will have good or bad lives (e.g., higher well-being from a good person acting prosocially), often no substantial causal links exist between a person's character, their random misfortune, and their ultimate fulfillment in life; or a victim's previous misdeeds and their current misfortune. That is, the worth of a person does not cause random, unrelated misfortunes and enduring a random misfortune does not necessarily mean that an individual's later life will be better. Despite this seeming irrationality, people might nonetheless engage in immanent and ultimate justice reasoning in response to suffering and misfortune because doing so enables them to maintain important, functional beliefs. We examined whether immanent and ultimate justice reasoning might be driven, in part, by the belief that the world is a just, fair, and nonrandom place where people get what they deserve—a world where an appropriate relation exists between the value of people (good or bad) and the value of their outcomes (good or bad) [20], [3], see also [21]. In other words, both the processes of causally linking a random misfortune to someone's prior misdeeds (immanent justice) and perceiving benefits in the later lives of victims of misfortune (ultimate justice) might be driven, in part, by a concern for upholding notions of *deservingness*.

Deservingness refers to the perceived congruence between the value of a person and the value of his or her outcomes. Therefore, something bad happening to a "good" person is often perceived as undeserved, whereas the same outcome occurring to a "bad" person is often considered deserved [11], [22], [21], [23], [24]. Several studies have confirmed that the perceived deservingness of a random outcome is an important mediator of the extent to which people are willing to adopt immanent justice accounts of the outcome see [14]. Less is known, however, about the processes underlying ultimate justice reasoning. If the proposed negative relation between immanent and ultimate justice reasoning is driven by the ultimate goal of perceiving people's fates as deserved

in a just world, we predict that perceived deservingness should underlie the endorsement of both types of justice reasoning.

This analysis is consistent with Kruglanski's discussion of the principle of equifinality [25], which suggests that different substitutable and equal means are capable of reaching the same goal. In the context of the current research, immanent and ultimate justice reasoning can both be considered equal means to achieving the goal of preserving a belief that the world is a fair and just place where people get what they deserve. People can accomplish this goal via immanent justice reasoning by attributing the cause of a misfortune to the victim's prior misdeeds. Alternatively, people who engage in ultimate justice reasoning can uphold their just-world beliefs by believing that a victim's misfortune will be ultimately compensated [7]. If participants engage in one type of reasoning because of their concerns about deservingness, utilizing an additional type of reasoning would be redundant. For example, linking an individual's current misfortune to their prior misdeeds satisfies a concern for deservingness because the victim "got what she deserved". Further rationalizations of misfortune, such as believing the victim will be ultimately compensated, are therefore less necessary and support our prediction of a negative correlation between ultimate and immanent justice reasoning.

The extent to which perceived deservingness underlies immanent and ultimate justice reasoning, however, should depend on the specific outcome people believe is deserved. With immanent justice reasoning, causal connections are drawn between people's previous deeds and their *recently experienced outcomes*, whereas ultimate justice reasoning entails believing in more *"long-term" positive outcomes* for a victim who is suffering. Thus, whether a concern for deservingness helps explain immanent and ultimate justice reasoning should depend on what people perceive as deserved—later life fulfillment or a recently experienced random outcome—given the value of the person experiencing the outcome. The idea that specific perceptions of deservingness might differentially predict immanent and ultimate justice reasoning resonates well with research showing greater congruency between constructs that are measured at the same level of specificity (e.g., values and behavior) [26]. Accordingly, we examined the degree to which perceptions of deserving later-life fulfillment and a recently experienced outcome underlie ultimate and immanent justice reasoning, respectively. We predicted that perceiving a misfortune as deserved should better predict immanent justice reasoning [14], whereas perceiving a victim as deserving of later fulfillment should better predict ultimate justice reasoning.

Immanent and ultimate justice reasoning for the self

Lerner argued that principles of justice and deservingness for others should be equivalent to the self, as observing deservingness in another's life should mean, by generalization, that one's own life is just and fair [3], [27]. Early work by Lerner and colleagues [28], [29] showed that people are more likely to work towards fairness for others when they themselves have received unfair treatment, suggesting that people are responsive to the fates of others because this determines the fairness of the world they live in. As a result, one's own fate "is intertwined emotionally and practically with the ability of others to get what they deserve" [28] (p. 177).

Consistent with this view, observer judgments of deservingness are often comparable to deservingness judgments made for the self. That is, research has shown that people judge others, and themselves, as deserving bad (good) outcomes if they are perceived as bad (good) people [11], [22], [30], [23], [24], [31], [32]. For example, Wood and colleagues found that individuals chronically and situationally lower (vs. higher) in self-esteem saw themselves as more deserving of negative emotions [31]. More recently, Callan and colleagues found that participants' beliefs about deserving bad outcomes in life mediated the relation between trait self-esteem and a variety of self-defeating thoughts and behaviors (e.g., self-handicapping, thoughts of self-harm) [22]. Although this research highlights the important role that perceptions of deservingness for the self play in a host of self-relevant outcomes, no research to our knowledge has examined the role that personal deservingness plays in people's immanent justice and ultimate justice reasoning for self-relevant outcomes. To this end, in Study 2 we examined whether people would causally attribute their random bad breaks to their personal worth or believe they would achieve a fulfilling life as a function of their self-esteem and perceptions of deservingness. In other words, we examined whether the same relation between immanent and ultimate justice reasoning, and the same underlying processes of deservingness, in response to the misfortune of others (Study 1) would replicate when individuals considered their own misfortune (Study 2).

Current research

Over two sets of studies we sought to investigate whether (1) there is a negative relation between immanent and ultimate justice reasoning, (2) perceived deservingness underlies this relation, and (3) the relation and processes involved in immanent and ultimate justice reasoning are similar for one's own misfortunes as they are for the misfortunes of others. To accomplish these aims we manipulated the worth of a victim (Study 1) or measured people's perceived self-worth (Study 2) before assessing judgments of deservingness and ultimate and immanent justice reasoning.

If there is a negative relation between immanent and ultimate justice reasoning in response to misfortune, then people should engage in significantly more ultimate than immanent justice reasoning for a victim who is a good person and significantly more immanent than ultimate justice reasoning for a victim who is a bad person. We also predicted that specific perceptions of deservingness would underlie this relation, such that perceiving a victim as deserving of their misfortune would more strongly mediate immanent justice reasoning and perceiving a victim as deserving of a fulfilling later life would more strongly mediate ultimate justice reasoning. Finally, we predicted that this pattern of findings should be similar when participants consider their own misfortunes (Study 2).

Study 1

In Study 1 we manipulated the value of a victim of misfortune before assessing participants' perceptions of the degree to which he deserved his misfortune and deserved ultimate compensation along with immanent and ultimate justice reasoning. We predicted that a "good" victim would encourage participants to engage in more ultimate than immanent justice reasoning, largely due to the victim being deserving of ultimate compensation following their ill fate. When faced with a "bad" victim, however, we predicted that participants would interpret the victim's fate as deserved and therefore engage in more immanent rather than ultimate justice reasoning.

Method

Participants. The study was administered online and approved by the Ethics Committee of the University of Essex. Consent was achieved by asking participants to click a button to begin the study and give their consent or to close their browser and withdraw consent. We recruited two samples of participants

($Ns = 168$ and 100; total $N = 268$, 48.9% females, 0.4% unreported; $M_{age} = 35.35$, $SD_{age} = 11.88$) via Amazon's Mechanical Turk [33] and CrowdFlower. Twelve participants (4.5%) who incorrectly answered a simple manipulation question ("Is Keith Murdoch awaiting trial for sexually assaulting a minor?") were excluded from further analysis. The samples differed only in the ordering of the items (see procedure below).

Materials and procedure. Participants were told they would be partaking in a study "investigating memory and impressions of events". Participants were first presented with an ostensibly real news article that described a freak accident where a volunteer swim coach, Keith Murdoch, was seriously injured following a tree collapsing on his vehicle during high winds see [15]. Next, we manipulated the worth of the victim by telling participants that the victim was either a pedophile ("bad" person) or a respected swim coach ("good" person). Specifically, participants in the "bad" person condition learned that "Keith Murdoch is awaiting trial for sexually assaulting a 14-year-old boy while he worked at the Bitterne Leisure Center as a volunteer swim coach and that other charges of sexual exploitation of minors are pending given recent evidence obtained by police since the original charge." Participants in the "good" person condition read that "Keith Murdoch volunteered as a swimming coach at the Bitterne Leisure Centre and is a valued and beloved member of the community." We predicted that this information about the victim's character should determine how deserving the victim was of his random misfortune and ultimate compensation and, as a result, the extent of participants' immanent and ultimate justice reasoning respectively.

As a manipulation check, participants rated the goodness of the victim's character with the item, "How would you rate Keith Murdoch as a person?" ($1 = very\ bad$ to $6 = very\ good$).

Ordering of items for Sample 1. In our first sample, participants were then asked two questions to assess their perceptions of deservingness of the accident: "To what extent do you feel Keith Murdoch deserved to be in this accident?" ($1 = not\ at\ all\ deserving$ to $6 = very\ deserving$) and "To what extent do you feel that this accident was a just and fair outcome for Keith Murdoch?" ($1 = not\ at\ all\ just\ and\ fair$ to $6 = very\ just\ and\ fair$).

Adapted from items used to measure beliefs in conspiracy theories [34], participants then answered four items that assessed their immanent justice attributions for the accident: "To what extent do you feel it is worth considering that this accident might have been a result of Keith Murdoch's conduct as a swim coach?" ($1 = not\ at\ all\ worth\ considering$ to $6 = worth\ considering$), "How possible do you feel it is that this accident was a result of Keith Murdoch's conduct as a swim coach?" ($1 = not\ at\ all\ possible$ to $6 = possible$), "How plausible do you feel it is that this accident was caused by Keith Murdoch's conduct as a swim coach?" ($1 = not\ at\ all\ plausible$ to $6 = plausible$), and "I feel that this accident was a result of Keith Murdoch's conduct as a swim coach." ($1 = strongly\ disagree$ to $6 = strongly\ agree$).

Following this, participants were asked two items to assess perceptions of how deserving the victim was of ultimate compensation: "I feel that Keith Murdoch deserves to experience his life as meaningful in the long run" and "I believe Keith Murdoch deserves to find purpose and fulfillment later in his life" ($1 = strongly\ disagree$ to $6 = strongly\ agree$). Finally, three items assessed ultimate justice reasoning see [7]: "To what extent do you think Keith Murdoch will find his existence fulfilling later in his life?", "To what extent do you believe that in the future, Keith Murdoch will experience his life as meaningful?", and "To what extent do you think that in the long run, Keith Murdoch will find purpose in his life?" ($1 = not\ at\ all\ fulfilling/meaningful/$

Table 1. Descriptive and inferential statistics for the measures employed in Studies 1 and 2.

Measures	Worth of Victim Manipulation		t	d	Intercorrelations				
	Pedophile	Volunteer			1.	2.	3.	4.	5.
Study 1	M (SD)	M (SD)							
1. Deservingness of misfortune	3.26 (1.65)	1.34 (0.71)	12.19**	1.51	[.86]	–	–	–	–
2. Immanent justice reasoning	1.98 (1.34)	1.27 (0.70)	5.28**	0.66	.56**	[.94]	–	–	–
3. Deservingness of later fulfillment	3.19 (1.29)	5.09 (0.73)	14.57**	1.81	-.67**	-.35**	[.86]	–	–
4. Ultimate justice reasoning	2.49 (1.08)	4.66 (0.97)	16.93**	2.11	-.64**	-.33**	.76**	[.97]	–
Study 2	M	SD							
1. Self-esteem	4.46	0.98	–	–	[.93]	–	–	–	–
2. Deservingness of misfortune	2.81	1.18	–	–	-.36**	[.69]	–	–	–
3. Immanent justice reasoning	2.93	1.22	–	–	-.39**	.66**	[.94]	–	–
4. Deservingness of later fulfillment	4.81	0.95	–	–	.38**	-.23**	-.16*	[.68]	–
5. Ultimate justice reasoning	4.70	1.05	–	–	.62**	-.25**	-.22**	.43**	[.94]

Note. Higher values indicate more of each construct. Where applicable, alpha reliabilities (or the bivariate correlations for the deservingness measures) are presented in brackets.
*p<.05
**p<.01

purposeful to 6 = *very fulfilling/meaningful/purposeful*). All items within each construct reached acceptable internal consistency (see Table 1), so were averaged to form measures of perceived deservingness of an accident, immanent justice attributions, perceived deservingness of ultimate compensation, and ultimate justice judgments.

Ordering of items for Sample 2. Because we were concerned that the fixed ordering of our items in Sample 1 may have biased participants toward the first opportunity they were given to resolve the injustice (i.e., immanent justice reasoning), we recruited another sample of participants and reversed the ordering of items from Sample 1. Sample 2, therefore, was identical to Sample 1, with the exception of the ordering of items. The questionnaire was structured so that after rating the goodness of the victim's character, participants answered the items regarding how deserving the victim was of ultimate compensation and deserving of the accident, followed by the ultimate justice reasoning items and finally the immanent justice reasoning items.

Results and Discussion

Preliminary analyses showed that there were no significant differences between the two samples in terms of the effect of the experimental manipulation on our dependent measures or the correlations among the measures (i.e., there were no significant interactions with sample/item order, all $ps>.05$), and the same patterns of results replicated across samples. Thus, the ordering of items did not appear to affect participants' responses. Accordingly, data from the two samples were collated and analyzed together.

Analysis of the manipulation check confirmed that participants who learned that the victim was a pedophile ($M = 1.64$, $SD = 0.76$) perceived him as less good than participants who learned that he was a respected volunteer ($M = 5.14$, $SD = 0.57$), $t(251) = 41.66$, $p<.001$, $d = 5.22$). Shown in Table 1, participants who were presented with a "bad" victim rated him as more deserving of his random bad outcome than participants who read about a "good" victim, conceptually replicating previous research [11], [35]. Also, participants who were presented with a "good" victim saw him as more deserving of later fulfillment than a "bad" victim. Table 1 also shows the correlations among the measures we employed in Study 1. Of note, both types of perceived deservingness correlated significantly with both types of justice judgments, and immanent and ultimate justice reasoning correlated negatively.

The interplay between immanent and ultimate justice reasoning. To examine the interplay between immanent and ultimate justice reasoning as a function of the value of the victim, we conducted a 2 (victim worth: good vs. bad) by 2 (type of justice reasoning: immanent justice vs. ultimate justice) mixed model ANOVA, with type of justice reasoning as the within-subjects factor. Because people are typically more willing to endorse ultimate justice than immanent justice in absolute terms, we standardized the data for comparisons across types of justice reasoning (the unstandardized data is presented in Table 1). Analyses revealed the predicted Victim Worth X Type of Reasoning interaction, $F(1, 254) = 176.09$, $p<.001$, $\eta_p^2 = .41$.

Shown in Figure 1, decomposing the interaction revealed that participants engaged in relatively more immanent justice than ultimate justice reasoning when the victim was a pedophile, $t(124) = 7.96$, $p<.001$, and more ultimate justice than immanent justice reasoning when he was a respected volunteer, $t(130) = 12.01$, $p<.001$.

Perceived Deservingness. We examined whether the perceived deservingness of the victim's fate accounts for the observed relation between participants' judgments of immanent justice and ultimate justice. That is, a concern for deservingness should

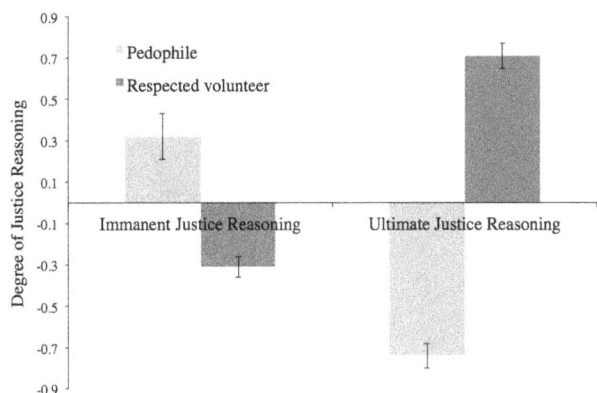

Figure 1. Mean level of immanent justice and ultimate justice reasoning from Study 1 (standardized) as a function of the victim's personal worth (pedophile versus respected volunteer). Error bars show standard errors of the means.

underpin the degree to which people engage in more or less immanent justice reasoning relative to ultimate justice reasoning as a function of the worth of the victim. More specifically, perceiving a victim as deserving of his fate should better underlie immanent justice judgments and perceiving a victim as deserving of later life fulfillment should better predict ultimate justice reasoning, as a function of the victim's worth.

To test this hypothesis, we conducted multiple mediation analyses with Preacher and Hayes's (2008) bootstrapping procedure (10,000 resamples; see Figure 2) [36]. As predicted, bootstrapping analyses revealed that perceived deservingness of the accident mediated the effect of the victim's worth on immanent justice reasoning (indirect effect $= -0.81$, BCa CI $= -1.13$ to -0.56), but perceived deservingness of later fulfillment did not (indirect effect $= 0.06$, BCa CI $= -0.19$ to 0.31). The same analysis conducted with ultimate justice reasoning showed both types of deservingness mediated the effect of the victim's worth on justice reasoning, but perceived deservingness of later fulfillment (indirect effect $= .88$, BCa CI $= 0.63$ to 1.15) was a stronger mediator than perceived deservingness of the accident (indirect effect $= .23$, BCa CI $= .06$ to 0.45). The same mediation pattern was observed for both samples separately. The exception being that for the second sample, perceived deservingness of the accident did not mediate the effect of the manipulation on ultimate justice reasoning (cf. Study 2; indirect effect $= -0.02$, BCa CI $= -0.24$ to 0.25). In sum, the value of a victim affects whether people view the misfortune or later life fulfillment as deserved, which in turn predicts the extent of immanent justice reasoning over ultimate justice reasoning and vice versa.

Study 2

In Study 2, we sought to conceptually replicate our Study 1 findings in the context of participants' considerations of their *own* misfortunes. Study 1 found that participants perceived greater immanent justice for a victim with negative (vs. positive) worth and greater ultimate justice reasoning for a victim of positive (vs. negative) worth. In Study 2, we predicted that people's perceived self-worth should similarly influence the extent of justice reasoning for their own outcomes. Specifically, we assessed whether people are more likely to engage in immanent or ultimate justice reasoning for the self after considering their own misfortunes as a function of their perceptions of personal deservingness. To test

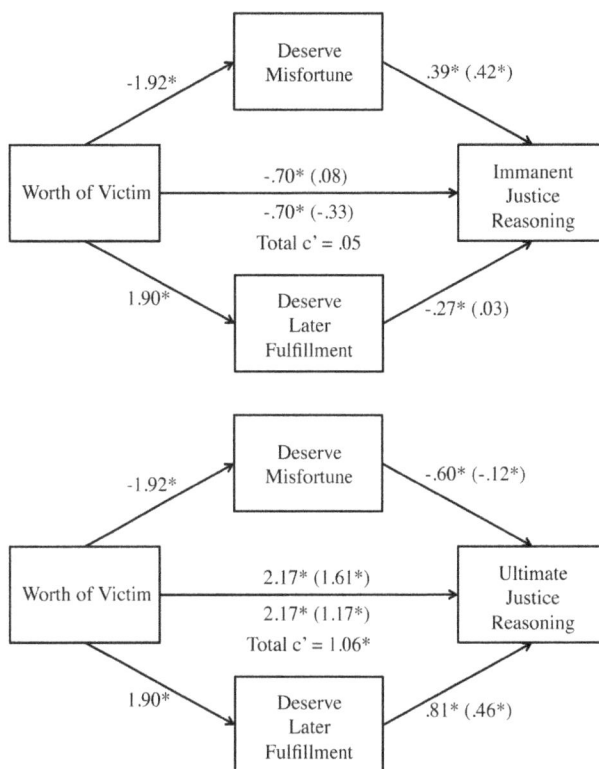

Figure 2. Mediational model from Study 1, predicting immanent justice and ultimate justice reasoning from the worth of a victim, beliefs about deserving bad outcomes, and beliefs about deserving later fulfillment. The victim of negative worth (pedophile) was coded as 1 and the victim of positive worth (respected volunteer) was coded as 2. Values show unstandardized path coefficients. * $p<.05$.

this notion, we measured participants' self-esteem before asking them to respond to deservingness, immanent, and ultimate justice items in relation to their own recent bad breaks. Paralleling our Study 1 effects, we predicted that self-esteem would correlate negatively with immanent justice reasoning and positively with ultimate justice reasoning. Crucially, we predicted that perceived deservingness would underlie the relations between self-esteem and justice reasoning for the self. Per our Study 1 findings, we predicted that perceiving a bad break as deserved would better predict immanent justice reasoning for the self and perceiving oneself as deserving of later life fulfillment should be a better predictor of ultimate justice judgments for the self.

Method

Participants. Participants were recruited online via Amazon's Mechanical Turk for a nominal payment ($N = 102$) or the University of Essex volunteer e-mail list for the chance to win a £20 gift voucher ($N = 100$; total $N = 202$, 56.9% females; $M_{age} = 27.64$, $SD_{age} = 9.58$). One participant was excluded from further analysis because he/she only answered one item from the self-esteem measure. Ethical approval and informed consent was obtained in the same way as Study 1.

Materials and procedure. Participants took part in a study that was ostensibly about "people's perceptions of their personal experiences." We first assessed participant's self-esteem via Rosenberg's 10-item self-esteem scale ($1 = $ *strongly disagree* to $6 = $ *strongly agree*) [37]. We then asked participants to think about their recent random "bad breaks." Bad breaks were described to participants as "those sorts of negative experiences we have that we do not intend, expect, or plan to occur—they just happen to us."

Next, participants answered a questionnaire similar to that of Study 1, although the questions were framed around participants' personal random bad breaks and in more general terms, due to the recalled "bad breaks" being general events rather than a specific incident of victimization. First, participants answered two items that aimed to assess their perceived deservingness of general bad outcomes: "I often feel that I deserve the bad breaks that happen to me" and "When I've experienced bad breaks in my life, I've sometimes thought that I deserved them" (1 = *strongly disagree* to 6 = *strongly agree*). Similar items from Study 1 were used to assess immanent justice reasoning (e.g., "How possible do you feel it is that your bad breaks were a result of the kind of person you are?"). Next, we presented participants with two items that assessed how deserving they felt of greater life fulfillment and meaningfulness (e.g., "I feel that I deserve to experience my life as meaningful in the long run") and three ultimate justice items based on those from Study 1 (e.g., "To what extent do you think you will find your existence fulfilling later in life?"). Table 1 shows that each of these measures achieved acceptable internal consistency.

Results and Discussion

Shown in Table 1, participants' self-esteem was negatively related to immanent justice judgments, showing that the lower their self-esteem, the more participants felt their bad breaks were caused by the kind of person they were. Self-esteem and ultimate justice reasoning were positively related, indicating that the higher participants' self-esteem, the more they engaged in ultimate justice reasoning for themselves. These findings replicate our Study 1 results, but do so in the context of participants considering their *own* bad breaks rather than the misfortune of someone else. Indeed, reflecting the interaction pattern shown in Figure 1, a test of the difference between overlapping correlations [38] showed that the correlation between self-esteem and immanent justice reasoning was significantly different from the correlation between self-esteem and ultimate justice reasoning (95% confidence interval: −1.16, −.85).

Of particular importance was the mediating role of deservingness beliefs in these relations, which we specified into two forms: (1) the deservingness of past bad breaks and (2) the deservingness of later life fulfillment. We again conducted multiple mediation analyses with Preacher and Hayes's (2008) bootstrapping procedure (10,000 resamples) [36]. When entering both deservingness of bad breaks and deservingness of later fulfillment as possible mediators of the relation between self-esteem and immanent justice reasoning, only the former provided a significant indirect effect. In other words, perceived deservingness of bad breaks significantly mediated the relation between self-esteem and immanent justice reasoning (indirect effect = −0.27, BCa CI = −0.41 to −0.14) but perceived deservingness of later fulfillment did not (indirect effect = 0.03, BCa CI = −0.04 to 0.08). Conducting the same analysis for ultimate justice reasoning revealed that perceived deservingness of bad breaks did not mediate the relation between self-esteem and ultimate justice reasoning (indirect effect = 0.003, BCa CI = −0.05 to 0.06) but perceived deservingness of later life fulfillment did (indirect effect = 0.09, BCa CI = 0.03 to 0.19).

Therefore, only deservingness of bad breaks mediated the relation between self-esteem and immanent justice reasoning, whereas only deservingness of later life fulfillment mediated the relation between self-esteem and ultimate justice reasoning for the self (see Figure 3).

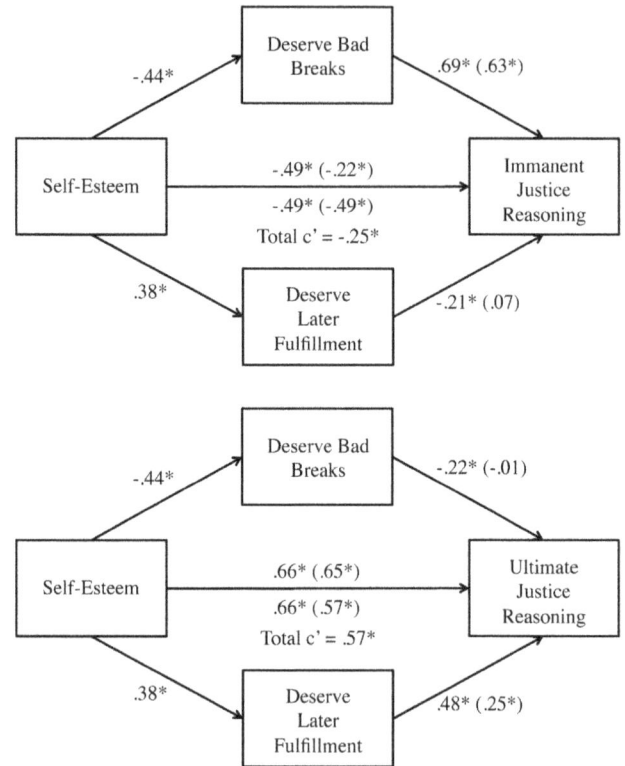

Figure 3. Mediational model from Study 2, predicting immanent justice and ultimate justice reasoning from self-esteem, beliefs about deserving bad outcomes, and beliefs about deserving later fulfillment. Values show unstandardized path coefficients. * $p<.05$.

General Discussion

Over two studies we sought to determine (1) the relation between immanent justice and ultimate justice reasoning, (2) the underlying mechanism responsible for this relation, and (3) if the relation between immanent and ultimate justice reasoning not only applies to the misfortunes of others, but also to one's own misfortunes. Study 1 showed that participants engaged in immanent justice reasoning to a greater extent when they learned that a victim was a "bad" (vs. "good") person, whereas they perceived more ultimate justice reasoning when the victim was a "good" (vs. "bad") person. When people are given to making immanent justice attributions (i.e., when a victim is of low worth), ultimate justice judgments are lower. However, when individuals are prone to ultimate justice reasoning (i.e., when a victim is of high worth), immanent justice reasoning is reduced.

Importantly, perceived deservingness mediated these effects. When confronted with a "good" person who experienced a random ill-fate, participants saw the victim as deserving of later life fulfillment and therefore, rejected an immanent justice account of the event in favor of perceiving benefits in the later life of the victim. When the victim was considered in negative terms, however, participants were more willing to see the misfortune as deserved and causally attribute the freak accident to the victim's past behavior, as well as reducing their ultimate justice judgments accordingly. As a result, participants engaged in immanent and ultimate justice reasoning as a function of their concerns for deservingness. The type of perceived deservingness that best predicted the extent of justice reasoning was that which was the most compatible on specificity. In other words, perceived deservingness of the current misfortune was more specific to immanent justice reasoning and proved to be the strongest predictor. However, perceptions of deservingness in later life outcomes was more congruent with ultimate justice reasoning and therefore best predicted people's ultimate justice judgments.

Study 2 extended these findings into the domain of considering one's own bad breaks and future fulfillment in life. After thinking about their own bad breaks, ultimate justice reasoning for the self was greater among participants higher in self-esteem, whereas immanent justice reasoning was more pronounced among participants lower in self-esteem. Study 2 also mirrored Study 1's effects of deservingness as underling these reactions to one's own outcomes. The perceived deservingness of bad breaks mediated the negative relation between self-esteem and immanent justice attributions, whereas only perceived deservingness of future life fulfillment mediated the positive relation between self-esteem and ultimate justice reasoning for the self.

These findings contribute to the literature in two important and novel ways: First, we examined how people try to make sense out of the misfortunes of others by engaging in both immanent and ultimate justice reasoning at once. We showed that these two types of justice reasoning are negatively related to one another and perceived deservingness plays an important role in the interplay between immanent and ultimate justice reasoning in response to the misfortunes of others. These findings therefore contribute to the limited literature examining when, and for whom, different reactions to instances of misfortune are apparent [1], [9], [17], [39], [40], [10]. As Hafer and Bègue argued, no one response is dominant across situations or individuals, and therefore multiple reactions should be assessed to gain a more comprehensive knowledge of how people make sense out of and find meaning in suffering and misfortune [1], also see [41]. Our work takes one step in that direction by suggesting the worth of a victim is key to determining perceptions of deservingness, which in turn influences the extent of both immanent and ultimate justice reasoning.

Of course, responding in terms of immanent and ultimate justice are by no means the only ways people make sense of misfortune and suffering. Interestingly, our manipulation of victim worth in Study 1 could be considered a manipulation of "just-world" threat, presumably because the "good" victim poses a larger threat to participants' just-world beliefs than the "bad" victim. Research has shown that people perceive the suffering of "good" victims as more unfair than the suffering of "bad" victims (e.g., when a physically attractive vs. an unattractive person is harmed) [42], [43], [44], [45]. Therefore, the interplay between other known responses to just-world threat, such as victim blaming see [1], and the responses to misfortune we measured here have yet to be investigated. It is therefore important for future research to examine perceptions of immanent and ultimate justice alongside other means by which people might maintain a perception of justice in the face of threat.

Second, the interactive pattern between the worth of a victim and type of justice reasoning we observed in Study 1 was replicated in Study 2 in the context of participants considering their own misfortunes. Of particular intrigue, we found that participants lower in self-esteem saw themselves as more deserving of their negative outcomes and were willing to adopt immanent justice attributions for their own fortuitous bad breaks. Although research into immanent justice reasoning has almost exclusively focused on people's causal attributions for the random misfortunes occurring to *others* [14], we found that the same processes operate when people entertain the causes of their own random bad breaks, and personal deservingness plays a crucial mediating role in this

relation. In addition, we found that participants with higher self-esteem believed they were more deserving of, and would therefore receive, a fulfilling and meaningful life. These findings add to the existing literature on how people make sense of their misfortunes [46] by suggesting that perceived deservingness of ultimate compensation plays an important meditational role.

Further, our findings may be important and applicable to our understanding of people's coping and resilience in the face of personal suffering and misfortune. Some research has shown that sufferers of illnesses engage in thought processes akin to ultimate and immanent justice reasoning, and these types of reasoning can be either beneficial or detrimental to their health [47], [48], [49], [50]. Our findings suggest that deservingness—either in the form of deserving one's recent bad breaks or deserving fulfillment later in life—might be underlying these types of responses to misfortune and as a result, may determine the trajectory of patient's well-being and recovery. For example, believing that one contracted an illness because they were a bad person deserving of bad outcomes may lead to heightened anxiety, lower levels of life-satisfaction, and a reduced likelihood of recovery cf. [48]. In a similar vein, Callan and colleagues found that individuals who held stronger beliefs that they deserved bad outcomes engaged in more self-defeating behaviors, including self-handicapping, wanting close others to evaluate them negatively, and seeking negative feedback about their performance during an intelligence test [22]. On the other hand, adopting the belief that one deserves a fulfilling and meaningful life in the future may lead to greater general well-being in the face of illness cf. [47]. Of course, more research is needed on the role that these deservingness beliefs might play in people's responses to their own misfortunes, but our work offers a theoretical perspective and empirical findings that point to their potential importance.

Finally, the present research encourages related lines of future research. We considered immanent and ultimate justice as reactions to undeserved *negative* outcomes, but both of these types of justice reasoning might also be adopted when people make sense of undeserved *positive* outcomes e.g., [11]. Therefore, it is important for future research to extend these findings in the context of positive outcomes. Although some research has examined the effects of undeserved positive outcomes on immanent justice reasoning (e.g., a man won the lottery *because* he was pleasant and hard working) [11], to our knowledge no research has considered ultimate justice reasoning in response to undeserved positive outcomes. We speculate that observing a good person experiencing a good outcome should result in individuals perceiving the two as causally connected (i.e., immanent justice reasoning) cf. [11], but observing the same outcome occurring to a bad person should encourage individuals to believe that the lucky individual will receive their comeuppance in the future (i.e., ultimate justice reasoning). Although much of just-world research has been concerned with victims of misfortune see [1], Lerner suggested that any injustice, good or bad, threatens our commitment to a just world [27]. Therefore, to further our understanding of how responses to misfortune operate, it is important for future research to consider both sides of the coin—people's responses to undeserved positive outcomes as a well as undeserved negative outcomes.

Supporting Information

Dataset S1   Study 1. Raw data and composite scores from Study 1 in SPSS.

Dataset S2   Study 2. Raw data and composite scores from Study 2 in SPSS.

Author Contributions

Conceived and designed the experiments: AH MC. Performed the experiments: AH MC. Analyzed the data: AH MC. Contributed reagents/materials/analysis tools: AH MC. Contributed to the writing of the manuscript: AH MC.

References

1. Hafer CL, Bègue L (2005) Experimental research on just-world theory: Problems, developments, and future challenges. Psychol Bull 131: 128–167.
2. Jost JT, Kay AC (2010) Social justice: History, theory, and research. In: Fiske ST, Gilbert D, Lindzey G, editors. Handbook of social psychology (5th edition, Vol. 2). Hoboken, NJ: Wiley. pp. 1122–1165.
3. Lerner MJ (1980) The belief in a just world: A fundamental delusion. New York: Plenum Press.
4. Callan MJ, Ellard JH (2010) Beyond victim derogation and blame: Just world dynamics in everyday life. In: Bobocel DR, Kay AC, Zanna MP, Olson JM, editors. The psychology of justice and legitimacy: The Ontario Symposium (Vol. 11) New York: Psychology Press. pp. 53–77.
5. Ellard J, Harvey AJ, Callan MJ (in press) The Justice Motive. In: Sabbagh C, Schmitt M, editors, Handbook of Social Justice Theory and Research.
6. Hafer CL, Gosse L (2010) Preserving the belief in a just world: When and for whom are different strategies preferred? In: Bobocel DR, Kay AC, Zanna MP, Olson JM, editors. The psychology of justice and legitimacy: The Ontario symposium. (Vol. 11). New York: Psychology Press. pp. 79–102.
7. Anderson JE, Kay AC, Fitzsimons GM (2010) In search of the silver lining: The justice motive fosters perceptions of benefits in the later lives of tragedy victims. Psychol Sci 21: 1599–1604.
8. Anderson JE, Kay AC, Fitzsimons GM (2013) Finding silver linings: Meaning-making as a compensatory response to negative experiences. In: Markman KD, Prolux T, Lindberg MJ, editors. The Psychology of Meaning. Washington, DC: American Psychological Association. pp. 279–296.
9. Hafer CL, Gosse L (2011) Predicting alternative strategies for preserving a belief in a just world: The case of repressive coping style. Eur J Soc Psychol 41: 730–739.
10. Warner RH, Vandeursen MJ, Pope ARD (2012) Temporal distance as a determinant of just world strategy. Eur J Soc Psychol 42: 276–284.
11. Callan MJ, Ellard JH, Nicol JE (2006) The belief in a just world and immanent justice reasoning in adults. Pers Soc Psychol Bull 32: 1646–1658.
12. Piaget J (1965) The moral judgment of the child. London: Kegan, Paul, Trench, Trubner, & Co. (Original work published 1932).
13. Raman L, Winer GA (2002) Children's and adults' understanding of illness: Evidence in support of a coexistence model. Genet Soc Gen Psychol Monogr 128: 325–355.
14. Callan MJ, Sutton RM, Harvey AJ, Dawtry RJ (2014) Immanent justice reasoning: Theory, research, and current directions. In: Olson JM, Zanna MP, editors. Adv Exp Soc Psychol (Vol. 49) London: Academic Press. pp. 105–161.
15. Callan MJ, Harvey AJ, Dawtry RJ, Sutton RM (2013) Through the looking glass: Long-term goal focus increases immanent justice reasoning. Br J Soc Psychol 52: 377–385.
16. Hafer CL (2000) Investment in long-term goals and commitment to just means drive the need to believe in a just world. Pers Soc Psychol Bull 26: 1059–1073.
17. Harvey AJ, Callan MJ (2014) The role of religiosity in ultimate and immanent justice reasoning. Pers Individ Dif 56: 193–196.
18. Maes J (1998) Immanent and ultimate justice: Two ways of believing in justice. In: Lerner MJ, Montada L, editors. Responses to victimizations and belief in a just world. New York: Plenum. pp. 9–40.
19. Maes J, Schmitt M (1999) More on ultimate and immanent justice: Results from the research project "justice as a problem within reunified Germany." Soc Justice Res 12: 65–78.
20. Lerner MJ, Miller DT, Holmes JG (1976) Deserving and the emergence of forms of justice. Adv Exp Soc Psychol 9: 133–162.
21. Feather NT (1999) Values, achievement, and justice: Studies in the psychology of deservingness. New York: Plenum.
22. Callan MJ, Kay AC, Dawtry RJ (2014) Making sense of misfortune: Deservingness, self-esteem, and patterns of self-defeat. J Pers Soc Psychol 107: 175–208.
23. Pepitone A, L'Armand K (1996) The justice and injustice of life events. Eur J Soc Psychol 26: 581–597.
24. Rice S, Trafimow D (2011) It's a just world no matter which way you look at it. J Gen Psychol 138: 229–242.
25. Kruglanski AW (1996) Motivated social cognition: Principles of the interface. In: Higgins ET, Kruglanski AW, editors. Social Psychology Handbook of Basic Principles. New York: Guilford Press. pp. 493–520.
26. Maio GR (2010) Mental representations of social values. In Zanna M, editor. Adv Exp Soc Psychol (Vol 42).Amsterdam: Elsevier.pp. 1–43.
27. Lerner MJ (1977) The justice motive. Some hypotheses as to its origins and forms. J Pers 45: 1–32.
28. Braband J, Lerner MJ (1974) "A little time and effort"…Who deserves what from whom? Pers Soc Psychol Bull 1: 177–179.
29. Simmons CH, Lerner MJ (1968) Altruism as a search for justice. J Pers Soc Psychol 9: 216–225.
30. Callan MJ, Kay AC, Davidenko N, Ellard JH (2009) The effects of justice motivation on memory for self- and other- relevant events. J Exp Soc Psychol 45: 614–623.
31. Wood JV, Heimpel SA, Manell LA, Whittington EJ (2009) This mood is familiar and I don't deserve to feel better anyway: Mechanisms underlying self-esteem differences in motivation to repair bad moods. J Pers Soc Psychol 96: 363–380.
32. Heuer L, Blumenthal E, Douglas A, Weinblatt T (1999) A deservingness approach to respect as a relationally based fairness judgment. Pers Soc Psychol Bull 25: 1279–1292.
33. Buhrmester M, Kwang T, Gosling SD (2011) Amazon's mechanical Turk: A new source of inexpensive, yet high-quality, data? Perspect Psychol Sci 6: 3–5.
34. Wood MJ, Douglas KM, Sutton RM (2012) Dead and alive: Belief in contradictory conspiracy theories. Soc Psychol Personal Sci 3: 767–773.
35. Callan MJ, Sutton RM, Dovale C (2010) When deserving translates into causing: The effect of cognitive load on immanent justice reasoning. J Exp Soc Psychol 46: 1097–1100.
36. Preacher KJ, Hayes AF (2008) Asymptotic and resampling strategies for assessing and comparing indirect effects in multiple mediator models. Behav Res Methods 40: 879–891.
37. Rosenberg N (1965) Society and the adolescent self-image. Princeton, NJ: Princeton University Press.
38. Zou GY (2007) Toward using confidence intervals to compare correlations. Psychol Methods 12: 399–413.
39. Karuza JJr, Carey TO (1984) Relative preference and adaptiveness of behavioral blame for observers of rape victims. J Pers 52: 249–260.
40. Miller DT (1977) Altruism and threat to a belief in a just world. J Exp Soc Psychol 13: 113–124.
41. Harvey AJ, Callan MJ, Matthews WJ (2014) How much does effortful thinking underlie observers' reactions to victimization? Soc Justice Res 27: 175–208.
42. Jones C, Aronson E (1973) Attribution of fault to a rape victim as a function of respectability of the victim. J Pers Soc Psychol 26: 415–419.
43. Correia I, Vala J, Aguiar P (2007) Victims innocence, social categorization, and the threat to the belief in a just world. J Exp Soc Psychol 43: 31–38.
44. Callan MJ, Powell NG, Ellard JH (2007) The consequences of victim's physical attractiveness on reactions to injustice: The role of observers' belief in a just world. Soc Justice Res 4: 433–456.
45. Callan MJ, Dawtry RJ, Olson JM (2012) Justice motive effects in ageism: The effects of a victim's age on observer perceptions of injustice and punishment judgments. J Exp Soc Psychol 48: 1343–1349.
46. Janoff-Bulman R (1985) The aftermath of victimization: Rebuilding shattered assumptions. In: Figley CR, editor. Trauma and its Wake: The study and treatment of post-traumatic stress disorder (Vol. 1). Bristol, PA: Taylor & Francis Group. pp. 15–35.
47. Affleck G, Tennen H, Croog S (1987) Causal attribution, perceived benefits, and morbidity after a heart attack: An 8 year study. J Consult Clin Psychol 55: 29–35.
48. Büssing A, Fischer J (2009) Interpretation of illness in cancer survivors is associated with health-related variables and adaptive coping styles. BMC Womens Health 2: 1–11.
49. Caress AL, Luker KA, Owens RG (2001) A descriptive study of meaning of illness in chronic renal disease. J Adv Nurs 33: 716–727.
50. Taylor S, Lichtman RR, Wood JV (1984) Attributions, beliefs about control, and adjustment to breast cancer. J Pers Soc Psychol 46: 489–502.

21

The Importance of *Acacia* Trees for Insectivorous Bats and Arthropods in the Arava Desert

Talya D. Hackett[1], Carmi Korine[2,3], Marc W. Holderied[1]*

1 Department of Biological Sciences, University of Bristol, Bristol, United Kingdom, **2** Mitrani Department of Desert Ecology, Swiss Institute for Dryland Environmental and Energy Research, Jacob Blaustein Institutes for Desert Research, Ben-Gurion University of the Negev, Midreshet Ben-Gurion, Israel, **3** The Dead Sea and the Arava Science Center, Tamar Regional Council, Neveh Zohar, Israel

Abstract

Anthropogenic habitat modification often has a profound negative impact on the flora and fauna of an ecosystem. In parts of the Middle East, ephemeral rivers (wadis) are characterised by stands of acacia trees. Green, flourishing assemblages of these trees are in decline in several countries, most likely due to human-induced water stress and habitat changes. We examined the importance of healthy acacia stands for bats and their arthropod prey in comparison to other natural and artificial habitats available in the Arava desert of Israel. We assessed bat activity and species richness through acoustic monitoring for entire nights and concurrently collected arthropods using light and pit traps. Dense green stands of acacia trees were the most important natural desert habitat for insectivorous bats. Irrigated gardens and parks in villages and fields of date palms had high arthropod levels but only village sites rivalled acacia trees in bat activity level. We confirmed up to 13 bat species around a single patch of acacia trees; one of the richest sites in any natural desert habitat in Israel. Some bat species utilised artificial sites; others were found almost exclusively in natural habitats. Two rare species (*Barbastella leucomelas* and *Nycteris thebaica*) were identified solely around acacia trees. We provide strong evidence that acacia trees are of unique importance to the community of insectivorous desert-dwelling bats, and that the health of the trees is crucial to their value as a foraging resource. Consequently, conservation efforts for acacia habitats, and in particular for the green more densely packed stands of trees, need to increase to protect this vital habitat for an entire community of protected bats.

Editor: Ben J. Mans, Onderstepoort Veterinary Institute, South Africa

Funding: This study was supported by the Israeli Ministry of Science and Technology (to C.K.): US $13333; The Explorers Club Exploration Fund (to T.D.H.): US $750; European Commission Dryland Research Specific Support Action Plan (to T.D.H.): £9,000. The funders had no role in study design, data collection and analysis, decision to publish, or preparation of the manuscript.

Competing Interests: The authors have declared that no competing interests exist.

* E-mail: Marc.Holderied@bristol.ac.uk

Introduction

Desert habitats are resource limited by definition, putting flora and fauna under particular constraints [1]. Anthropogenic disturbance of such extreme natural habitats can have long-lasting deleterious effects [2]. Within mammals, bats are the second most species rich order [3], provide valuable ecosystem services [4], are abundant in many habitats, can easily be monitored through recordings of their powerful sonar vocalisations and are good bioindicators of habitat quality [5]. In desert areas of Israel (e.g. Negev, Arava and Judean) there are 17 species of insectivorous bats, representing more than half of the country's desert mammals [6,7]. All insectivorous bats are protected by Israeli law and are either 'vulnerable', 'near-threatened' or 'endangered' on the International Union for Conservation of Nature (IUCN) red list for Israel [8].

Acacia trees are widely regarded as a keystone species with most desert fauna depending on them, either directly or indirectly, for food and shade [9–11]. They have an established positive impact on soil chemistry as nitrogen fixers [12] and increase herbaceous understory productivity [13]. Acacias hold crucial links to arthropods [14–16], such as ants, which live on acacias [17–20], bees which rely on acacia pollen [21] and bruchid beetles that infest seed pods [10,17,22,23]. Gazelle (*Gazella dorcas*), Arabian oryx (*Oryx leucoryx*), small nocturnal omnivorous rodents (*Mastomys natalensis, Saccostomus campestris* and *Aethomys chrysophilus*) [24,25], ostriches (*Struthio camelus*) and giraffes (*Giraffa camelopardalis*) [24] all consume the seeds of acacia, disperse and then fertilize pods aiding in germination while reducing the effect of seed parasites [23]. Three species, *Acacia tortilis, A. raddiana,* and *A. pachyceras,* provide the majority of wooded habitats in the Arava [26].

Acacia trees, particularly *A. raddiana* are in decline [27,28]; the total mortality of acacia trees in the Arava Rift Valley may be as high as 61% over 14 years [9,27]. This is primarily due to water stress, low recruitment of young acacia seedling and loss/change of habitat and water flow patterns [9,27]. As acacia trees rely predominantly on surface water, the latter factor is of great concern [29]. Additionally, there is a significant decline in annual precipitation, which is likely to have a negative effect on mortality and recruitment of acacias [30]. Rohner and Ward [23] predict that loss of acacia trees in the Middle East would lead to a significant loss of biodiversity in the region.Despite the wealth of research on the ecology of acacia trees it is almost completely unknown how and to what degree bats and their nocturnal arthropod prey might utilize acacia trees. Vaughan and Vaughan [31] found that the central African bat *Lavia frons* uses *A. tortilis* and occasionally *A. elatior* as a night roost from which to forage, and suggest that the bats are feeding on insects that are attracted to acacia trees. In Australia *Vespadelus pumilus* selectively roost in *A. melanoxylon* despite their relative rarity in the area [32]. Moreover, surveys of bats in the Sinai [33], Kenya [34] and Swaziland [35] found bat foraging activity at sites that contained acacia trees. None of these papers examined a specific interaction between bats and acacia trees, nor was there any explicit comparison to other available foraging habitats.

Here we examine activity levels and species richness of insectivorous desert bats and the abundance and richness of their arthropod prey in available natural and artificial desert habitats, including irrigated agricultural sites (date palms) and villages where desert-dwelling species are attracted to artificial light sources [6,36]. We hypothesise that acacias are a keystone genus in the nocturnal food web and therefore predict that nocturnal arthropods are diverse and abundant around acacia trees, and that bats are attracted to this foraging resource. Concurrently, we further hypothesise that the declining health of acacia habitats would negatively influence the bat and arthropod community and therefore predict that arthropod abundance and richness as well as bat activity and richness will be greater at dense green acacia stands than other available acacia habitats. Because artificial irrigation increases productivity in water limited ecosystems, we further predict that bat activity and arthropod abundance will be high in man-made habitats. We also predict that bat communities will differ between natural and man-made habitats, with a higher proportion of species that are typically recorded in the desert in natural habitats and more generalist synantropic species in the latter.

Materials and Methods

2.1 Study Site

The Arava rift valley, which connects the Dead Sea to the Red Sea, is an extremely arid desert with approximately 25–50 mm of annual rainfall and an average summer temperature of 31°C [37]. The area is characterised by ephemeral rivers that flood briefly after occasional, often distant, rains in most winters but otherwise remain dry (wadis). Scattered small settlements with irrigated parks, gardens and agricultural fields exist along the entire length of the valley. Potential foraging habitats for insectivorous bats in the Arava thus range from open desert with scarce vegetation that is typically dry in the summer months, through wadis with shrubs or trees to artificially irrigated and lit settlements and agricultural fields (e.g. date palms). The most stable biomass producers in the desert resource web are trees belonging to the genus *Acacia*. They occur in scattered lines along some wadis and, less frequently, in dense assemblages which are typically near small, often seasonal, springs, that are drying up due to aquifer pumping and climatic changes [38].

2.2 Habitats

We studied six habitat types; four natural: (1) densely packed green acacia trees (predominantly *A. tortilis* and *A. raddiana*) with trees clustered less than 30 m apart, (2) sparsely distributed green acacia trees with trees separated by greater than 50 m, (3) brown/barren acacia trees and (4) desert sites without acacia trees, as well as two modified by human habitation: (5) agriculture in the form of date plantations, and (6) irrigated vegetation at walkways or gardens in villages. We selected five different locations for each of the six habitat types giving a total of 30 sites in a 20×15 km area between the villages of Idan and Ein Yahav and using, from north to south, the accessible wadis Bitaron, En Zach, Masor, Shehaq

and Dohan. There were only five dense green acacia stands in the research area; one in each of the five wadis. All other natural sites were selected to span the same north-south range around these (numbered 1–5 from north to south) preferentially with one site of each habitat within each respective wadi. Because adequate habitats were not always available in the same wadi, one barren acacia (B1), one sparse acacia (S5) and one no acacia (N5) site had to be located outside of the five wadis, and two barren acacia sites (B4 and B5) were in one wadi. There were five settlements in the area so artificial sites were selected in and around each settlement again spanning the area in the north-south range (Fig. 1). No sites were within 100 m of water available for drinking and most were >500 m away. All natural sites were pristine and away from public illumination, and town sites were the only artificial habitats that had any non-natural lighting. Potential roosts in the form of caves and crevices were plentiful throughout the study region; buildings, occupied or abandoned, were located within likely bat commuting distance of all sites (<5 km). We collected data at the 30 sites from April to August 2009. We visited sites randomly but made sure to visit one site for all six habitat types before starting with the next set of six different sites. After all 30 sites had been sampled once, we repeated this two more times but in a different random order. To minimise any potential lunar effect, we did not sample for five days around the full moon.

2.3 Arthropod Sampling

We used a fluorescent light trap (Sylvania 15 W black light actinic bulb, 350 nm) suspended in front of a white cotton sheet to sample arthropods at each study site (for review of light trapping see [39]). Starting at 30 min after sunset, the light was turned on every hour for 30 minutes. All arthropods on the sheet were then collected during the following 10 minutes, and the light was then turned off for 20 minutes to avoid cumulative effects. We repeated this cycle four additional times and again once more beginning 90 min before sunrise. In villages, where ambient light might bias the attractiveness of our light trap, we placed the trap in darker areas, or those shielded from light. A pit trap was located less than 1 m from the sheet and checked every hour. We collected large arthropods in vials and smaller ones in pooters. This combination of methods has a bias against those arthropods not attracted to light but alternative methods were not suitable to the habitat/situation. Sweep netting was unviable as the net would get caught in acacia trees' thorns damaging both the tree and the net while allowing all arthropods to escape; sticky traps quickly became covered in sand carried by the persistent winds; and, as many sites were in a national park, use of pesticides was not permitted. We identified all specimens at least to order and classified them as morphospecies. To create a reference collection of morphospecies hard-bodied specimens were collected, frozen and then pinned. Soft-bodied arthropods (Araneae, Scorpionidae and Solifugae) were stored in 70% ethanol. We analysed arthropod abundance per hour to account for periods of equipment failure in the light trap.

2.4 Acoustic Monitoring and Species Identification

At each site we used a full spectrum, direct recording automatic acoustic monitoring device with an omnidirectional microphone (BatCorder, EcoObs, Nuremberg, Germany) to record bat echolocation calls (@ 500 kHz and 16 bit) following the general approach outlined in Hayes et al. [40]. This device was hung from the edge of a tree 1–2 m from the ground. At sites where no trees were suitable a 1 m-high artificial stand was used. To avoid influencing recorded bat activity, the BatCorder was set at least 25 m from the arthropod trap. Once set, the BatCorder

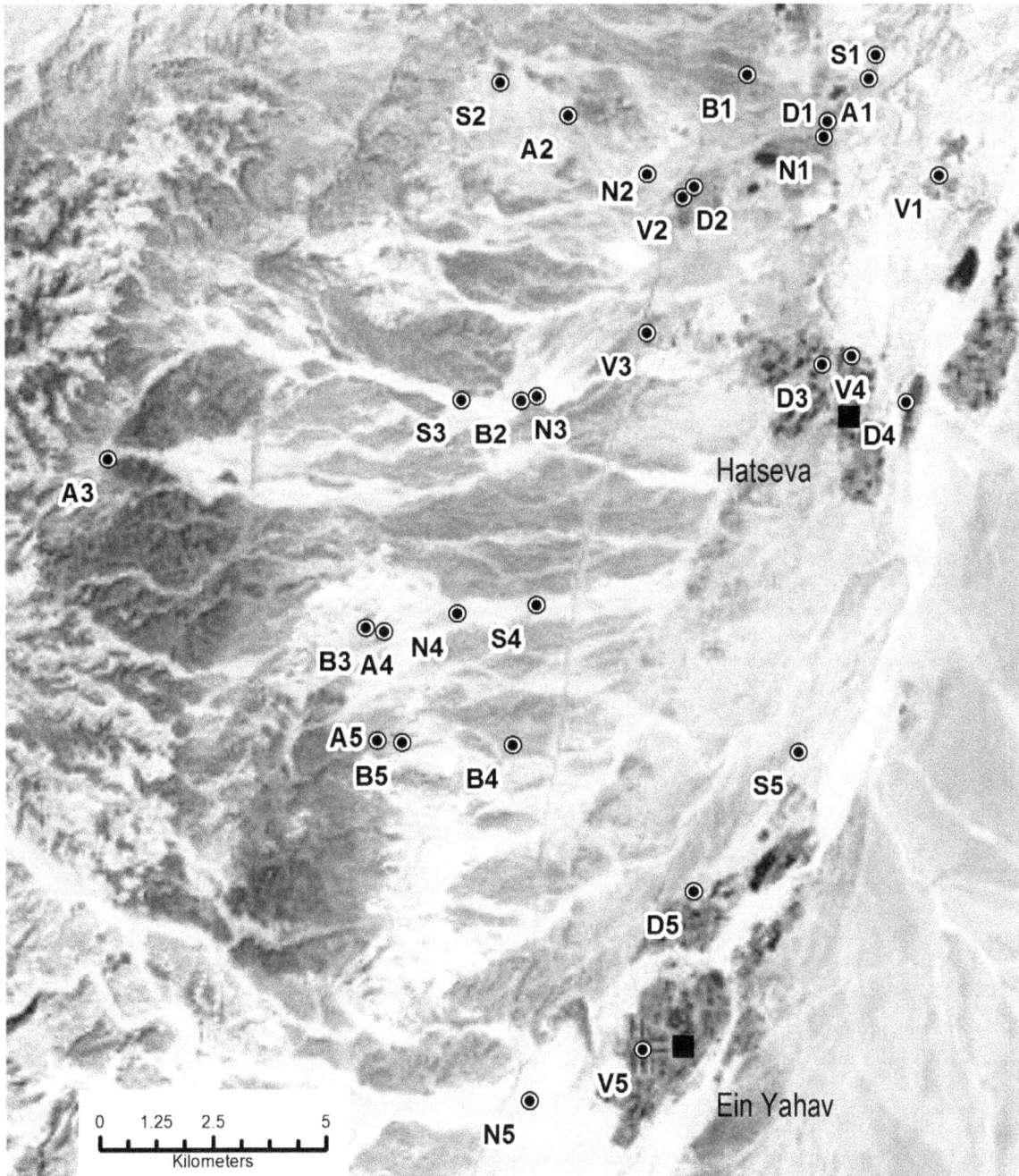

Figure 1. Satellite map of sites. A is dense acacia stands, S is sparse acacia stands, B is barren acacia stands, N is non-acacia desert sites, V is village sites and D is date plantations. The five replicates of each habitat type are numbered one to five from north to south.

automatically records upon detection of a bat call and continues recording as long as bat calls are detected. After 800 ms of silence it ceases recording until triggered by a new call which starts a new file. Detection was assumed to be equal in all sites since desert habitats are all acoustically transmissible, with little canopy cover even at dense acacia sites. Within villages, sites were also open and there were no tall buildings obstructing bat flight. Agricultural sites were the most densely covered, but trees were still spaced 8–10 meters apart with crown diameters that leave gaps of 2–4 m between trees. There is currently no quantitative data on the

transmissibility of habitat types for ultrasound, but large gaps in foliage in all habitats and the omnidirectional microphone of the BatCorder [41] mean that this is unlikely to have been a strong effect in this region. Due to differences in the source levels of different species' echolocation calls, "loud" aerial hawkers are likely to be recorded over greater distances than "whispering" gleaning species [41]; this bias could not be eliminated but was equal across all sites. Activity was measured as number of bat passes per night and each recording file was defined as one pass. This is a conservative measure of bat activity if two passes of the

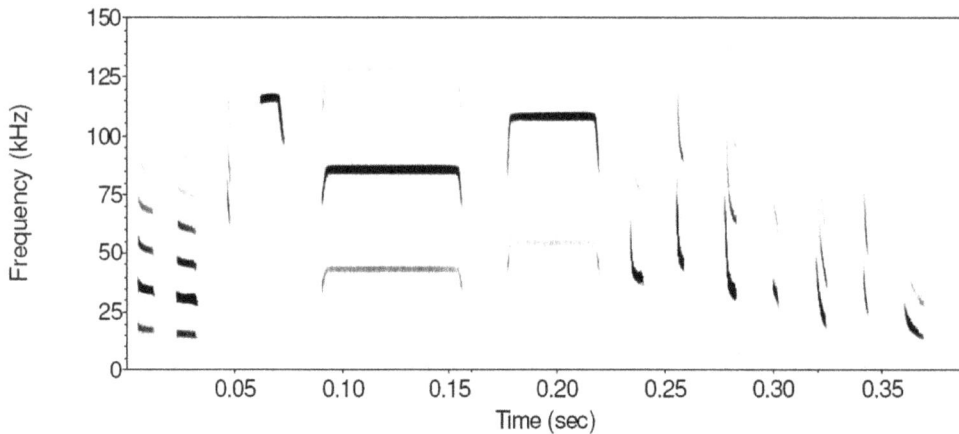

Figure 2. Spectrogram of one typical echolocation call from each identified species. From left to right: *Rhinopoma hardwickii, R. microphyllum, Nycteris thebaica, Asellia tridens, Rhinolophus hipposideros, R. clivosus, Pipistrellus kuhlii, Hypsugo bodenheimeri, Eptesicus bottae, Barbastella leucomelas, Otonycteris hemprichii, Plecotus christii* and *Tadarida teniotis*. Spectrogram parameters: FFT 1024, frame 100%, overlap 98.43%, window flat top.

same species are separated by less than 800 ms of silence. Multiple species present in the same file were defined as separate passes [42]. In a pilot survey in summer 2008, we identified the bat species foraging around acacia trees in the Arava using a combination of recordings from hand-released bats and descriptions of echolocation calls from studies in the broader region [33,43]. We established that desert bats in Israel can be identified to species level based on species/specific echolocation call design (see Fig. 2).

We used a weather monitoring device (Silva ADC Pro, Silva Sweden AB, Sweden) to record temperature, humidity and wind speed at the position of the BatCorder, with measurements taken one hour after sunset. At the beginning the night, wind was often blowing constantly with speeds of 5–10 km/h and occasionally as high as 20 km/hr. At a variable time, typically before midnight, this wind stopped abruptly and conditions remained calm for the rest of the night. Because of this pattern, wind was recorded as either present or absent one hour after sunset coinciding with the usual peak foraging activity. To standardise bat species identification and efficiently process the large number of recordings, we developed an automatic classification algorithm in SasLab Pro v. 4.40 (Avisoft Bioacoustics, Berlin, Germany). Peak frequencies at the start, end and maximum amplitude were measured for each echolocation call, which was then classified to species using defined frequency ranges per species. When compared to manual species identification the automatic classification correctly identified 95% of bat passes across all species (695 passes over 3 nights from pilot data in 2008); errors occurred when the recorded calls were too faint for the automatic classification to pick up (13 out of 115 passes for *Rhinopoma hardwickii*, 7 out of 537 passes for *Hypsugo bodenheimeri*), and when a *H. bodenheimeri* call overlapped with a *R. hardwickii* call it was mistakenly classified as *Eptesicus bottae* (5 passes); the passes of all other species were correctly identified in all cases. In order to reduce any further errors, some files had to be checked manually for potential misclassifications. This was necessary for recordings where (a) no call was classified/detected, or where (b) all calls classified as *Otonycteris hemprichii, Plecotus christii* or *Tadarida teniotis*, because low frequency noise was sometimes mistakenly classified one of these species. Echolocation calls from *O. hemprichii* and *P. christii* differ characteristically in the end frequencies, duration and the amount of spectral overlap between the first and second harmonic, but they had to be separated

manually because automatic classification was unreliable. There were also rare misclassifications between solitary calls of three species with peak frequencies of approximately 30 kHz (*E. bottae, R. hardwickii*, and *R. microphyllum*). Hence, all files containing two or fewer such calls were also checked manually. Mist nets were routinely placed at each site, with the intention of confirming activity estimates, but due to the open nature of the desert habitats we rarely caught bats, stressing the advantages of acoustic monitoring for bat surveys. Bat captures and surveys were conducted under license #34615 given to CK by the Israel Nature and Park Authority, and all sites were visited with permission from land owners or the Israel Nature and Park Authority.

2.5 Statistical Analysis

Bat passes, arthropod abundance and the total number of bat species recorded at each site were heteroscedastic, therefore a $\log_{10}(x+1)$ transformation was applied to the data enabling the use of parametric tests. A mixed-model analysis of variance (ANOVA) with one between-subjects factor (habitat) and one within-subjects factor (visit number) was performed on bat passes and arthropod abundance/hour. Green, dense acacia habitats were then individually compared to all other habitats using pairwise t-tests with sequential Bonferroni adjustment. Bat and arthropod species richness was measured as the number of species or morphospecies present at each site during the period of study. An ANOVA was performed on the number of bat species and arthropod morphospecies across habitats. A Pearson correlation test was performed to determine the relationship between bat passes and arthropod abundance. To test for any confounding effect of abiotic factors (temperature, humidity and wind) we performed a multivariate ANOVA for each visit cycle on the total number of bat passes per night and arthropod abundance per hour per night; we used a Bonferroni correction to control for multiple tests. All statistics were computed and graphs created using R-2.7.1 statistical environment (The R Foundation for Statistical Computing, 2008).

Results

We collected a total of 46,471 arthropods almost exclusively at the light trap over 533 hours with only 10 arthropods in pit traps.

We identified 733 arthropod morphospecies in the following systematic groups: Lepidoptera (234), Coleoptera (100), Orthoptera (31), Mantoidea (12), Diptera (109), Neuroptera (30), Hymenoptera (39), Hemiptera (144), Blattaria (5), Odonata (4), Dermaptera (5), Isopoda (2), Ixodida (1), Pseudoscorpionida (1), Aranae (13), Solifugae (1) and Scorpiones (2).

Arthropod abundance was affected by both habitat type ($F_{5,24} = 3.94$; p $= 0.009$; fig. 3a) and visit number ($F_{2,48} = 5.26$, p $= 0.009$; Fig. 3a). Arthropod abundance was lower at dense green acacia trees than at date sites ($t_{21} = 3.42$; p $= 0.012$), but after correcting for multiple testing there was no difference compared to other habitats (sparse acacia trees: $t_{24} = 0.29$, p $= 0.77$; barren acacia trees $t_{27} = 2.55$, p $= 0.067$; no acacia trees: $t_{26} = 0.96$, p $= 0.69$, village sites $t_{24} = -2.19$, p $= 0.11$). The number of arthropod morphospecies did not change significantly between habitat ($F_{5,24} = 2.27$, p $= 0.079$; Fig. 3c).

We identified 13 bat species by their echolocation calls (Fig. 2): *Rhinopoma hardwickii*, *R. microphyllum*, *Nycteris thebaica*, *Asellia tridens*, *Rhinolophus hipposideros*, *R. clivosus*, *Pipistrellus kuhlii*, *Hypsugo bodenheimeri*, *Eptesicus bottae*, *Barbastella leucomelas*, *Otonycteris hemprichii*, *Plecotus christii* and *Tadarida teniotis*. We caught seven of these species in mist nets: *R. hardwickii*, *A. tridens*, *R. clivosus*, *H. bodenheimeri*, *E. bottae*, *O. hemprichii* and *P. christii*.

Over a total duration of 963 hours on 72 recording nights we recorded 6,575 bat passes in 5,586 files. Typically each file contained a single pass by a single individual, but passes of different species sometimes occurred in the same file indicating that multiple species were foraging at the same place and time. The number of bat passes was affected by both habitat type ($F_{5,24} = 3.55$; p $= 0.015$; Fig. 3b) and visit number ($F_{2,48} = 3.84$, p $= 0.028$; Fig. 3b). The number of bat passes was significantly greater in dense green acacia trees than in all other natural desert habitats (sparse acacia trees $t_{27} = 4.05$, p $= 0.001$; barren acacia trees $t_{20} = 5.88$, p<0.001; no acacia trees: $t_{28} = 4.23$, p $= 0.001$) and date fields ($t_{22} = 2.43$, p $= 0.05$); but was not significantly different from village sites ($t_{27} = 1.47$, p $= 0.15$).

The number of bat species present was affected by habitat type ($F_{5,24} = 2.80$, p $= 0.042$; Fig. 3d), with dense green acacia trees having more species than barren acacia trees ($t_8 = 3.50$, p $= 0.04$), but not significantly different from any other habitat (sparse acacia

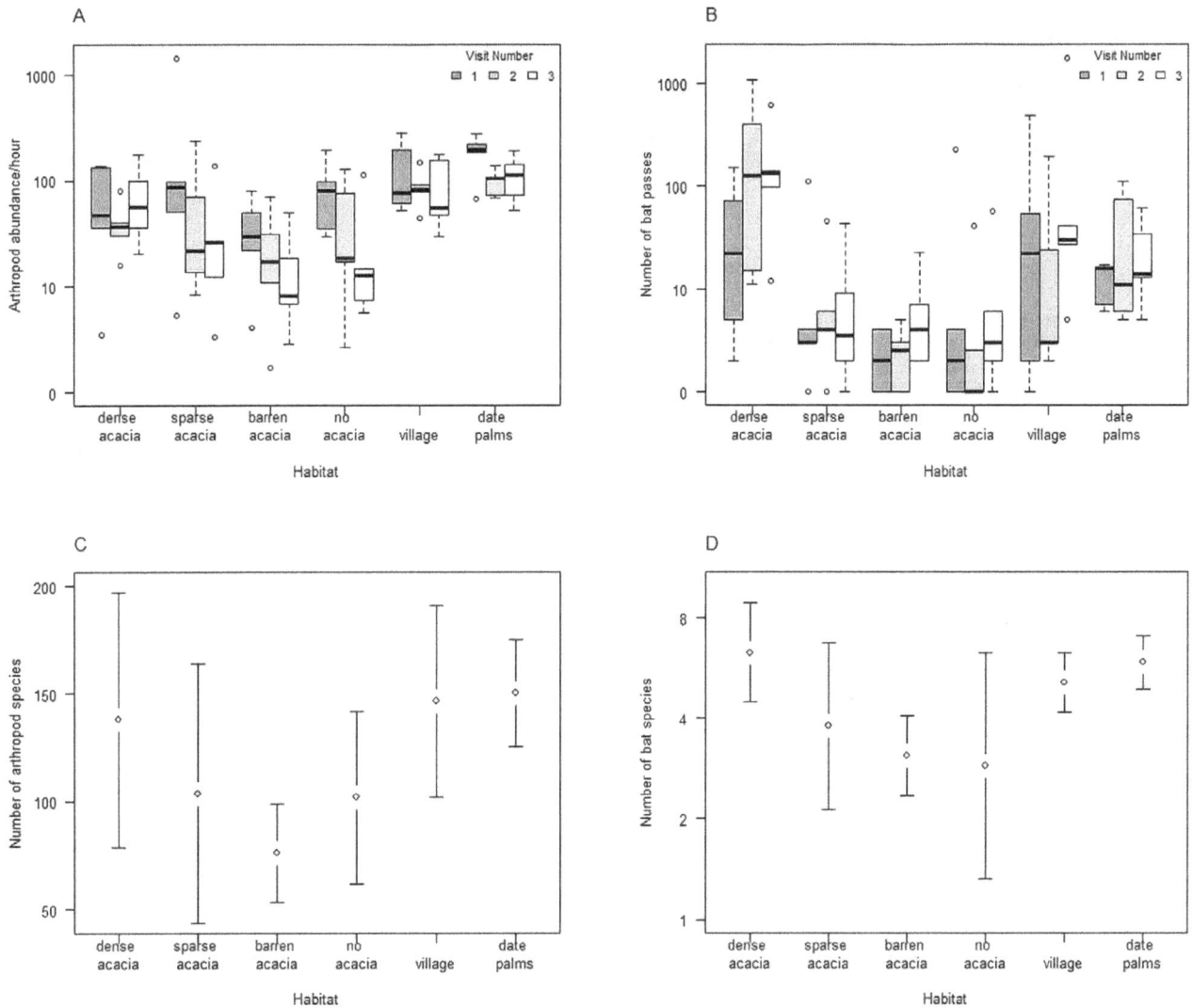

Figure 3. Arthropod abundance, bat activity and species richness for each habitat. a: box plot of arthropod abundance per hour for three consecutive repeats (visits). **b**: box plot of total number of bat passes for three consecutive repeats (visits). **c**: total number of arthropod morphospecies (mean ± standard deviation. **d**: total number of bat species recorded (mean ± standard deviation).

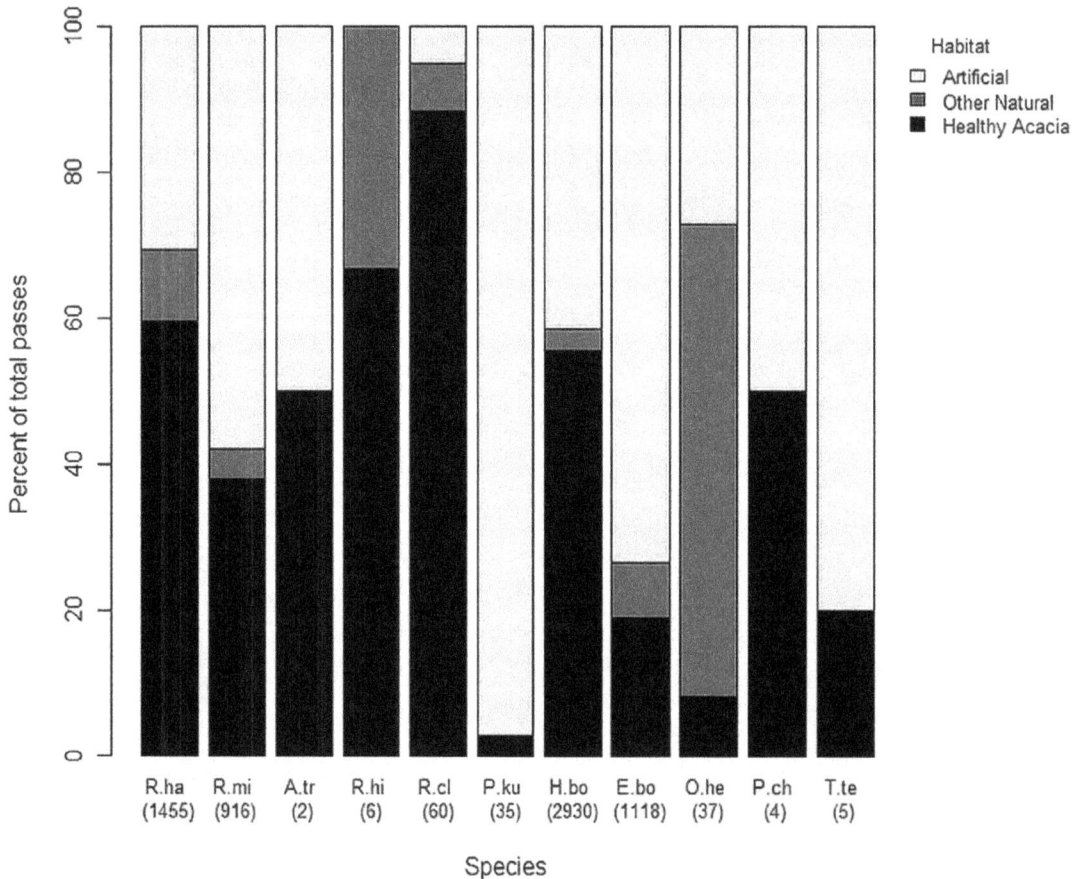

Figure 4. The percentage of recorded bat passes in each habitat per species. R.ha: *Rhinopoma hardwickii*, R.mi: *R. microphyllum*, A.tr: *Asellia tridens*, R.hi: *Rhinolophus hipposideros*, R.cl: *R. clivosus*, P.ku: *Pipistrellus kuhlii*, H.bo: *Hypsugo bodenheimeri*, E.bo: *Eptesicus bottae*, O.he: *Otonycteris hemprichii*, P.ch: *Plecotus christii*, T.te: *Tadarida teniotis*. Numbers in brackets indicate the total number of passes for that species.

trees: $t_8 = 1.64$, $p = 0.44$; no acacia trees: $t_8 = 5.54$, $p = 0.36$; village sites: $t_8 = 6.40$, $p = 0.60$ and date palms: $t_8 = 6.25$, $p = 0.75$). Some species of bat were recorded almost exclusively in desert habitats and, within them, mostly in healthy acacia stands while others were more likely to be recorded in non-natural habitats (Fig. 4). For instance Rhinolophid species were recorded almost exclusively in natural habitats while *P. kuhlii* and *T. teniotis* were almost exclusively found in artificial sites. Other species (e.g. *R. hardwickii* and *H. bodenheimeri*) were more equally distributed between habitats.

There was a positive correlation between the number of bat passes and the arthropod abundance across all habitats but only 20.8% of the variation is accounted for by this relationship ($R^2 = 0.21$; $t_{88} = 4.81$, $p<0.001$; Fig. 5). Wind had an effect on arthropod abundance in each visit cycle (1st visit: $F_{1,28} = 12.73$, $p = 0.011$; 2nd visit: $F_{1,28} = 13.81$, $p = 0.008$; 3rd visit: $F_{1,28} = 9.20$, $p = 0.048$) but not on bat activity (all visits $F_{1,28}<1.92$, $p>0.18$). No other abiotic factor affected either arthropod abundance or bat activity (all $F_{1,28}<8.20$, $p>0.07$).

Discussion

In accordance with our hypothesis, insectivorous bat activity was higher in dense green acacia stands than any other natural habitat, and species richness was high at habitats with dense green acacia trees. While dense green acacia trees only differed significantly from barren acacia trees in terms of bat species

richness it was only at green acacia sites that we recorded all 13 species, which make up 76% of insectivorous bat species known from the deserts of Israel and Jordan [33,44,45]. Since natural sites were located along wadis, they may all be used as commuting routes. Thus the difference in bat activity levels, but not in species richness, between dense green acacias and the other natural desert habitats could be a result of bats flying through sparse acacia and no acacia sites en route to dense green acacias. These results indicate that healthy stands of acacia trees are key natural foraging resources for desert-dwelling insectivorous bats.

There was however only a weak link between habitat type and arthropods; no natural habitat differed in arthropod abundance or richness from green, dense acacias. One possible explanation for this is that our arthropod trapping method is biased towards light-attracted species, thus we are likely under sampling arthropods not attracted to light in all habitats. There might be a bias if these species' abundance differed between habitats. While we attempted other sampling methods, these were not effective or not viable. The effect of visit number on both the number of bat passes and abundance of arthropods indicates that there is a potential seasonal component to habitat profitability and use. Healthy acacias remain green all year but partition flowering seasonally [46], and were in full flower during the 3rd visit. There is a general trend for arthropod abundance to decrease across the visits, particularly the 3rd visit, in the more barren sites (barren acacia trees and natural non-acacia sites). However, at green acacia trees

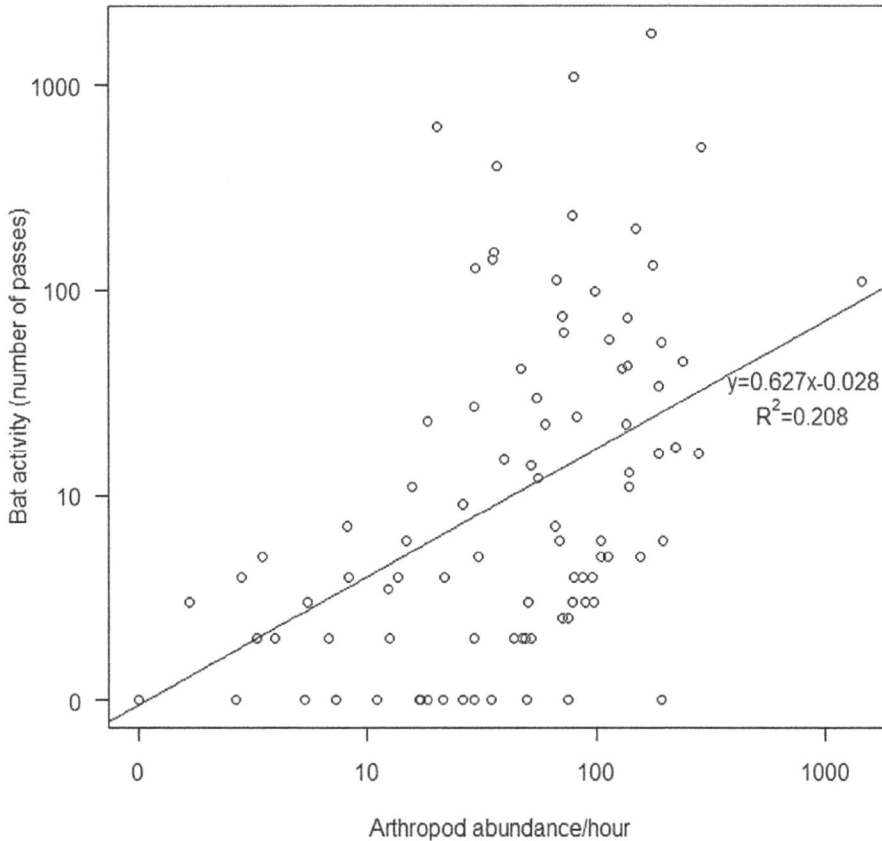

Figure 5. The relationship between arthropod abundance and bat activity. Each data point represents one entire night of sampling. Solid line is a linear regression (y = 0.627 × −0.028; R^2 = 0.21; t_{88} = 4.81, p<0.001).

(both dense and sparse) the level remains high during the 3rd visit in midsummer. It is therefore likely that green flowering acacias become even more important for bats and nocturnal arthropods as summer progresses and other habitats get less productive.

As we predicted, artificially irrigated and lit man-made habitats did have high arthropod abundance and bat activity. Date palms supported a greater abundance of arthropods than dense green acacias, while village sites and dense green acacias had equally high levels of bat activity. Moreover, arthropod and bat species richness for both date palms and village sites did not differ significantly from dense green acacias. These findings support the observation that for some species of bat, man-made habitats can, in fact, act as an alternative foraging resource [6].

Our results are consistent with previous studies that found bat activity correlated with arthropod abundance [47,48]. We recorded a range of bat species with different dietary niches [49], thus the activity of some species would likely correlate better with specific species of arthropods than others. Further studies into the diet of the bats in this area are needed to clarify this.

As predicted, some bat species relied more heavily on acacia trees than others; all species recorded here are listed at least as regionally vulnerable [8] (Table 1). Use of green acacia habitats was strongest in species typically recorded in deserts: *R. clivosus*, which mainly catches flying Coleoptera near vegetation, where echoes may come from objects that are not the target (cluttered environment) [6,44,49]; and *O. hemprichii*, which gleans terrestrial arthropods from surfaces [49–52] and tends to forage in xeric, sparsely vegetated, rocky environments [53] that are usually

cluttered [6]. Conversely, *P. kuhlii*, a generalist in terms of prey and habitat selection that favours habitats with street lights [49,54], and *T. teniotis*, which hunts for flying insects in open spaces but is a generalist in terms of prey selection [6,33,49,54] were encountered mainly in non-natural habitats. Four species were recorded approximately equally in all habitats, both natural and artificial. All of these are aerial insectivores foraging in background cluttered space: *H. bodenheimeri*, is known to be a generalist in terms of both habitat and prey selection [6,49,55,56]; *R. hardwickii* forages on aerial Coleoptera and swarming Hymenoptera in open habitats [49,57]; *R. microphyllum* predominantly consumes Coleoptera and is often found in sympatry with *R. hardwickii* [57,58]; and *E. bottae*, is a background cluttered space aerial insectivore [6]. Thus, artificial habitats created by the settlements appear to at least partially compensate for the habitat loss, but only for some of the desert species. Moreover, *P. kuhlii*, a species which has only expanded its range into desert areas of Israel following human habitation, could compete for resources with desert specialists in the vicinity of settlements [36,44,59,60].

Three species were rarely recorded (Fig. 4): *A. tridens* and *R. hipposideros* tend to forage in highly cluttered environments [6,49,61–63], and *P. christii* is a recently isolated whispering bat presumed to consume mainly Lepidoptera [64]. *A. tridens* and *P. christii* were sampled equally at healthy acacia and artificial sites while *R. hipposideros* was recorded mostly at healthy acacia sites and never in artificial habitats. Artificial light has been shown to negatively influence activity levels of *R. hipposideros* so it would not be expected in, or very near to, villages [65].

Table 1. Regional and global conservation status of recorded bat species.

Species	Regional status [8]	Global status [83]
Rhinopoma hardwickii	Vulnerable	Least concern
Rhinopoma microphyllum	Vulnerable	Least concern
Nycteris thebaica	Endangered	Least concern
Asellia tridens	Vulnerable	Least concern
Rhinolophus hipposideros	Vulnerable	Least concern
Rhinolophus clivosus	Vulnerable	Least concern
Pipistrellus kuhlii	Near threatened	Least concern
Hypsugo bodenheimeri	Endangered	Data deficient
Eptesicus bottae	Vulnerable	Least concern
Barbastella leucomelas	Endangered	Least concern
Otonycteris hemprichii	Vulnerable	Least concern
Plecotus christii	Endangered	Least concern
Tadarida teniotis	Near threatened	Least concern

Of particular interest are two additional rare species that were only recorded outside our analysed sampling period, but exclusively at dense green acacia sites: *B. leucomelas* and *N. thebaica*. *B. leucomelas* has been caught only five times before in Israel. We have recorded it five times at three different dense green acacia tree sites. As they are so rare, nothing is known about habitat selection of *B. leucomelas* and this is the first occurrence of consistent recordings in the region. *N. thebaica* is a generalist/opportunistic feeder [49,66] found foraging in open savannah woodland areas [66]. It is also a whispering bat, hunting in flight or from a perch [66,67]. As these two as well as *P. christii* and *O. hemprichii* are presumed whispering bats with low intensity calls, they will have been under sampled and in fact be more prevalent in the area than determined by acoustic monitoring [68].

Many studies have found increased bat activity at sites with water [6,69,70], but that is not likely to be the explanation for the site-dependant differences we recorded. Sites of different habitats were all an approximately equal flight distance from standing water. Moreover, during the summer months the natural pools and springs often dry up completely, yet activity levels remained high.

Environmental factors are suggested to play a role in where bats forage. Temperature is negatively correlated with activity levels [71–76], while heavy rains can stop all foraging [71,72,77], and relative humidity is positively correlated to bat activity [78]. Environmental conditions remained similar during the period of each cycle of 30 sites, with a temperature range of less than $\pm 5\,^{\circ}\mathrm{C}$ and without any precipitation. We found no significant effect of

either temperature or humidity on either arthropod abundance or bat activity. The abiotic factor most likely to influence bat activity in the Arava is strong wind, because this increases energy expenditure for powered flight [79]. The effect of wind on bats is somewhat ambiguous, with evidence of both no change in bat activity [71] as well as a decrease in activity [54,78]. We did not find an effect of the presence of wind on bat activity over the whole night, but did on arthropod abundance, possibly because the wind moved the collection sheet thereby preventing insects from landing. Since wind was discontinuous and stopped abruptly during the night it is possible that bats shifted their activity to calm periods; thus leaving bat activity levels over the entire night unaffected.

Kunz et al. [4] reviewed the ecosystem services provided by bats, concluding that insectivorous bats potentially exercise a top down control of arthropods in both natural and agricultural ecosystems. The use of exclusion nets to determine the relative effects of predation by birds and bats on arthropods indicates that there is an equal or stronger effect of bats on arthropod abundance [80–82]. Moreover, bat predation of arthropods had an indirect effect on herbivory, providing a strong case for bats as biological agents of pest control [81,82]. Thus, the high activity level and diversity of insectivorous bats we found around dense healthy assemblages of acacia trees and also in irrigated agriculture might indicate that bats act as a biological control agent in both natural and agricultural habitats in the Arava.

Our findings provide evidence that acacia habitats are keystone foraging sites especially for rare bat species and desert specialists. Irrigated habitats in deserts are frequented by a selection of desert species, but synantropic species might increase resource competition [60]. We also give evidence that the health of the tree has a strong influence on activity level. Acacia trees' further decline will have a significant impact on bats that forage in these areas. There is a need for better conservation and protection, particularly given the protected status of insectivorous bats in Israel and the ecosystem services they can provide.

Acknowledgments

Data was collected with the aid of Lauren Holt, Helen Hedworth, Orly Razgour, Shai Pilosof and especially Melia Nafus. Rangers from the Israel Nature and Park Authority were very helpful and friendly, particularly Yoram Hemo, Harel Ben Shahar, Roy Talbi and Asaf Tsoar as well as Eran Levin. We wish to thank Elizabeth Clare, Holger Goerlitz and two anonymous reviewers for comments on earlier versions of the manuscript. This is publication no. 789 of the Mitrani Department of Desert Ecology.

Author Contributions

Conceived and designed the experiments: TDH CK MWH. Performed the experiments: TDH. Analyzed the data: TDH MWH. Wrote the paper: TDH CK MWH. N/A.

References

1. Noy-Meir I (1973) Desert ecosystems: environment and producers. Annual Review of Ecology and Systematics 4: 25–51.
2. Lovich JE, Bainbridge D (1999) Anthropogenic degradation of the southern California desert ecosystem and prospects for natural recovery and restoration. Environmental Management 24: 309–326.
3. Simmons NB (2005) Order Chiroptera. In: Wilson DE, Reeder DM, editors. Mammal Species of the World: A Taxonomic and Geographic Reference. third ed: Johns Hopkins University Press. 312–529.
4. Kunz TH, Braun de Torrez E, Bauer D, Lobova T, Fleming TH (2011) Ecosystem services provided by bats. Annals of the New York Academy of Sciences 1223: 1–38.
5. Jones G, Jacobs DS, Kunz TH, Willig MR, Racey PA (2009) Carpe noctem: the importance of bats as bioindicators. Endangered Species Research 8: 93–115.

6. Korine C, Pinshow B (2004) Guild structure, foraging space use, and distribution in a community of insectivorous bats in the Negev Desert. Journal of Zoology 262: 187–196.
7. Yom-Tov Y (1993) Character displacement among the insectivorous bats of the Dead Sea area. Journal of Zoology 230: 347–356.
8. Dolev A, Perevolotsky A (2004) The Red Book: Vertebrates in Israel; Israel TINaPATSftPoNi, editor.
9. Ward D, Rohner C (1997) Anthropogenic causes of high mortality and low recruitment in three acacia tree taxa in the Negev desert, Israel. Biodiversity and Conservation 6: 877–893.
10. Or K, Ward D (2003) Three-way interactions between *Acacia*, large mammalian herbivores and bruchid beetles - a review. African Journal of Ecology 41: 257–265.

11. Munzbergova Z, Ward D (2002) Acacia trees as keystone species in Negev desert ecosystems. Journal of Vegetation Science 13: 227–236.

12. Belsky AJ, Amundson RG, Duxbury JM, Riha SJ, Ali AR, et al. (1989) The effects of trees on their physical, chemical, and biological environments in a semi-arid savanna in Kenya. Journal of Applied Ecology 26: 1005–1024.

13. Weltzin JF, Coughenour MB (1990) Savanna tree influence on understory vegetation and soil nutrients in Northwestern Kenya. Journal of Vegetation Science 1: 325–334.

14. Stone G, Willmer P, Nee S (1996) Daily partitioning of pollinators in an African acacia community. Proceedings of the Royal Society of London Series B-Biological Sciences 263: 1389–1393.

15. Kruger O, McGavin GC (1998) Insect diversity of Acacia canopies in Mkomazi game reserve, north-east Tanzania. Ecography 21: 261–268.

16. Kruger O, McGavin GC (2000) Macroecology of local insect communities. Acta Oecologica-International Journal of Ecology 21: 21–28.

17. Ernst WHO, Tolsma DJ, Decelle JE (1989) Predation of seeds of *Acacia tortilis* by insects. Oikos 54: 294–300.

18. Young TP, Stubblefield CH, Isbell LA (1997) Ants on swollen thorn acacias: Species coexistence in a simple system. Oecologia 109: 98–107.

19. Kruger O, McGavin GC (1998) The influences of ants on the guild structure of *Acacia* insect communities in Mkomazi Game Reserve, north-east Tanzania. African Journal of Ecology 36: 213–220.

20. Palmer TM (2003) Spatial habitat heterogeneity influences competition and coexistence in an African Acacia ant guild. Ecology 84: 2843–2855.

21. Martins DJ (2004) Foraging patterns of managed honeybees and wild bee species in an arid African environment: ecology, biodiversity and competition. International Journal of Tropical Insect Science 24: 105–115.

22. Hauser TP (1994) Germination, predation and dispersal of *Acacia albida* seeds. Oikos 70: 421–426.

23. Rohner C, Ward D (1999) Large mammalian herbivores and the conservation of arid *Acacia* stands in the Middle East. Conservation Biology 13: 1162–1171.

24. Miller MF (1995) Acacia seed survival, seed-germination and seedling growth following pod consumption by large herbivores and seed chewing by rodents. African Journal of Ecology 33: 194–210.

25. Downs CT, McDonald PM, Brown K, Ward D (2003) Effects of acacia condensed tannins on urinary parameters, body mass, and diet choice of an acacia specialist rodent, *Thallomys nicricauda*. Journal of Chemical Ecology 29: 845–858.

26. Horovitz A, Danin A (1983) Relatives of ornamental plants in the flora of Israel. Israel Journal of Botany 32: 75–95.

27. Shrestha MK, Stock WD, Ward D, Golan-Goldhirsh A (2003) Water status of isolated Negev desert populations of *Acacia raddiana* with different mortality levels. Plant Ecology 168: 297–307.

28. Ashkenazi S (1995) *Acacia* trees in the Negev, Israel, following reported large-scale mortality Jerusalem: Land Development Authority (HaKaren Ha Keyemet L'Israel).

29. Sher AA, Wiegand K, Ward D (2010) Do Acacia and Tamarix trees compete for water in the Negev desert. Journal of Arid Environments 74: 338–343.

30. Ginat H, Shlomi Y, Batarshe S, Vogel J (2011) Reduction in precipitation levels in the Arava valley. Journal of Dead-Sea and Arava Research 1: 1–7.

31. Vaughan TA, Vaughan RP (1986) Seasonality and the behavior of the African yellow-winged bat. Journal of Mammalogy 67: 91–102.

32. Law BS, Anderson J (2000) Roost preferences and foraging ranges of the eastern forest bat *Vespadelus pumilus* under two disturbance histories in northern New South Wales, Australia. Austral Ecology 25: 352–367.

33. Benda P, Dietz C, Andreas M, Hotovy J, Lucan RK, et al. (2008) Bats (Mammalia: Chiroptera) of the Eastern Mediterranean and Middle East. Part 6. Bats of Sinai (Egypt) with some taxonomic, ecological and echolocation data on that fauna. Acta Societas Zoologicae Bohemicae 72: 1–103.

34. Webala PW, Oguge NO, Bekele A (2001) Bat species diversity and distribution in three vegetation communities of Meu National Park, Kenya. African Journal of Ecology 42: 171–179.

35. Monadjem A, Reside A (2008) The influence of riparian vegetation on the distribution and abundance of bats in an African savanna. Acta Chiropterologica 10: 339–348.

36. Polak T, Korine C, Yair S, Holderied MW (2011) Differential effects of artificial lighting on flight and foraging behaviour of two sympatric bat species in a desert. Journal of Zoology 285: 21–27.

37. Goldreich Y, Karni O (2001) Climate and precipitation regime in the Arava Valley, Israel. Israel Journal of Earth Sciences 50: 53–59.

38. Bruins HJ, Sherzer Z, Ginat H, Batarseh S (2012) Degredation of springs in the Arava Valley: anthropogenic and climatic factors. Land Degradation & Development 23: 365–383.

39. Young M (2005) Insects in flight; Leather SR, editor: Blackwell Science Publ, Osney Mead, Oxford Ox2 0el, Uk. 116–145 p.

40. Hayes JP, Ober HK, Sherwin RE (2009) Survey and Monitoring of Bats. In: Kunz TH, Parsons S, editors. Ecological and Behavioral Methods for the Study of Bats. 2nd ed. Baltimore, MD USA: The Johns Hopkins University Press. 112–129.

41. Adams AM, Jantzen MK, Hamilton RM, Fenton MB (2012) Do you hear what I hear? Implications of detector selection for acoustic monitoring of bats. Methods in Ecology and Evolution.

42. Fenton MB (1970) A technique for monitoring bat activity with results obtained from different environments in southern Ontario. Canadian Journal of Zoology 48: 847–851.

43. Dietz C, von Helversen O (2004) Illustrated identification key to the bats of Europe. 1.0 ed. Germany: Tuebingen & Erlangen.

44. Yom-Tov Y, Kadmon R (1998) Analysis of the distribution of insectivorous bats in Israel. Diversity and Distributions 4: 63–70.

45. Shalmon B, Kofyan T, Hadad E (1993) A field guide to the land Mammals of Israel: their tracks and signs. Jerusalem, Israel: Keter Publishing House, Ltd.

46. Stone GN, Willmer P, Rowe JA (1998) Partitioning of pollinators during flowering in an African Acacia community. Ecology 79: 2808–2827.

47. Anthony ELP, Stack MH, Kunz TH (1981) Night roosting and the nocturnal time budget of the little brown bat, *Myotis lucifugus* - effects of reproductive status, prey density, and environmental conditions. Oecologia 51: 151–156.

48. Hayes JP (1997) Temporal variation in activity of bats and the design of echolocation-monitoring studies. Journal of Mammalogy 78: 514–524.

49. Feldman R, Whitaker JOJ, Yom-Tov Y (2000) Dietary composition and habitat use in a desert insectivorous bat community in Israel. Acta Chiropterologica 2: 15–22.

50. Holderied M, Korine C, Moritz T (2011) Hemprich's long-eared bat (*Otonycteris hemprichii*) as a predator of scorpions: whispering echolocation, passive gleaning and prey selection. Journal of Comparative Physiology A Neuroethology Sensory Neural and Behavioral Physiology 197: 425–433.

51. Arlettaz R, Dandliker G, Kasybekov E, Pillet JM, Rybin S, et al. (1995) Feeding-Habits of the Long-Eared Desert Bat, *Otonycteris hemprichi* (Chiroptera, Vespertilionidae). Journal of Mammalogy 76: 873–876.

52. Fenton MB, Shalmon B, Makin D (1999) Roost switching, foraging behavior, and diet of the vespertilionid bat, *Otonycteris hemprichii*. Israel Journal of Zoology 45: 501–506.

53. Gharaibeh BM, Qumsiyeh MB (1995) *Otonycteris hemprichii*. Mammalian Species 0: 1–4.

54. Russo D, Jones G (2003) Use of foraging habitats by bats in a Mediterranean area determined by acoustic surveys: conservation implications. Ecography 26: 197–209.

55. Riskin DK (2001) *Pipistrellus bodenheimeri*. Mammalian Species: American Society of Mammalogists. 1–3.

56. Whitaker JO, Shalmon B, Kunz TH (1994) Food and Feeding-Habits of Insectivorous Bats from Israel. Zeitschrift Fur Saugetierkunde-International Journal of Mammalian Biology 59: 74–81.

57. Whitaker JOJ, Yom-Tov Y (2002) The diet of some insectivorous bats from northern Israel. Mammalian Biology 67: 378–380.

58. Schlitter DA, Qumsiyeh MB (1996) *Rhinopoma microphyllum*. Mammalian Species: 1–5.

59. Mendelssohn H, Yom-Tov Y (1999) Fauna Palaestina: Mammalia of Israel. Jerusalem, Israel: The Israel Academy of Science and Humanities.

60. Razgour O, Korine C, Saltz D (2011) Does interspecific competition drive patterns of habitat use in desert bat communities? Oecologia 167: 493–502.

61. Bontadina F, Schofield H, Naef-Daenzer B (2002) Radio-tracking reveals that lesser horseshoe bats (*Rhinolophus hipposideros*) forage in woodland. Journal of Zoology 258: 281–290.

62. Jones G, Rayner JMV (1989) Foraging Behavior and Echolocation of Wild Horseshoe Bats *Rhinolophus ferrumequinum* and *Rhinolophus hipposideros* (Chiroptera, Rhinolophidae). Behavioral Ecology and Sociobiology 25: 183–191.

63. Zahn A, Holzhaider J, Kriner E, Maier A, Kayikcioglu A (2008) Foraging activity of *Rhinolophus hipposideros* on the island of Herrenchiemsee, Upper Bavaria. Mammalian Biology 73: 222–229.

64. Spitzenberger F, Strelkov PP, Winkler H, Haring E (2006) A preliminary revision of the genus *Plecotus* (Chiroptera, Vespertilionidae) based on genetic and morphological results. Zoologica Scripta 35: 187–230.

65. Stone EL, Jones G, Harris S (2009) Street Lighting Disturbs Commuting Bats. Current Biology 19: 1123–1127.

66. Gray PA, Fenton MB, Cakenberghe VV (1999) *Nycteris thebaica*. Mamalian Species: The American Society of Mammalogists. 1–8.

67. Fenton MB, Gaudet CL, Leonard ML (1983) Feeding behaviour of the bats *Nycteris grandis* and *Nycteris thebaica* (Nycteridae) in captivity. Journal of Zoology 200: 347–354.

68. Barclay RMR (1999) Bats are not birds - A cautionary note on using echolocation calls to identify bats: A comment. Journal of Mammalogy 80: 290–296.

69. Russ JM, Montgomery WI (2002) Habitat associations of bats in Northern Ireland: implications for conservation. Biological Conservation 108: 49–58.

70. Rebelo H, Brito JC (2006) Bat guild structure and habitat use in the Sahara desert. African Journal of Ecology 45: 228–230.

71. Kunz TH (1973) Resource utilization - temporal and spatial components of bat activity in central Iowa. Journal of Mammalogy 54: 14–32.

72. Rydell J (1991) Seasonal Use of Illuminated Areas by Foraging Northern Bats *Eptesicus nilssoni*. Holarctic Ecology 14: 203–207.

73. O'Donnell CFJ (2000) Influence of season, habitat, temperature, and invertebrate availability on nocturnal activity of the New Zealand long-tailed bat (*Chalinolobus tuberculatus*). New Zealand Journal of Zoology 27: 207–221.

74. Meyer CFJ, Schwarz CJ, Fahr J (2004) Activity patterns and habitat preferences of insectivorous bats in a West African forest-savanna mosaic. Journal of Tropical Ecology 20: 397–407.

75. Milne DJ, Fisher A, Rainey I, Pavey CR (2005) Temporal patterns of bats in the Top End of the Northern Territory, Australia. Journal of Mammalogy 86: 909–920.

76. Ciechanowski M, Zajac T, Bitas A, Dunajski R (2007) Spatiotemporal variation in activity of bat species differing in hunting tactics: effects of weather, moonlight, food abundance, and structural clutter. Canadian Journal of Zoology-Revue Canadienne De Zoologie 85: 1249–1263.

77. Fenton MB, Boyle NGH, Harrison TM, Oxley DJ (1977) Activity patterns, habitat use, and prey selection by some African insectivorous bats. Biotropica 9: 73–85.

78. Adam MD, Lacki MJ, Barnes TG (1994) Foraging areas and habitat use of the Virginia big-eared bat in Kentucky. Journal of Wildlife Management 58: 462–469.

79. Schnitzler HU (1971) Bats in the wind tunnel. Zeitschrift fuer Vergleichende Physiologie 73: 209–221.

80. Williams-Guillén K, Perfecto I, Vandermeer J (2008) Bats Limit Insects in a Neotropical Agroforestry System. Science 320: 70.

81. Kalka MB, Smith AR, Kalko EKV (2008) Bats limit arthropods and herbivory in a tropical forest. Science 320: 71.

82. Böhm SM, Wells K, Kalko EKV (2011) Top-down control of herbivory by birds and bats in the canopy of temperate broad-leaved okas (*Quercus robur*). PLoS ONE 6: 1–8.

83. IUCN (2011) The IUCN Red List of Threatened Species. Version 2011.2. Available: http://www.iucnredlist.org. Downloaded 29 May 2012.

PERMISSIONS

The contributors of this book come from diverse backgrounds, making this book a truly international effort. This book will bring forth new frontiers with its revolutionizing research information and detailed analysis of the nascent developments around the world.

We would like to thank all the contributing authors for lending their expertise to make the book truly unique. They have played a crucial role in the development of this book. Without their invaluable contributions this book wouldn't have been possible. They have made vital efforts to compile up to date information on the varied aspects of this subject to make this book a valuable addition to the collection of many professionals and students.

This book was conceptualized with the vision of imparting up-to-date information and advanced data in this field. To ensure the same, a matchless editorial board was set up. Every individual on the board went through rigorous rounds of assessment to prove their worth. After which they invested a large part of their time researching and compiling the most relevant data for our readers.

The editorial board has been involved in producing this book since its inception. They have spent rigorous hours researching and exploring the diverse topics which have resulted in the successful publishing of this book. They have passed on their knowledge of decades through this book. To expedite this challenging task, the publisher supported the team at every step. A small team of assistant editors was also appointed to further simplify the editing procedure and attain best results for the readers.

Apart from the editorial board, the designing team has also invested a significant amount of their time in understanding the subject and creating the most relevant covers. They scrutinized every image to scout for the most suitable representation of the subject and create an appropriate cover for the book.

The publishing team has been an ardent support to the editorial, designing and production team. Their endless efforts to recruit the best for this project, has resulted in the accomplishment of this book. They are a veteran in the field of academics and their pool of knowledge is as vast as their experience in printing. Their expertise and guidance has proved useful at every step. Their uncompromising quality standards have made this book an exceptional effort. Their encouragement from time to time has been an inspiration for everyone.

The publisher and the editorial board hope that this book will prove to be a valuable piece of knowledge for researchers, students, practitioners and scholars across the globe.

LIST OF CONTRIBUTORS

Zhuofei Xu, Martin Asser Hansen, Lars H. Hansen, Samuel Jacquiod and Søren J. Sørensen
Section of Microbiology, Department of Biology, University of Copenhagen, Copenhagen, Denmark

Min Wang
Linze Inland River Basin Research Station, Chinese Ecosystem Network Research, Cold and Arid Regions Environmental and Engineering Research Institute, Chinese Academy of Sciences, Lanzhou, Gansu, China
University of Chinese Academy of Sciences, Beijing, China

Yongzhong Su and Xiao Yang
Linze Inland River Basin Research Station, Chinese Ecosystem Network Research, Cold and Arid Regions Environmental and Engineering Research Institute, Chinese Academy of Sciences, Lanzhou, Gansu, China

Tomislav Hengl, Jorge Mendes de Jesus, Niels H. Batjes, Eloi Ribeiro, Bas Kempen, Johan G. B. Leenaars and Maria Ruiperez Gonzalez
ISRIC — World Soil Information, Wageningen, the Netherlands

Robert A. MacMillan
LandMapper Environmental Solutions Inc., Edmonton, Canada

Gerard B. M. Heuvelink
ISRIC — World Soil Information, Wageningen, the Netherlands
Wageningen University, Wageningen, the Netherlands

Alessandro Samuel-Rosa
Federal Rural University of Rio de Janeiro, Rio de Janeiro, Brazil

Markus G. Walsh
The Earth Institute, Columbia University, New York, New York, United States of America, and Selian Agricultural Research Inst., Arusha, Tanzania

Pingping Zhang
State Key Laboratory of Soil Erosion and Dryland Farming on the Loess Plateau, Northwest A & F University, Yangling, China

Ming'an Shao
State Key Laboratory of Soil Erosion and Dryland Farming on the Loess Plateau, Northwest A & F University, Yangling, China
Key Laboratory of Ecosystem Network Observation and Modeling, Institute of Geographical Science and Natural Resources, Chinese Academy of Sciences, Beijing, China

Eric D. Freeman, Tiffanny R. Sharp and Brock R. McMillan
Brigham Young University, Department of Plant and Wildlife Sciences, Provo, Utah, United States of America

Robert N. Knight
United States Army Dugway Proving Ground, Environmental Programs, Dugway, Utah, United States of America

Steven J. Slater
HawkWatch International, Conservation Director, Salt Lake City, Utah, United States of America

Randy T. Larsen
Brigham Young University, Department of Plant and Wildlife Sciences, Provo, Utah, United States of America
Monte L. Bean Life Science Museum, Provo, Utah, United States of America

Benjamin L. Allen
Robert Wicks Pest Animal Research Centre, Biosecurity Queensland, Toowoomba, Queensland, Australia
School of Agriculture and Food Sciences, the University of Queensland, Gatton, Queensland, Australia

Luke K.-P. Leung
School of Agriculture and Food Sciences, the University of Queensland, Gatton, Queensland, Australia

Roberto O. Chávez, Jan G. P. W. Clevers, Jan Verbesselt and Martin Herold
Laboratory of Geo-Information Science and Remote Sensing, Wageningen University, Wageningen, The Netherlands

Paulette I. Naulin
Laboratorio de Biología de Plantas, Departamento Silvicultura y Conservación de la Naturaleza, Universidad de Chile, Santiago, Chile

Zhang Zhaoyong
State Key Laboratory of Desert and Oasis Ecology, Xinjiang Institute of Ecology and Geography, Chinese Academy of Sciences, Urumqi, China
University of the Chinese Academy of Sciences, Beijing, China

Jilili Abuduwaili
State Key Laboratory of Desert and Oasis Ecology, Xinjiang Institute of Ecology and Geography, Chinese Academy of Sciences, Urumqi, China

Hamid Yimit
Key Laboratory of Xingjiang Arid Land Lake Environment and Resource, Xinjiang Normal University, Urumqi, China

Veerle Linseele and Wim Van Neer
Laboratory of Biodiversity and Evolutionary Genomics, Katholieke Universiteit Leuven, Leuven, Belgium
Royal Belgian Institute of Natural Sciences, Brussels, Belgium

Sofie Thys
Royal Belgian Institute of Natural Sciences, Brussels, Belgium

Rebecca Phillipps and Simon Holdaway
Anthropology Department, The University of Auckland, Auckland, New Zealand

René Cappers
Groningen Institute of Archaeology (GIA), Rijksuniversiteit Groningen, Groningen, The Netherlands

Willeke Wendrich
Cotsen Institute of Archaeology, University of California Los Angeles, Los Angeles, California, United States of America

Steven J. Dempsey and Bryan M. Kluever
Department of Wildland Resources, Utah State University, Logan, Utah, United States of America

Eric M. Gese
United States Department of Agriculture, Animal and Plant Health Inspection Service, Wildlife Services, National Wildlife Research Center, Department of Wildland Resources, Utah State University, Logan, Utah, United States of America

Michele T. Jay-Russell, Alexis F. Hake, Yingjia Bengson, Anyarat Thiptara and Tran Nguyen
Western Center for Food Safety, University of California Davis, Davis, California, United States of America

Jean-Baptiste Ramond, Annelize Pienaar, Alacia Armstrong and Don A. Cowan
Center for Microbial Ecology and Genomics (CMEG), Genomic Research Institute, University of Pretoria, Pretoria, South Africa

Mary Seely
Gobabeb Research and Training Center (GRTC), Walvis Bay, Namibia

Da Yong Yang
Southern Medical University, Guangzhou, Guangdong, China
Department of Ophthalmology, Guangdong General Hospital/Guangdong Academy of Medical Sciences, Guangzhou, Guangdong, China

Kai Guo, Yan Wang, Yuan Yuan Guo, Xian Rong Yang, Xin Xia Jing and Dan Zhu
The Affiliated Hospital of Inner Mongolia Medical University, Hohhot, Inner Mongolia, China

Hai Ke Guo
Department of Ophthalmology, Guangdong General Hospital/Guangdong Academy of Medical Sciences, Guangzhou, Guangdong, China

Yong Tao
Department of Ophthalmology, People's Hospital, Peking University, & Key Laboratory of Vision Loss and Restoration, Ministry of Education, Beijing, China

Jost B. Jonas
Department of Ophthalmology, Medical Faculty Mannheim of the Ruprecht-Karls-University Heidelberg, Mannheim, Germany

Rachele Gallisai and Francesc Peters
Departament de Biologia Marina i Oceanografia, Institut de Ciències del Mar, CSIC, Barcelona, Spain

Gianluca Volpe
Istituto di Scienze dell'Atmosfera e del Clima, Roma, Italy

Sara Basart
Earth Sciences Department, Barcelona Supercomputing Center-Centro Nacional de Supercomputación, BSC-CNS, Barcelona, Spain

José Maria Baldasano
Earth Sciences Department, Barcelona Supercomputing Center-Centro Nacional de Supercomputación, BSC-CNS, Barcelona, Spain Environmental Modelling Laboratory, Technical University of Catalonia, Barcelona, Spain

Qing Zhang and Jianming Niu
School of Life Sciences, Inner Mongolia University, Hohhot, China
Sino-US Center for Conservation, Energy and Sustainability Science, Inner Mongolia University, Hohhot, China

Xiangyang Hou and Yong Ding
Grassland Research Institute of Chinese Academic of Agricultural Science, Hohhot, China

Frank Yonghong Li
AgResearch Grasslands Research Centre, Palmerston North, New Zealand

Xin Li
Sino-US Center for Conservation, Energy and Sustainability Science, Inner Mongolia University, Hohhot, China

Yanlin Zhou, Liqing Zhao, Wenjing Ma and Sarula Kang
School of Life Sciences, Inner Mongolia University, Hohhot, China

Paulo C. Manzig
Centro Paleontoló gico da UnC (CENPÁ LEO), Universidade do Contestado, Mafra, Santa Catarina, Brazil
Programa de Pós-Graduação IEL-Labjor, Universidade Estadual de Campinas (UNICAMP), Campinas, São Paulo, Brazil

Alexander W. A. Kellner
Laboratory of Systematics and Taphonomy of Fossil Vertebrates, Departamento de Geologia e Paleontologia, Museu Nacional/Universidade Federal do Rio de Janeiro, Rio de Janeiro, Brazil

Luiz C. Weinschütz, João H. Z. Ricetti and Camila C. de Moura
Centro Paleontoló gico da UnC (CENPÁ LEO), Universidade do Contestado, Mafra, Santa Catarina, Brazil

Carlos E. Fragoso
Universidade Estadual de Ponta Grossa, Ponta Grossa, Paraná, Brazil

Cristina S. Vega
Departamento de Geologia, Universidade Federal do Paraná, Curitiba, Paraná, Brazil

Gilson B. Guimarães, Luiz C. Godoy and Antonio Liccardo
Departamento de Geociê\cias, Universidade Estadual de Ponta Grossa, Ponta Grossa, Paraná, Brazil

Wei Feng, Yuqing Zhang, Xin Jia, Bin Wu, Tianshan Zha, Shugao Qin, Ben Wang, Chenxi Shao, Jiabin Liu and Keyu Fa
Yanchi Research Station, College of Soil and Water Conservation, Beijing Forestry University, Beijing, China

Matina Donaldson-Matasci
Department of Biology, Harvey Mudd College, Claremont, California, United States of America
Department of Ecology & Evolutionary Biology, University of Arizona, Tucson, Arizona, United States of America

Anna Dornhaus
Department of Ecology & Evolutionary Biology, University of Arizona, Tucson, Arizona, United States of America

Annelie J. Harvey and Mitchell J. Callan
Department of Psychology, University of Essex, Essex, United Kingdom

Index

www.ingramcontent.com/pod-product-compliance
Lightning Source LLC
Chambersburg PA
CBHW061248190326
41458CB00011B/3614